REACTIVE INTERMEDIATES
IN
ORGANIC CHEMISTRY

REACTIVE INTERMEDIATES
IN
ORGANIC CHEMISTRY

N. S. ISAACS

Department of Chemistry,
University of Reading

JOHN WILEY & SONS

London · New York · Sydney · Toronto

Library of Congress Cataloging in Publication Data:

Isaacs, Neil S.   1934–
Reactive intermediates in organic chemistry.

1. Chemistry, Physical organic. I. Title.

QD476.I85   547′.2   73–8194
ISBN 0 471 42861 2

Printed in Great Britain by William Clowes and Sons Limited, London, Colchester and Beccles

To
Diane

Preface

Physical organic chemistry is a remarkably ill-defined subject; witness the almost complete lack of overlap in the subject matter covered in four recent text books with that title. The physical principles of organic chemistry are now universally taught at the university level; nevertheless it would seem that there is a wide diversity in the approach to the subject.

This book is an attempt to present physical organic chemistry from the point of view of reactions, but with strong emphasis on physical principles which are separately discussed. Furthermore the organization is such as to draw together reactions taking place via a common type of reactive intermediate and thus the emphasis is upon the chemistry of these important and often controversial species. This has the advantage of enabling topics such as aromatic electrophilic substitution, olefin addition and superacid chemistry to be treated under one heading, and to discuss their common characteristics as carbonium ion reactions.

All methods of subdividing an integrated subject such as organic chemistry must have disadvantages, however, and in the present treatment the reader will have to seek information on substitution at the aromatic ring under some five separate headings according to the type of intermediate involved. The inconvenience thus caused should be compensated by thorough indexing and full cross-referencing.

This book treats only reactions which proceed by way of reactive intermediates; single-step reactions are included where appropriate (e.g. nucleophilic aliphatic substitution and polar eliminations) to contrast the characteristics with related multistep processes. In scope it is intended for the undergraduate rather than the specialist but includes material somewhat beyond most degree courses for those interested in the subject.

1973   N. S. Isaacs

vii

Contents

Chapter 1

Physical Principles of Organic Chemistry

1.1   THE VALENCE STATES OF CARBON

The normal state of combination of carbon is tetracovalent and this is brought about by the pairing of each of the valence electrons ($2s^2$ and $2p^2$) with four electrons provided by the ligands. In the saturated, tetrahedral form characteristic, say of methane, this is accomplished by the 'hybridization' of the carbon atomic orbitals (**1a**) to a set of four equivalent orbitals, denoted sp^3 (**1b**), each of which overlaps with a $1s$ orbital of a hydrogen atom (**1c**). The result is four localized two-electron bonds set as far apart from each other as is

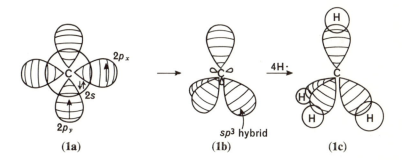

(1a)   (1b)   (1c)

geometrically possible. By alternative hybridization schemes, sp^2 and sp, trigonal and digonal bonding arrangements may be realized, one and two $2p$ orbitals being retained for π-bond formation as in ethylene (**2**), and acetylene (**3**), respectively. In each of these states, carbon exhibits a formal tetracovalency

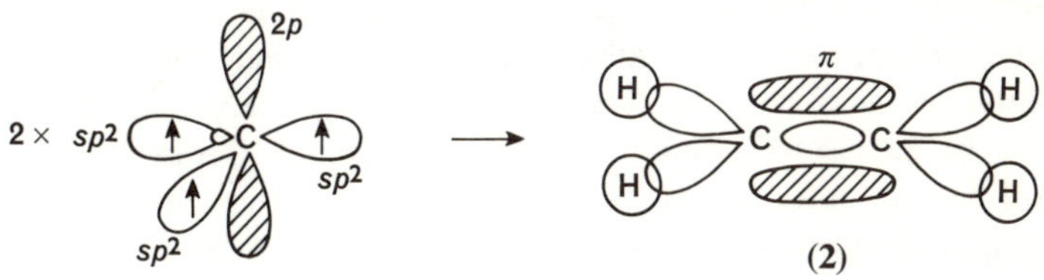

(2)

1

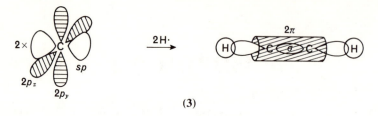

(3)

and a filled octet of valence electrons. Other, less usual, valence states of carbon may be formed in reactions, but are transient high-energy species which tend to revert rapidly to the tetracovalent state. Species containing these unusual valence states are some of the reactive intermediates which form the subject of this book.

Trivalent carbon is well established. If three ligands are attached to a central carbon atom by two-electron σ-bonds, the remaining orbital may hold two, one or no electrons; no other possibilities are permitted. This results in the formation of a carbanion (4), a free radical (5), and a carbonium ion (6),

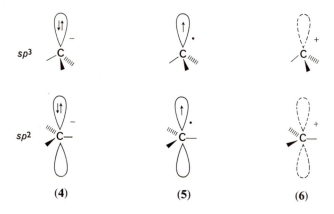

(4)   (5)   (6)

respectively, with formal charges of -1, 0 and $+1$. In principle we may describe the structures of these species using sp^3 or sp^2 hybrid orbitals, which will produce either a pyramidal or a planar geometry. Of these three types of trivalent carbon, only the carbanion has a filled octet. The radical and the carbonium ion are both electron deficient. In its chemistry, the carbanion tends to be an electron-pair donor (a nucleophile), the carbonium ion an electron-pair acceptor (an electrophile), and the radical tends to react with other unpaired electron species and form a new two-electron bond.

$$\text{\scriptsize}\;C\colon\curvearrowright H\!\!-\!\!B \quad\longrightarrow\quad C\!\!-\!\!H \;+\; :B^-$$

$$\overset{|}{\underset{|}{C^+}} + \;:O\!\!\begin{smallmatrix}H\\ \\H\end{smallmatrix} \quad\longrightarrow\quad C\!\!-\!\!OH \;+(H^+)$$

$$\overset{|}{\underset{|}{C\cdot}} + \;\cdot C \quad\longrightarrow\quad C\!\!-\!\!C$$

The double-barbed arrow ⌢ signifies the movement of an electronpair, and the singly-barbed arrow ⌢ the movement of an unpaired electron.

Divalent carbon compounds are also known. Some of the possible forms these could take are shown in **7**. However, it seems that only the neutral form

(=)	(−)		(+)	$\left(\begin{smallmatrix}+\\+\end{smallmatrix}\right)$
$\overset{\cdot\cdot}{\underset{\cdot\diagdown\;\cdot\diagup}{C}}\cdot\cdot$	$\overset{\cdot\cdot}{\underset{\cdot\diagdown\;\cdot\diagup}{C}}\cdot$	$\overset{\cdot\cdot}{\underset{\diagup\cdot\;\cdot\diagdown}{C}}\cdot$	$\overset{\cdot}{\underset{\cdot\diagup\;\cdot\diagdown}{C}}\cdot$	$\underset{\cdot\diagup\;\cdot\diagdown}{C}$
a	**b**	**c**	**d**	**e**

(**7**)

(**7c**), has any existence which has so far been discovered. Compounds with two covalent bonds and an unshared pair at carbon are known as carbenes (methylenes) and many members of this family are believed to be formed very transiently in certain reactions (Chapter 6). Carbenes will react with almost anything and, if nothing better is available, will dimerize;

$$Cl_2C\colon + \;:CCl_2 \quad\longrightarrow\quad Cl_2C\!\!=\!\!CCl_2$$

Monovalent carbon compounds such as R—C: may be formed very transiently in high-temperature reactions (e.g. flames, electric discharges) but not apparently in reactions occurring under more mild conditions.

Pentavalent carbon is traditionally considered to be energetically impossible since the bonding would require bringing in high-lying carbon $3s$ orbitals. Very recently this view has had to be modified since it now appears that pentavalent cations of carbon, e.g. CH_5^+ (**8a**), may be involved in very acidic media. The subject is discussed further (Section 2.30). It is unlikely that bonding involves higher atomic orbitals of carbon, and a three-centre two-electron bond as in B_2H_6 (**8b**) is assumed.

(**8a**)   (**8b**)

The bonding discussed above refers to the lowest (ground) electronic states. All molecules may in principle be promoted to an excited electronic state by absorption of a quantum of radiation of the appropriate energy. This results in the reorganization, in the main, of the highest lying valence electrons.

1.2   THE ENERGETICS OF REACTIONS

Chemical reactions are the spatial reorganizations of atoms and of the valence electrons which constitute covalent bonds. Except in the case of photo-chemical reactions, only the lowest (ground) electronic state of the reagents is involved and the passage from the reagents to the products may be considered to take place in a continuous fashion. Thus at any intermediate stage during the progress of a reaction, it is possible in principle to describe the reacting system in terms of its energy, the coordinates of all the atoms and so on. The progress of the reaction can then be mapped on a multidimensional 'surface'; the most favourable pathway, the 'reaction coordinate', will be that which requires the system to acquire the least amount of potential energy along the route. In real terms, the problem of determining (by calculation) the potential energy surface of a multi-atom system is an immensely difficult task in computation and only recently have solutions for quite small systems been attempted. However, the concept of a reaction coordinate is qualitatively useful and can aid our understanding of reaction mechanisms.

Consider the case of the simple exchange reaction:

$$A + B—C \longrightarrow A—B + C$$

If the three atoms retain a linear arrangement throughout, the only coordinate variables are the internuclear distances A, B and B, C. A three-dimensional graph of these two variables plotted against the third, potential energy, can be represented in two dimensions by drawing contours of equal energy (Figure 1a). It will be noticed that there is a relatively easy route from initial to final states along the energy 'valley' a–b, over the saddle-point b, and finally by descent through the valley b–c, in analogy with the map of a mountain pass. If we plot the potential energy of this route, which is the reaction coordinate, against some measure of progression along it, a curve such as Figure 1b is obtained. This indicates that even in the most favourable circumstances the reacting system must acquire potential energy in order to react, this process being known as 'activation'. The minimum amount of energy required, the 'activation energy', is the difference between the energies of the initial or reagent state, a, and the saddle-point, b, which is known as the 'transition state'. This energy is supplied to the reacting molecules as thermal energy—kinetic,

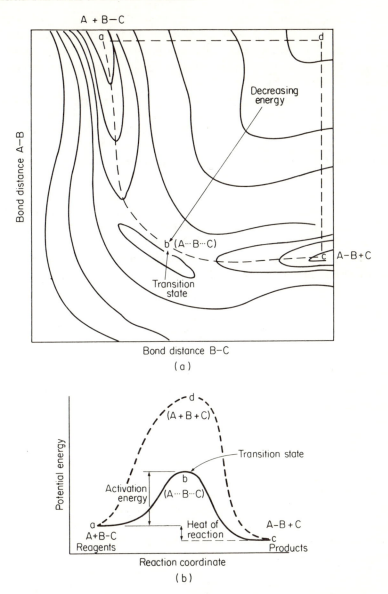

Figure 1. (a) Schematic potential energy surface for a simple displacement reaction; (b) reaction pathways for the concerted reaction (solid line) and non-concerted reaction (dashed line).

vibrational and rotational energy of the reactants. At any instant, only a small fraction of molecules will possess sufficient energy to surmount the potential barrier to reaction. It can be seen that any pathway other than the reaction coordinate between reagents and products will require an even greater amount of activation and will therefore be less favourable. For instance, compare the pathway a–d–c (Figure 1b) by which the B—C bond is first broken and then the A—B bond formed in two separate steps. The activation energy is much larger. It appears, in this case at least, that reaction occurs by a *concerted* bond making and bond breaking, that the energy released by the former in part compensates for the energy required by the latter. The net reaction is exothermic, the heat of reaction being the difference in energy between reagents and products. Another point to note is that the most favourable pathway from between *products* back to *reagents* is the same reaction coordinate; information on the course of a reaction in one direction is therefore applicable for the reverse reaction. This is known as the 'Principle of microscopic reversibility'. A potential energy diagram of this type would be applicable to a displacement reaction at a saturated carbon, such as:

transition state

We know that activation energy is required for almost all chemical reactions to occur, since their rates increase with temperature. At higher temperatures, a greater proportion of the reactants will possess the necessary kinetic energy to pass over the activation barrier.

1.3   SIMPLE AND COMPLEX REACTIONS

The reaction discussed above was characterized by a single activation barrier along the reaction coordinate and may be designated a 'concerted' reaction. Many common reactions are of this type even though they may involve large molecules and the formation and breaking of several covalent bonds, for example in β-elimination:

transition state

and in the Diels–Alder reaction:

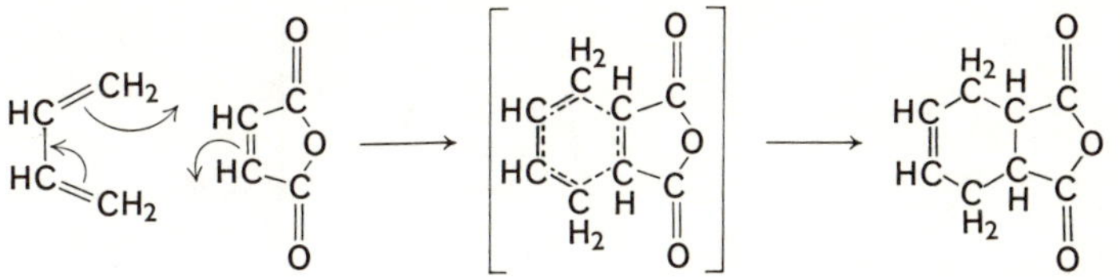

transition state

On the other hand, many organic reactions are more complex in the sense that the reaction coordinate has more than one energy maximum and consequently one or more *reactive intermediates* represented by an energy minimum (Figure 2d–i), but nonetheless a high-energy species with respect to reagents and products. A reactive intermediate has some stability with respect to small displacements along the reaction coordinate unlike a transition state which is unstable. An intermediate may be represented by a conventional valence-bond structure (or series of structures) whereas a transition state must often be represented with partly formed bonds. Intermediates have a finite lifetime which under appropriate conditions may extend to isolation as a stable species. The lifetime of the transition state of a reaction is exceedingly short, probably of the order of a period of vibration (ca 10^{-13} s). Some examples of reactions which proceed via one or more reactive intermediates are given below and will be discussed in more detail in later chapters. Further schematic reaction coordinate diagrams are given in Figure 2.

Tertiary halide hydrolysis

$$(CH_3)_3C{-}Br + H_2O \xrightleftharpoons{\text{slow}} \left[(CH_3)_3\overset{\delta+}{C}{---}\overset{\delta-}{Br}\right] \longrightarrow (CH_3)_3\overset{+}{C}\ Br^-$$

transition state — intermediate carbonium ion

$$(CH_3)_3\overset{+}{C}\ Br^- \xrightarrow[\text{fast}]{H_2O} (CH_3)_3C{-}OH + (H^+)$$

Base decomposition of chloroform

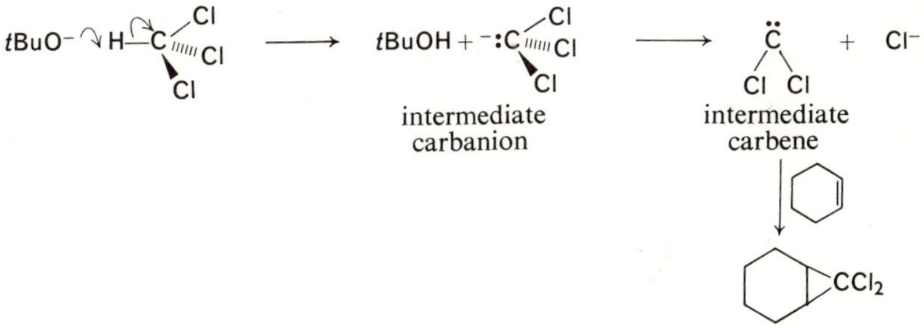

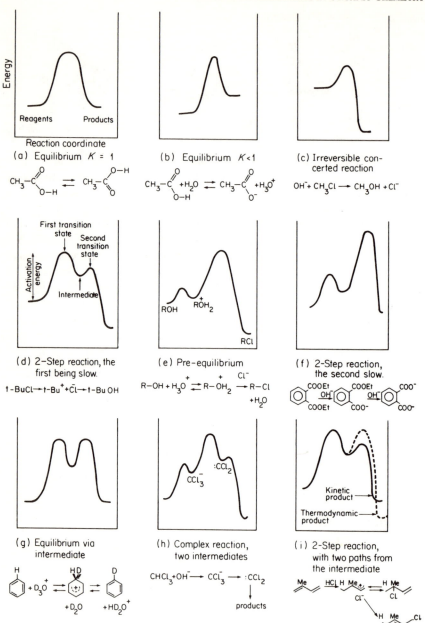

Figure 2. Schematic reaction pathways for various reaction types.

In these two cases, the intermediates are consumed very rapidly and cannot be detected directly. There are examples of intermediates which may be isolated during the normal course of the reaction such as the epoxide (9), from the

chlorohydrin   (9)

hydrolysis of a chlorohydrin, or the half ester (10), from the esterification of a

(10)

dibasic acid. Such stable species will only be mentioned where their presence illustrates an appropriate principle. Also omitted from our discussion of intermediates are those stable compounds which are *intermediate* in some desirable synthesis but which are formed and reacted further under entirely different conditions. For instance, acetylene is an intermediate in this sense in the preparation of polyvinyl chloride, PVC. However the formation and reactions of acetylene are entirely separate steps in the production of the polymer:

$$CaC_2 + H_2O \longrightarrow \underset{\text{acetylene}}{HC\!\equiv\!CH} \xrightarrow{HCl} \underset{\substack{\text{vinyl} \\ \text{chloride}}}{CH_2\!=\!CHCl} \xrightarrow{\text{catalyst}} \underset{\substack{\text{polyvinyl} \\ \text{chloride}}}{(CH_2\!-\!CHCl)_n}$$

1.4   THE DETECTION OF REACTIVE INTERMEDIATES

The elucidation of a reaction mechanism is an exercise which may call for the interplay of a wide range of physical techniques with the processes of inductive reasoning. Some of the techniques which have proved especially fruitful and to which frequent reference will be made, are the following:

(a) Kinetic studies, including the measurement of reaction rates, activation parameters, the effects of substituents and structure, solvents and isotopic substitution on rate; also acid–base catalysis.
(b) The use of isotopic tracers or other means of marking an atom in the reagent.
(c) Direct observation of the intermediate by spectroscopic means.

(d) Isolation of the intermediate even when this entails using conditions far removed from those in the reaction of interest.

(e) 'Trapping' the intermediate by the addition of a suitable reagent with which it forms a stable and characteristic compound.

(f) Determination of the stereochemistry of the reaction.

Where two or more possible mechanisms for a reaction may be considered, experimental evidence will be sought from criteria whose predicted results may afford a distinction to be made. There must be an adequate theoretical understanding of the interpretation and limitations of each technique before its results can be applied with confidence. These principles will now be considered in more detail. Examples of their use will be found throughout the text.

1.5   REACTION KINETICS AND THERMODYNAMICS

The rate of a homogeneous chemical reaction, i.e. one which occurs in a single phase, is given by an equation of the form:

$$\text{rate} = \frac{-d}{dt}[\text{reagent}] = \frac{d}{dt}[\text{product}] = k \cdot \Pi[\text{reagent}]$$

where k is a specific rate constant, and the product of the reagent concentrations includes terms for *all species which take part in the transition state of the rate-determining (slow) step of the reaction.* The rate equation is an empirical statement expressing the dependence of reaction rate, a measurable quantity, on the concentrations of all relevant species. The aim of kinetic measurements is to determine a rate equation to fit the observed rate dependence upon concentration and this in turn will give valuable information concerning the composition of the transition state of the slow step, as the following examples show:

(a) $PhCH_2Br + NMe_3 \longrightarrow PhCH_2^+NMe_3\ Br^-$

$$\text{rate} = k_2[PhCH_2Br][NMe_3]$$

The rate is found by experiment to be proportional both to benzyl bromide and to trimethylamine concentrations. It is inferred from the observed second-order kinetics, that the reaction is *bimolecular*, i.e. one molecule of each reagent is involved in the transition state. This is clearly a reasonable deduction; the reaction is initiated by the collision of suitably activated reagent molecules and the transition state is depicted as **11**, although other evidence is needed to distinguish other possible structures such as **12**.

$$
\begin{array}{cc}
\underset{\delta-}{Br}\text{---}\underset{\overset{|}{H}\ H}{\overset{Ph}{C}}\text{---}\underset{\delta+}{NMe_3} & \underset{\overset{|}{H}}{\overset{Ph}{C}}\overset{\delta-}{\text{--}Br}\ \ \underset{\delta+}{NMe_3} \\
(11) & (12)
\end{array}
$$

(b)

$$
2\,Me_2C{=}C{=}O \longrightarrow
\begin{array}{c}
Me_2C\text{---}C{=}O \\
\mid\qquad\ \ \mid \\
O{=}C\text{---}CMe_2
\end{array}
\qquad
\begin{array}{c}
Me_2C{=\!=\!=}C{=}O \\
\vdots\qquad\quad\vdots \\
O{=}C{=\!=\!=}CMe_2
\end{array}
$$

$$(14)\qquad\qquad\qquad (13)$$

$$\text{rate} = k_2[Me_2C{=}C{=}O]^2$$

The squared concentration term again indicates that the reaction is bimolecular; this time two molecules of dimethylketen (14) are required in the composition of the transition state, depicted as 13 though the kinetic evidence alone says nothing of the bonding in the transition state.

(c) $$Me_3C\text{---}Br + OH^- \longrightarrow Me_3C\text{---}OH + Br^-$$

$$\text{rate} = k_1[Me_3C\text{---}Br]$$

The rate is found to be independent of the hydroxide-ion concentration although this is manifestly consumed in the reaction. One must therefore conclude that the hydroxide ion reacts in a fast step which is subsequent to the slow step of the reaction. It follows that the transition state of the slow step involves only a molecule of alkyl bromide and that since there are two steps to the reaction, there must be a reactive intermediate formed between them. These results are compatible with the following mechanism. One thing

$$
\underset{Me}{\overset{Me}{Me}}C\text{---}Br \ \xrightarrow[\text{slow}]{H_2O}\
\left[\ \underset{Me}{\overset{Me}{Me}}\overset{\delta+}{C}{\text{---}}\overset{\delta-}{Br}\ \right] \longrightarrow
$$

transition state

$$
\underset{Me}{\overset{Me}{\underset{\ \ }{C^+}}}\ \overset{Br^-}{} \xrightarrow{OH^-} \xrightarrow[\text{fast}]{OH^-} \underset{Me}{\overset{Me}{C}}\text{---}OH
$$

intermediate

further must be mentioned: the rate equation will *not* be found to contain concentration terms of substances which are in large excess, since their concentrations change negligibly during the course of the reaction. In the last example it is quite possible and indeed likely that water is involved in the transition state, but since it is in vast excess we are unable to measure the kinetic order with respect to this component of the reaction medium. As far as the kinetics are concerned, the following mechanism for the hydrolysis of *t*-butyl bromide is equally compatible with the data:

transition state

On chemical grounds we would think the latter mechanism unlikely since it is known that hydroxide ion is a far more powerful nucleophile than water and it would be unlikely that it would not take part in a bimolecular reaction in preference to water, even allowing for the large concentration difference of the two species. Other information such as substituent effects also allows us to rule out this mechanism.

Similarly, in strictly neutral aqueous solution, the bromination of acetone is independent of bromine concentration:

(d) $CH_3COCH_3 + Br_2 \longrightarrow CH_3COCH_2Br + HBr$

$$\text{rate} = k_1[CH_3COCH_3]$$

This is interpreted by the following mechanism in which the ketone is converted in the slow step to a reactive species, the enol (15), which in turn reacts rapidly with bromine:

(15)

In general, wherever the observed rate is independent of the concentration of an obvious reagent species, and one which is not present in large excess, or when fractional orders are revealed, a multi-step reaction probably occurs and at least one reactive intermediate is implicated. In such cases one must conceive of as many reaction schemes as are allowable on chemical and other grounds, deduce the rate law for each and compare the results with the experimental rate law.

Higher (e.g. third-) order reactions are not uncommon but in solution usually reveal hidden complexities. The simultaneous collision of three particles is relatively unlikely:

$$A + B + C \xrightarrow[k_3]{\text{slow}} [A...B...C] \longrightarrow \text{products}$$

In the gas phase, termolecular processes involving two reacting species and one 'third body' to carry away the energy of reaction are fairly usual, but in solution the third body would be a solvent molecule and hence would not appear in the kinetics. The usual situation which leads to third-order kinetics is a pre-equilibrium before the slow step,

$$A + B \rightleftharpoons C$$

$$C + D \xrightarrow{\text{slow}} \text{products}$$

$$\text{rate} = [A][B][D]$$

If the equilibrium lies well to the left, the intermediate, C, may not be detectable but its concentration (dependent upon A and B) directly affects the rate. The reaction between ethylene oxide (16) and aqueous HCl or HBr is of third order,

$$\text{rate} = k_3 \left[\begin{array}{c} CH_2\!-\!CH_2 \\ \diagdown\!\diagup \\ O \end{array} \right] \left[H_3O^+ \right] \left[\text{hal}^- \right]$$

and points to an equilibrium protonation of the epoxide ring and rate-determining attack upon the conjugate acid of ethylene oxide (17). The form of

$$\begin{array}{ccc} CH_2\!-\!CH_2 + H_3O^+ & \rightleftharpoons & CH_2\!-\!CH_2 \xrightarrow[\text{hal}^-]{\text{slow}} CH_2\!-\!CH_2\!-\!\text{hal} \\ \diagdown\!\diagup & & \diagdown\!\diagup \qquad\qquad\quad | \\ O & & O^+ \qquad\qquad\quad OH \\ & & | \\ & & H \end{array}$$

(16)   (17)

the kinetics would be inconsistent with alternative mechanisms such as,

$$\begin{array}{ccc} CH_2\!-\!CH_2 + \text{hal}^- \xrightarrow{\text{slow}} & CH_2\!-\!CH_2\!-\!\text{hal} \xrightarrow[\text{fast}]{H_3O^+} & CH_2\!-\!CH_2\!-\!\text{hal} \\ \diagdown\!\diagup & \quad | & \quad | \\ O & \quad O^- & \quad OH \end{array}$$

$$\text{rate} = k' \left[\begin{array}{c} CH_2\!-\!CH_2 \\ \diagdown\!\diagup \\ O \end{array} \right] \left[\text{hal}^- \right]$$

or

$$CH_2\!\!-\!\!CH_2 + H_3O^+ \xrightarrow{\text{slow}} \overset{+}{CH_2}\!\!-\!\!CH_2 \quad \xrightarrow[\text{fast}]{\text{hal}^-} \quad CH_2\!\!-\!\!CH_2\!\!-\!\!hal$$
$$\underset{O}{\diagdown\diagup} \qquad\qquad\qquad |\;\;\;\; \qquad\qquad\qquad\qquad |$$
$$\qquad\qquad\qquad\qquad OH \qquad\qquad\qquad\qquad\quad OH$$

$$\text{rate} = k'' \left[\underset{O}{\overset{CH_2\!-\!CH_2}{\diagdown\diagup}} \right] \left[H_3O^+ \right]$$

But it would be consistent with a termolecular process, though this is considered

$$CH_2\!\!-\!\!CH_2 + H_3O^+ + hal^- \longrightarrow \left[\begin{array}{c} \overset{\delta-}{hal} \\ CH_2\!\!-\!\!CH_2 \\ O \\ H \\ \underset{H-O}{\overset{\delta+}{}} \\ H \end{array} \right] \longrightarrow \begin{array}{c} hal \\ CH_2\!\!-\!\!CH_2 \\ OH \\ H-O \\ H \end{array}$$

unlikely for reasons given. The fact that a reaction obeys a kinetic law which contains the concentration of a reagent does not necessarily imply that this species is the reacting entity in the slow step. For instance, the nitration of

$$HNO_3 + \underset{(18)}{\overset{NO_2}{\bigodot}} \xrightarrow{H_2SO_4} \overset{NO_2}{\underset{NO_2}{\bigodot}}$$

(18)

nitrobenzene (18), under certain conditions obeys the rate law, but we would be

$$\text{rate} = k_2[\text{PhNO}_2][\text{HNO}_3]$$

wrong to ascribe this to the rate-determining attack of a nitric acid molecule on nitrobenzene. Non-kinetic evidence indicates that the reacting species in nitration is the nitronium ion NO_2^+ which may be produced from nitric acid by the action of sulphuric acid,

$$HNO_3 + 2H_2SO_4 \rightleftharpoons NO_2^+ + H_3O^+ + 2HSO_4^-$$

The concentration of the nitronium ion, depends on the HNO_3 and hence this mechanism is *kinetically equivalent* to direct attack by HNO_3. Kinetic evidence for the correctness of this interpretation is suggested by the rate equation which is obeyed by a very reactive substrate such as anisole (19),

$$\text{rate} = k_1[HNO_3]$$

or, with nitric acid in large excess,

$$\text{rate} = k_0$$

Now it is unlikely that a radical change of mechanism has occurred, yet the reaction rates are notably faster. It is consistent with the kinetics and an economy of assumptions that the attack, as before, is by NO_2^+ on the anisole but this occurs at a rate which is *faster than the rate of production of* NO_2^+. The nitrating agent therefore is consumed as fast as it is generated and adding more anisole will not make the reaction go faster. It has also been shown that an additional limit to the rate of this reaction is the rate at which the reacting species can diffuse together. With such reactive aromatic molecules, *diffusion control* also limits the rate of nitration which becomes similar for many molecules of high, but nonetheless different, intrinsic reactivity. An intermediate situation may be encountered in the nitration of benzene in which both processes are of comparable rate, with the result that the order of reaction with respect to benzene is non-integral, between 0 and 1. Much more complex situations can arise if a reaction proceeds by a series of steps, some of whose rates are comparable. Each case must be considered individually and analysed by comparing experimental concentration dependence with that predicted on the basis of possible reaction schemes. The decomposition of some dialkyl peroxides in solution might be imagined to be a simple unimolecular process, but the kinetics reveal complexities, e.g.

$$Bu\!-\!O\!-\!O\!-\!Bu \xrightarrow{\text{slow}} BuO\!\cdot + \cdot OBu \xrightarrow{\text{fast}} \text{products}$$

$$\text{rate} = k[ROOR] + k'[ROOR]^{1.5}$$

The decomposition contains a first-order term but, in addition, a higher order dependence shows that some other decomposition pathway is occurring. The following sequence is proposed:

(a) $\text{ROOR} \xrightarrow{k_1} 2\text{RO·}$

(b) $\text{RO·} + \text{HS} \xrightarrow{k_2} \text{ROH} + \text{S·}$

(c) $\text{S·} + \text{ROOR} \xrightarrow{k_3} \text{ROS} + \text{RO·}$

(d) $2\text{S·} \xrightarrow{k_4} \text{S—S}$

(e) $\text{RO·} + \text{S·} \xrightarrow{k_5} \text{ROS}$

where HS is a protic solvent, e.g. if ethanol, then S = OEt. The reaction scheme is a simple unimolecular decomposition of peroxide (a), and an induced decomposition (b) and (c) by a radical chain process in which the products of reaction themselves cause further decomposition. Steps (d) and (e) are possible chain-terminating reactions which remove the chain-carrying radicals.

If we assume that step (e) is negligible we may derive a rate equation thus:

$$\text{rate} = -\frac{d}{dt}[\text{ROOR}] = k_1[\text{ROOR}] + k_3[\text{ROOR}][\text{S·}] \tag{1.1}$$

we now assume that the concentrations of the reactive species, S· and RO·, are stationary, that is the rate of formation equals the rate of destruction (stationary state approximation), whence [S·] may be obtained from

$$k_2[\text{RO}][\text{HS}] = k_3[\text{S·}][\text{ROOR}] + k_4[\text{S·}]^2 \tag{1.2}$$
$$\text{(production)} \qquad\qquad \text{(destruction)}$$

and [RO·] from

$$k_2[\text{RO·}][\text{HS}] = \tfrac{1}{2}k_1[\text{ROOR}] + k_3[\text{ROOR}][\text{S·}] \tag{1.3}$$
$$\text{(destruction)} \qquad\qquad \text{(production)}$$

Combining equations 1.2 and 1.3

$$k_3[\text{S·}][\text{ROOR}] + k_4[\text{S·}]^2 = k_3[\text{ROOR}][\text{S·}] + k_1[\text{ROOR}] \tag{1.4}$$

$$k_4[\text{S·}]^2 = k_1[\text{ROOR}] \tag{1.5}$$

$$[\text{S·}] = (k_1/k_4[\text{ROOR}])^{0.5} \tag{1.6}$$

Substituting for [S·] in equation 1.1,

$$\text{rate} = k_1[\text{ROOR}] + (k_3 k_1/k_4)^{0.5}[\text{ROOR}]^{1.5}$$
$$= k_1[\text{ROOR}] + k'[\text{ROOR}]^{1.5} \tag{1.7}$$

Hence this scheme accords with the observed kinetic dependence. Had the termination step been reaction (e), with (d) negligible, quite a different result would be obtained. Assuming

$$\frac{d}{dt}[S\cdot] = 0$$

$$k_2[RO\cdot][HS] = k_3[S\cdot][ROOR] + k_5[RO\cdot][S\cdot] \qquad (1.8)$$

and assuming

$$\frac{d}{dt}[RO\cdot] = 0$$

$$\tfrac{1}{2}k_1[ROOR] + k_3[ROOR][S\cdot] = k_2[RO\cdot][HS] + k_5[RO\cdot][S\cdot] \qquad (1.9)$$

Combining equations 1.8 and 1.9,

$$\tfrac{1}{2}k_1[ROOR] = 2k_2[RO\cdot][HS] - 2k_3[S\cdot][ROOR]$$

whence

$$[S\cdot] = \frac{k_2[RO\cdot][SH]}{k_3[ROOR]} - \frac{k_1}{4k_3} \qquad (1.10)$$

But [RO·] is constant, and [HS] is in large excess (constant), hence

$$[S\cdot] = \frac{k''}{[ROOR]} - c$$

Substituting in equation 1.1

$$\text{rate} = k_1[ROOR] + k_3 k'' \frac{[ROOR]}{[ROOR]} - c$$

$$= k_1[ROOR] + C \qquad (1.11)$$

i.e. first order kinetics would be observed although the more complex scheme operates.

By manipulating the temperature (the range $-80°$ to $400°$ is relatively easily available), the rates of a great many organic reactions can be adjusted so that they occur over periods ranging from minutes to days and can thereby be followed by appropriate sampling techniques. Some processes occur at very much faster rates which, if they are not equilibrium processes, require very special apparatus for their measurement. The recombination of oppositely charged ions, for example, often occurs at rates governed by the frequency of their collisions, each collision resulting in reaction. This is known as a *diffusion-controlled reaction* and bimolecular rate constants may be as high as 10^{11} l mol^{-1} s^{-1}. Acid–base proton transfers in water are almost as fast (often ca 10^{10} l mol^{-1} s^{-1}). Proton transfer in a protic solvent can occur by a co-operative mechanism by which the acid relays a proton through solvent

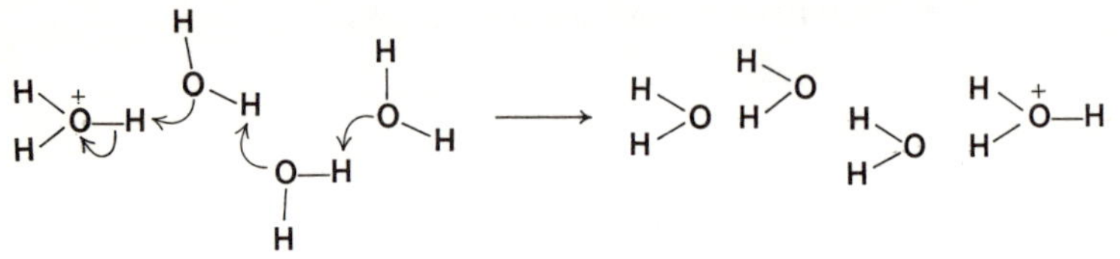

molecules. Consequently, proton transfers are not likely to be rate-determining steps (though see Section 4.3) and, if they occur subsequent to a slow reaction, are not kinetically detectable. A proton transfer prior to the slow step will however be revealed by kinetics and the acid (or base) will appear in the rate law. The acid-catalysed hydrolysis of an ester illustrates these points. The

$$R \cdot COOR' + H_3O^+ \longrightarrow R \cdot COOH + R'OH + (H^+)$$
$$\text{rate} = k_2[R \cdot COOR'][H_3O^+]$$

kinetics can only tell us that one molecule of ester and a proton are involved in the slow step. By a combination of other evidence (see Section 8.3) it appears that the mechanism can be formulated as follows:

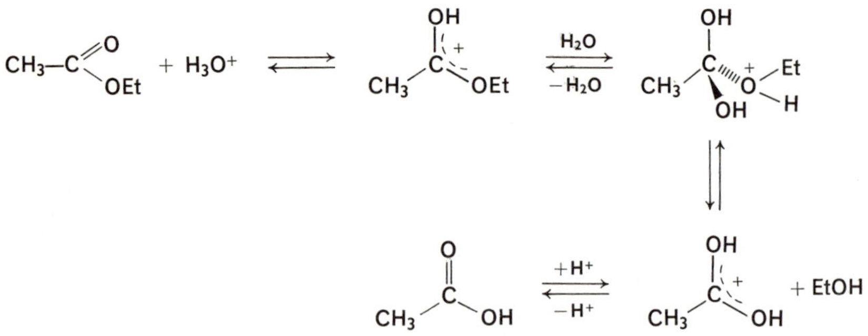

Several proton transfers are involved and the whole scheme is reversible. Some of the intermediates have been detected as stable species in highly acidic solution.

1.6 RELATIVE RATE STUDIES

In addition to determining reaction order—the concentration terms which appear in the rate equation—kinetic measurements can serve a completely different purpose if specific rate constants are measured while a systematic change in the reaction variables is made. For instance, the effects on rates of change in structure, solvent, isotopic substitution as well as temperature and pressure, may give unequivocal information on reaction mechanisms and the presence of reactive intermediates. Before discussing examples of these techniques, some precautions must be entered.

It is frequently implied that the effects of structural change on reactions of a series of compounds with similar functional groups are mirrored in their rates of reaction, measured under identical conditions. This may be true; for example the bromination of benzene and its homologues is uniformly facilitated by methyl substitution (which we know is in turn due to the electron-donating capabilities of the methyl group).

$ArH + Br_2 \longrightarrow ArBr + HBr$	Relative rate, k_{rel}
Benzene	1.0
Methylbenzene	600
1,4-Dimethylbenzene	2500
1,3-Dimethylbenzene	5×10^5
1,2,4-Trimethylbenzene	1.2×10^6
1,2,3,5-Tetramethylbenzene	1×10^7
Pentamethylbenzene	1×10^9

It would however be erroneous to conclude that the same effect is predominating in the hydrolysis of methyl bromide and its methyl-substituted homologues:

$R—Br + OH^- \longrightarrow R—OH + Br^-$		
	Alkyl group	k_{rel}
Bimolecular mechanism	CH_3-	1.0
$\downarrow$	CH_3CH_2-	0.08
	$(CH_3)_2CH-$	4.5×10^{-3}
Unimolecular mechanism	$(CH_3)_3C-$	48.3

The explanation of this reactivity lies in the gradual change of mechanism between the methyl and *t*-butyl compounds (Section 2.3). *In comparing rates one must be sure that the same mechanism is operating.* Compare also the relative rates of hydrolysis of the following alkyl bromides in which successive methyl substitution is made on the β-carbon;

$$R\!-\!Br + OH^- \longrightarrow R\!-\!OH + Br^-$$

Alkyl group	k_{rel}
$CH_3\!-\!CH_2\!-$	1.0
$CH_3CH_2\!-\!CH_2\!-$	0.28
$(CH_3)_2CH\!-\!CH_2\!-$	0.03
$(CH_3)_3C\!-\!CH_2\!-$	4×10^{-6}

On the basis of the observed effects upon aromatic bromination we might be tempted to argue that methyl substitution here disfavours the reaction by its electronic effect (i.e. by electron donation). This would be unreasonable on several grounds; firstly the effect is large especially for the last member of the series, neopentyl bromide, and from other experience we would not expect the fairly mild electronic effect of a methyl group to be felt so strongly when three bonds are removed from the reaction centre in a saturated system. Secondly, the effect of the third methyl group is much larger than that of either the second or the first; now increments in electronic perturbations usually *fall off* in magnitude as one increases the number of perturbing groups. This is a so-called 'saturation' effect, e.g.

Acid	pK_a	Increment
CH_3COOH	4.786	
		1.964
$ClCH_2COOH$	2.824	
		1.457
$Cl_2CHCOOH$	1.366	
		0.701
Cl_3CCOOH	0.665	

The fact that methyl substitution does not show a saturation effect suggests that it is not operating by electronic perturbation. In fact the explanation of these hydrolysis rates is in the steric effects of the methyl groups which effectively prevent the approach of the reagent to the substrate. One must therefore be careful in the interpretation of rate data in case one is observing different types of effect in different situations, by a given type of structural change.

Yet another pitfall may arise when comparing two reactions which have significantly different activation energies, E_a, (Section 1.8). An Arrhenius plot of $\log k$ for each system against the reciprocal temperature (Figure 3) shows that at high temperatures reaction A goes faster than B, while at low temperatures the opposite is true. A comparison of rates at one temperature would

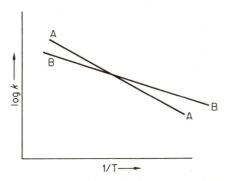

Figure 3. Arrhenius plot for two reactions
of differing activation energy.

therefore not afford a basis for any conclusions on the effects of structural change and one should properly compare the activation energies. Furthermore, to be meaningful, rate comparisons should also be limited to reactions which have comparable entropies of activation (pre-exponential factors, Section 1.7), otherwise differences in rate may be ascribable to a combination of steric and electronic effects which are difficult to disentangle. For example, the effects of a given set of substituent groups upon a reaction at or near an aromatic nucleus fall into a rational pattern provided that the locations of the substituents are limited to the *meta* and *para* positions. *Ortho* substituents often behave abnormally since they exert a steric as well as an electronic effect on the reaction site, reflected in differences in the activation entropy. Provided that

steric repulsion
between X and NO_2

these limitations are understood, comparisons of reaction rate provide an excellent and sensitive probe into transition-state stability and hence into structure-reactivity relationships.

1.7   TRANSITION STATE THEORY

This assumes that the rate of reaction is the rate at which reacting systems pass over the activation barrier. It assumes further that the transition state is an entity which may be ascribed all the normal thermodynamic properties (free energy, enthalpy, entropy, etc.) and is in equilibrium with the starting materials. In order for an equilibrium to be established it is necessary that only a small fraction of reacting systems which reach the transition state actually pass on to products, the majority returning to the starting materials. If the activation barriers on either side of the transition state are of comparable magnitude, then the reaction as a whole will be reversible if it goes at all (Figure 1). If the product state is much more stable (of lower energy) than the initial state then a non-reversible reaction may occur if the temperature is such that the forward reaction is rapid and the reverse reaction relatively very slow.

An approximate treatment of a simple process may be considered as follows and concerns the situation in which a covalent bond breaks to form products, its vibrational energy being transformed into kinetic energy of the fragments.

The vibrational energy of the bond, E, is related to its frequency, v, by the Planck equation,

$$E = hv$$

and at temperatures around room temperature by the Boltzmann equation,

$$E \approx kT \qquad k = \text{Boltzmann constant}$$

It follows that

$$v \approx \frac{kT}{h}$$

Now v may be considered proportional to the frequency at which the transition state decomposes since this occurs by a continuation of the vibrational motion in the stretching direction. The proportionality is required since, as mentioned, most transition states decompose to starting materials. Hence,

$$\text{rate of reaction} = v[AB]^{+}$$

$$= \kappa v K^{+}[A][B]$$

where the constant of proportionality, κ, is known as the Transmission coefficient.* Since

$$-\Delta G = RT \ln K^{\pm}, \qquad K^{\pm} = e^{-\Delta G^{\pm}/RT}$$

$$\text{rate of reaction} = kT/h \, e^{-\Delta G^{\pm}/RT} [\text{A}][\text{B}] \qquad (1.12)$$

and, since the Gibbs function, G, can be partitioned into enthalpy, H, and entropy, S, components,

$$\Delta G^{\pm} = \Delta H^{\pm} - T\Delta S^{\pm},$$

$$\text{rate of reaction} = \kappa \frac{kT}{h} e^{-\Delta H^{\pm}/RT} e^{\Delta S^{\pm}/R} [\text{A}][\text{B}] \qquad (1.13)$$

also, since for a bimolecular reaction

$$\text{rate} = k_2 [\text{A}][\text{B}] \qquad k_2 = \text{specific rate constant}$$

$$k_2 = \kappa \frac{kT}{h} e^{\Delta S^{\pm}/R} e^{-\Delta H^{\pm}/RT} \qquad (1.14)$$

this may be compared directly with the classical Arrhenius equation which expresses the effect of temperature upon reaction rate,

$$k_2 = A \, e^{-E_a/RT} \qquad (1.15)$$

in which the 'Arrhenius activation energy', E_a, replaces the enthalpy of activation, $\Delta H^{\pm}$, and the pre-exponential term, A, embodies the entropy of activation, $\Delta S^{\pm}$.

Both $\Delta G^{\pm}$ and $\Delta S^{\pm}$ are fundamental quantities which characterize a reaction and from which much may be learned the reaction mechanism. In general, the effect of structural change upon the rate of a reaction may be assessed by a comparison of the set of values of $\Delta G^{\pm}$ for the reacting systems with the proviso that the entropies of activation of each are similar. Under these conditions, one may equate the change in activation energy, $\Delta G^{\pm}$ to the structural change and attempt to rationalize the result in terms of stabilization or destabilization of the transition state.

1.8   FREE ENERGIES AND ENTROPIES OF ACTIVATION

Equation 1.14 shows that the rate of a reaction is determined by both the changes in free energy and in entropy in going from the initial to the transition

---

* 'Primed' symbols, e.g. $\Delta H^{\pm}$, refer to the transition state or to the activation process.

state, i.e. by the terms $\Delta G^{\ne}$ and $\Delta S^{\ne}$. The changes in rate brought about by substituent, solvent and structural change are the result of an interplay between these terms. Figure 4 indicates the effects of these two factors independently upon the rate. It will be seen that a thousandfold increase in rate is brought about by either an *increase* in $\Delta S^{\ne}$ of 57 J K^{-1} or a *decrease* in $\Delta H^{\ne}$ (at 273.2 K) of 16.2 kJ mol^{-1}.

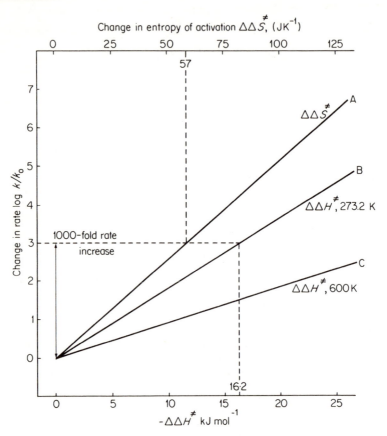

Figure 4. Effects on reaction rate ($\log k/k_0$) produced by a change in entropy of activation (at constant enthalpy), curve A, and by change in enthalpy of activation (at constant entropy), curves B and C.

Entropies of activation refer to changes in vibrational, rotational and translational freedom of all species involved in the transition state, including solvent molecules which are a part of the solvation sphere and not necessarily covalently bound at any time. Absolute values of $\Delta S^{\ne}$ are often quoted in support of reaction mechanisms since the combination of two species together or the coordination of solvent molecules can be expected to produce a negative ΔS^{+} while a positive value is characteristic of dissociative and solvation-releasing processes. For example, the Diels–Alder reaction between a conjugated diene and an olefin commonly has ΔS^{+} in the range -100 to -150 J K^{-1}

mol^{-1} consistent with a 'tight' transition state in which the two components are bound cyclically. Essentially three degrees of translational freedom are lost during the reaction, although compensated to some extent by a gain in

vibrational freedom. At the other extreme, an 'open' or 'dissociative' transition state in which extra degrees of freedom are acquired, has a positive ΔS^{+}. It

$$\Delta S^{+} = +51 \text{ JK}^{-1}$$

$$\Delta S^{+} = +37 \text{ JK}^{-1}$$

must be realized that non-stoichiometric involvement of solvent in the transition state will contribute to, and may even dominate ΔS^{+}, making interpretation of this quantity difficult. The interpretation of absolute values of ΔS^{+} requires there to be some estimate of the expected value for a given mechanism. This may be relatively qualitative (i.e. 'large, negative' for the Diels–Alder reaction) or may, particularly for gas-phase reactions, be calculated by more or less sophisticated methods. Changes in ΔS^{+} for a related series of reactions are somewhat easier to interpret. For example, the series of reactions of 2-substituted pyridines with methyl iodide show the following trend:

| | | $\Delta\Delta S^+$ |
R	R'	(relative to pyridine)
H	H	0
Me	H	−1.3
Et	H	−1.9
iso Pr	H	−2.1
Me	Me	−2.8
t-Bu	H	−4.6

The enthalpies of activation are similar and hence there is no doubt that these values reflect increasing steric hindrance towards the attack of the reagent and a transition state which becomes more rigid and compressed the larger the flanking groups. Interpretations of substituent effects should always be made with the knowledge of which thermodynamic parameter is being affected. It would be easy to ascribe the slow rate of hydrolysis of acetate esters compared with the analogous formates in terms of electron release by the methyl group, a factor which is known to slow the reaction down. However, experiment shows that the rate change is due to entropy changes, the free energies of activation being very similar. The effect of the methyl group is to block the approach of the reagent by its greater size, as also occurs in alkyl-halide hydrolysis. The

nature of the solvent markedly affects entropies of activation and this provides valuable information concerning solvent involvement in the transition state. Solvents of lower 'polarity' (Section 1.15) generally bring about a more negative ΔS^+ in a reaction with an ionic transition state since the *change* in freedom of movement of a solvent molecule is more marked than with a polar solvent which is usually associated and rather restricted. For example, the solvolysis of t-butyl chloride;

$$t\text{Bu—Cl} \quad \longrightarrow \quad (t\overset{+}{\text{Bu}}....\text{Cl}^-) \quad \longrightarrow \quad t\text{Bu}^+ \ \text{Cl}^-$$

	Solvent	$\Delta S^{\ddagger}$
	Water	$+51$ J K^{-1}
Decreasing	Methanol	-19
polarity	Ethanol	-20
	90% acetone/10% water	-71
	90% dioxan/10% water	-78

Entropy studies can reveal whether a transition state resembles reagents or products. The enolization of cyclohexanedione (**20**) is associated with very different values of $\Delta S^{\ddagger}$ for the forward and reverse reactions:

keto

(**20**)

enol

$\Delta S_0 = -88$

Now the entropy difference between the keto and enol forms, ΔS_0, is -88 J K^{-1} which implies that the absolute entropies of the keto form and the transition state are similar and hence their solvation is similar.

1.9   VOLUMES OF ACTIVATION

The net spatial requirements of a reacting system may increase or decrease during passage to the transition state. Hence the effect on rate of decreasing the volume of the system by the application of pressure may be inferred by Le Chatelier's principle as in gas phase reactions. If the volume of the system decreases on activation, external compression will facilitate the reaction, and vice versa. The analogy with a gaseous reaction is rather superficial, however. Liquids, being rather incompressible, require far higher pressures to show a noticeable effect upon rate; in fact the useful pressure range for such studies is measured in kilobars (1 kilobar = 1000 atmospheres approx.). Relatively simple apparatus may be used to study rates at up to 10 kbar, but studies have been extended up to 30–40 kbar for some reactions. To date, the pressure dependence of some 500 systems has been measured.

The change in volume upon activation (volume of activation), ΔV_0^+ is defined by the equation,

$$\left(\frac{\delta \ln k}{\delta P}\right)_T = \frac{-\Delta V_0^+}{RT} \qquad \text{where } P = \text{pressure}$$

Typical values of ΔV_0^+ vary between +25 and −25 cm³, a negative value indicating an increase in rate with pressure, and a positive value, a decrease. Thus, many reactions show a rate dependence of up to three-fold per kbar applied pressure, either positive or negative. In this context, the reacting system includes all solvent molecules which are associated with the reacting molecules, hence volumes of activation can give information concerning changes in solvation on activation. On the other hand this can make interpretation of data difficult since solvation is such an imprecise concept. However, the magnitudes and signs of ΔV_0^+ in general obey the following principles and hence can be useful in mechanistic studies.

	Nature of the slow step	Average ΔV_0^+	$\dfrac{k_{p=1000\,atm}}{k_{p=1\,atm}}$
Bond fission (homolytic)	$A\!-\!B \rightarrow A \cdots B$	+5 to 10	0.8–0.7
Bond fission (heterolytic)	$A\!-\!B \rightarrow A^+ \cdots B^-$	−15 to −20	2.0–3.0
Neutral associations	$A + B \rightarrow A \cdots B$	−10 to −15 per bond formed	1.5–1.9
Ion-forming associations	$A + B \rightarrow A^+ \cdots B^-$	−10 to −30	1.5–3.6
Ion-destroying associations	$A^+ + B^- \rightarrow A^+ \cdots B^-$	+10 to 20	0.7–0.4
Displacement reactions	$(S_N2)\, A + B\!-\!C \rightarrow$ $A \cdots B \cdots C$	−20 to −40	2.0–4.0

The above ranges of values are only approximate and individual structures and conditions will largely determine ΔV_0^+. The magnitude of the volume change will in general be greater the larger the molecules involved in the reaction; ring-closure reactions tend to increase activation volumes while the effect of solvents is profound. The above table indicates that negative volume changes are associated with the generation of charge, which is due to the increasing attraction of the transition state for solvent molecules or 'electrostriction'.

The following reactions exemplify these principles. Diels–Alder reactions and other concerted processes show large negative values of ΔV_0^+ consistent with simultaneous bond formation at both ends of the system and not consistent with non-concerted reactions which would show more positive values.

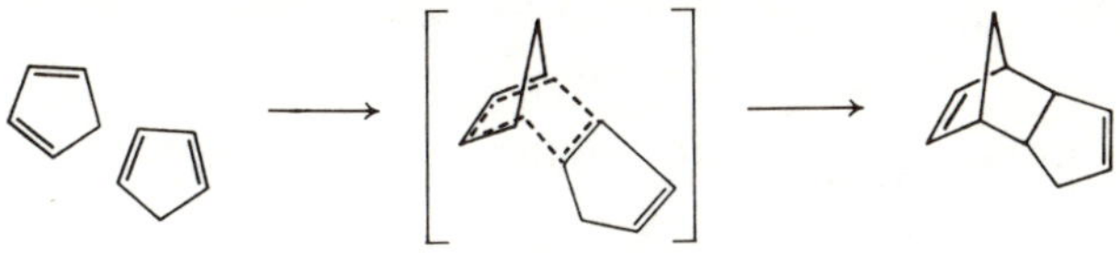

$$\Delta V_0^+ = -24 \text{ cm}^3$$

Diels–Alder reaction (dimerization of cyclopentadiene)

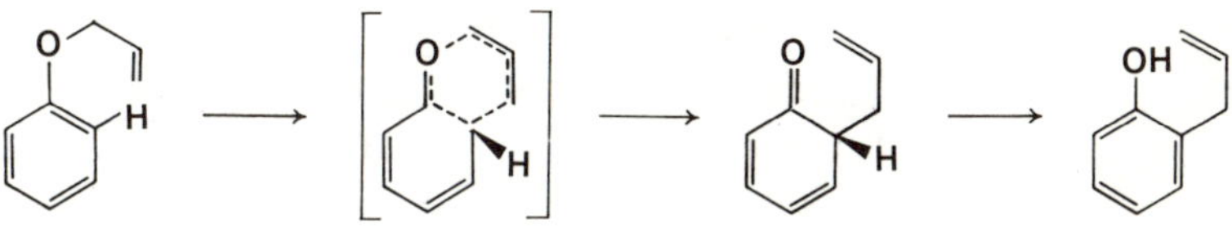

$$\Delta V_0^+ = -15 \text{ cm}^3$$

Claisen rearrangement (of allyl phenyl ether)

Nucleophilic displacements at saturated carbon both by S_N1 and S_N2 mechanisms are accelerated by pressure, little distinction between the two being made,

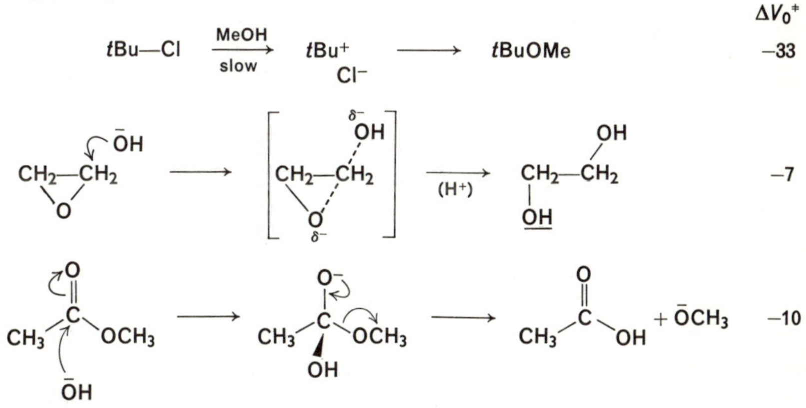

$$\Delta V_0^+$$

tBu—Cl $\xrightarrow[\text{slow}]{\text{MeOH}}$ tBu^+ $\longrightarrow$ $tBuOMe$ -33
                                    Cl^-

-7

-10

unless the charge is destroyed or neutralized:

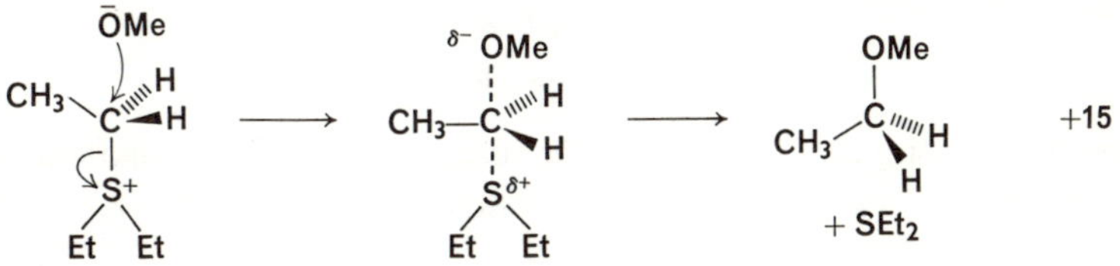

$+15$

Dissociative processes are associated with a positive $\Delta V_0^{\ddagger}$ also:

$$Ar\text{—}N_2^{+} \longrightarrow Ar^{+}....N_2^{+} \longrightarrow Ar^{+} + N_2 \qquad +10$$

$$Me_2C\overset{+}{\underset{CH_2CH_3}{\overset{SMe_2}{\diagdown}}} \longrightarrow Me_2\overset{+}{C}\text{—}CH_2CH_3 + SMe_2 \qquad +14$$

$$tBuO\text{—}OtBu \xrightarrow[C_6H_6]{\Delta} tBuO....OtBu \longrightarrow 2tBuO\text{·} \qquad +13$$

But the nature of the solvent is always important, even in a homolytic reaction:

$$2tBuO\text{—}OtBu \xrightarrow{\Delta} 2tBuO\text{·}$$

Solvent	$\Delta V_0^{\ddagger}$
Carbon tetrachloride	13.3
Benzene	12.6
Cyclohexane	6.7
Toluene	5.4

It is likely that carbon tetrachloride and benzene form more efficient 'cage' structures under pressure which, by effectively preventing the escape of radicals to form products, reduce the rate of decomposition of the peroxide.

1.10   SUBSTITUENT EFFECTS ON REACTION RATE

Valuable information could potentially be obtained if we were able to perturb the valence electrons in the vicinity of a reaction centre in some predictable way and observe the effect upon the reaction rate. In practice this can be done by the introduction of substituent groups which do not themselves take part in the reaction but by supplying or withdrawing electrons according to their nature may affect the stability of the transition state.

That substituent effects may be very significant is apparent from qualitative experiments on the nitration of some aromatic compounds (see also Section 1.6). Benzene is nitrated fairly readily by concentrated nitric–sulphuric acids while toluene reacts much more readily under the same conditions. Phenol is nitrated, even in dilute aqueous nitric acid, while nitrobenzene requires nitric acid–oleum at 100°C to form the dinitro compound. We can say that the order of ease of nitration of C_6H_5X, is

$$X = OH \gg Me > H \gg NO_2$$

However, the point to realize is that the interpretation of these data demands that we know *either* the electronic demands of the reaction *or* the effects produced by these groups. Knowing one we can deduce the other. Thus the above series might indicate an order of increasing electron-withdrawing power and a reaction that is facilitated by electron release, or the converse; an order of increasing electron release and a reaction which demands electron withdrawal. Further information is necessary to distinguish between these two possibilities; as will be seen, the former is the correct one. It is necessary further to ascertain that substituent groups behave consistently, that a strongly electron-withdrawing group will behave as such in every situation. These questions were examined by L. P. Hammett, around 1935, who chose as a primary measure of substituent action their effect on the dissociation of benzoic acid, the substituents being placed in *meta* and *para* positions. Changes

in the dissociation constant brought about by *meta* or *para* substitution should affect only the free-energy change ($\Delta\Delta G^+$ since this is an equilibrium) since, as has been mentioned, the rigid aromatic ring prevents steric interactions between carboxyl group and substituent. Furthermore, since dissociation involves the loss of a proton from the carboxyl group we can *predict* reliably that the process is assisted by electron withdrawal. This is in accordance with other series of acids, for instance the hydrides of first-row elements whose acidity increases strongly with electronegativity:

$$\longrightarrow \quad \text{increasing acidity}$$
$$\longrightarrow \quad \text{increasing electronegativity}$$

Hammett defined the Substituent Constant, σ, as

$$\sigma = \log \frac{K}{K_0}$$

where K_0 is the dissociation constant of benzoic acid in water at 25° and K that of the substituted benzoic acid under the same conditions: σ is appropriate to a given substituent in a definite position (*meta* or *para*). The logarithmic relationship arises since we are comparing effects of the substituent on the free energy of dissociation for which

$$\Delta\Delta G^{+} = RT\log K$$

For example, the dissociation constants of benzoic and *para*-chlorobenzoic acids are 6.25×10^{-5} and 10.0×10^{-5} respectively:

$$\sigma(p\text{-Cl}) = \log\left(\frac{10.0 \times 10^{-5}}{6.25 \times 10^{-5}}\right) = 0.204$$

Table 1. Dissociation constants of substituted benzoic acids and substituent constants

Substituent	Effect	$10^5\,K^a$		Hammett substituent constant σ^b ($= \log K/K_0$)		Augmented substituent constant σ^+ (*para*)
		meta	*para*	*meta*	*para*	
$-NH_2$	$+M > -I$	3.07	5.13	-0.16	-0.66	-1.7
$-OH$	$+M, -I$	8.33	2.62	$+0.12$	-0.37	
$-OMe$	$+M, -I$	8.17	3.38	$+0.115$	-0.27	-0.8
$-t\text{Bu}$	$+M$	5.2	3.98	-0.10	-0.20	
$-Me$	$+M$	5.35	4.24	-0.07	-0.17	-0.32
H	—		$6.25 = K_0$	0	0	
$-F$	$+M < -I$	14.0	7.22	$+0.34$	$+0.06$	
$-Cl$	$+M < -I$	15.0	10.0	$+0.37$	$+0.23$	$+0.11$
$-Br$	$+M < -I$	15.2	10.7	$+0.39$	$+0.23$	
$-COOH$	$-M, -I$	24.0	29.0	$+0.35$	$+0.41$	σ^- (*para*)
$-COOEt$	$-M, -I$	17.6		$+0.37$	$+0.45$	
$-CN$	$-M, -I$	25.2	28.0	$+0.56$	$+0.66$	$+0.92$
$-IO_2$	$-M, -I$	25.2	28.8	$+0.70$	$+0.76$	
$-NO_2$	$-M, -I$	35.5	36.1	$+0.71$	$+0.78$	$+1.24$
$-NMe$	$-I$			$+0.88$	$+0.82$	

[a] Values of K are taken from G. Kortum, W. Vogel and K. Andrussow, *Dissociation Constants of Organic Acids in Aqueous Solution*, Butterworth, London, 1961.
[b] Values of σ are averaged from the measured substituent effects upon a large number of reactions and will not in all cases agree exactly with values obtained from the benzoic acid acidities.

Similarly values of σ are obtained for all other substituent groups which it is desired to consider. The *para*-NO_2 group is clearly acid strengthening and therefore must be electron withdrawing *relative to the corresponding substituent in benzoic acid* (H). Electron-withdrawing groups will have a positive value of σ and electron-donating ones, a negative one, with hydrogen being zero by definition. Values of σ are set out in Table 1 for a number of important groups. It should be noticed that values of σ are not the same for *meta* and *para* positions; thus, the value of K for *meta*-chlorobenzoic acid is 15×10^{-5} whence σ (*meta*-chloro) = 0.380.

The values of substituent constants may be rationalized if one considers two contributing factors operating. The *Inductive* effect results from polarization of mainly the σ-bond framework between substituent and reaction centres due to differences in electronegativity. The effect is abbreviated I, $-$ or $+$ according to whether withdrawing or donating in direction, and is usually more strongly felt from the nearer *meta* position. The *Mesomeric* (M) effect operates by

conjugation between substituent and aromatic ring or other unsaturated system. Atoms which bear a lone pair of electrons adjacent to the ring (N, O, S, halogen) are able to supply electrons by this mechanism particularly to positions *ortho* and *para* to the substituent as shown by the valence bond canonical structures for anisole (methoxybenzene):

On the other hand, carbonyl-containing substituents, the nitro, sulphonyl and cyano groups, *withdraw* electrons by a mesomeric effect ($-M$) again affecting mainly the positions *ortho* and *para*. A summary of these effects is given in Table 1.

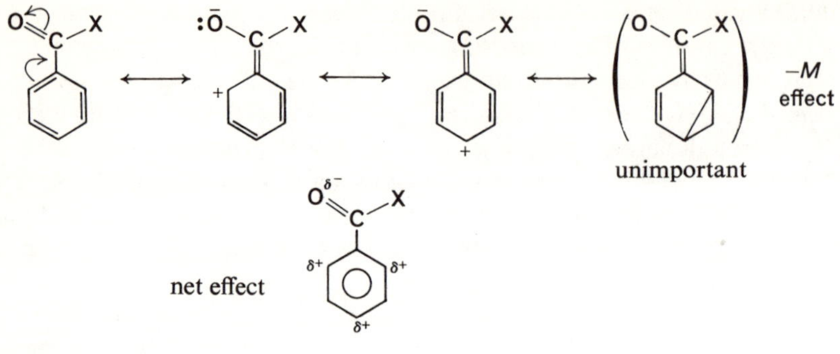

The values of σ obtained from benzoic-acid dissociations describe substituent effects quantitatively for this equilibrium and enable a series of substituent groups to be placed in a regular order. For such a series to be useful for investigating other reaction mechanisms one must have confidence that the same order applies to other reactions. As an example, the rates of base-catalysed hydrolysis of *meta-* and *para*-substituted ethyl benzoates (**21**), may be compared. It is found that electron-withdrawing groups uniformly assist

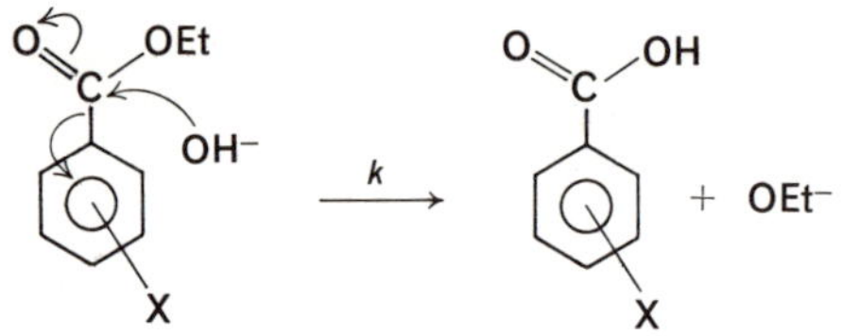

(**21**)

this reaction (Section 8.2). A plot of $\log k/k_0$ for this reaction (k_0 refers to the rate of hydrolysis of ethyl benzoate ($X = H$) and k to the substituted analogue) against σ is a very reasonable straight line proving that substituent effects in the two reactions are, to a very good approximation, parallel (Figure 5). The equation of the line is known as the Hammett equation, and is the best known example of a *linear free-energy relationship*.

$$\log \frac{k}{k_0} = \rho\sigma$$

The slope, ρ, is a constant which is characteristic of the two reactions being compared. In general ρ, the *Reaction Constant*, expresses the degree to which the reaction under examination responds to substituent effects, compared to

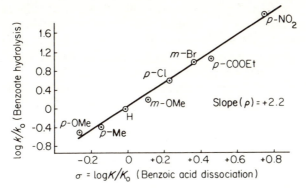

Figure 5. Hammett plot for alkaline hydrolysis of ethyl
benzoates (*meta-* and *para-*substituted).

the standard reaction, benzoic acid dissociation. In the example above,
benzoate ester hydrolysis may be said to be 2.2 times more sensitive to electronic
perturbation than the benzoic acid equilibrium.

The universality of this substituent order has been tested and found to hold
for many hundreds of reactions and equilibria which occur at or near an
aromatic nucleus. A reaction constant may be positive or negative meaning
that the reaction is accelerated by electron withdrawal or electron donation,
respectively. An example of the latter is the reaction of 2,4-dinitrochloro-
benzene (**22**), with a series of substituted anilines:

$$\rho = -3.98$$

$$(22)$$

Not only does this figure confirm a demand for electrons at the reaction centre,
as expressed in the equation, but the very large magnitude of the effect means
that a great deal of charge transfer from the aromatic ring occurs in the
transition state. This type of information is of great value in mechanistic
studies and many examples will be met in later chapters.

Some extensions of the Hammett equation are important. When a reaction creates an electron-deficient centre adjacent to the aromatic ring, substituents with a large $+M$ effect are found to give significant deviations in a simple Hammett plot (Figure 6a). It is as if the amount of electron donation which occurs is far greater than normal, as expressed by the value of σ. An example is the hydrolysis of α,α-dimethylbenzyl halides (23), which leads to an intermediate cation (24) in which direct conjugation with such groups as –OMe, –NMe$_2$ is

(23)   (24)
                                    intermediate cation
                                    (resonance stabilized)

σ^+ (p-OMe) $= -0.8$
(normal $\sigma = -0.17$)

possible from the *para* position and an abnormally large substituent effect is observed. In order to bring these aberrant points back on to the linear plot, new values of σ are computed for use in this situation. In order to distinguish these values from the normal Hammett constants they are denoted σ^+. Similarly, direct conjugation to a negative centre adjacent to the ring can occur with p-NO$_2$, p-CN, p-COOH, etc., necessitating a new set of values of substituent constants, σ^-, for this type of reaction, e.g. the dissociation of phenols

(25)

σ^- (p-NO$_2$) $= +1.24$
(normal $\sigma = +0.78$)

(25), see Figure 6b. Some values for these augmented substituent constants are given in Table 1. It may seem that the arbitrary adjustment of the parameters to fit a preconceived relationship would have no fundamental value; however, this is a convenient method of determining the existence of deviations from the Hammett relationship. If the plot of $\log k/k_0$ for a reaction is linearly related to σ^+, rather than σ, it implies that a vacant orbital adjacent to the ring is being created in the transition state and if correlation with σ^- is found, the creation

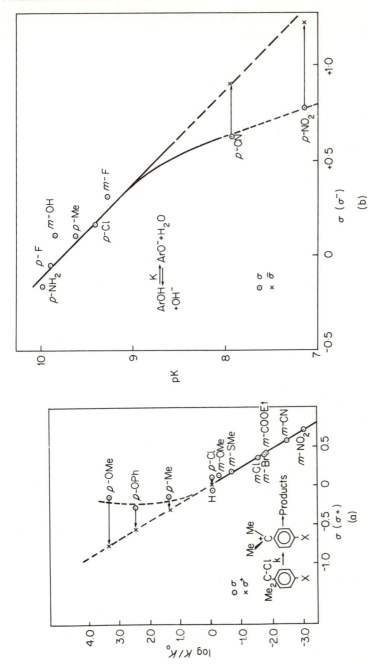

Figure 6. (a) Modified Hammett plot for reaction involving cationic conjugation with the aromatic ring; (b) modified Hammett plot for reaction involving anionic conjugation with the aromatic ring.

of a filled p-orbital, conjugated with the ring is inferred. In this way a great deal of information is obtained concerning the movement of charge in the slow step of the reaction.

1.11   ISOTOPIC EFFECTS UPON RATE

Isotopic substitution may affect reaction rate in a number of ways, all of which can yield valuable mechanistic evidence. All isotope effects ultimately depend upon relative masses of the two nuclides involved and this ratio is greatest for the hydrogen isotopes as are the observed effects. By far the largest number of studies have involved a comparison of a deuterium-substituted reagent with the normal ^1H analogue, but the principles can be extended to other atoms.

Primary Kinetic Isotope Effects

The dissociation energy of a covalent bond depends, other things being equal, upon the nuclear masses which are linked. Since a covalent bond behaves to a good approximation as a harmonic oscillator, its frequency of oscillation, v, is governed by a relationship analogous to Hooke's law for a spring,

$$v = \frac{1}{2}\left(\frac{k}{\mu}\right)^{0.5}$$

where μ is the reduced mass of the system. It equals $m_1 m_2/(m_1 + m_2)$, in which m_1 and m_2 are the separate masses linked by the bond.

   The analogy with a spring is not complete, for the permitted energies of vibration are not unlimited but quantized into discrete vibrational energy states, V_n. The lowest energy state, V_0, which is that populated at low temperatures, even down to absolute zero, is associated with energy E_0, known as zero-point energy for which

$$E_0 = hv = \frac{h}{4\pi}\left(\frac{k}{\mu}\right)^{0.5}$$

The relative zero-point energies associated with a C—H bond and a C—D bond in a large molecule for which m_1 is large and $m_2 = 1$ and 2 respectively, are given by,

$$\frac{E_0(H)}{E_0(D)} \approx \left(\frac{2}{1}\right)^{0.5}$$

This ratio approaches unity as the bond becomes broken but it is clear that more energy will have to be put in to break a C—D bond than a C—H since the former lies at a lower level in the initial state (Figure 7). It follows that if a C—H bond is being broken in the slow step of a reaction, the corresponding C—D compound will react more slowly under the same conditions. This is

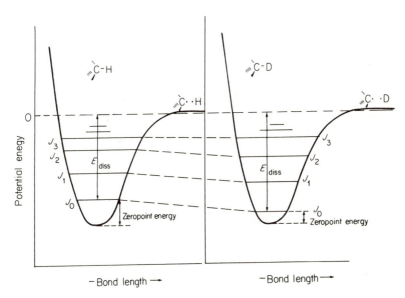

Figure 7. Potential curves for a C—H and a C—D bond.

known as a *Primary Kinetic Isotope Effect*. The same considerations apply to any pair of light and heavy isotopes, e.g. ^{14}C, ^{12}C, and ^{34}S, ^{32}S, although the magnitude of the effect is much smaller. The observation of a primary isotope effect is diagnostic of bond breaking to that isotope in the slow step of the reaction. The magnitudes of the various isotope effects vary with the degree of bond breaking in the transition state, reaching a maximum value if this process is far advanced (Table 2). The isotope effect diminishes with temperature since more of the reacting systems are initially in higher vibrational states in which isotopic differences are smaller. Some examples of reactions showing primary isotope effects follow.

The abstraction of hydrogen from toluene by a bromine atom in solution has $k_H/k_D = 4.6$ around room temperature, less than the maximum value, but still a large effect. The full value is found for the oxidation of isopropanol by

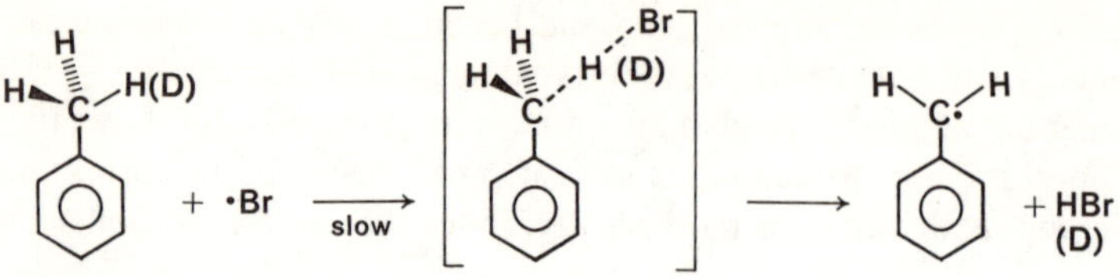

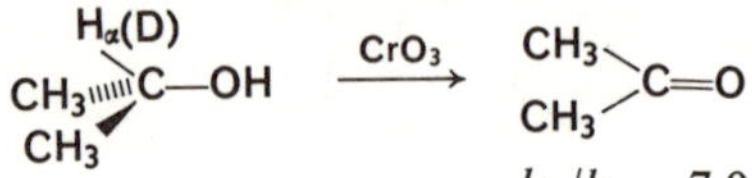

$$k_H/k_D = 7.0$$

chromic acid at 25°. This is diagnostic that the C—H$_\alpha$ bond is broken in the slow step of the reaction. It is probable that it is removed as hydride ion from a preformed chromate ester (Section 9.1).

Table 2. Maximum values of primary isotope effects

Nuclide		k_{light}/k_{heavy}
C—H/C—D	0°	8.3
	25°	6.9
	100°	4.7
	200°	3.4
	500°	2.1
C—H/C—T		~13
$^{12}C/^{13}C$		1.04
$^{12}C/^{14}C$		1.07
$^{15}N/^{14}N$		1.03
$^{16}O/^{18}O$		1.02
$^{32}S/^{34}S$		1.01
$^{35}Cl/^{37}Cl$		1.01

A distinction between the two mechanisms of elimination may be made on the basis of primary deuterium isotope effects; the E2 mechanism (26) occurs

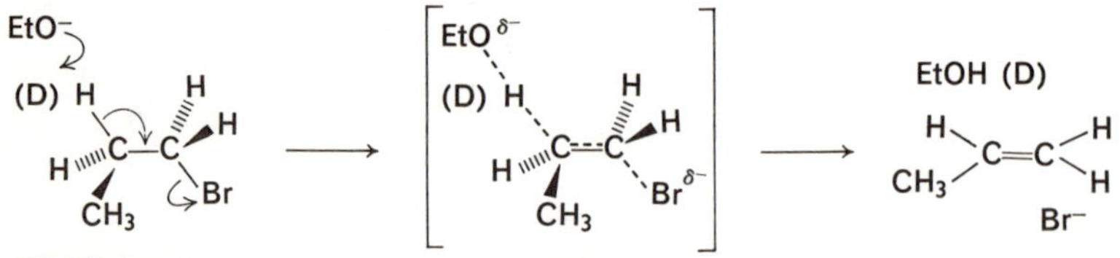

E2 Mechanism   (26) $k_H/k_D = 6.7$

by attack of base on the β-hydrogen atom and the concerted loss of proton and leaving group. On the other hand, the $E1$ mechanism (27) proceeds via a carbonium ion formed in the slow step of the reaction, the loss of β-hydrogen

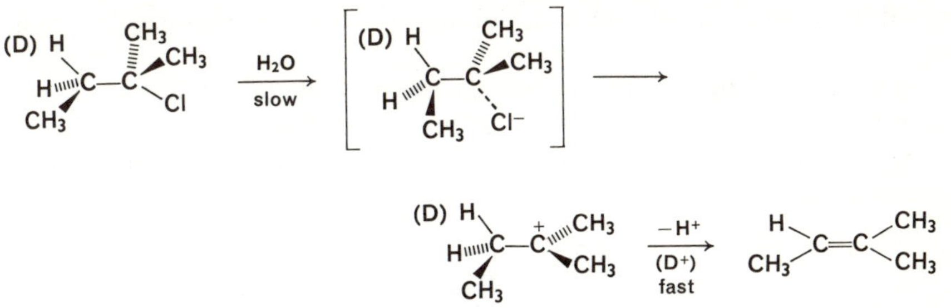

E1 Mechanism   (27) $k_H/k_D = 1.40$

being fast. The small observed isotope effect is probably secondary in nature (Section 1.12). For a transfer reaction, such as is almost invariably the case with the proton (deuteron), a maximum isotope effect is found when hydrogen is bonded equally to two atoms in the transition state. For example, the value

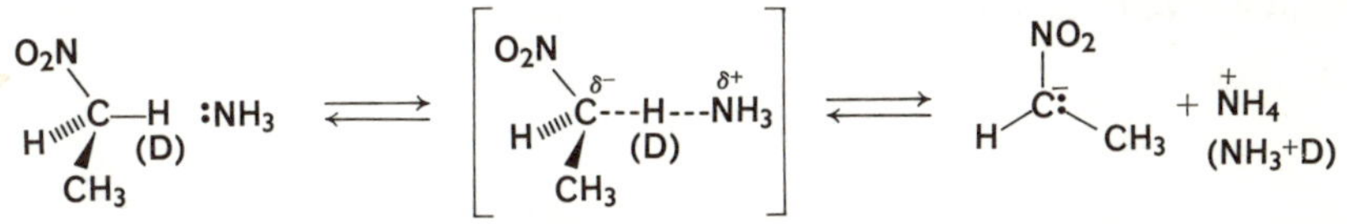

of k_H/k_D is about 10 for the ionization of nitroethane by ammonia, whereas smaller values are found for both stronger and weaker bases with which a less symmetrical transition state would form, being more 'product-like' and 'reagent-like', respectively. Abnormally high isotope effects sometimes occur. That for proton exchange between 2-nitropropane and *s*-trimethylpyridine (28), amounts to 24.2, far greater than the maximum based on a single

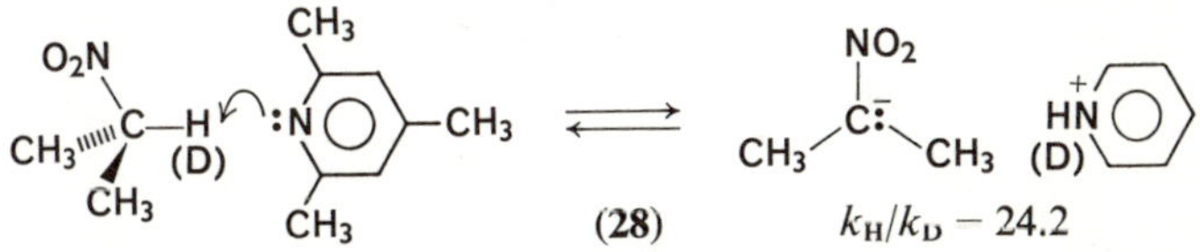

(28) $k_H/k_D - 24.2$

vibrational frequency. The reason for such a large value is not completely understood but quantum-mechanical 'tunnelling' is implicated. There is a finite probability that the reaction system may pass to products without going 'over' the activation barrier. This probability increases the narrower is the activation barrier, i.e. the smaller the atomic displacement which results in conversion of starting material to products. Furthermore, 'tunnelling' in this way is less probable for deuterium than for hydrogen (^{1}H). Hence if this amounts to a significant fraction of the reaction path, further retardation of the deuterium-containing substance on top of the normal isotope effect will occur.

Although hydrogen isotopes are by far the most important, isotope effects of other elements have been observed and serve diagnostically to indicate bond-breaking processes. On account of the much smaller mass differences, these effects are far less than is exhibited by hydrogen (Table 2), e.g.

$$\frac{E_0\,(^{14}N)}{E_0\,(^{15}N)} \approx \left(\frac{15}{14}\right)^{0.5} = 1.035$$

The decomposition of azo compounds shows a nitrogen isotope effect,

$$R-N=N-R \longrightarrow \left[R...N\equiv N...R\right] \longrightarrow R\cdot \; N\equiv N \; R\cdot$$

$$\frac{k_{14}}{k_{15}} = 1.02 \; (2\%)^*$$

A similar value is found for the Hofmann elimination, (*E*2 mechanism)

$$CH_3-CH_2-\overset{+}{N}Me_3 \xrightarrow{\;OH^-\;} \cdots \longrightarrow CH_2=CH_2 + NMe_3 + H_2O$$

$$\frac{k_{14}}{k_{15}} = 1.0186 \qquad (1.86\%)$$

In strong acid solution, the decarboxylation of azulene-1-carboxylic acid (**29**), shows a carbon isotope effect confirming that C—C bond fission occurs in the slow step. The reaction probably involves pre-equilibria which are fast:

*^{12}C or ^{13}C

$\longleftrightarrow$ etc.

(**29**)

$$\frac{k_{12}}{k_{13}} = 1.04 \; (4\%)$$

* These small effects are often given in terms of a percentage $\left(\dfrac{k_1}{k_2} - 1\right) \times 100$.

A sulphur isotope effect is found in the decomposition of sulphonium ions in an $E2$ elimination (Hofmann elimination); C—S bond fission therefore occurring in the slow step.

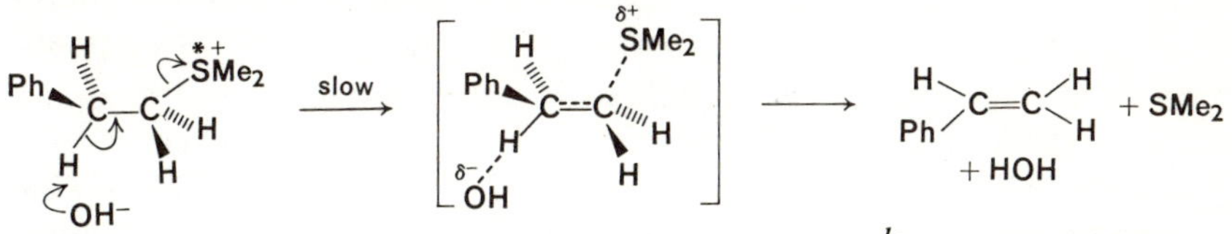

$$\frac{k_{32}}{k_{34}} = 1.072 \ (0.72\%)$$

* 32S or 34S

1.12   SECONDARY ISOTOPE EFFECTS

The substitution of deuterium for hydrogen may produce a kinetic effect even when the isotopic substitution is remote from the reaction centre and the C—H bond is not being broken at all in the reaction. The ionization constants for benzhydrol (30) in sulphuric acid are significantly altered by isotopic substitution at the α-position, though no C—H bond fission occurs. Deuterium

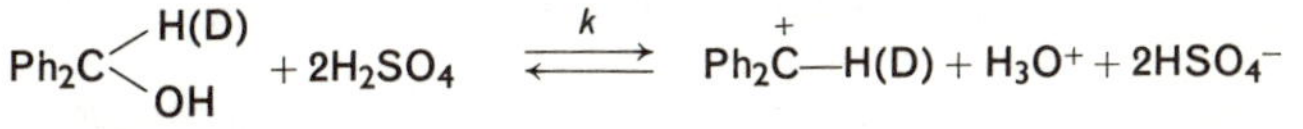

$$Ph_2C\begin{smallmatrix}H(D)\\ \\OH\end{smallmatrix} + 2H_2SO_4 \ \underset{\longleftarrow}{\overset{k}{\longrightarrow}} \ Ph_2\overset{+}{C}-H(D) + H_3O^+ + 2HSO_4^-$$

(30)

substitution both at the reaction centre and also on adjacent carbon produces a comparable kinetic effect, e.g. acetolyses of cyclopentyltoluene sulphonate (31), by the S_N1 mechanism:

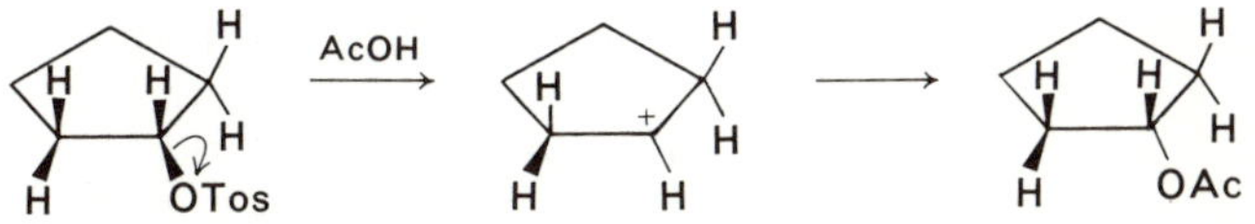

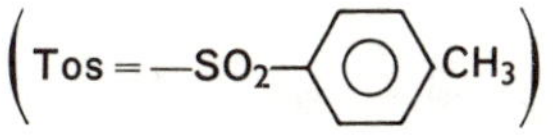

$$\left(Tos = -SO_2-\!\!\left\langle\!\bigcirc\!\right\rangle\!-CH_3\right) \quad (31)$$

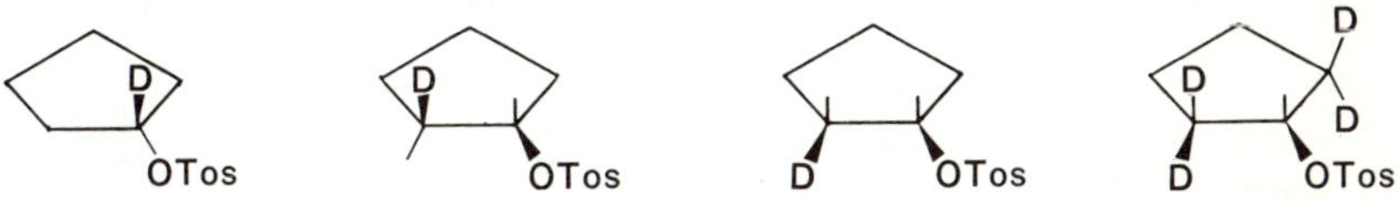

$\dfrac{k_H}{k_D} =$	1.15	1.16	1.22	2.06
	(15%)	(16%)	(22%)	(106%)

These small kinetic effects in systems in which the C—H bond is not being broken at any stage in the reaction, are termed *Secondary Kinetic Isotope Effects*. Their origin lies in the perturbation which isotopic substitution produces on various vibrational modes of the molecule. On passing from the initial to the transition state, complex vibrational modes of the molecule, involving the motions of many atoms, will suffer a change. Such changes will clearly occur if a change in hybridization occurs at the reaction centre. The prediction of the magnitude or even the sign of the isotope effect requires taking into account changes in zero-point energy of all the vibrational modes affected and hence is usually very complex. The observation of secondary isotope effects, however, can be used diagnostically. Thus it seems to be a general rule that in solvolytic reactions proceeding by the S_N1 mechanism (Section 2.3), the α-deuterium isotope effect is approximately 15% and the β-effect a little larger, 20–22%. On the other hand, solvolyses occurring by the S_N2 route (direct displacement by solvent) show a very small, often inverse (<1) effect in the range $k_H/k_D = 0.96$–1.02, and thus a distinction can often be made by this criterion. Typical values for secondary isotope effects are given in Table 7, Section 2.10.

1.13   SOLVENT ISOTOPE EFFECTS

The autoprotolysis constant of D_2O is smaller than that of water by a factor

$$2D_2O \underset{\longleftarrow}{\overset{k_D}{\longrightarrow}} D_3O^+ + OD^-$$

of 5, which means that D_2O (and OD^-) are weaker bases than H_2O and OH^-, respectively, and also that D_2O and D_3O^+ are stronger acids than H_2O and H_3O^+. It follows that a reaction which occurs by an equilibrium pre-proto-nation will be favoured in D_2O as compared to H_2O as solvent, i.e. an *inverse*

$$B + H_2O \underset{\longleftarrow}{\overset{k_{H_2O}}{\longrightarrow}} BH^+ + OH^-$$
$$B + D_2O \underset{\longleftarrow}{\overset{k_{D_2O}}{\longrightarrow}} BD^+ + OD^- \qquad \frac{k_{D_2O}}{k_{H_2O}} \sim 3\text{–}4$$

isotope effect will be observed. This effect arises from differences in vibrational frequencies of the species on either side of the equation. The main contributing modes are bending frequencies involving the $\overset{+}{B}$—H bond in the conjugate acid. However, the situation is complex and, indeed, some systems show a value less than unity making the interpretation of the effect more equivocal than is desirable for a diagnostic tool. The bromination of acetone, catalysed by weak acids, proceeds twice as fast in D_2O as in H_2O; a pre-equilibrium protonation

$$CH_3\text{—}\underset{\parallel}{\overset{O}{C}}\text{—}CH_3 + \underset{(DA)}{HA} \rightleftharpoons CH_3\text{—}\underset{\parallel}{\overset{+O\text{—}H}{C}}\text{—}CH_3 \xrightarrow[Br_2]{slow} \longrightarrow CH_3\text{—}\underset{\parallel}{\overset{O}{C}}\text{—}CH_2Br$$

$$\frac{k_{D_2O}}{k_{H_2O}} = 2$$

is implied. In its acid-catalysed mutarotation, α-D-glucose (32) undergoes isomerization in water to an equilibrium mixture of the α- and β-anomers* (32 and 33). The reaction is acid catalysed and is believed to proceed by protonation on the cyclic oxygen, ring opening to the aldehyde (open) form and reclosure

(32)
α-D-glucose
(pyranose form)

solvent
isotope
effect

(D⁺)
H⁺

primary
isotope
effect

open form
of glucose

β-D-glucose
(33)

to either the α- or β-anomer. This situation is complicated by the presence of a primary isotope effect since the slow step requires the loss of the hydroxyl proton from C-1 (all hydroxyl protons exchange very rapidly with deuterium in D_2O). This itself would produce a retardation in D_2O by approximately a factor of 7. Coupled with the acceleration by solvent D_2O of about three, the net expected isotope effect is:

$$\frac{K_{D_2O}}{K_{D_2O}} = 3 \times \tfrac{1}{7} = 0.43$$

which is approximately the experimental value. On the other hand, base-catalysed reactions will be slower in D_2O than H_2O. The decomposition of nitramide is characterized by a large deuterium solvent effect ($k_H/k_D > 1$).

* 'anomers' = stereoisomers at C_1 of the sugar.

$$\frac{k_{D_2O}}{k_{H_2O}} = \frac{1}{5.2}$$

1.14   SOLVENT EFFECTS ON RATES

Most of the reactions discussed in this book occur in solution. In many instances the solvent plays a vital role which nonetheless tends to be overlooked since it is very difficult to obtain detailed information on this aspect of the mechanism. One of the most obvious manifestations of medium effects is that on rate. Some reactions are found to be very sensitive to the nature of the solvent while others are much less so. The response of a reaction to changes in the solvent can often throw light on its mechanism since all such effects ultimately derive from interactions, or a lack of them, between solvent molecules and the reagents and transition state. These interactions are of several types, hence the complexity of the subject.

Interactions with Ions; Solvent Polarity

The most frequently encountered solvent effects are those which are attributable to interactions with ionic solutes. The formation of two ions from neutral species occurs much more rapidly in acetone than in hexane. Most of

$$Et_2N + Et—I \longrightarrow Et_3N^+ \; I^-$$
$$k_2(\text{acetone}) = 4.2 \times 10^{-3} \; \text{mol s}^{-1}$$
$$k_2(\text{hexane}) = 5.0 \times 10^{-6} \; \text{mol s}^{-1}$$

this is attributable to a much reduced activation energy in the former solvent.

The activity coefficients of ions are found to be markedly dependent upon the medium. The absolute values are difficult to measure but values for the chloride ion relative to Ph_4B^- (which is almost unaffected by solvent, being very large and non-nucleophilic) in a number of solvents, related in turn to dimethylformamide, are given below.

$$\log(\text{Standard Activity}) \text{ in, } a_{Cl^-}/a_{Ph_4B^-}$$

Solvent:	$\underset{0}{\overset{\overset{\displaystyle O}{\overset{\displaystyle \parallel}{}}}{HC—NMe_2}}$	H_2O	MeOH	$Me_2{=}O$	MeCN
	0	−18.4	−11.9	−3.0	−2.4

The more negative the figure, the more interaction with solvent is indicated. The differences are very large indeed and this is only within the range of fairly 'active' or 'polar' solvents.

The Nature of Solvation

A solute may be pictured as surrounded by solvent molecules, continually interchanging and under the influence of considerable attractive forces, particularly by those nearest to the solute molecule or ion. These forces are of several kinds, mainly electrostatic in nature, and may be summarized as follows:

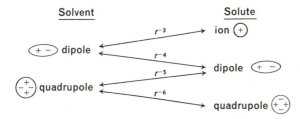

The powers of r, the distance of separation, indicate the dependence of the force on separation. Clearly ion–dipole and dipole–dipole forces are the most important, the others may be ignored as only applicable at short range when the former are present.

In addition, and perhaps the most important interaction for hydroxylic solvents, hydrogen bonding occurs when the solute has an atom with lone-pair electrons or other 'donor' features and may be described by contributions from valence-bond structures such as **34**. The weakest forces, van der Waals' or

Hydrogen bond

Solute Solvent   (34)

dispersion forces, are ubiquitous and important principally in the absence of the stronger forces, e.g. in hydrocarbon solvents. They arise on account of the tendency of the valence electrons on one molecule to 'avoid' those on a neighbouring one leading to a weak attraction. Interactions which depend on dipoles and quadrupoles are directional, as also are hydrogen-bonding and other specific interactions which may occur for a particular solvent–solute pair, such as donor–acceptor forces (Chapter 10) and polarizability.

The term *Solvent polarity* is used to describe the sum of these interactions between solvent and an ionic or at least dipolar solute. Conceptually, the term implies a measure of the solvation forces or the free energy of transfer of an ionic (or dipolar) solute from the gas phase to solution. A highly 'polar' solvent will therefore tend to stabilize ions in solution as free (but solvated) species rather than cation–anion pairs; it will also favour the conversion of a dipole to two ions since for the latter longer-range solvation forces are able to act ($\propto r^{-3}$ rather than r^{-4}). That is, a highly polar solvent will favour heterolysis,

$$
\begin{array}{ccccc}
\text{H—Cl} & \underset{\substack{\text{'non-polar'}\\ \text{solvent}}}{\overset{\substack{\text{'polar'}\\ \text{solvent}}}{\rightleftharpoons}} & \text{H}^+ & + & \text{Cl}^- \\[2pt]
\substack{\text{strong}\\ \text{covalent bond}} & \substack{\text{weak}\\ \text{solvation}} & \substack{\text{strong}\\ \text{solvation}} & & \substack{\text{strong}\\ \text{solvation}}
\end{array}
$$

and conversely. At the same time, solvation of ions will tend to supress their tendency to react since the solvent molecules have to be at least partly removed in order for the reagent to approach. This will not be apparent from kinetic studies if the reaction of the ion is fast compared to ionization.

1.15   MEASUREMENT OF SOLVENT POLARITY

Many chemical and physical systems which respond to the effects of the medium have been suggested as empirical measures of solvent polarity. The test for their reliability would lie in their providing a universal order of solvent effects. This however, is unlikely to be found since solvation is such a complex phenomenon. The measurements discussed below produce reasonable agreement.

Dielectric Constant, D

This is a property of the bulk solvent and measures the energy stored in the medium under the influence of an electric field, due to the partial ordering of the molecules. Dielectric constant provides a rough guide to polar solvent effects in chemistry and is of limited use, Table 3. The function $(D - 1)/(2D + 1)$ has more significance in this respect since it may be shown (Kirkwood equation), that the free-energy change accompanying transfer of a molecule of dipole moment, μ, from a medium of $D = 1$ to a medium of $D = D$ is given by

$$\Delta G^{\ast} = \frac{\mu^2}{r^3}\frac{(D-1)}{(2D+1)}, \qquad (r = \text{radius of solute})$$

whence the change in free energy of activation for the same solvent change $(D=1$ to $D=D)$ is proportional to the second term:

$$\Delta\Delta G^{\ast} = \log\frac{k_{(D=D)}}{k_{(D=1)}} = \frac{(D-1)}{(2D+1)}$$

In practice, this approach works reasonably well for a number of reactions which proceed by way of an ionization in the slow step.

Table 3. Solvent polarity parameters

	Solvent	Dielectric constant	Y	Z	E_T	Ω
	Water	78.5	3.49	94.6	63.1	
	Formic Acid	57.9	2.05			
	Methanol	32.6	−1.09	83.6	55.5	0.845
	Ethanol	24.3	−2.03	79.6	51.9	0.718
Decreasing polarity	Acetic acid	6.19	−1.64	79.2	51.9	0.83
	Isopropanol	18.3	−2.73	76.3	48.6	
	Acetonitrile	37.5		71.3	46.0	0.69
	Acetone	20.5		65.7	42.2	0.62
	Dichloromethane	8.9		64.2	41.1	
	Chloroform	4.7		63.2	39.1	
	Diethyl ether	4.22			34.6	
	Benzene	2.27			34.5	
	Carbon tetrachloride	2.23			32.5	
	Hexane	1.90			30.9	

The Winstein–Grunwald Relationship

This and subsequent solvation scales illustrate the use of operational definitions of 'polarity'. The solvolysis of t-butyl chloride (**35**) occurs almost exclusively by the ionization mechanism (S_N1) hence the transition state is more dipolar than the initial state. The rate of reaction therefore is increased by solvents of high 'polarity' (as are all reactions involving ionization), which interact more with the transition state than with the reagent, and lower the

(35)

products

energy of activation. Winstein and Grunwald defined a *Solvent polarity parameter*, Y, in terms of the rates of this reaction in a given solvent, k, and in water at 25°, k_0,

$$\log k/k_0 = m_s\, Y(+m_n\, N)*$$

This will be recognized as a linear free-energy expression. The constant m_s is unity for the solvolysis of t-butyl chloride but may be determined for other substrates and indicates their relative susceptibilities to solvent effects in the solvolytic reaction. Values of Y for a number of solvents are given in Table 3. This scale is necessarily limited to protic solvents (or their mixtures) in which solvolysis is possible, and includes water, alcohols and carboxylic acids. The polarity scale based on solvolytic rates is extended to media more polar than water in Chapter 2 (p. 105).

Isomer Ratios in a Diels–Alder Reaction

The addition of cyclopentadiene to methyl acrylate yields two isomeric products, the *exo*, **(36)** and *endo*-bicycloheptene carboxylic esters **(37)**. While the rate of a cycloaddition such as this is not markedly solvent dependent, the isomer ratio is. The transition state leading to the *endo* product **(38)** must have a higher dipole moment than that leading to the *exo* **(39)**. Hence the ratio of *endo* to *exo* rates or proportions should increase the more polar the solvent.

(38)   **(37)**

* The term $(m_n N)$ refers to the nucleophilic power, N, of the two solvents being compared and the susceptibility of the reaction to nucleophilicity, m_n. It is usually ignored—i.e. solvents are (erroneously) assumed to have similar N.

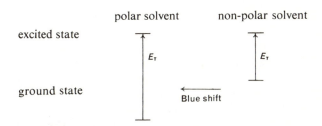

(39)   (36)

This reasoning was used by Berson and coworkers who defined the solvent polarity constant, Ω:

$$\Omega = \log k_N/k_x = \log[N]/[X]$$

where N and X refer to *endo* and *exo* rates and products. Some values of Ω are given in Table 3.

Solvatochromic Dyes

The absorption spectra of certain compounds change markedly according to the solvent in which they are dissolved. The absorption maxima, or the transition energies E_T at the maximum wavelength, provide an index of solvent polarity which is very easily determined and not limited to any type of medium. Solvatochromic materials, as they are called, are usually species which have a highly dipolar ground state and a much less polar excited state. If, to a first approximation, we regard the free energy of the excited state in any solvent as being approximately constant, that of the ground state will vary widely being lower the more polar the solvent. Thus, the energy required for excitation, E_T, will be greater the more polar the solvent, and the absorption maximum will

	polar solvent	non-polar solvent
excited state		
	E_T	E_T
ground state		
	Blue shift	

shift towards the blue end of the spectrum. Two dyes which exhibit marked solvatochromism are the pyridinium iodide, (40), and the betaine, (41). In the latter, this property is especially developed: it is yellow in methanol, red in isopropanol and blue in chloroform. Kosower defined a solvent polarity parameter, Z, as

$$Z = E_T \text{ of } \mathbf{40} \text{ in solvent X}$$

$$= hc/\lambda_{max} \qquad \text{where } h = \text{Planck's constant}$$

$$c = \text{velocity of light}$$

$$= \frac{(2.859 \times 10^4)}{\lambda_{max}(nm)} \qquad \text{kcal mol}^{-1}$$

An exactly parallel set of values, denoted E_T, was determined by Dimroth and coworkers (**41**). Values of Z and of E_T are given in Table 3.

1.16   POLARIZABILITY OF SOLVENT MOLECULES

Refractive index, n, or rather n^2, measures the interaction of the molecules in a bulk material with the electric vector of a light beam. This is, in effect, an extremely high frequency (10^{17} Hz) alternating electric field. Now no molecules are able to rotate and align their dipoles with this field at such a frequency, so the interaction is by distortion of the electron distribution of the molecule which is able to respond much more rapidly. Compounds which have highly mobile electrons (e.g. aromatic compounds) usually have high refractive indices. Now suppose a relatively non-polar solute molecule absorbs light and is transformed to a highly polar excited state. The excited state is formed in a very short period of time (ca 10^{-16} s), hence it is not able to be stabilized by dipolar interactions with the solvent, oriented almost randomly, but only by the polarizability of the solvent molecules. The system would show solvatochromism but the variation of the excitation energy would correlate with

non-polar state   excited state

refractive index. Such appears to be the case with charge-transfer complexes (Chapter 10) which have polar excited states, and which are formed in the very short period of time characteristic of an electronic transition. The solvato-chromic behaviour of dimethyl aniline–tetracyanoethylene (TCNE) complexes

(42)

(42) is related to refractive index of the medium.

		λ_{max}/nm
	Refractive	(TCNE–dimethyl
Solvent	index	aniline)
Diethyl ether	1.350	490
Cyclohexane	1.429	500
Methyl iodide	1.529	536

It is likely that polarizability, or time-dependent polarity, comes into play in many situations but is frequently masked by more important contributions from permanent polarities of the solvent molecules. The different parameters are by no means parallel though there is a reasonable trend common to all. Solvents of higher polar character than water are discussed in connection with solvolytic reactions, (Section 2.7). Mixed solvents in general will exhibit properties which are approximately a linear function of their composition.

1.17   SOLVENT POLARITY CORRELATIONS

Having established an order of solvent polarity as in Table 3, kinetic measure-ments in a series of different solvents can be used to ascertain whether the solvent interacts more or less with the transition state than with the reagents and therefore whether the one is more polar (i.e. shows more charge separation) than the other. In general, if the transition state is more stabilized by solvent than the reagents, the rate will be accelerated by a more polar solvent, and conversely. This effect is usually very pronounced. If charge is not being created or destroyed but merely spread over more atoms (dissipated) in going to the transition state, the solvation energy is diminished but rather weakly. These considerations lead to the following predictions of the effects of solvents on any reaction. They were originally formulated by Ingold (see also Section 2.7):

	Effect of *increasing* polarity of solvent upon reaction rate
A. Transition state more polar than initial state (charge separation)	Strong acceleration
B. Transition state less polar than initial state (charge neutralization)	Strong retardation
C. Dispersion of charge in transition state	Weak retardation
D. No change of charge distribution between initial and transition states	Little effect

The following examples illustrate these categories.

A. A unimolecular solvolysis, neutral reagents

$$Me_3C-Cl + SH \longrightarrow Me_3\overset{+}{C}...\overset{-}{Cl} \underset{solv}{\longrightarrow} Me_3C-S + HCl$$

transition state

where SH is a protic solvent

Increasing polarity ↓	SH	k_{rel}
	MeOH	0.1
	alc. H_2O (60%)	1500

B. Bimolecular displacement, oppositely charged reagents

$$HO^- \quad \underset{H}{\overset{H}{\searrow}} C-\overset{+}{S}Me_2 \longrightarrow HO\overset{\delta^-}{---}\underset{H\ H}{\overset{H}{\underset{|}{C}}}\overset{\delta^+}{---}SMe_2 \longrightarrow HO-C\underset{H}{\overset{H}{\cdots}}_H + SMe_2$$

Solvent	k_{rel}
MeOH	1
H_2O	5×10^{-5}

C. Bimolecular displacement, one charged, one neutral reagent.

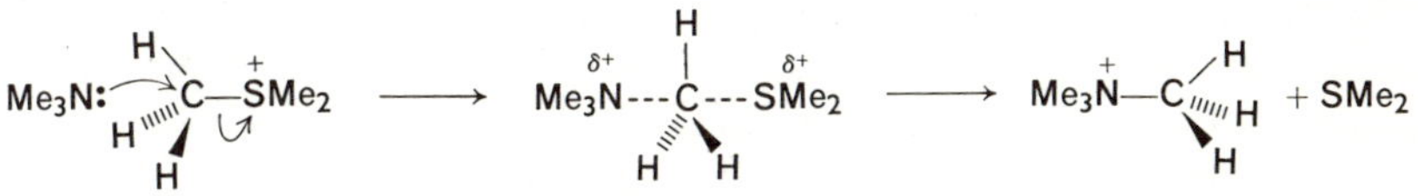

Solvent	k_{rel}
MeOH	1
H_2O	0.17

D. Pure homolytic reaction;

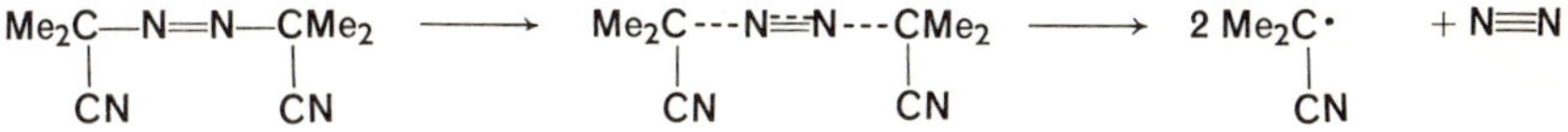

Solvent	k_{rel}
Benzene	1
Nitrobenzene	1
Isobutanol	1
Acetic acid	1

1.18   **DONOR–ACCEPTOR SOLVENT PROPERTIES**

Yet another solvent interaction parameter may be conceived by considering the donor–acceptor properties of the medium. This would be a measure of solvation by covalency formation rather than electrostatic interaction and would depend upon the availability of filled orbitals (unshared pairs) or vacant and accessible orbitals, either of a π or σ-type. A solvent scale based on these properties has been proposed by Gutmann who defined the donor number, DN, in terms of the enthalpy of interaction of solvents with antimony pentachloride, a strong Lewis acid or acceptor, both species being in dilute solution in dichloroethane. Drago and Wayland have proposed a two-parameter scale to measure this quantity,

$$-\Delta H = E_A E_B + C_A C_B$$

where ΔH is the enthalpy of complexation of a donor solvent with a given acceptor or an acceptor solvent with a donor species. In either case, the donor and the acceptor are characterized by two constants each, E_B, C_B and E_A, C_A, respectively. The E parameters express contributions to electrostatic interactions and the C values contributions to covalency interactions, thus expressing that neither are mutually exclusive. Some DN, E and C values are given in Table 4.

Table 4. Solvent donor–acceptor parameters

Solvent	Donor number (DN)	C_B	E_B
Methyl cyanide	14.1	1.77	0.553
Sulpholane	14.8	3.30	1.09
Ethylene carbonate	16.4		
Acetone	17.0	0.66	0.706
Ethyl acetate	17.1	2.42	0.639
Water	18.0		
Diethyl ether	19.2	3.55	0.654
Tetrahydrofuran	20.0	4.61	0.61
Dimethylformamide	26.6	3.00	1.00
Dimethylsulphoxide	29.8	3.42	0.97
Pyridine	33.1	6.92	0.88

Increasing donor character (↓ arrow along left of table)

Donor–acceptor character of the solvent becomes prominent in reactions such as complexation where for example, a solvent competes with a donor species for an acceptor molecule:

$$SbCl_5 + Cl^- \overset{k}{\rightleftharpoons} SbCl_6^-$$

$$\Updownarrow S:$$

$$S-SbCl_5$$
$$\log k \propto 1/DN\ (S:)$$

and also in many redox reactions (one-electron donor–acceptor) such as,

$$\overset{IV}{S:} + N_2O_4 \overset{k}{\rightleftharpoons} \overset{III}{S-NO} + \overset{V}{NO_3^-}$$

Rates of certain acid–base reactions may also correlate with these scales rather than polarity scales, e.g.

$$Ph_2C{=}\overset{+}{N}{=}\overset{-}{N} + RCOOH \xrightarrow[\text{slow}]{k} Ph_2CH{-}\overset{+}{N_2}, RCOO^- \xrightarrow{\text{fast}}$$

$$Ph_2CH{-}O{\cdot}COR + N_2$$

The rate order of this reaction in some common solvents is:

dimethylformamide < ether < acetone < benzene < chloroform

which clearly does not fit any of the 'polarity' scales (Table 3).

1.19   DIPOLAR APROTIC SOLVENTS

A number of solvents are known which have a substantial dielectric constant and dipole moment and which have no acidic hydrogen available for hydrogen bonding. Such compounds as acetone (**43**), dimethylformamide (**44**), dimethyl-acetamide (**45**), dimethylsulphoxide (**46**) and hexamethylphosphoric triamide (**47**), are good examples. The solvation properties of these compounds are

| (43) | (44) | (45) | (46) | (47) |

rather specific. Owing to their molecular shape they are much better able to

solvate cations than anions. This property is illustrated by measurements of free energies of transfer, ΔG_t, from methanol to dimethylsulphoxide (DMSO),

$$\text{Ion(MeOH)} \longrightarrow \text{Ion(DMSO)}; \quad \Delta G_t$$

for the potassium ion, $\Delta G_t = -25 \text{ kJ mol}^{-1}$, and for the chloride ion, $+30 \text{ kJ mol}^{-1}$. In other words, the cation is more strongly solvated in DMSO than in methanol while the opposite is true for the anion. The result is that while ionic compounds are rather soluble in these media, anions are much more 'free' and unsolvated than is the case in hydroxylic solvents, and consequently much more reactive. The relative standard activities of typical anions is much higher in dipolar aprotics than in hydroxylic solvents; e.g. for the chloride ion—

Solvent	H_2O	MeOH	DMSO	HMPT	DMA
			(43)	(47)	(45)
log (Standard activity)	−15.3	−9.2	−1.3	+0.2	+1.3

The result is that reactions of anions are greatly accelerated. Further examples

$$MeI + Cl^- \longrightarrow MeCl + I^-$$

$$k_{MeOH} = 1$$
$$k_{DMA} = 7.4 \times 10^6$$

are discussed in Section 4.4.

1.20   **KINETIC SALT EFFECTS**

The addition of salts (i.e. ions) to the reaction medium may also have a profound effect upon rate. There are three types of salt effect, each of which may operate separately or concurrently.

Ionic Strength Effects

A salt solution is a more polar medium than the solvent itself, hence reaction rates respond to the addition of non-reacting salts as for a change to a more polar medium (Section 1.14). The additional solvation energy arises from relatively strong interactions between the added ions and solute ions or dipoles. For a reaction between two ions (or its reversal) the effect on rate may be predicted.

$$A^\pm + B^\pm \longrightarrow (A\cdots B)^\pm \longrightarrow A-B^\pm$$

transition state

where $\pm$ refers to any charge species.

The free energy of activation, ΔG^+ in a medium of zero ionic strength is given by:

$$\Delta G^+ = G_{AB+} - (G_A + G_B)$$

where AB refers to the transition state and A, B to the reagents. If the system is now transferred to a medium of ionic strength $= I$, each free energy term changes so that:

$$\Delta\Delta G^+ = \Delta G_{AB+}^* - (\Delta G_A + \Delta G_B)$$

Applying the Arrhenius equation (1.7) and equating E_A with ΔG^+,

$$k_0 = A\,e^{-\Delta G^+/RT} \qquad \text{at } I = 0$$

$$k_I = A'\,e^{-(\Delta G^+ + \Delta\Delta G^+)/RT} \qquad \text{at } I = I$$

whence,

$$\log\left(\frac{k}{k_0}\right) = \frac{-\Delta\Delta G^+}{RT} = (\Delta G_A + \Delta G_B - \Delta G_{AB+})/RT$$

Assuming that the solutions behave ideally, at $I = 0$

$$G_A = RT\ln f_A$$

$$G_B = RT\ln f_B$$

$$G_{AB+} = RT\ln f_{AB+} \qquad \text{where the } f\text{s are activity coefficients}$$

$$\text{hence } \log\frac{k}{k_0} = \frac{f_A f_B}{f_{AB}}$$

Values of activity coefficients may be calculated from Debye–Hückel theory which in the limiting form gives,

$$-\log f = z^2 \mathscr{S} I^{0.5}$$

whence

$$\log\frac{k}{k_0} = 2z_A z_B \mathscr{S} I^{0.5*}$$

where z_A, z_B are the charges on A and B, $\mathscr{S}$ is a constant depending upon the solvent and temperature; for water at 25°, $\mathscr{S} = 0.509$. This expression is known as the Brønsted Salt Effect relationship. It will be apparent that if the charges on A and B are of opposite sign—an anion reacting with a cation— $k < k_0$ and added salt should retard the rate. If the signs are the same, two cations or two anions reacting, then $k > k_0$ and acceleration by added salt would be predicted. Generally good agreement between theory and experiment is found as, for instance, for the reaction between the Crystal Violet cation (48) and hydroxide ion, for which retardation proportional to $I^{0.5}$ is found. This

---

* A more accurate version uses the ionic strength function $(1 + 0.921 I^{0.5})$ for aqueous solutions.

$$\left(Me_2N\!-\!\!\left\langle\!\bigcirc\!\right\rangle\!\right)_3\!\!-\!C^+ + OH^- \longrightarrow \left(Me_2N\!-\!\!\left\langle\!\bigcirc\!\right\rangle\!\right)_3\!\!-\!C\!-\!OH$$

(48)

relationship enables a distinction to be made between reagents of different charge which might be involved, e.g. CrO_3 and CrO_4^{-2} in oxidations.

Reactions which generate ionic products may show autocatalysis due to increasing ionic strength, e.g. solvolytic reactions which proceed by an ionization mechanism,

$$R\!-\!Cl \longrightarrow (R^+\!\cdots Cl^-) \longrightarrow (R^+) \xrightarrow{Cl^-\ H_2O} ROH + Cl^- + H_3O^+$$

In order to observe good first-order kinetics of these reactions it is usual to add some inert salt to swamp the added effect of the product ions. Alkali-metal perchlorates are often used as inert salts.

Common Ion Effects

If the slow step of a reaction involves the reversible production of ions, addition of further quantities of these ions may force the equilibrium in favour of reagents and cause a decrease in rate. This is a common-ion or mass-law effect. For example, the apparent rate constant for the hydrolysis of dimethyl-benzhydryl chloride **(49)** falls with the progress of the reaction. Although accelerated by the increasing ionic strength, the reaction is retarded even more strongly by the reversibility of the slow step which is more apparent as the

$$\left(CH_3\!-\!\!\left\langle\!\bigcirc\!\right\rangle\!\right)_2\!\!CHCl \rightleftharpoons \left(CH_3\!\left\langle\!\bigcirc\!\right\rangle\!-\right)_2\!\overset{+}{C}H\ \overset{-}{Cl} \xrightarrow{H_2O} \left(CH_3\!\left\langle\!\bigcirc\!\right\rangle\!\right)_2\!\!CHOH$$

(49) $+ Cl^-$

concentration of chloride ion increases. The effect of added chloride ion is specific in causing retardation; an inert anion such as perchlorate would cause acceleration of the rate by the increased ionic strength, but would not combine with the carbonium ion.

Specific Salt Effects

These are rate accelerations, far greater than could be due to increased ionic strength, which are brought about by certain anions upon solvolytic reactions. The interpretation is in an ion–pair exchange which effectively prevents the

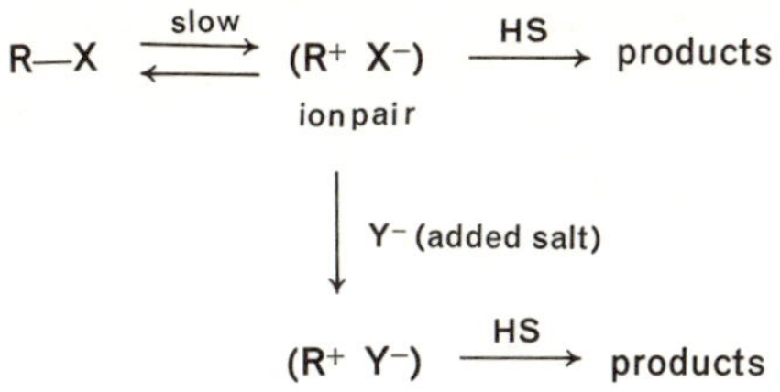

$$R\!-\!X \xrightleftharpoons{\text{slow}} (R^+ \; X^-) \xrightarrow{\text{HS}} \text{products}$$

ion pair

↓ Y⁻ (added salt)

$$(R^+ \; Y^-) \xrightarrow{\text{HS}} \text{products}$$

reversal of the rate-determining ionization. This phenomenon is discussed more fully in Section 2.9.

1.21 CATALYSIS BY ACIDS AND BY BASES—PROTONATED INTERMEDIATES

Some reactions require, or at least are greatly accelerated by, the addition of catalytic quantities of acid or of base, the catalyst not being consumed although its concentration may enter the rate equation. Catalysis by acids implies the existence of a protonated intermediate. For example, the hydrolysis of ethylene oxide (**49**) is acid catalysed. It is inferred that the reaction occurs at a more rapid rate at the conjugate acid of the epoxide (**50**).

Water reaction

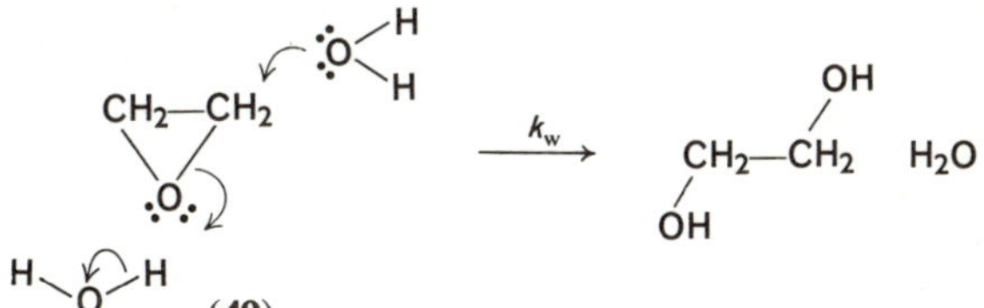

Acid-catalysed reaction

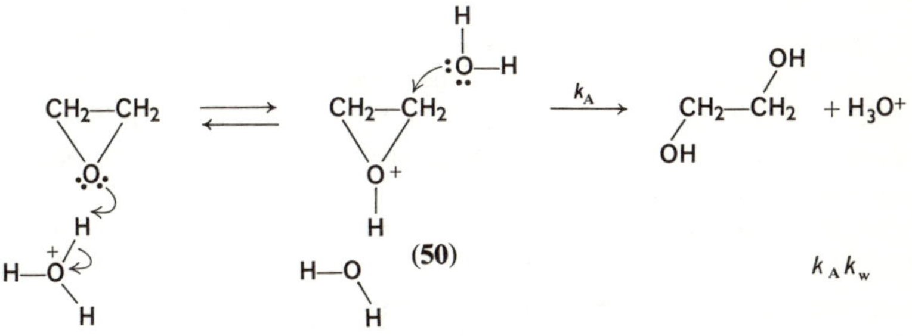

Catalysis by base is ambiguous in that it may imply a pre-equilibrium transfer of a proton from the substrate, i.e. true base catalysis as in the proton exchange of nitromethane. Reactions involving proton transfer from carbon form the subject of Chapter 4. Base catalysis may mean nucleophilic catalysis where the lyate ion (i.e. de-protonated solvent) is a stronger nucleophile than the solvent itself. The latter interpretation is undoubtedly correct in the base-catalysed hydrolysis of ethylene oxide,

In order to put acid–base catalysis upon a quantitative basis it is necessary first to discuss the measurement of acidity and basicity in solutions in general.

1.22   MEASUREMENT OF ACIDITY

In dilute aqueous solution, acidity or the tendency to supply a proton is adequately measured by the pH scale which is defined in terms of the standard potential of the hydrogen electrode and readily measured with accuracy sufficient for most purposes by the glass electrode. The pH of a solution under these conditions is equated with the concentration (or activity) of hydronium ions which are the sole source of protons in aqueous solutions of strong acids. Within the range that these assumptions are applicable (to about pH = 1), acid–base dissociation equilibria and hence rates of acid- or base-catalysed reactions are a function of pH.

   Similar measurements may be made in other media; the glass electrode will respond reversibly also in alcohols, acetic acid, acetonitrile or dimethyl-sulphoxide and in mixtures of polar solvents. Acidity measurements in these media are often internally consistent though the exact thermodynamic meaning of acidity here is uncertain.

   In solutions of high acidity, pH as measured by the glass electrode becomes increasingly meaningless as a measure of the protonating ability of the solution since species other than the hydronium ion are present and act as strong proton donors. For instance, in sulphuric acid the hydronium ion concentration increases only about one pH unit between 20% and 80% acid whereas the protonating capacity (due increasingly to the species H_2SO_4 and HSO_4^-) increases steadily to a maximum in 100% sulphuric acid. For these cases, another measure of acidity is required.

This problem was examined by Hammett (1932) who proposed the use of indicators as an operational measure of acidity. An indicator, *In*, for these acidic systems would be a weak base whose basic and protonated forms have distinct spectroscopic properties, allowing the indicator ratio ($[In]/[InH^+]$) to

$$In + H^+solv \underset{\longleftarrow}{\overset{K_0}{\longrightarrow}} IdH^+$$

be determined readily. In dilute aqueous solution the thermodynamic dissociation constant, K_0 can be determined from the relationship,

$$K_0 = [H_3O^+] \cdot [In]/[InH^+] \quad \text{or} \quad \log K_0 = -pH + \log([In]/[InH^+])$$

whence

$$pH = pK_0 + \log([In]/[InH^+]) \tag{1.16}$$

In analogy to equation 1.16, Hammett defined an acidity function, H_0, applicable to any acid solution,

$$H_0 = pK_0 + \log([In]/[InH^+]) \cdot -\log \frac{a_{H^+} \cdot a_I}{a_{IH^+}}$$

where *a* is activity. In terms of *p*-nitroaniline as the indicator,

$$In = \quad \begin{array}{c} NH_2 \\ | \\ \bigcirc \\ | \\ NO_2 \end{array}$$

The Hammett acidity function, H_0 (or h_0, where $H_0 = -\log h_0$) provides an operational measure of the protonating ability of the solution which is independent of concentration measurements of any of the protonating species. The function increases smoothly, though not linearly, throughout the concentration range of sulphuric acid (Figure 8). For practical purposes it is

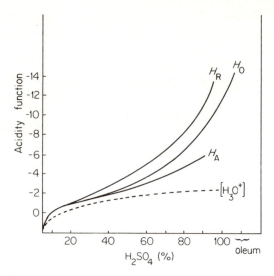

Figure 8. Some acidity functions of sulphuric acid solutions (from L. P. Hammett, *Physical Organic Chemistry*, McGraw-Hill, New York, 1970).

desirable to measure indicator ratios which are between the values 0.1 and 9.0. Since one indicator was insufficient for a wide range of acidity measurements, Hammett devised a series of related weakly basic amines such that at least one gave an indicator ratio between these limits at any desired acidity. Values of H_0 for strong acid solutions are based on the dissociations of a series of over-lapping indicators. The values of K_0 are obtained from measurements in dilute aqueous solution in which $H_0 = pH$.

It subsequently became apparent that the H_0 scale was appropriate only for dissociations of aromatic amine-type indicators (Hammett indicators). The dissociations of different chemical types of weak base did not necessarily follow H_0 and, in general, each different type required its own acidity function. The following equilibria have been used to define further acidity functions analogous to H_0.

Protonation of alcohols:

$$R\text{---}OH + H^+ \rightleftharpoons R^+ + HOH$$
$$H_R = pK + \log[ROH]/[R^+]$$

where $R = Ph_3C\text{--}$ and related groups (Section 2.24, and Chapter 2, Table 14).

Protonation of amides:

$$RC\overset{O}{\underset{NH_2}{\diagdown}} + H^+ \rightleftharpoons RC\overset{O}{\underset{NH_3^+}{\diagdown}}$$

$$H_A = pK + \log[R \cdot CONH_2]/[R \cdot CONH_3^+]$$

Protonation of positively charged bases:

$$H_+ = pK + \log[\overset{+}{A}rNH_2]/[\overset{+}{A}rNH_3^+]$$

The acidity functions H_R and H_A would be more appropriate for the description of protonation equilibria on oxygen bases and amides respectively, whereas H_0 refers to aromatic amines. Figure 7 illustrates the differences in these scales in sulphuric acid of varying concentration.

Acid-catalysed reactions in the high acidity range, therefore, will in general be found to be proportional to the appropriate acidity function of the medium. It should be understood that the concept of 'acidity' as a property of any medium as such is quantitatively meaningless since it depends upon the method of measurement. Thus, 55% sulphuric acid is more acidic than 80% phosphoric acid towards aromatic amines ($H_0 = -3.91$, -1.85, respectively). The opposite is found for carbinol dissociations ($H_R = -7.07$, -8.57, respectively). Basicity measurements pose the same problem. The pH scale is an adequate measure of basicity only to about pH 14 (1 M NaOH) but much more basic solutions are known such as hydroxide or methoxide ion in dimethylsulphoxide, or cyclohexylamide in cyclohexylamine (Section 4.2). Acidity functions, H_-, for these solutions which range up to 10–12 orders of magnitude greater than 1 M aqueous hydroxide, have been devised, based on the dissociation of weak acids, HA, where

$$H_- = pK_0 + \log([HA]/[A^-])$$

Scales based on the dissociation as acids of substituted anilines and fluorene

$$ArNH_2 + B \rightleftharpoons ArNH^- + BH^+$$

derivatives (51), have been suggested. Many base-catalysed reactions at these high basicities exhibit rates proportional to H_-. This would tend to confirm

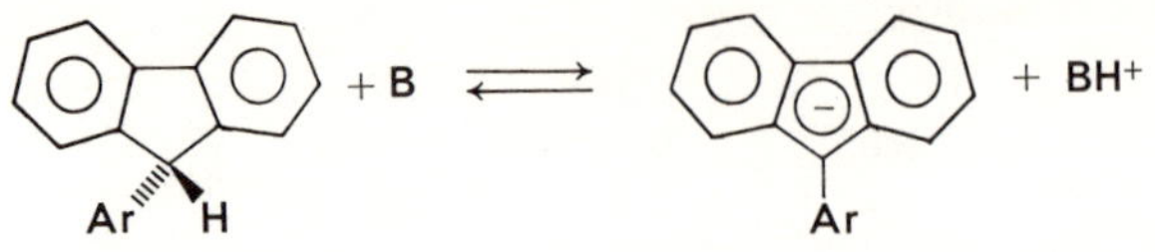

(51)

the role of the base as a proton remover since its action as a nucleophile to carbon or some other element would not in general parallel its basic behaviour (nucleophilicity to hydrogen). The H_- scale will differ to some extent depending upon the medium and base including the associated metal cation. Systems which have been frequently used include $NaOH-H_2O$, $NaOMe-MeOH$ and $NaOH-H_2O$-dimethylsulphoxide.

1.23   **MECHANISMS OF ACID CATALYSIS**

Two main types of acid catalysis may be distinguished. The $A1$ mechanism proceeds by the equilibrium protonation of the substrate, X, followed by a unimolecular decomposition of the conjugate acid, XH^+:

$$X + H^+ \underset{}{\overset{\text{fast, } K}{\rightleftarrows}} XH^+$$

$$XH^+ \xrightarrow[k_1]{\text{slow}} A \xrightarrow[H_2O]{\text{fast}} \text{products}$$

The solvent is not necessarily water.

$$\text{rate} = k_1[XH^+] = \frac{k_1}{K}[X][H^+] = k_{obs}[X + XH^+] \tag{1.17}$$

where k_{obs} is the experimentally determined first-order rate constant. Whence

$$k_{obs} = \frac{k_1}{K} \frac{[X][H^+]}{[X + XH^+]}$$

or, from equation 1.17

$$k_{obs} = \frac{k_1}{K} \frac{[X]}{[X + XH^+]} \frac{h_0 a_{IH^+}}{a_I}$$

The most common situation in acid-catalysed reactions has K very small, i.e. the equilibrium lies far to the left when $[X] \approx [X + XH^+]$. Therefore

$$k_{obs} = \frac{k_1}{K} h_0 \frac{a_{IH^+}}{a_I}$$

or

$$\log k_{obs} = -H_0 + \log(k_1/K) + \log(a_{IH^+}/a_1)$$

The second mechanism for acid catalysis is the $A2$ process in which the slow

$$X + H^+ \underset{}{\overset{K}{\rightleftharpoons}} XH^+$$

$$XH^+ + H_2O \xrightarrow[k_2]{slow} products$$

step requires water as reagent. Application of the same kinetic treatment yields the rate law (equation 1.18),

$$\log k_{obs} = -H_0 + \log(k_2/K) + \log(a_{IH^+}/a_1 + a_{H_2O}) \tag{1.18}$$

The two rate laws differ in the inclusion of the activity of water in the $A2$ case. Examples of $A1$ and $A2$ mechanisms are, respectively, the depolymerization of trioxan (**52**) and the iodination of acetone (in which the acid-catalysed process is enolization, Section 4.7).

$A1$ Mechanism

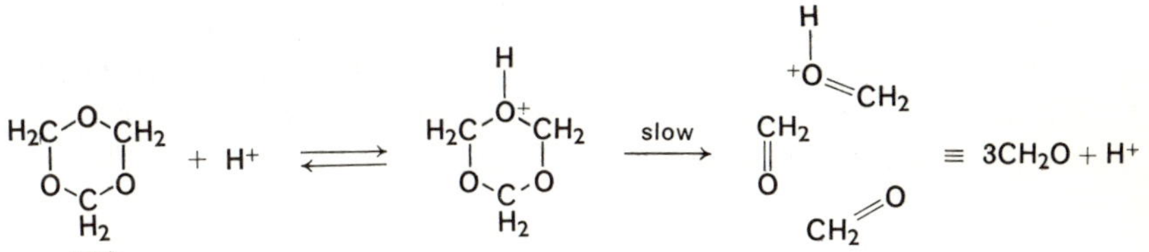

(**52**)

$A2$ Mechanism

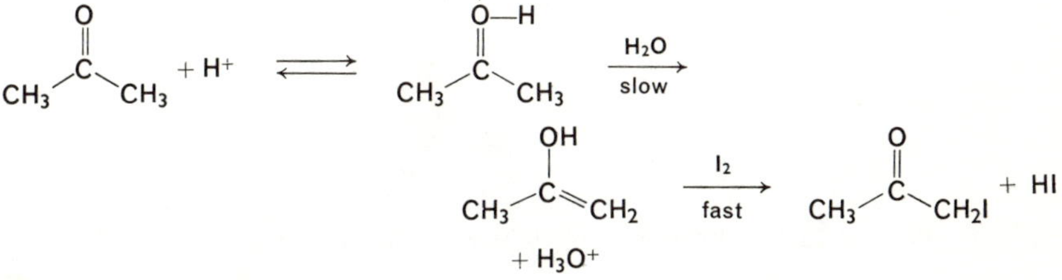

Entropies of activation may be used to distinguish the $A1$ and $A2$ mechanisms; in general, the range of values experienced for $A1$ processes is 0 to +45 J K^{-1} and for the $A2$, −60 to −120 J K^{-1}. The much more negative value of $\Delta S^{\ddagger}$ for the bimolecular mechanism is a result of the covalent binding of a water molecule in the transition state. Table 5 illustrates this principle.

Table 5. Entropies of activation of acid-catalysed reactions

Substrate	ΔS^+	ΔV_0^+	Supposed mechanism
Hydrolysis of methyl acetate	−90	−8.5	$A2$ ($A_{AC}2$)
Hydrolysis of acetamide	−155		$A2$
Hydrolysis of ethylene oxide	−26	−7.9	$A2$
Hydrolysis of benzoic acid	−120		$A2$
(Exchange of carbonyl oxygen)			
Hydrolysis of t-butyl mesitoate	+40		$A1$
Hydrolysis of acetaldehyde dimethylacetal	+55	0	$A1$
Trioxane depolymerization	+17		$A1$
Hydrolysis of t-butyl chloride	+51		S_N1

Volumes of activation also help to distinguish $A1$ and $A2$ mechanisms; the former are rather pressure insensitive ($\Delta V_0^+ \sim 0$) while the latter rates increase with pressure ($\Delta V_0^+ = -8$ to -11 cm³).

General and Specific Catalysis

In principle, a pre-equilibrium involving protonation of the substrate would be established with every proton-donating species in solution independently. Each acid would then contribute to the measured rate. Such a situation occurs in the hydrolysis of ethyl orthoacetate (**53**), for example,

When catalysed by aqueous benzoic acid, the rate equation takes the form,

$$\text{rate} = [CH_3C(OEt)_3](k_w + k_h[H_3O^+] + k_a[PhCOOH])$$

where k_w, k_h and k_a are the specific rate constants (catalytic constants) for the reaction components mediated by water, hydronium ion and undissociated benzoic acid, respectively. Such a reaction is said to be *General Acid Catalysed.*

On the other hand, the rate law may reveal that all but one of these rate constants are negligibly small and that the catalysis involves only one acidic species as proton donor. This is usually the hydronium ion, the strongest acid present in dilute aqueous solution (or in general the 'lyonium ion'-protonated solvent). An example of this *Specific Acid Catalysis* is the hydrolysis of

$$CH_3-C\overset{OEt}{\underset{H}{\overset{|}{C_{\prime\prime\prime\prime}OEt}}} + H_3O^+ \rightleftharpoons CH_3-C\overset{H\overset{+}{\diagdown}OEt}{\underset{H}{\overset{|}{C_{\prime\prime\prime}OEt}}} \xrightarrow{\text{slow}} CH_3-C\overset{OH}{\underset{H}{\overset{|}{C_{\prime\prime\prime}OEt}}} + H_3O^+ + EtOH$$

(54) $+ H_2O$

$$\text{rate} = k_2[\text{acetal}][H_3O^+]$$

fast $\downarrow$

$$CH_3CHO$$

acetaldehyde diethylacetal (54). In the same way, base catalysis may be general or specific (e.g. to the hydroxide ion); where catalysis by acids and by bases occurs, a reaction may show general catalysis by one and specific to the other. Many combinations are possible and prediction is difficult.

1.24   **RATES OF PROTON TRANSFER**

It has been mentioned that proton transfers are usually very fast (except where carbon acids are involved) and occur frequently at the encounter rate; it is only fairly recently that it has been possible to measure the rates directly. The 'relaxation' methods pioneered by Eigen involve perturbing an acid–base equilibrium by the sudden application of a change in temperature, pressure, electric field, etc., and the observation by a very fast spectrophotometric method of the reestablishment of equilibrium as a function of time. The individual rate constants involved in the acetic acid dissociation were found in

$$CH_3COOH + H_2O \underset{k_{-1}}{\overset{k_1}{\rightleftharpoons}} CH_3COO^- + H_3O$$

$$\log k_1 = 5.9, \; \log k_{-1} = 10.7$$
$$(pK = \log k_{-1} - \log k_1 = 4.8)$$
$$= \text{normal equilibrium constant.}$$

this way. The broadening of a proton resonance absorption line is related to the lifetime of the species and is also used to study these fast reactions. Thus, the OH-resonance line in very pure ethanol is less sharp than the C–H resonances and indicates an exchange rate of the order of 1 s. The addition of either 10^{-4} M alkali or 10^{-5} M acid increases the rate by a factor of 100. The rate of recombination of hydroxide and hydronium ions is enormously fast, $k \sim 10^{11}$ s^{-1}, which is due to the cooperative proton transfer which occurs (page 18). Further examples of acid catalysis will be met in Chapters 8 and 10.

1.25   **THE BRØNSTED CATALYSIS LAW**

Studying the general base-catalysed decomposition of nitramide (**55**), Brønsted

$$H_2N\text{—}NO_2 \xrightarrow{\text{ B: }} H_2O + N_2O$$
$$(\mathbf{55})$$

and Pedersen (1924) found that the rate of reaction was dependent on the base strength of the catalyst employed—which included varied types of base, anilines, carboxylate ions, etc.—according to the equation,

$$k_B = C_B K_b^{\beta} = C_B (1/K_a)^{\beta}$$

or

$$\log k_B = \beta \log(1/K_a) + C_B$$

where K_b and K_a are the dissociation constants of the base and the corresponding conjugate acid, respectively. C_B and β are constants for the reaction. An analogous relationship was found for acid-catalysed reactions where

$$\log k_A = \alpha \log k_a + C_A$$

This relationship holds well for the dehydration of acetaldehyde hydrate (**56**), which was investigated extensively by Bell using as catalysts some sixty acids

$$CH_3\text{—}CH\begin{smallmatrix}OH\\OH\end{smallmatrix} + H^+ \rightleftharpoons CH_3CH\begin{smallmatrix}OH\\OH_2^+\end{smallmatrix} \rightleftharpoons$$
$$(\mathbf{56})$$

$$CH_3\overset{OH^+}{\underset{+\,H_2O}{C\text{—}H}} \rightleftharpoons CH_3CHO + H^+ + H_2O$$

with a span of acidities of 10^{10} (Figure 9). Many general acid- and base-catalysed reactions obey this Brønsted catalysis law, which is a linear free-energy relationship; written in the form,

$$\log k_a/k_a^0 = \alpha \log K/\mathbf{R}_0$$

where k_a, K and k_a^0, K_0 refer to two different catalysts (the values of k are catalytic constants, p. 68) the analogy with the Hammett equation (Section 1.10) becomes obvious. The Brønsted constant α expresses the relative ability

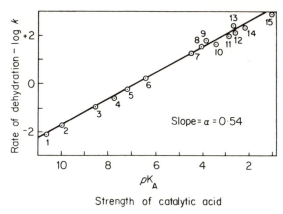

Strength of catalytic acid

Key

Acid	
1 hydroquinone	8 phenylacetic acid
2 phenol	9 *m*-toluic acid
3 *o*-chlorophenol	10 formic acid
4 2,4-dichlorophenol	11 phenoxyacetic acid
5 *p*-nitrophenol	12 chloracetic acid
6 2,4,6-trichlorophenol	13 salicyclic acid
7 acetic acid	14 cyanoacetic acid
	15 dichloracetic acid

Figure 9. Brønsted plot for the dehydration of acetaldehyde hydrate, catalysed by a variety of organic acids.

$$CH_3CH(OH)_2 \xrightarrow[k]{HA} CH_3CHO + H_2O$$

(From R. P. Bell and W. C. E. Higginson, *Proc. Roy. Soc. (A)*, **197**, 141 (1949)).

of the acid to protonate the substrate and to protonate water, and β similarly for deprotonation. The values almost invariably lie between 0 and 1. The value of α (β) has also been interpreted as a measure of the extent of proton transfer in the transition state, a value near zero indicating little transfer and a value near unity indicating a product-like transition state. It should be mentioned that base- and acid-catalysed processes are known which do not obey the

Brønsted Law. For example, the hydrogen-exchange reaction (conveniently followed by loss of tritium):

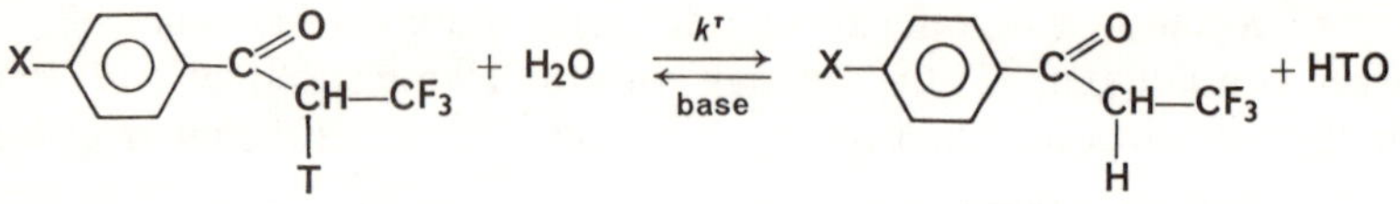

a plot of $\log k^T$ against the pK of the base is a curve rather than a straight line. Furthermore, values of α greater than unity have been found. The precise meaning of deviations from the Brønsted law is controversial, so caution should be exercised in its use.

1.26   TRACER METHODS

The principle of using a substituent group or an isotopic atom to mark a given atom during a reaction is straightforward and is a most valuable technique for obtaining evidence of mechanism and the presence of reactive intermediates. The following examples will illustrate the scope of this method.

The Claisen rearrangement of allyl ethers (57) to *ortho*-allylphenols (58) proceeds without the dissociation of the allyl group from the system, as shown by the invariable attachment of C-3 to the aromatic ring in the product. This can be demonstrated using either a methyl label or ^{14}C. If the *ortho*

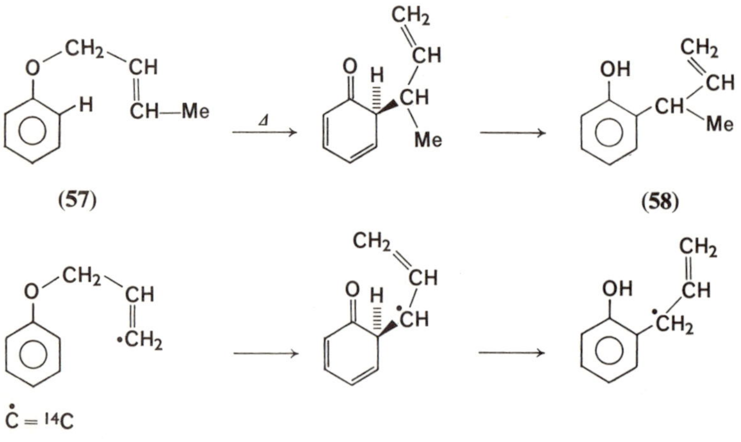

positions are blocked, rearrangement to *para*-allylphenol (59) may occur. Now the product has C-1 attached to the ring. The following mechanism provides

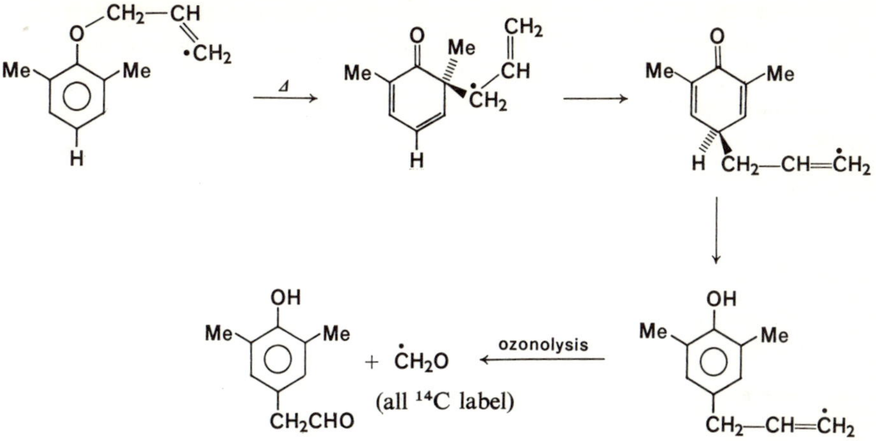

(60)   **(59)**

an explanation. The point to note here is that if an allyl radical or cation had become free at some stage **(60)**, the products would be expected to have *both* C-1 and C-3 attached to the ring. The ^{14}C experiment in particular demonstrates that these two carbons never become equivalent during the reaction. The

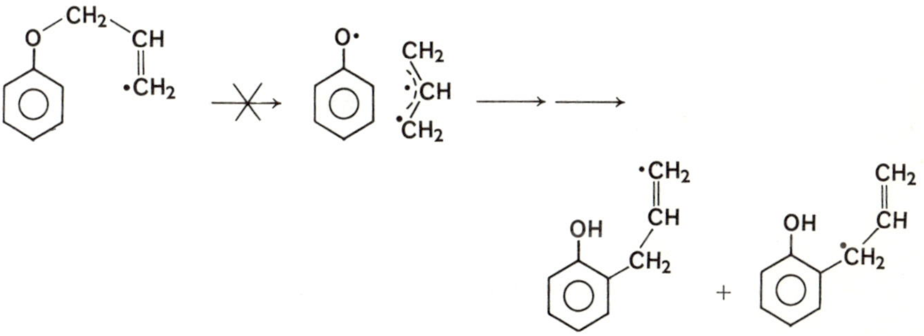

Wilgerodt reaction, the treatment of acetophenone with sulphur and ammonia in pyridine, yields phenylacetamide **(61)** and phenylacetic acid **(62)**. When the reaction was carried out on acetophenone labelled with ^{14}C at the carbonyl position the following label positions were found:

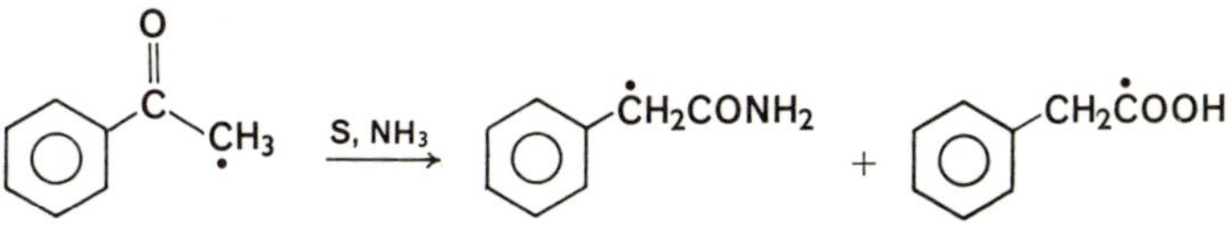

(61) 80%   **(62)**

The phenylacetic acid must have been formed with a skeletal rearrangement but the phenylacetamide has not. This demonstrated that the two products were formed by different routes.

The acetolysis of the toluenesulphonate ester (63) with ^{14}C at the α-carbon gives an acetate (64) in which both α- and β-carbons are labelled. This suggests an intermediate in which phenyl group migration is occurring before reaction

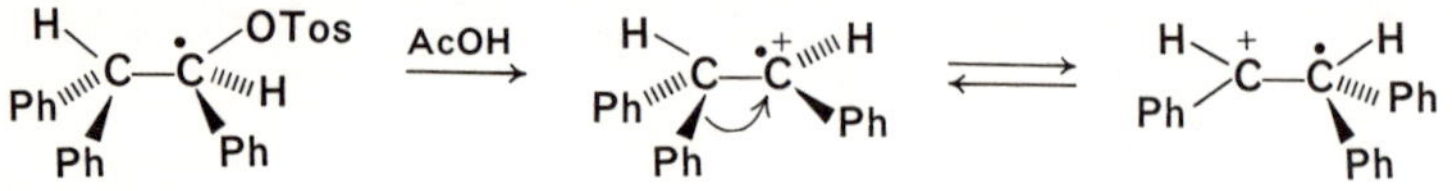

(63)   (65)

with acetic acid (Section 3.3) (65). Pre-equilibria may be revealed by tracer

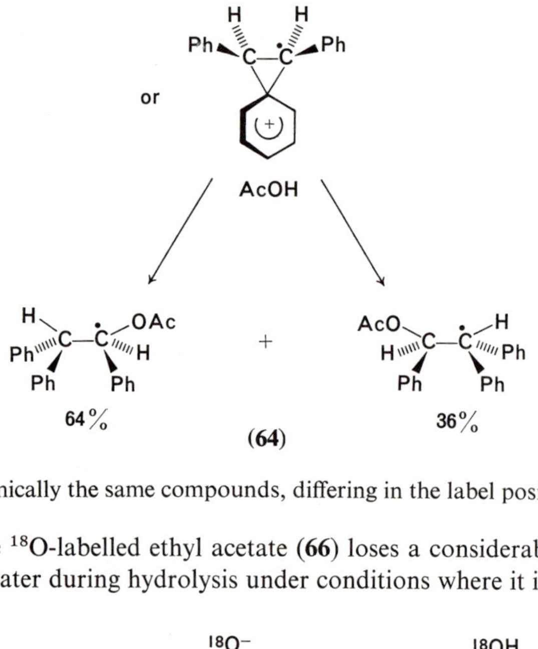

or

AcOH

64%   +   36%

(64)

(Chemically the same compounds, differing in the label position)

methods. The ^{18}O-labelled ethyl acetate (66) loses a considerable amount of the label to water during hydrolysis under conditions where it is stable in the

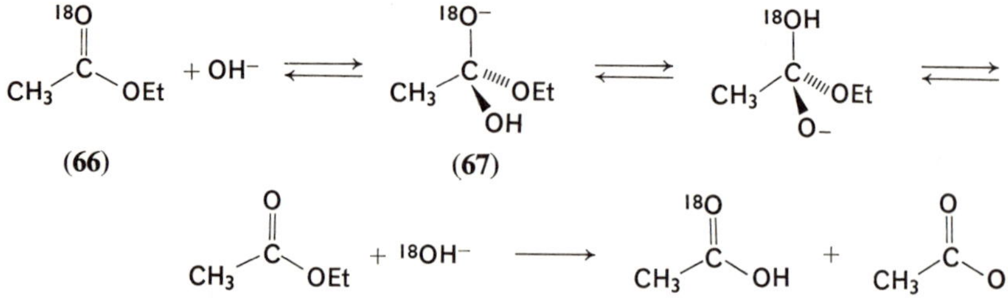

(66)   (67)

acetic acid. Exchange occurs from a tetrahedral intermediate (67) formed during the reaction.

Isotopic labelling is often used for kinetic studies of exchange reactions where the symmetry of the transition state is a simplifying feature. Further

$$^{131}I^- + \quad {}^{H}_{H}\!\!\diagdown\!\!{}^{\diagup}_{\diagdown}C\!-\!{}^{127}I \quad \longrightarrow \quad {}^{131}I\!-\!C\!\!\diagup\!\!{}^{H}_{H} \; + \; {}^{127}I$$

examples of the use of isotopic labelling will be found in subsequent chapters.

1.27   DIRECT OBSERVATION OF REACTIVE INTERMEDIATES

Many forms of spectroscopy are now available and there are considerable prospects of detecting directly certain short-lived intermediates by their characteristic absorption of radiation. Two conditions must be met, namely that the concentration of the reactive species be sufficient for detection to be possible and secondly that the species sought be sufficiently long-lived to be able to absorb the appropriate radiation. The latter depends upon the frequency of the radiation being used. Absorption of ultraviolet–visible light is very fast (ca 10^{-16} s) and infrared absorption requires only that the species shall be longer lived than molecular vibration times (ca 10^{-11} s). Problems begin to appear when using radiofrequencies (10^6–10^9 Hz). Electron spin resonance and nuclear magnetic resonance require the species under examination to be stable for periods in excess of 10^{-6} and ca 10^{-3} s, respectively. If a species has a lifetime appreciably shorter than these times, absorption lines will be broadened and may not be visible. A combination of these two circumstances usually means that detection of an intermediate during a reaction is only possible if it is not too transient, since short lifetimes mean that low concentrations are inevitable.

1.28   ISOLATION OF REACTIVE INTERMEDIATES

Reactive intermediates are, by their nature, unstable species under normal reaction conditions. Nevertheless it may be possible to isolate such species under more favourable conditions either in the pure state or at least in solution by suppressing their further reaction. While this does not necessarily prove the existence of the compounds in a reaction path, it lends plausibility by demonstrating that the proposed intermediates can have an independent existence.

Alkyl radicals in a matrix of sodium and sodium halide at low temperature are stable, unable to react with the surrounding medium or to dimerize. Many

$$R\text{—}Cl + Na \xrightarrow{-180°} R\text{·} + NaCl$$

alkyl cations have now been prepared in solution in extremely strongly acidic

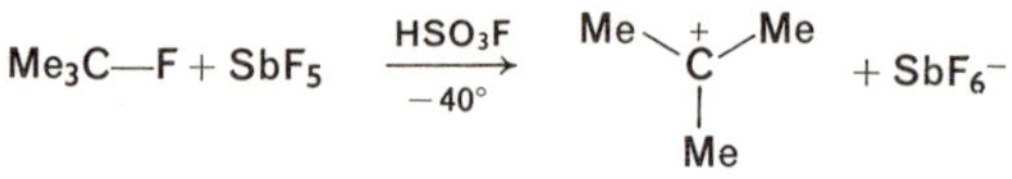

$$Me_3C\text{—}F + SbF_5 \xrightarrow[-40°]{HSO_3F} \underset{Me}{\overset{Me \quad Me}{\underset{|}{C^+}}} + SbF_6^-$$

media such as fluorosulphonic acid. In other instances, actual reaction intermediates cannot be isolated, but related species can. The nitrobenze-nonium ion (68) and its homologues, which are supposed intermediates in aromatic nitration, have not been isolated (though it has recently been observed in solution) but alkylated benzenonium ions such as 69 have long

(68) 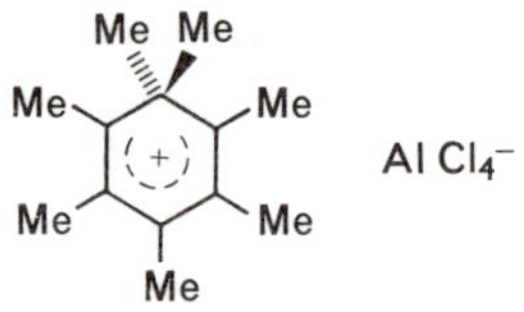   (69)

been known as stable crystalline materials. Similarly, the intermediate 70, believed to lie on the reaction path for the nucleophilic displacement reaction between fluorodinitrobenzene and an amine, is not isolable but a related species (71, a Meisenheimer complex) has long been known. Cyclic chloronium

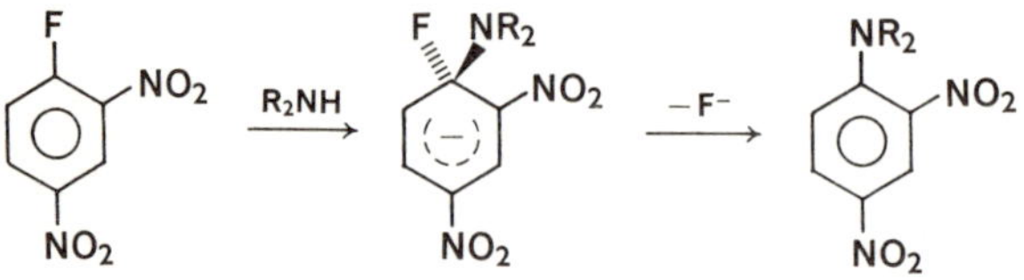

(70) 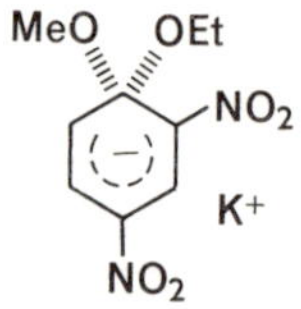   (71)

(72) and bromonium ions (73) are inferred as intermediates, e.g. in the halogenation of olefins. Recently, a number of non-cyclic analogues have been obtained crystalline (74).

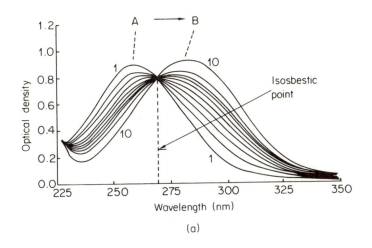

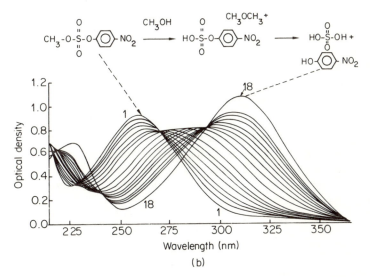

Figure 10. Spectroscopic scans for reactions (a) with, and (b) without, an isosbestic point (from E. Buncel, A. Raoult, and J. F. Wiltshire, *J. Amer. Chem. Soc.*, **95**, 799 (1973)).

R–CH=CH₂ (with Br–Br) $\longrightarrow$ (73) $R\text{-CH-CH}_2$ with Br⁻ and Br⁺ bridged

(74) $Me-\overset{+}{Br}-Me$ SbF_6^-

dimethylbromonium
hexafluoroantimonate

(72) $R\text{-CH-CH}_2$ with ⁺Cl bridge

Isosbestic Points

Even if an intermediate is not isolable, its presence may sometimes be inferred from spectra run during the course of a reaction. When one species is transformed directly into another, both having characteristic spectra, a spectrum run at intervals during the reaction will show the disappearance of one peak and its replacement by another (Figure 10a). There is one wavelength at which the absorption does not change with time, known as the isosbestic point.

If, on the other hand, the reaction proceeds via a fairly long-lived intermediate whose concentration becomes significant during the reaction, there will be no isosbestic point. Figure 10b shows that in the methanolysis of methyl *p*-nitrophenyl sulphate, under neutral conditions, there is an intermediate, *p*-nitrophenyl hydrogen sulphate, *en route* to the final products, *p*-nitrophenol and sulphuric acid.

1.29   'TRAPPING' REACTIVE INTERMEDIATES

This term implies adding reagents or changing the conditions such that a reactive intermediate goes on to form a product which is particularly characteristic of the species, in contrast to the 'normal' reaction product which may not be unambiguous in this respect. For example, the displacement of halogen by amide ion at an aromatic ring might conceivably occur by either of the two mechanisms:

(75)

In the first an anionic species similar to **70** is intermediate, while in the second a 'benzyne' (**75**) is produced (see Chapter 7). If on adding some anthracene to the reaction mixture the adduct triptycene (**76**) is isolated, this points to the latter intermediate which is known to add in this way while the anion is known not to. The formation of the dimer, diphenylene (**77**) is also characteristic of benzyne.

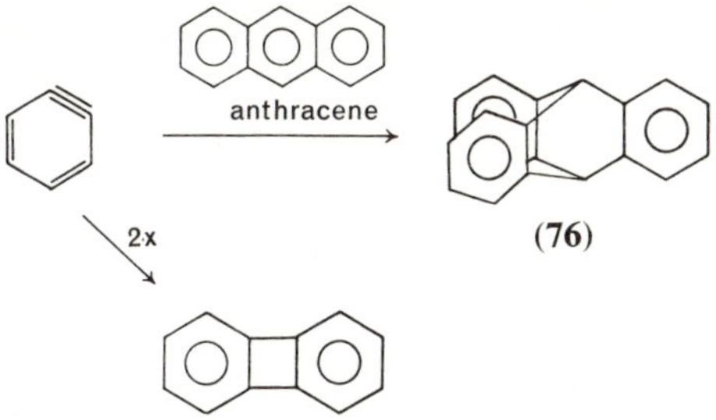

(76)

(77)

Deuterium exchange can be used in a similar manner. Suppose a reactant is first converted to its conjugate base by removal of a proton before this, in turn, forms products. The proton transfer is likely to be reversible and, in a deuterated solvent, deuterium incorporation will occur in the *unreacted starting*

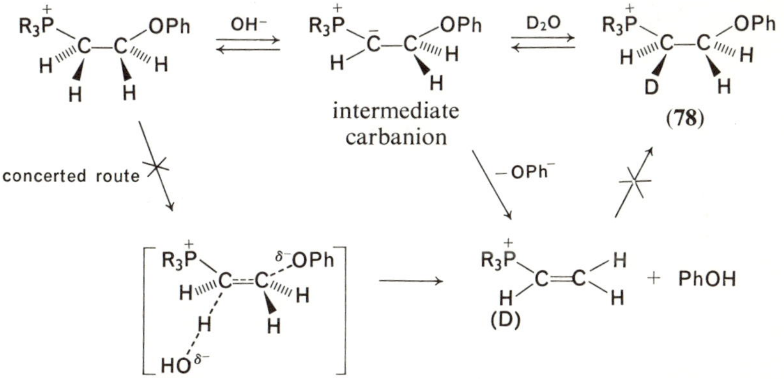

material. In the elimination reaction above, the concerted and the stepwise mechanisms can be distinguished by the fact that during the reaction unreacted starting material shows an increasing amount of deuterium incorporation (**78**). This is incompatible with elimination to the observed products in a single step. The products were separately shown not to reform the starting material.

1.30 **STEREOSPECIFICITY IN REACTION**

When two or more steric courses are open to a reaction, it may or may not be specific in the products which are formed, but in either case it is often possible to make deductions concerning the reaction mechanism and possibly the involvement of intermediates.

When two molecules undergo a cycloaddition, i.e. combination at two positions with the formation of a ring, the formation of the two new bonds may in principle occur sequentially or simultaneously. If the former, a diradical or zwitterion may be involved as intermediate while the concerted

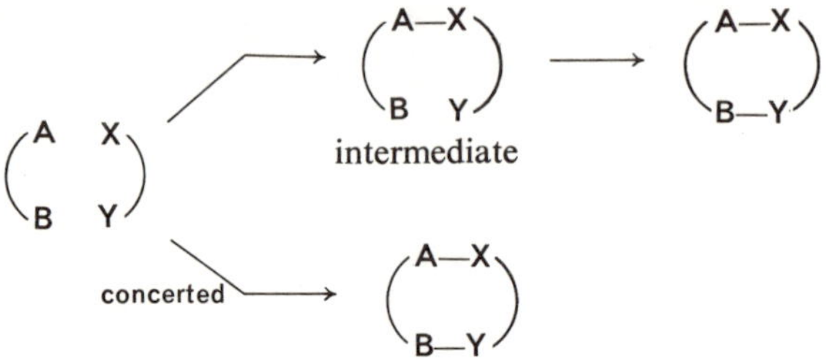

mechanism requires no intermediate. A familiar example of a cycloaddition reaction is the Diels–Alder synthesis of a cyclohexene from 1,3-diene and olefin components. It is found that the geometry of the substituents on the olefin remains unchanged in the products. For example, diethyl maleate (**79**) will add to butadiene to give cleanly the *cis* product (**80**), and diethyl fumarate

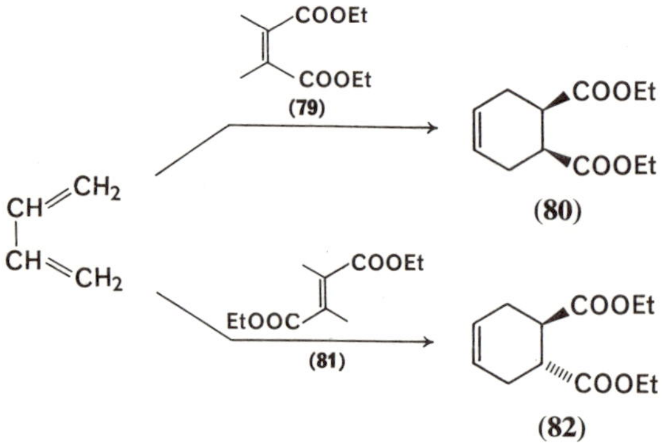

(**81**) only the *trans* (**82**). It is clear that no rotation around the C—C bond of the olefin component has been able to occur during the reaction. This is a *required* condition for a concerted reaction. It is unlikely to be the result of a truly two-step cycloaddition since an intermediate (**83**), if existing for a period comparable with molecular vibration times, would be able to rotate about the C—C bond and give at least some of the other stereoisomer in the product. On

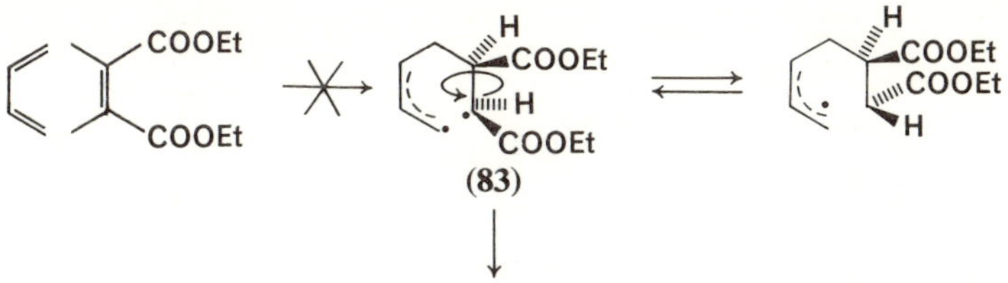

(83)

↓

(80)   +   (82)

the other hand, the addition of either *cis* or *trans*-2-butenes to an olefin such as cyclohexenone under photochemical conditions leads to the same mixture of the two isomeric adducts, **84** and **85**. Once it is established that neither the olefin nor the products can be isomerized under the reaction conditions, this result indicates conclusively the presence of an intermediate in which rotation

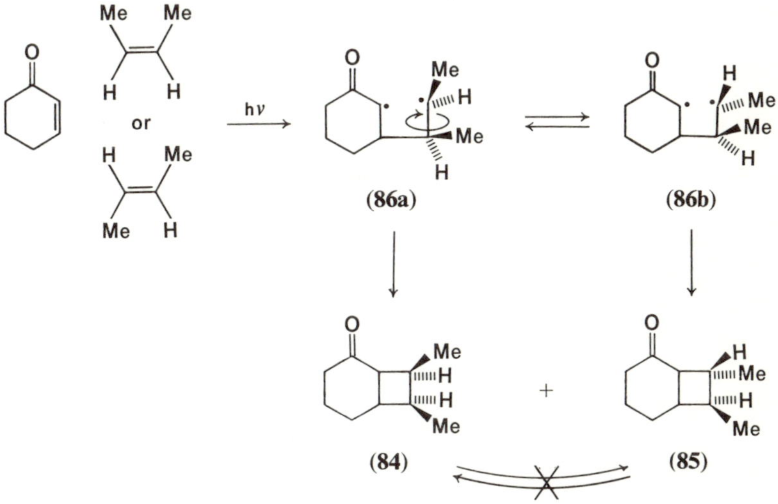

(86a)   (86b)

↓   ↓

(84)   +   (85)

of the carbon–carbon bond can occur (**86**). Stereochemical criteria may also be applied where the loss or retention of asymmetry of a molecule during reaction is significant. The conversion of tetrahedral carbon, potentially a centre of asymmetry (chirality), to trivalent carbon (cation, radical or anion) is invariably accompanied by loss of that asymmetry (racemization) if the latter has an independent existence. The trivalent carbon species may be planar (carbonium ions, **86**, and radicals, **87**, for instance) and have a plane of symmetry or may be pyramidal and rapidly inverting (carbanions, **88**) and thus effectively are symmetrical. Loss of asymmetry, as measured by optical activity, may indicate that a trivalent intermediate is formed and conversely chiral

(86, 87) achiral (88)

the retention of asymmetry, either by retention of *configuration* or inversion, may mean their absence although special circumstances may require these generalizations to be modified to some extent. Some examples from the field of substitution reactions will illustrate these points.

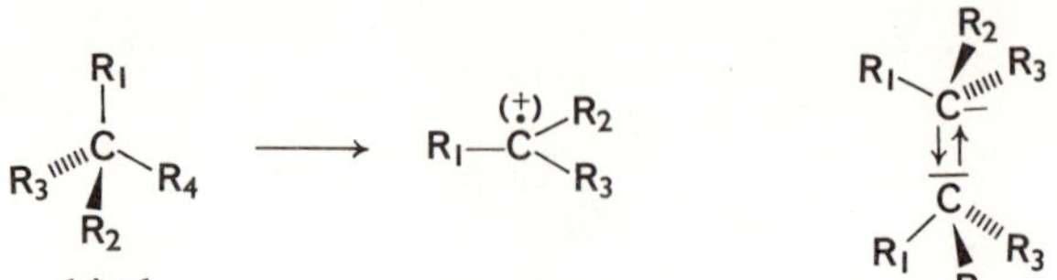

The reaction between optically active 2-octyl bromide (**89**) and sodium azide gives 100% 2-octyl azide (**90**) having the inverted configuration but otherwise

(89) (90)

inversion

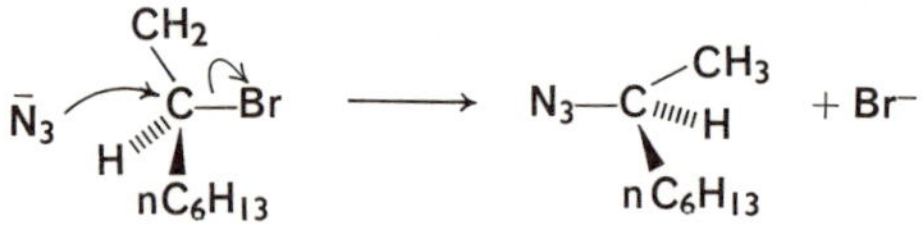

completely asymmetric. It is clear that no symmetric intermediate is involved and the azide has replaced bromide from the opposite side of the carbon atom.

The decomposition of the alkoxide (**91**) in the presence of *t*-butanol as a proton donor, results in the displacement of the butanone fragment by a proton. In this case, the proton enters on the *same* side as the leaving group (retention of configuration) but again, no symmetric intermediate has been

(91) retention

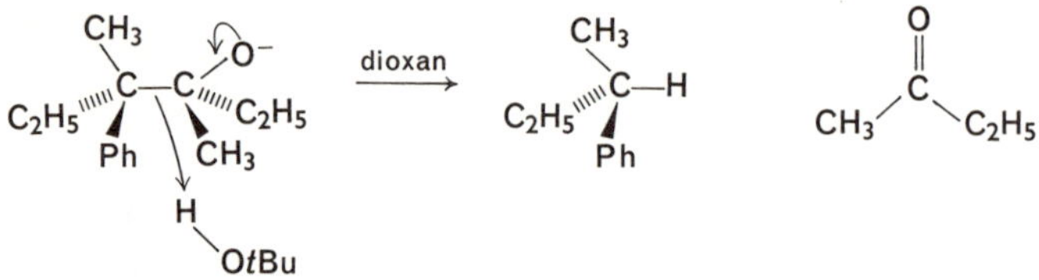

formed. The reaction of α-phenethyl chloride (**92**) in methanol, leading to the corresponding methyl ether, causes the loss almost entirely of all optical activity which the reagent possessed. It is likely on these and other grounds that a carbonium ion is formed as intermediate which reacts with solvent equally from either side.

$$\underset{\underset{Me}{H^{\prime\prime\prime\prime}}}{\overset{Ph}{\underset{|}{C}}}\!\!-\!Cl \quad \xrightarrow{MeOH} \quad \underset{\underset{H}{}}{\overset{Ph}{\underset{|}{C^+}}}\!\!\underset{Me}{\diagdown} \quad Cl^- \quad \longrightarrow \quad \underset{\underset{Me}{H^{\prime\prime\prime\prime}}}{\overset{Ph}{\underset{|}{C}}}\!\!-\!OMe$$

(92)

+

$$MeO\!-\!\underset{\underset{Me}{}}{\overset{Ph}{\underset{|}{C^{\prime\prime\prime\prime}H}}}$$

racemization

1.31   ORBITAL SYMMETRY CONSIDERATIONS

It has become apparent in recent years that an overriding factor in determining whether or not a reaction will occur is to be found in the symmetry properties of the reacting molecular orbitals. The principle was explicitly stated by Woodward and Hoffmann in 1965 that, if the reacting molecular orbitals of the reagent molecules transform into a set of orbitals in the product such that the two sets possess the same symmetry properties, then the reaction is greatly facilitated ('permitted') compared to a reaction in which no such correlation exists (a 'forbidden' process). An example will make this clearer. Consider the joining of two ethylene molecules to form a molecule of cyclobutane. Only the two pairs of π-electrons are required to convert into two σ-bonds. The system possesses two planes of symmetry at right angles to the plane of the paper and Figure 11 indicates the symmetries with respect to each, of each reacting orbital, and the corresponding antibonding orbitals. In each case, the pairs of orbitals are considered as combinations of the simple, localized orbitals. Now it will be observed that the correlation of identical symmetry properties between reagent and product orbitals requires that one of the bonding π-orbitals (ψ_2) transforms into an *antibonding* σ-orbital and also one of the antibonding π-orbitals of the reagents transforms to a bonding σ-orbital in the product. By the Woodward–Hoffmann principle of orbital-symmetry conservation it would be predicted that the conversion of two ethylenes into cyclobutane by a single-step, concerted, process would be energetically very difficult since it would tend to produce an electronically excited product and the same consideration applies to the reverse reaction. This means in practice that the concerted transformation in either direction is forbidden and that, if the reaction occurs at all, it will do so in two stages, thus:

$$\begin{array}{ccc} H_2C\!=\!CH_2 & H_2\overset{\cdot}{C}\!-\!CH_2 & H_2C\!-\!CH_2 \\ & | & | \quad | \\ H_2C\!=\!CH_2 & H_2\overset{\cdot}{C}\!-\!CH_2 & H_2C\!-\!CH_2 \end{array}$$

(93)

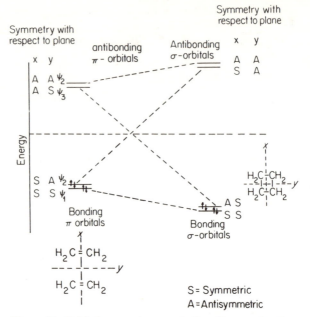

Figure 11. Orbital symmetry correlation diagram for the dimerization of ethylene to cyclobutane.

The orbital-symmetry principle tells us in effect that an intermediate should be expected on theoretical grounds, a diradical for instance (**93**). If one were to promote a bonding electron in one of the reagents to an antibonding orbital (i.e. to an excited state), then the reaction could proceed downhill in energy and with correct symmetries direct to the bonding orbitals of the product. This reaction is therefore 'permitted' when photochemically excited.

While the dimerization of ethylene cannot in practice be achieved, the dissociation of cyclobutane can occur at a sufficiently high temperature. It has been shown that the reaction takes place in two steps via **93** and requires an activation energy of 260 kJ mol⁻¹, despite the assistance to reaction arising from the relief of steric strain in the four-membered ring, approximately 110 kJ mol⁻¹.

Compare this reaction with the cycloaddition of butadiene to ethylene to form cyclohexene, and its regression:

This is comparable in the sense that two π-bonds are converted into two σ-bonds. The system has only one symmetry plane at right angles to the paper and the symmetries of the reacting orbitals with respect to it are indicated in Figure 12. Here there is a difference which is immediately apparent; the bonding orbitals on one side correlate with bonding orbitals of matching symmetry on the other and there is no need to involve high-energy antibonding

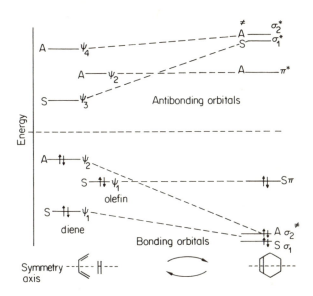

Figure 12. Orbital-symmetry correlation diagram for the reaction of butadiene with ethylene.

orbitals. This reaction is said to be 'orbital symmetry permitted' and is the well-known Diels–Alder reaction. Many examples are known whereby a diene and olefin react together at room temperature to give the cyclohexene. The thermal decomposition of cyclohexene requires an estimated 240 kJ mol^{-1} in activation energy with no compensating energy release from steric strain, so is seen to be very much more facile than the decomposition of cyclobutane. Both the forward and reverse reactions occur by concerted processes with no intermediate.

Four–centre reactions are frequently 'forbidden' by symmetry; for example, the addition of a chlorine molecule to a double bond does not correlate in the ground state:

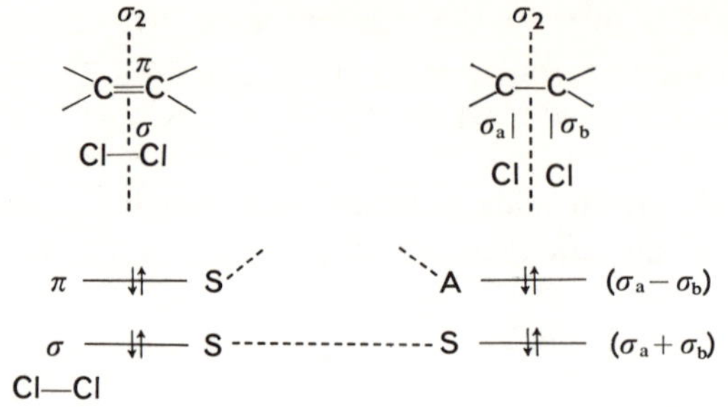

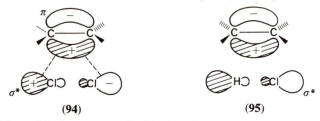

To look at this reaction in another way, the two orbitals which must interact are the π-orbital of the alkene and the lowest vacant orbital of chlorine—$\sigma*$ (an antibonding σ-orbital). These have zero net overlap due to symmetry and are non-interacting, **94**. These considerations provide an explanation as to why

(94)   (95)

the addition of halogen to a double bond does not occur by a concerted mechanism but goes through an intermediate carbonium or chloronium ion (Section 2.13). The addition of HCl to the olefin, strictly speaking, has no symmetry plane for a concerted reaction and the interaction of the $\sigma*$ and π orbitals would have non-zero overlap since the latter is polarized. However, one can still postulate that the concerted addition is unfavourable on account of the 'pseudosymmetry' (**95**).

For symmetry reasons the following reactions are forbidden (to be concerted); if they are observed we would expect an intermediate to be required.

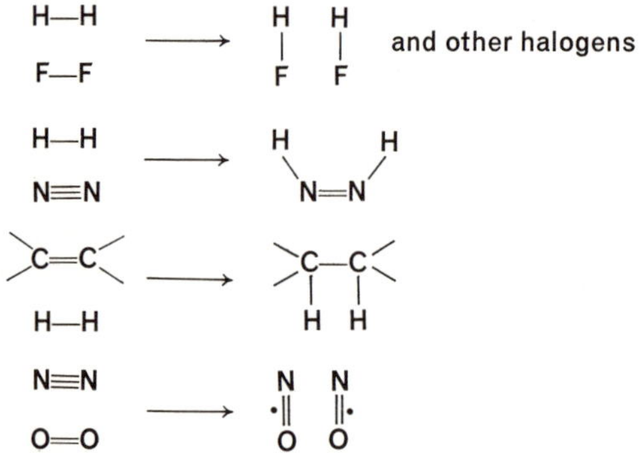

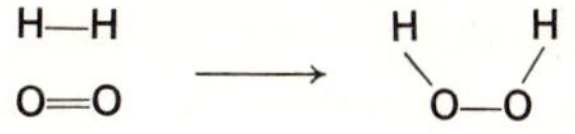

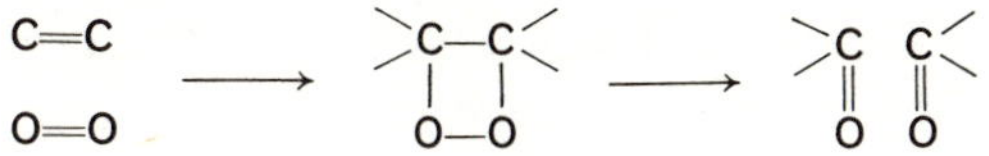

(see Section 10.8)

Even when the enthalpy of reaction is very favourable (e.g. $2NO \rightarrow N_2 + O_2$, $\Delta H = -180$ kJ mol^{-1}), the activation energy is very high. In the example, $\Delta G^{\neq} = 210$ kJ mol^{-1}.

By contrast, the following processes are symmetry allowed. It is probable that these reactions occur by concerted processes with no intermediates. While

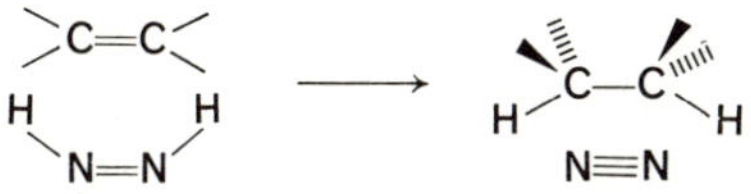

(Section 10.5)

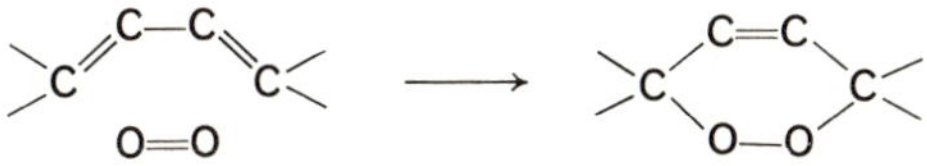

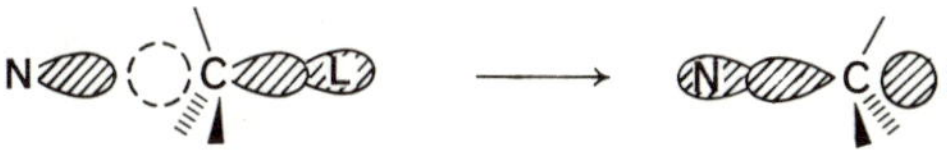

S_N2 reactions
(N = nucleophile)
(Section 2.3)

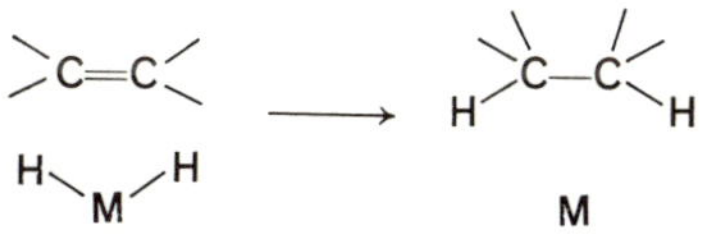

Catalytic hydrogenation
(M = metal)
(Section 10.7)

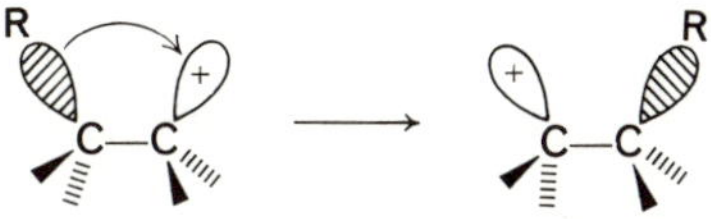

Wagner–Meerwein
rearrangement
(Section 2.29c)

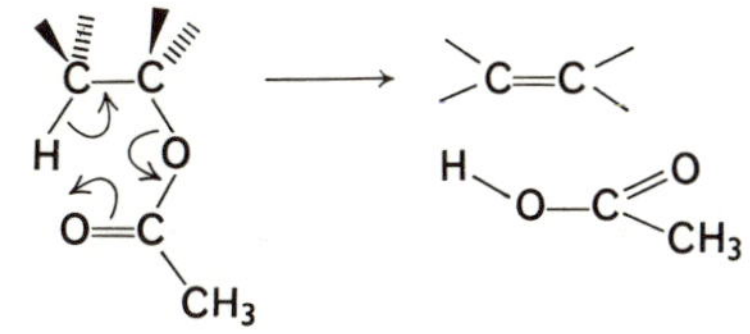

Ester pyrolysis

symmetry permittedness cannot invariably predict whether a reaction will occur, since there may be adverse energy considerations or alternative and more favourable paths, the orbital-symmetry correlation principle can certainly predict the concerted processes which will not occur. The principle is so fundamental that exceptions are not to be expected.

Orbital-symmetry restrictions can enforce stereospecificity upon a reaction. Cyclobutenes and butadienes can be relatively easily interconverted and this requires the transformation of a σ-bond of the cyclobutene into a π-bond, in fact the highest-energy π-orbital, of the diene. It may be seen that in order to

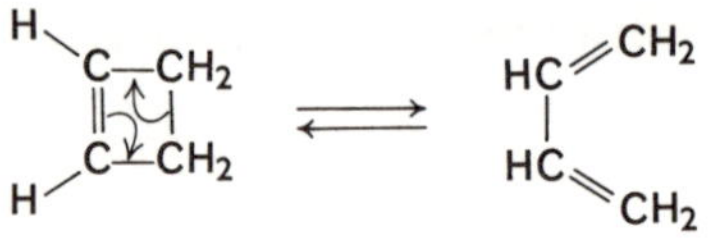

obtain positive overlap between the ends of this π-orbital the carbons must twist in the same direction, since only in this way do the combining ends present regions with the same sign of the wave function. This is known as a 'conrotatory' process; the other possibility, each terminus turning in the

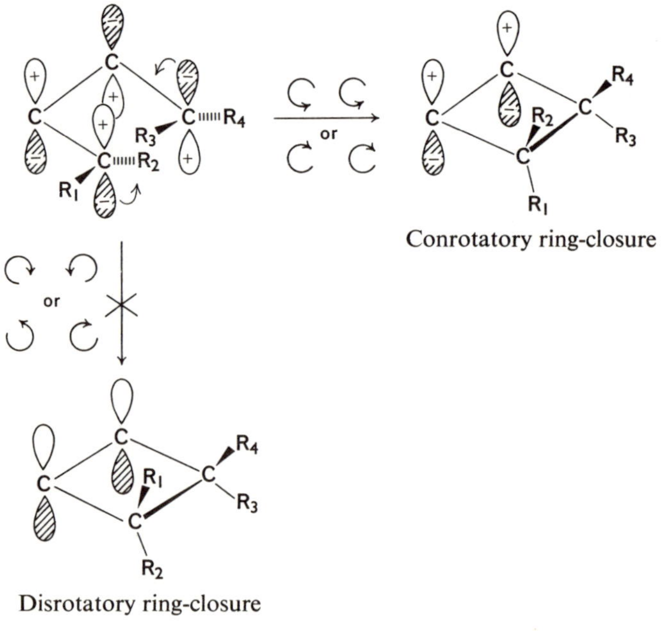

Conrotatory ring-closure

Disrotatory ring-closure

opposite direction, is known as 'disrotatory'. As a result, the *trans, trans* (**96**), or *cis, cis* (**97**), diene is converted to *trans*-substituted cyclobutene, and *cis,* *trans* diene (**98**) to the *cis*-substituted cyclobutene, and these steric relationships are quite specific. The operation of symmetry-controlled reactions is very

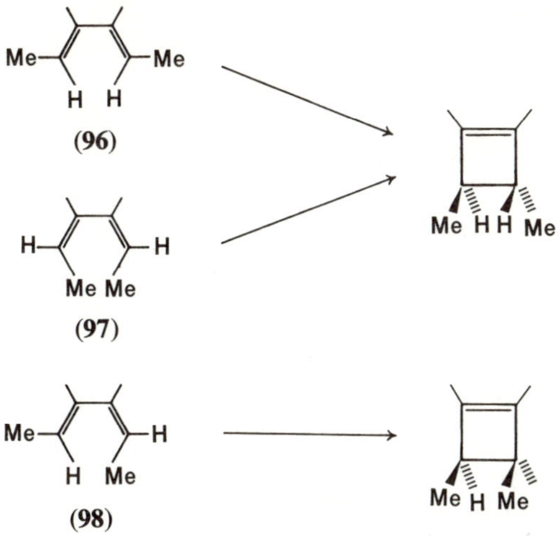

widespread and their recognition is a vital necessity which in many instances immediately sets out those reactions expected to be concerted and those which are multistep processes.

SUGGESTIONS FOR FURTHER READING

The following list of general texts on physical organic chemistry contain discussions of material relevant to this and to subsequent chapters.

General Texts

L. Ferguson, *Modern Structural Theory of Organic Chemistry*, Prentice–Hall, New York, 1963.

R. D. Gilliom, *Introduction to Physical Organic Chemistry*, Addison–Wesley, New York, 1970.

E. S. Gould, *Mechanism and Structure in Organic Chemistry*, Holt, Rinehart and Winston, New York, 1959.

L. P. Hammett, *Physical Organic Chemistry*, 2nd edition, McGraw-Hill, New York, 1970.

J. Hine, *Physical Organic Chemistry*, McGraw-Hill, New York, 1962.

C. K. Ingold, *Structure and Mechanism in Organic Chemistry*, 2nd edition, Bell, London, 1969.

E. M. Kosower, *Physical Organic Chemistry*, Wiley, New York, 1968.

A. L. Liberles, *Introduction to Theoretical Organic Chemistry*, Macmillan, New York, 1968.

K. B. Wiberg, *Physical Organic Chemistry*, Wiley, New York, 1964.

The following monographs and reviews are recommended for further information on topics discussed in Chapter 1.

Kinetics

S. W. Benson, *The Foundations of Chemical Kinetics*, McGraw-Hill, New York, 1960.
L. L. Schaleger and F. A. Long, 'Entropies of Activation and Mechanisms of Reactions in Solution', *Advances in Physical Organic Chemistry*, **1**, 1 (1963).

Pressure Effects

C. A. Eckert, *Rev. Phys. Chem.*, 239 (1972).
G. Kohnstam, 'Heat Capacities of Activation and their Uses in Mechanistic Studies', *Adv. Phys. Org. Chem.*, **5**, 121 (1967).
G. Kohnstam, 'The Kinetic Effects of Pressure', *Progr. Reaction Kinetics*, **5**, 335 (1970).
W. J. Le Noble, 'Kinetics of Reaction of Solutions under Pressure', *Progr. Phys. Org. Chem.*, **5**, 207 (1967).
E. Whalley, 'Use of Volumes of Activation for Determing Reaction Mechanism', *Adv. Phys. Org. Chem.*, **2**, 93 (1964).

Substituent Effects

H. H. Jaffe, *Chem. Rev.*, **53**, 191 (1953).
G. Kortum, W. Vogel, and K. Andrussow, *Dissociation Constants of Organic Acids in Aqueous Solution*, Butterworth, London, 1961.
L. M. Stock and H. C. Brown, 'A Quantitative Treatment of Directive Effects in Aromatic Substitution', *Progr. Phys. Org. Chem.*, **1**, 35 (1963).
P. R. Wells, *Linear Free Energy Relationships*, Academic Press, New York, 1968.
P. R. Wells, *Progr. Phys. Org. Chem.*, **6**, (1969).

Isotope Effects

C. J. Collins and N. S. Bowman, *Isotope Effects in Chemical Reactions*, Amer. Chem. Soc. Monograph, 167 (1971).
E. A. Halevi, 'Secondary Isotope Effects', *Progr. Phys. Org. Chem.*, **1**, 109 (1963).
L. Melander, *Isotope Effects on Reaction Rates*, Ronald Press, New York, 1960.
F. H. Westheimer, *Chem. Rev.*, **61**, 265 (1961).
K. B. Wiberg (see above under General Texts), and 'The Deuterium Isotope Effect', *Chem. Rev.*, **55**, 713 (1955).

Acid–Base Catalysis

R. P. Bell, *Acid–Base Catalysis*, Oxford University Press, 1941.
R. P. Bell, *The Proton in Chemistry*, Methuen, London, 1959.
M. Eigen, 'Relaxation Methods for the Study of Fast Reactions', *Disc. Farad. Soc.*, **17**, 194 (1954).

E. Grunwald and E. K. Ralph, 'Proton Exchange Mechanisms', *Accounts Chem. Res.*, **4**, 107 (1971).

M. A. Paul and F. A. Long, *Chem. Rev.*, **57**, 1 (1957).

C. H. Rochester, *Acidity Functions*, Academic Press, New York, 1970.

C. H. Rochester, 'Salt and Medium Effects on Reaction Rates', *Progr. Reaction Kinetics*, **6**, 144 (1971).

Orbital–Symmetry Effects

N. Entwistle, *Introduction to Orbital Symmetry Correlations in Organic Chemistry*, Van Nostrand–Reinhold, New York, 1971.

R. B. Woodward and R. Hoffmann, *The Conservation of Orbital Symmetry*, Academic Press, New York, 1970.

Chapter 2

Carbonium Ions as Reaction Intermediates

2.1   **INTRODUCTION**

Species which contain a trivalent carbon atom with a formal positive charge are known as carbonium ions. The positive carbon is therefore electron deficient (having six valence electrons) and could in principle make use of either sp^2 or sp^3 hybrid orbitals for bonding, being thereby a planar or a pyramidal species, respectively (**1a, 1b**). In either case, a low-lying vacant

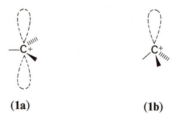

(**1a**)   (**1b**)

orbital is available. Simple examples of carbonium ions are the methyl cation (**2**), which may be regarded as the parent compound, the trimethylcarbonium ion or the *t*-butyl cation* (**3**), the benzyl cation (**4**) and the allyl cation (**5**). Structures **4** and **5** illustrate how the carbonium carbon may conjugate with unsaturated structures using its vacant orbital and thus cause the positive

---

* Two systems of nomenclature are in common use for carbonium ions: either the species may be named as a derivative of 'carbonium ion', i.e. CH_3^+, or it may be named as an alkyl (or aryl) cation. For example Me_3C^+ may be called the trimethylcarbonium ion or the *t*-butyl cation. It is incorrect to name this ion the *t*-butyl carbonium ion as this would refer to $Me_3\overset{+}{C}CH_2$—the neopentyl cation. Similarly, $Ph\overset{+}{C}HCH_3$ would be either the methylphenyl carbonium ion or the 1-phenylethyl cation. The name 'carbenium ion' has been proposed for trivalent cations on the grounds that the termination '-onium ion' implies an increase in coordination number and hence should apply to CH_5^+ and derivatives (cf. NH_4^+, ammonium). Since intermediates with pentavalent carbon may indeed have some existence (Section 2.30), this name may become common in future, but is not used in this book.

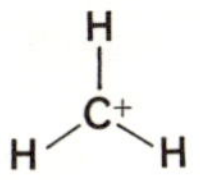

(2) 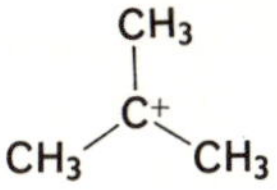   (3) 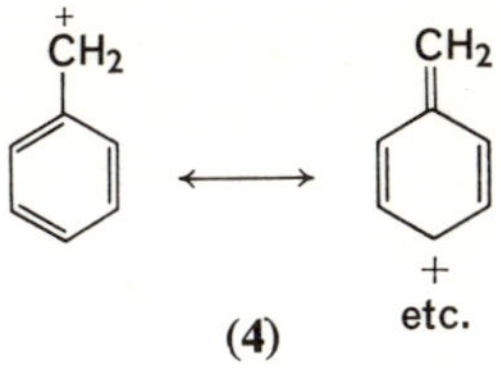   (4)   etc.

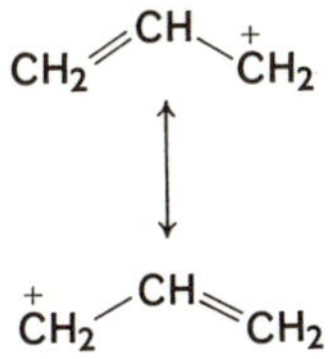

(5)

charge to become delocalized. Conjugation with adjacent atoms which possess a lone pair is also possible as in the oxo-carbonium ion (6) and the

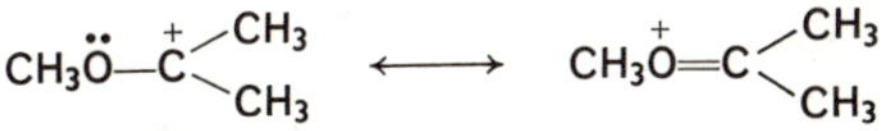

(6a)   (6b)

(7a)   (7b)

iminium ion (7). It is a matter of discretion whether such species should be regarded as carbonium ions since the positive charge is resident mainly upon the heteroatom, especially when it is the less electronegative nitrogen, rather than oxygen. Thus there is little contribution from **7a** to the overall electronic structure of the iminium ion.

2.2   THE RECOGNITION OF CARBONIUM IONS

The formation of deep yellow colours on solution of triphenylmethyl halides in certain solvents was reported in 1902 by Gomberg and Norris, among other workers. They attributed these reversible changes to what would today be called 'ionic dissociation', the ions being intensely coloured. The name

$$Ph_3C—Cl \underset{\text{or } H_2SO_4}{\overset{\text{liquid } SO_2}{\rightleftharpoons}} Ph_3C^+ + Cl^-$$

colourless   yellow

'carbonium ion' was applied to these species by von Baeyer. Dissociation of the covalent halide was confirmed by physical studies such as the conductivities of the yellow solutions and the twofold freezing point depression* brought about by solution of triphenylmethyl chloride in sulphuric acid (Hantzsch).

In the twenty years following their discovery, many carbonium ions containing three or two aryl groups were prepared and characterized. Some were found to have valuable dye or indicator properties, e.g. Crystal violet (**8**) and Malachite green (**9**), although these are best described as iminium ions (**8b, 9b**).

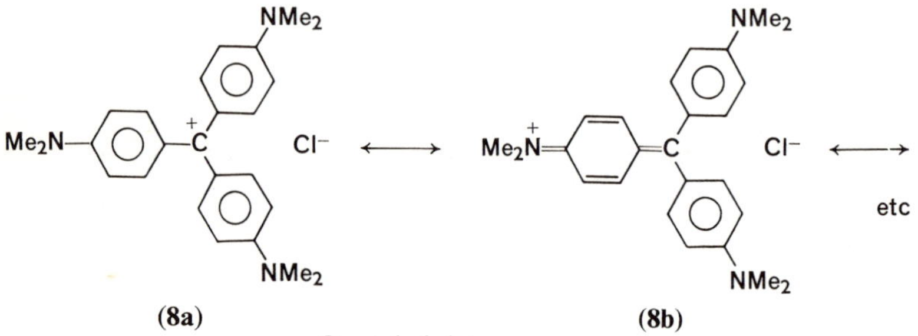

(**8a**)   (**8b**)

Crystal violet

(**9a**)   (**9b**)

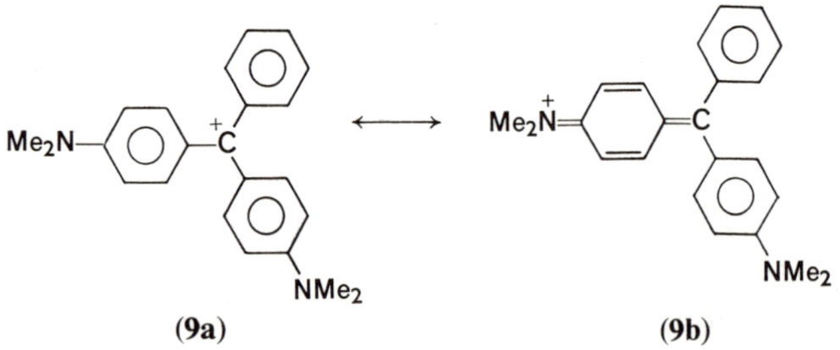

Malachite green

The next phase in the development of carbonium-ion theory began in 1922 with the suggestion by Meerwein that cationic intermediates were present in the rearrangement of camphene hydrochloride (**10**) to isobornyl chloride (**11**) which occurs in polar solution. Much of the baffling and complex chemistry of

---

* The depression of freezing point of a solvent in which is dissolved a solute is proportional to the number of moles of each separate solute species. Thus if one molar quantity of a solute is dissolved and it dissociates into *two* ions, the depression of freezing point is twice that which would have been observed had no dissociation occurred.

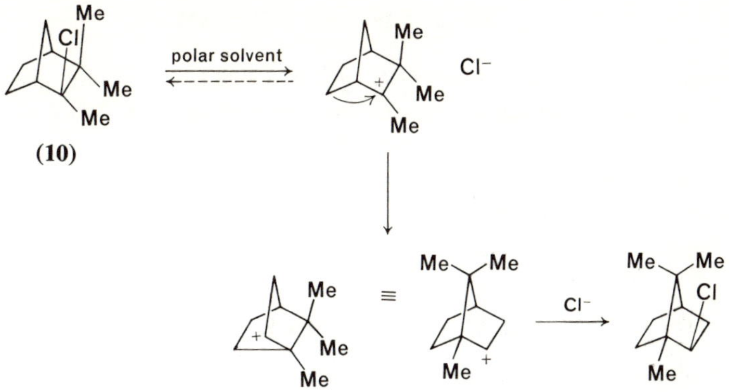

(10)

(11)

the terpenes was found to be explicable once it was recognized that skeletal changes involving the migration of adjacent groups to positive carbon were a

common occurrence. In the 1930s, the pioneering research of Whitmore, Hughes and Ingold, led to the postulation of carbonium ion intermediates in certain substitution reactions, e.g. the hydrolysis of *t*-butyl chloride (**12**) and

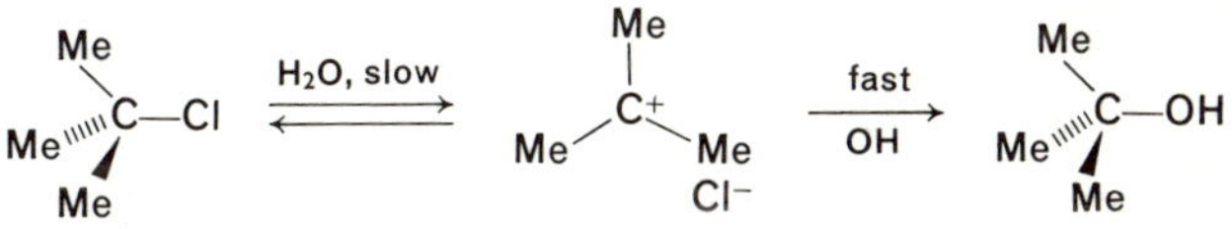

(12)

also in polar additions to multiple bonds, for example:

$$PhCH{=}CH_2 + H{-}Cl \longrightarrow \overset{+}{PhCH}{-}CH_3 \longrightarrow PhCH{-}CH_3$$
$$Cl^- \qquad\qquad Cl$$

Today, carbonium ions are recognized as transient intermediates in many solvolytic and acid-catalysed reactions while, since about 1965, great advances have been made in techniques for the stabilization and study of species previously thought to be capable only of a fleeting existence. There can be no doubt as to the real existence of these intermediates in a wide variety of reactions.

2.3   SOLVOLYTIC REACTIONS

This term refers to reactions in which a protic solvent, HS, exchanges its lyate ion, S^- for a suitable anionic group in a substrate in solution, R—X. Solvolytic

$$R—X + SH \longrightarrow R—S + H—X$$

reactions are commonly carried out in water (hydrolysis), acetic acid (acetolysis), ethanol (ethanolysis) and related solvents.

The Hydrolysis of Alkyl Halides; the Mechanistic Dichotomy

Alkyl halides are converted to the corresponding alcohols by aqueous alkali.

$$R—Cl + H_2O \longrightarrow R—OH + HCl$$

Hughes, Ingold and coworkers, around 1935, studied many such displacement reactions and measured their kinetics. They revealed two fundamentally different types of behaviour, illustrated by the hydrolysis of n- and t-butyl chlorides. The kinetic rate law for the hydrolysis of the primary halide was found to be second order, while that for hydrolysis of the tertiary isomer was

$$\text{rate} = -d/dt\,[nBuCl] = k_2[nBuCl][OH^-]$$

first order, the rate being independent of the hydroxide ion concentration. It will be shown that this reaction proceeds via a carbonium ion.

$$\text{rate} = -d/dt\,[tBuCl] = k_1[tBuCl]$$

It is apparent that two mechanisms of distinct character are operating. That of second order requires one molecule each of alkyl halide and hydroxide ion to be involved in the transition state.

S_N2 Mechanism

transition state

This direct displacement of chloride by hydroxide ion in a single step is referred to as an S_N2 process (Substitution, Nucleophilic, Second order). By contrast the rate-determining step in the hydrolysis of t-butyl chloride requires no hydroxide ion (although OH$^-$ is consumed in the reaction). The kinetic result is explained if the halide undergoes ionization in the slow step of the reaction forming the *t*-butyl cation which then in a subsequent and rapid step co-ordinates with the hydroxide ion (or a water molecule).

S_N1 Mechanism

transition state

intermediate

This is termed an S_N1 process (Substitution, Nucleophilic, Unimolecular). The entropies of activation (Section 1.8) lend support to these mechanisms: for S_N2 reactions, ΔS^* is usually small or negative (the value for the hydrolysis of CH_3Cl is -17 J K^{-1}) while S_N1 reactions have typically large, positive values of ΔS^* (e.g. *t*-butyl chloride hydrolysis, $+51$ J K^{-1}), implying an increase in the number of degrees of freedom of the system in going towards the transition state.

In general, methyl and primary halides tend to react by the S_N2 route and tertiary compounds by the S_N1. Secondary substrates occupy an intermediate position and may tend towards either, or a mixture, of mechanisms depending upon the circumstances. Factors which tend to stabilize a carbonium ion (e.g. highly solvating solvents, Section 1.15, and conjugating substituents) will favour the S_N1 mechanism. In extreme cases, a combination of these favourable circumstances may make the ionization route preferred even for primary halides. Although the S_N2 mechanism does not involve a reactive intermediate, some of its characteristics will be mentioned where these contrast with the carbonium-ion route.

Scope of the S_N1 Reaction

The basic requirements for S_N1 reactions are, an easy leaving group, a solvating solvent and a structure which leads to a reasonably stable carbonium ion.

The nature of the leaving group affords considerable variation. It should be, ideally, a very weak nucleophile to carbon. Chloride, bromide and iodide fulfil the requirements, but a number of sulphonate groups, particularly those with electron-withdrawing groups, are much better leaving groups. They also have the advantage of being conveniently prepared from the alcohol. The

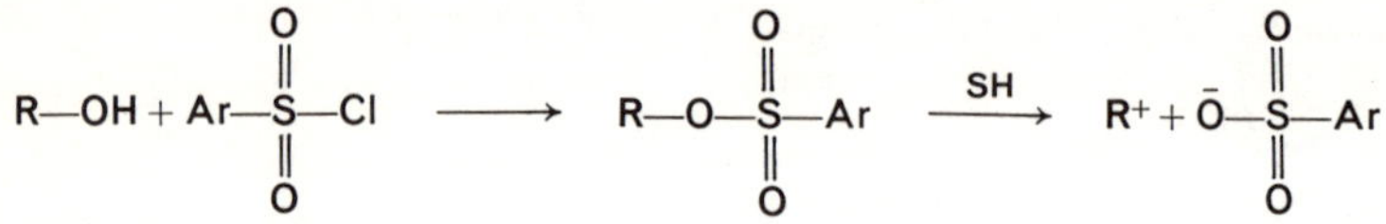

$$R{-}OH + Ar{-}\overset{\overset{O}{\|}}{\underset{\underset{O}{\|}}{S}}{-}Cl \longrightarrow R{-}O{-}\overset{\overset{O}{\|}}{\underset{\underset{O}{\|}}{S}}{-}Ar \xrightarrow{\ SH\ } R^+ + \bar{O}{-}\overset{\overset{O}{\|}}{\underset{\underset{O}{\|}}{S}}{-}Ar$$

following leaving groups are often encountered in solvolytic reactions:

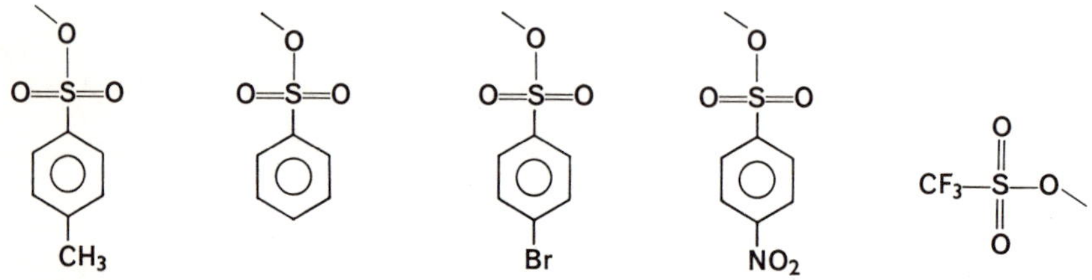

| p-toluene-sulphonate ('tosylate') = —OTos | benzene-sulphonate | p-bromo-benzene-sulphonate ('brosylate') = —OBros | p-nitro-benzene-sulphonate ('nosylate') | trifluoro-methane-sulphonate ('triflate') = —OTf |

Carboxylate groups, $R{-}C\overset{O}{\underset{}{\diagup}}O{-}$, do not ionize readily from carbon unless heavily substituted by electron-withdrawing groups; the same applies to phenolate ions. The following derivatives will take part in ionization reactions

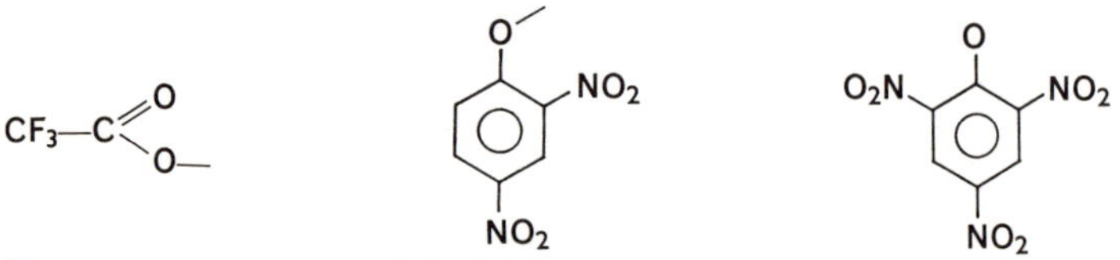

| trifluoroacetate | 2,4-dinitrobenzoate (DNB) | picrate |

from a suitable carbon. Alcohol and ether groups, normally difficult to displace (since OH⁻ and OR⁻ are strong nucleophiles), can be facilitated towards ionization by protonation: the same consideration applies to sulphur

$$R\text{—}OH \;\underset{\longleftarrow}{\overset{H^+}{\rightleftharpoons}}\; R\text{—}\overset{+}{O}\overset{H}{\underset{H}{\diagup}} \;\longrightarrow\; R^+ + H_2O$$

$$\underset{R}{\overset{O}{\diagup}}\!\!\diagdown_{CH\text{—}CH_2} \;\underset{\longleftarrow}{\overset{H^+}{\rightleftharpoons}}\; \underset{R}{\overset{+O}{\diagup}}\!\!\diagdown_{CH\text{—}CH_2} \;\longrightarrow\; \underset{R}{\overset{+}{CH}\text{—}CH_2OH}$$

and nitrogen groups. In a general sense this effect is an electrophilic catalysis
by the proton. Other examples of electrophilic catalysis in carbonium-ion

$$R\!\!\overset{\curvearrowleft}{\text{—}}Cl \;\;\; Ag^+ \;\longrightarrow\; R^+ + AgCl$$

formation include the silver-ion catalysed reaction of halides, and the racemization of α-phenethyl chloride (**13**) in the presence of a trace of stannic

(13)

chloride. The diazonium ion is also an excellent leaving group; indeed,
aliphatic diazonium ions have no stable existence, even in solution, but rapidly
lose nitrogen. They are easily generated by treatment of a primary amine with

$$R\text{—}NH_2 + HO\text{—}N\!\!=\!\!O \;\longrightarrow\; \underset{\underset{\text{N-nitrosoamine}}{H}}{R\text{—}N\text{—}N\!\!=\!\!O} + H_2O \;\;\overset{\text{'enolization'}}{\longrightarrow}$$

$$R\text{—}N\!\!=\!\!N\text{—}OH \;\longrightarrow\; R\text{—}\overset{+}{N}\!\!\equiv\!\!N + OH^- \;\longrightarrow\; \overset{+}{R} + N_2$$

diazotic acid   diazonium ion

nitrous acid or, in non-aqueous media, by an alkyl nitrite. Aromatic diazonium
ions are much more stable and frequently can be isolated. Their thermal

decomposition provides a route to aryl cations. The best possible leaving group may be considered to be a ^{3}He atom formed by radioactive decay of tritium, ^{3}H. Being unable to bond to carbon it provides a source of carbonium

$$H_3C\text{—}^3H \xrightarrow{\ \beta\text{-decay}\ } \begin{array}{c} H_3C\text{—}^3He^+ \\ +\ e^- \end{array} \longrightarrow H_3C^+ + {}^3He$$

ions in the gas phase. Leaving group facility is quite distinct in S_N1 and S_N2 processes and can often be used to gain mechanistic evidence; the relative rates of solvolysis of alkyl tosylates compared to the corresponding bromides is quite small in the bimolecular reaction and usually rather large in the unimolecular (Table 1). The *para*-substituted benzenesulphonates naturally act as leaving groups in order of σ-values of the substituents, $\rho = 1.2\text{–}1.5$.

A useful source of carbonium ions which is applicable in even non-protic solvents is the acid-catalysed decomposition of triazenes:

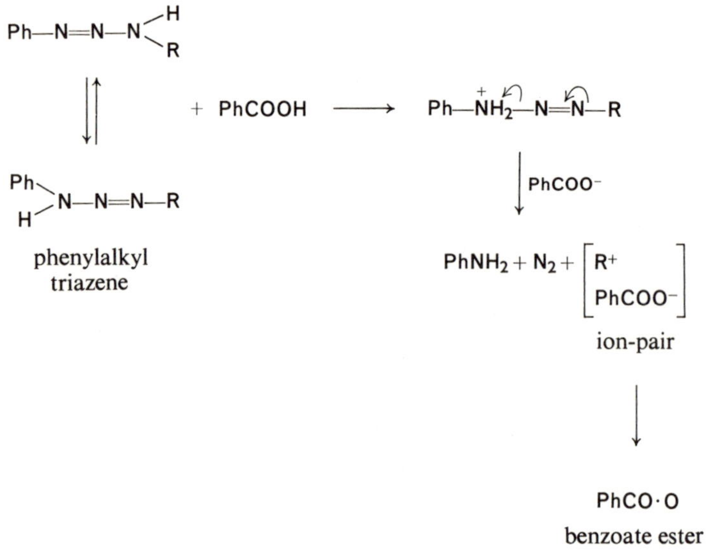

phenylalkyl
triazene

ion-pair

benzoate ester

Rates of reaction increase with chain branching of $R : R = Me : Et : isoPr$ $k_{rel} = 1:1.6:7$, the order for carbonium-ion stability. Also, the isopropyl analogue ($R = iPr$) gives products containing significant proportions of isopropyl ester. Again, this is the expected behaviour of a carbonium ion (see p. 167).

Table 1. Relative ethanolysis rates of alkyl tosylates and bromides

$$R—OTos + EtOH \xrightarrow{k_{OTos}} R—OEt + TosOH$$

$$R—Br + EtOH \xrightarrow{k_{Br}} R—OEt + HBr$$

R–	$\dfrac{k_{OTos}}{k_{Br}}$	Mechanism
Me-	11	S_N2
Et-	10	S_N2
isoPr-	40	S_N2
t-Bu-	4000	S_N1
1-Bicyclo[2,2,2]octyl, 	5000	S_N1

2.4 THE PROBLEM OF MECHANISM IN THE INTERMEDIATE REGION

While extreme types of nucleophilic substitution, denoted S_N1 and S_N2, occur in appropriate cases and are relatively well understood, the solvolysis of secondary substrates in moderately ionizing solvents and in the absence of strong bases may occupy a mechanistic position which is not represented by either extreme but is in the 'borderline' region. There are two ways of interpreting the mechanism in such cases. One may either postulate the existence of a continuity of transition states between extremes, following Ingold and co-workers:

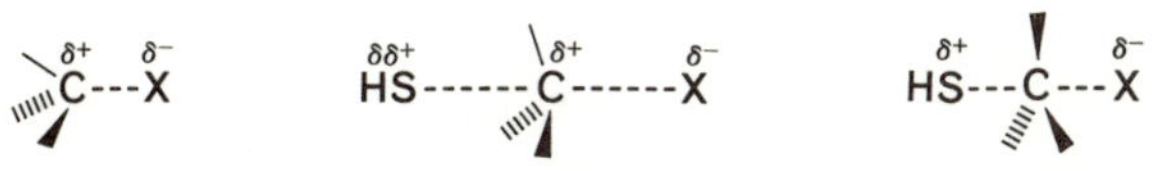

Or one may regard a borderline reaction as being partly composed of a true S_N1 component and partly of an S_N2 component (attack at the rearside by solvent), each being independent and competitive. This approach has been discussed by Winstein, Schleyer and others.

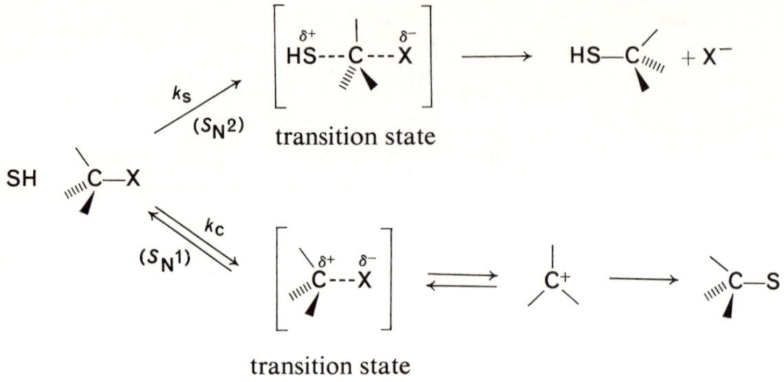

transition state

$$k_{total} = k_s + f k_c$$

where f = fraction of dissociations which lead to product rather than return. This scheme explains stereochemical results in which partial inversion and partial racemization occur; in certain cases, the two components can be dissected (Section 3.3). However, it should be realized that the subject is controversial.

2.5   KINETICS

The ionization of the leaving group from carbon must, in principle, be a potentially reversible process. The reaction sequence then may be written:

$$R\text{---}L \underset{k_{-1}}{\overset{k_1}{\rightleftarrows}} R^+ + L^-$$

$$R^+ + SH \xrightarrow{k_2} R\text{---}S + H^+ \quad \text{(or other products)}$$

$$\text{rate} = k_2[R^+]$$

If we assume that the concentration of carbonium ion is small and *constant* throughout the reaction (steady state approximation), i.e.

$$d/dt\,[R^+] = 0$$

then rate of formation of R^+ equals its rate of destruction;

$$k_1[RL] = k_{-1}[R^+][L^-]$$

$$[R^+] = \frac{k_1[RL]}{k_{-1}[L^-] + k_2}$$

$$\text{rate of reaction} = \frac{k_2 k_1[RL]}{k_{-1}[L^-] + k_2} = \frac{k_1[RL]}{\dfrac{k_{-1}}{k_2}[L^-] + 1} \tag{2.1}$$

or, for $k_{-1} \ll k_2$,

$$\text{rate} = k_1[RL] \tag{2.2}$$

Thus a full kinetic analysis would predict that first-order kinetics would only be found when the ionization is not appreciably reversible, (equation 2.2), but if that condition is not fulfilled, the rate should obey the more complex expression, equation 2.1. The ratio k_{-1}/k_2 may be evaluated from kinetic measurements and shows a wide variation in magnitude with structure and conditions. In general, a low value of this ratio implies a high reactivity and short lifetime for the carbonium ion, and kinetics which obey the limiting expression, equation 2.2 (Table 2). A high value of k_{-1}/k_2 indicates a rather stable carbonium ion and kinetics which obey the rigorous expression, equation 2.1.

Table 2. Reversibility of ionization in S_N1 reactions

$$RCl \xrightarrow{\text{aqueous acetone}} ROH + HCl$$

	R	k_{-1}/k_2
	Ph_3C	400
Increasing	$(p\text{-}MeC_6H_4)_2CH$	70
stability	$(p\text{-}MeC_6H_4)CHPh$	30
of R^+	Ph_2CH	15
	Me_3C	ca 2

2.6   SUBSTITUENT EFFECTS AND LINEAR FREE ENERGY RELATIONSHIPS

Ionization of the leaving group should be facilitated by electron-donating substituents, X, in the alkyl residue:

This is particularly true if the carbonium ion can be stabilized by conjugation with a group which will accept positive charge. The Hammett relationship (equation 1.10) has been applied to many S_N1 reactions occurring adjacent to an aromatic ring and good correlations are found with large, negative ρ-values (Table 3). Furthermore, the 'enhanced' values of the substituent constant, σ^+, must be used when a benzylic cation is formed as intermediate, since through-conjugation with $+M$ substituents (e.g. –OMe, –Me) in the *para* position is

(14)

possible (14). Intermediates such as 14 are especially stabilized.

Table 3. Reaction constants for some solvolytic reactions

Reaction	Mechanism	ρ	Correlation of k_{rel} with
Ar—CHCl + EtOH → Ar—CHOEt 　|　　　　　　　| 　Ph　　　　　　　Ph	S_N1	−5.09	σ^+
Ar—CHCl + EtOH → Ar—CHOEt 　|　　　　　　　| 　Me　　　　　　Me	S_N1	−5.7	σ^+
Ar—O—PNB* ... Ar—O—Ac + AcOH ⟶	S_N1	−5.1	σ^+
Ar—C(Ph₂)-Cl + EtOH → Ar—C(Ph₂)OEt	S_N1	−2.9	σ^+
ArCH₂Cl + OH⁻ → ArCH₂OH	S_N2	−0.14	σ
ArCH₂Cl + I⁻ → ArCH₂I	S_N2	+0.81	σ

* *p*-nitrobenzoate.

The correlation of rates with σ^+ rather than with σ affords excellent evidence for the intermediacy of cationic species in such reactions as these. By contrast, solvolyses occurring by the S_N2 mechanism exhibit small and often positive values of ρ and correlated with σ. An electron-donating group will still facilitate the departure of the leaving group but at the same time will retard the co-ordination of the incoming nucleophile, and conversely for an electron-withdrawing substituent (15). Since attachment of the nucleophile and detachment of the leaving group occur concurrently in the S_N2 process, electronic effects of substituents tend to cancel out and be rather unpredictable for this type of reaction.

facilitated

retarded

N:⇀C—L

retarded   facilitated
electron-donating   electron-withdrawing
substituent   substituent

(15)

2.7   **SOLVENT EFFECTS**

The ionic dissociation of a covalent bond is a process which requires considerably more energy than homolysis, since work has to be done against an electrostatic field, unless this is compensated by attractive interactions between the ions and solvent molecules (solvation energy). Consequently, ionic reactions tend almost exclusively to occur only in solution and, moreover, such reactions are understandably very sensitive to the nature of the solvent. The greater the ability of the solvent to solvate ions (i.e. its polar character, Section 1.14), the lower will be the activation energy for a heterolytic reaction

$$\text{C—L} \longrightarrow \text{C}^+ + \text{L}^- \qquad -\Delta H_{\text{heterolysis}}$$

$$\text{C}^+ + \text{L}^- + \text{solvent} \longrightarrow \text{C}^+_{\text{solv}} + \text{L}^-_{\text{solv}} + \Delta H_{\text{solvation}}$$

$$\Delta H_{\text{net}} = \Delta H_{\text{solvation}} - \Delta H_{\text{heterolysis}}$$

of an initially *neutral* substrate. S_N1 reactions are typically very dependent upon the nature of the solvent and tend only to occur in those media best able to solvate ions. The most powerful of the solvation forces is hydrogen bonding, followed by electrostatic charge–dipole interactions as somewhat shorter-range forces. Furthermore, the solvent should have little tendency to bond covalently to carbon, i.e. should be of low nucleophilic power, otherwise solvolysis products will form by the direct displacement reaction k_s (S_N2) rather than via solvated ions. This limits the range of solvents in which carbonium ions may be generated in solvolytic reactions to sulphonic and carboxylic acids, sulphuric acid, water and alcohols. The dipolar aprotic solvents such as acetone, dimethylsulphoxide, Me_2SO, and dimethylform-amide, $HC(=O)NMe_2$ (Section 1.18), although capable of solvating cations, are too nucleophilic to favour the S_N1 reaction in general.

Nonetheless, even in these 'polar' solvents a solvolytic reaction may occur by a combination of the two mechanisms, the solvent acting partly as a nucleophile attacking carbon and partly as a solvating medium. The rates of

the individual pathways may be referred to as k_s and k_c, respectively. Hence the total rate of a solvolysis is given by,

$$k_{obs} = k_s + fk_c$$

where f is the fraction of substrate molecules which, upon ionization, go on to form products (the fraction which returns to starting material $= 1 - f$). This approach to solvolytic reactions was first discussed by Winstein. For the purpose of studying the properties of solvolytic reactions via the S_N1 mechanism it is necessary to choose a solvent in which the ratio fk_c/k_s is as large as possible. This ratio is said to denote the 'limiting' character of a solvent since one strives to work under conditions at the limit where $k_s \approx 0$. A large body of data indicates that the following sequence of solvents increase in limiting character. Rates of ionization of a given substrate also tend to increase as the solvent is

$$BuOH < EtOH < MeOH < H_2O < AcOH < HCOOH < CF_3CH_2OH < CF_3COOH < H_2SO_4 < FSO_3H$$

—— Greater polar (limiting) character ——→

varied in this order. As one proceeds along this series, the likelihood of a purely carbonium ion reaction occurring is increased. Mixed solvents behave roughly as the mean of the individual components.

The effect of polar solvents upon displacement reactions of either mechanism in more general terms is summarized quantitatively in the Hughes–Ingold theory of solvent effects (Section 1.17) in which the effect upon the rate of reaction in going from a less to a more polar solvent depends upon the change brought about on the charge on the reagent(s) upon passing from the initial to the transition state. The following examples will illustrate the point that the solvent effect depends upon the charge type of the reaction.

Charge type	Charge on nucleophile	Leaving group	Transition state	Charge movement	Effect on rate of a more polar solvent
S_N1 type					
a		neutral	 $-C-Br \longrightarrow \overset{\delta+}{C}\cdots\overset{\delta-}{Br}$	Separation	Large acceleration
b		$+$	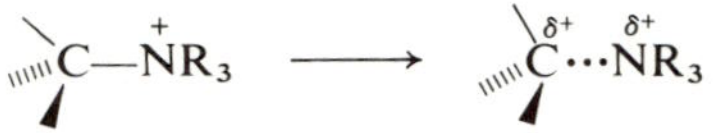 $-C-\overset{+}{N}R_3 \longrightarrow \overset{\delta+}{C}\cdots\overset{\delta+}{N}R_3$	Dissipation	Slight retardation
S_N1 type					
a	$-$	neutral	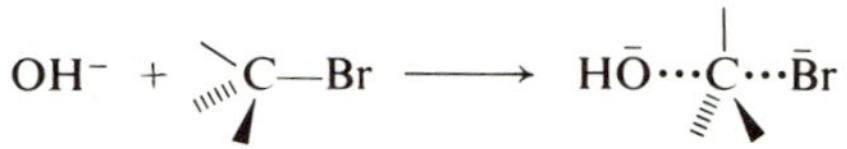 $OH^- + -C-Br \longrightarrow H\overset{-}{O}\cdots C\cdots \overset{-}{Br}$	Dissipation	Slight retardation
b	neutral	neutral	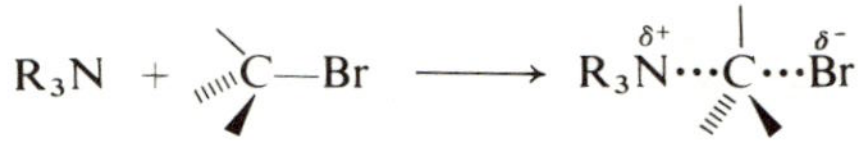 $R_3N + -C-Br \longrightarrow \overset{\delta+}{R_3N}\cdots C\cdots \overset{\delta-}{Br}$	Separation	Large acceleration
c	$-$	$+$	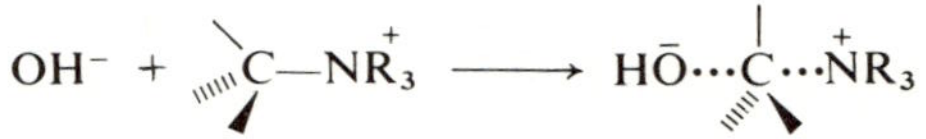 $OH^- + -C-\overset{+}{N}R_3 \longrightarrow H\overset{-}{O}\cdots C\cdots NR_3$	Neutralization	Large retardation
d	neutral	$+$	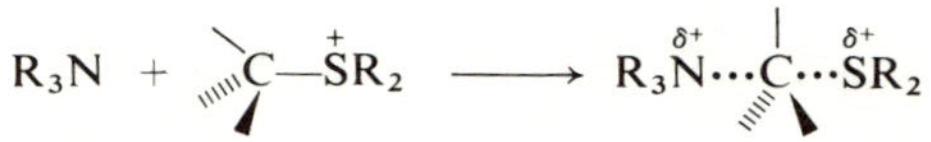 $R_3N + -C-\overset{+}{S}R_2 \longrightarrow \overset{\delta+}{R_3N}\cdots C\cdots \overset{\delta+}{S}R_2$	Dissipation	Slight retardation

These principles, which are applicable to a wide range of polar reactions, are further illustrated by the data of Table 5. It should be noticed that large accelerations are predicted for the reactions of neutral substrates with neutral nucleophile (e.g. solvent) by both the S_N1 and the S_N2 mechanisms. Thus the solvent effect would not be an effective technique for distinguishing these; however, in practice a distinction can be made by adding lyate ion (i.e. the anion derived from solvent by removal of a proton) and noting the solvent

Table 4. Grunwald–Winstein Solvent-susceptibility constants, m_s

Substrate	m_s	Solvolytic mechanism
t-Butyl chloride	1.00	S_N1
t-Pentyl chloride	0.90	S_N1
Benzhydryl chloride	0.757	S_N1
2-Adamantyl tosylate (**16**)	0.91	S_N1
Cyclohexyl tosylate (**17**)	0.44	S_N1 in very polar solvent / S_N2 in less polar solvent
2-Propyl tosylate (**18**)	0.42	S_N1 in very polar solvent / S_N2 in less polar solvent
Ethyl bromide	0.34	S_N2 almost entirely
Methyl bromide	0.22	S_N2

Table 5. Effects of solvent upon rates of solvolytic reactions

t-BuBr + SH → t-Bu-S + MBr   Charge type S_N1(a)

SH:	EtOH	80% aq. EtOH	50% aq. EtOH	H_2O	HCOOH
k_{rel}	1	10	29	1450	1.2×10^5

t-Bu-$\overset{+}{S}Me_2$ + SH → t-Bu-S + Me_2S Charge type S_N1(b)

SH:	EtOH	80% aq. EtOH	H_2O
k_{rel}	1	0.65	0.32

i-Pr-Cl + OH^- → i-Pr-OH + Cl^- Charge type S_N2(a)

solvent:	EtOH	20% aq. EtOH	40% aq. EtOH
k_{rel}	1	0.82	0.50

Me_3S^+ + OH^- → Me_2S + MeOH   Charge type S_N2(c)

solvent:	EtOH	20% aq. EtOH	40% aq. EtOH	H_2O
k_{rel}	1	0.025	0.0021	5×10^{-5}

i-Pr—Cl + SH → i-Pr—S + HCl   Charge type S_N2(b)

SH:	EtOH	20% aq. EtOH	40% aq. EtOH
k_{rel}	1	13.5	39

effects. Thus, the reaction of n-butyl chloride in aqueous ethanolic sodium hydroxide is slightly retarded the higher the water content of the medium (S_N2 reaction, OH^- a weaker nucleophile than OEt^-). By contrast under the same conditions (with or without added OH^-), the reaction of t-butyl chloride is greatly accelerated by increasing the water content (S_N1 reaction).

As mentioned earlier (Section 1.15), the rates of solvolysis of t-butyl chloride in a wide range of protic solvents (relative to the rate in 80% aqueous ethanol) have been used as an empirical measure of solvent polarity (Y-value), where

$$Y = m_s \log \left(\frac{k(\text{solvent})}{k_0(80\% \text{ aq. EtOH})} \right)$$

By definition, the constant of proportionality, m_s, which measures susceptibility of the solvolytic rate to solvent polarity, is unity for t-butyl chloride and should be of similar magnitude for any neutral substrate which reacts by the S_N1 mechanism. Table 4 gives some values of m_s and their interpretation. Some attempts have been made to separate the 'solvent-assisted' and 'ionization'

| (16) | (17) | (18) |

components, k_s and fk_c of solvolytic reactions in secondary substrates which react by a mixture of mechanisms.

2-Adamantyl tosylate (16), by virtue of its geometry, is incapable of undergoing the S_N2 reaction or of receiving nucleophilic assistance at the rearside opposite the leaving group. It may be argued that its reaction rates in any solvent should be those of the pure ionization component, fk_c. This is substantiated by a comparison of rate characteristics for this compound with values typical of compounds undergoing S_N1 and S_N2 reactions:

	2-Adamantyl	Typical S_N1	Typical S_N2
Winstein–Grunwald m_s value (Table 4)	0.91	1.0	0.2
Reactivity ratio, $k_{R\text{-OTos}}/k_{R\text{-Br}}$ (in AcOH)	16×10^3	10^5	500
Effect of α-methyl	$10^{7.5}$	10^8	10^{-1}–10^{-4}

Now, it may be supposed that the rate of reaction of the adamantyl ester is equal to the *ionization component*, fk_c, of other secondary tosylates with similar electronic and steric environments such as isopropyl tosylate (**18**). The rate ratio, k_{obs}(isopropyl)$/k_{obs}$(2-adamantyl) may therefore be taken as a

measure of the amount of the nucleophilic assistance due to solvent. This ratio is found to increase dramatically as one progresses from a highly limiting solvent (CF_3COOH) to a rather poor but nucleophilic one (EtOH), in which the component k_s has become dominant (Table 6). These figures illustrate the progression of mechanism from pure S_N1 in CF_3COOH to almost pure S_N2 in ethanol.

Table 6. Nucleophilic solvent assistance in solvolytic reactions of isopropyl tosylate

Solvent	$\log \dfrac{k_{obs}(\text{i-PrOTos})}{k_{obs}(\text{Ad-OTos})^b}$	$k_s/k_c{}^a$ (Isopropyl)	
CF_3COOH	−2.25	<1	S_N1
HCOOH	0.5	560	
CH_3COOH	1.1	2240	
50% aq. EtOH	1.3	7400	
EtOH	3.0	4×10^5	S_N2

^{a} assuming essentially limiting solvolysis of both substrates in CF_3COOH.
^{b} Ad = 2-adamantyl (**16**).

2.8   SALT EFFECTS UPON UNIMOLECULAR SOLVOLYSES

Added salts can have a profound effect upon a solvolytic reaction and can afford a further distinction between the two mechanistic types. There are, in fact, three ways in which a salt may act (Section 1.19).

Ionic Strength Effects

The solvation energy of an ion is increased by ions of opposite charge already in the solution which therefore have the effect of increasing the polarity of the medium. The effect of an added inert salt to the medium therefore can be predicted qualitatively from the principles set out on p. 58, and will depend upon the charge type of the reaction. Thus, the ionization rate of a *neutral* substrate will be enhanced by a high ionic strength. Quantitatively, the effect of ionic strength may be equated by the Brønsted equation (Brønsted Salt Effect),

$$\log(k/k_0) = 2\mathscr{S} z_a z_b \, F(I)$$

where k and k_0 are the specific rate constants at ionic strength $I = I$ and $I = 0$ respectively; z_a, z_b are the charges on the ions formed; $\mathscr{S}$ is the Brønsted coefficient, a property of the solvent and the particular function of ionic strength, $F(I)$, which is appropriate depends upon the salt concentration. At $I \sim 0.01$, the limiting value, $I^{0.5}$, may be used; at higher ionic concentrations, the more complex expression, $(1 + BI^{0.5})$, where B is a constant, becomes appropriate. It is generally assumed that ion pairs do not contribute to the ionic strength of the medium but the point is difficult to test.

Many unimolecular solvolytic reactions have been tested and found to obey the Brønsted equation with varying degrees of precision, though departures from ideal behaviour occur at relatively high ionic strength.

In order to study ionic strengths, the salt chosen must not be capable of reaction with either the substrate or the intermediate carbonium ion. An alkali metal perchlorate or tetrafluoroborate is suitable since the anions are of very low nucleophilic power.

As a consequence of the ionic strength effect, the apparent rate constant of a solvolytic reaction of a neutral substrate such as t-butyl bromide increases during the reaction due to the ions Br^- and H_3O^+ which are products of the

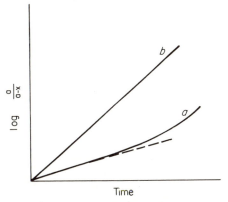

Figure 1. First-order rate plot of the solvolysis of R-Br, (a) in pure solvent, (b) in solvent containing 0.5 M $NaClO_4$.

reaction. To obtain the true solvolytic rate constant one must therefore extrapolate the value to zero time (Figure 1). This is not usually a procedure which can be accomplished with great accuracy, however, and since relative rates are usually as significant as absolute values in physical organic chemistry an alternative procedure is to add a large amount of inert salt to the medium so that additional salt effects due to products are 'swamped'. The rates will, of course, be higher than in pure solvent but will obey the first order law, and a series of substrates can be compared under identical conditions.

Common and Non-common Salt Effects

The ionization of the substrate to carbonium ion and nucleophile is a reversible process and an equilibrium is set up:

$$R-X \; \underset{}{\overset{K}{\rightleftharpoons}} \; R^+ + X^-$$

It follows that the higher the concentration of X^-, the lower is that of R^+. But the kinetic analysis (Section 1.5) is based on the steady state assumption that R^+ remains constant. Since the concentration of X^- necessarily increases during the reaction, because it is a product, the assumption cannot be strictly correct hence further deviations from the ideal rate law are manifest. In general, the presence of the common ion (i.e. that corresponding to the leaving group in the substrate) tends to retard the reaction by increasing the reversibility of the slow step of the reaction. This effect is, of course, superimposed upon the ionic-strength effect of the common ion, and is usually smaller than the latter.

The common-ion effect is just one manifestation of the principle that all nucleophiles present in solution compete for the carbonium ion once formed. These include solvent, leaving group and any other 'foreign' nucleophiles which may be added (non-common ions or neutral species). Thus, suppose t-butyl bromide was allowed to solvolyse in aqueous ethanol in the presence of sodium azide, the following scheme of reactions would take place:

Three substitution products are formed and starting material is reformed by the common ion in a series of nucleophilic attacks on the carbonium ion the amounts of each depending on the specific rate constants for the individual reactions; in addition, olefin (isobutene) can form by removal of the β-proton of the carbonium ion by any of the nucleophiles present ($E1$ process). All these processes can be regarded as examples of 'trapping' of the carbonium ion, conversion to recognizable stable products.

Note that the addition of non-common ion (azide) has no effect on the rate of reaction other than that due to increasing ionic strength of the medium.

These results may be contrasted with the effects of common and non-common ions on S_N2 reactions. The displacement of one (weak) nucleophile by another (stronger) one by the concerted mechanism is not usually reversible. Therefore common ion formed during the reaction or added independently will have no effect on the rate apart from a possible ionic strength effect. Non-common ions, if sufficiently nucleophilic, will be able to react with the substrate by completely independent S_N2 routes and therefore will increase the rate of disappearance of the substrate. Elimination can also occur by a

$$\text{rate} = \frac{-d[\text{RCH}_2\text{Br}]}{dt} = k[\text{RCH}_2\text{Br}][\text{OH}^-] + k'[\text{RCH}_2\text{Br}][\text{N}_3^-] + k_e[\text{RCH}_2\text{Br}][\text{OH}^-]$$

concurrent reaction ($E2$ process)—the simultaneous removal of H$^+$ and Br$^-$ initiated by attack of a nucleophile on the β-hydrogen.

A further consequence of reversibility arises when the substrate is optically active, since recombination of the carbonium ion with the leaving group may in principle lead to racemization of the substrate *before it is converted to product*. In this instance the return of the leaving group leads to formation, at least in part, of a molecule which differs from the starting material. The rate constant obtained from the change in optical rotation, k_α would be larger than that obtained by determination of the rate of acid produced, k_{titr}, and is nearer to

$$\frac{d}{dt}[HBr] = k_{titr}[RBr]$$

$$\frac{d}{dt}\alpha = k_{\alpha}[RBr] \qquad \alpha = \text{specific rotation}$$

$$k_{\alpha} > k_{titr}$$

the true ionization rate, $k_1(+)$ (though not necessarily identical with it, since return, $(k_{-1}(+))$ may be faster than inversion $(k_{-1}(-))$ if the ion pair has not time to become completely symmetrical).

The fact that polarimetrically obtained rate constants, k_{α}, are often found to be larger than those determined titrimetrically for the same reaction suggests that the above scheme is correct. However, it would be unlikely that a free carbonium ion would have much chance in ever recombining with its leaving group anion once they had become separated, since solvent is in such vast excess and is known to react with the carbonium ion at virtually diffusion-controlled rates. This suggests that racemization, and consequently other return reactions, take place from an ion pair in which the two ions remain throughout within their sphere of mutual electrostatic attraction. Yet another example of the internal return of ion pairs is the rearrangement of the starting material which often accompanies the solvolysis of an allylic compound.

3,3-Dimethylallyl chloride (**19**) is hydrolysed to a mixture of the isomeric alcohols, **20a** and **20b** via the allylic cation, **21**. If the chloride is reisolated before solvolysis is complete it is found to have partly rearranged to 1,1-dimethylallyl chloride (**22**). This must have arisen by the recapture of chloride ion in an intimate ion pair such as **21**.

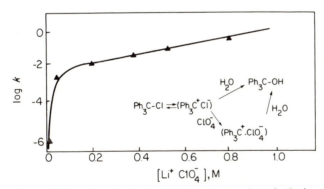

2.9 'SPECIAL' SALT EFFECTS

Occasionally, the addition of a very small concentration of an inert salt to a solvolysis can increase the rate by a factor much larger than could possibly be due to an ionic strength effect. For example, the addition of 0.1 M LiClO$_4$ can enhance the ionization rate of trityl chloride a thousandfold (Figure 2).

Figure 2. Special salt effect of Li$^+$ClO$_4^-$ upon the solvolysis of triphenylmethyl chloride in wet ether.

The interpretation of this 'special' effect suggests that the reaction in the absence of salt is highly reversible and that only a small fraction of ionizations lead ultimately to product. The anion of the added salt, though non-nucleophilic, is capable of displacing the leaving group ion from the intimate ion pair (23) which must exist. The new ion pair (24) always goes on to form products since there is little chance of the reformation of starting material once the carbonium ion and leaving group have become separated. The special salt

$$R-Br \; \rightleftharpoons \; (R^+ \; Br^-)_{solv} \; \xrightarrow{ClO_4^-} \; (R^+ \; ClO_4^-)_{solv} \; \xrightarrow{SH} \; R-S$$

$$\searrow SH \qquad\qquad + Br^- \qquad\qquad + H^+ + ClO_4^-$$

$$RS \qquad\qquad (24)$$

$$(23)$$

effect therefore is the suppression of ion-pair return allowing the rate of formation of product to be more closely a measure of the rate of ionization. This is good evidence for the existence of ion pairs.

2.10   ISOTOPE EFFECTS UPON SOLVOLYTIC REACTIONS

As explained previously (Section 1.11), breaking of a covalent bond to a light isotope of an element is easier than breaking the same type of bond to a heavier isotope (primary kinetic isotope effect) and the observation of this effect can provide evidence of which bond is breaking in the slow step.

A few primary isotope effects in nucleophilic-substitution reactions have been observed, but this is a difficult study since the effects are very small when relatively heavy elements are involved (the effect is approximately proportional to $(m_2/m_1)^{0.5}$ where m_1 and m_2 are the isotopic masses of the light and heavy isotopes respectively).

Sulphur isotope effects have been observed in substitutions of sulphonium salts, using mass spectrometry to determine the ratios of the naturally occurring

$$t\text{Bu}-{}^{32}\text{SMe}_2 \; \xrightarrow[\text{EtOH}]{k_{32}} \; t\text{Bu}^+ + {}^{32}\text{SMe}_2$$

$$t\text{Bu}-{}^{34}\text{SMe}_2 \; \xrightarrow{k_{34}} \; t\text{Bu}^+ + {}^{34}\text{SMe}_2$$

$$k_{32}/k_{34} = 1.0103 \qquad \left(\frac{34}{32}\right)^{0.5} = 1.031$$

isotopes ${}^{32}\text{S}$ and ${}^{34}\text{S}$ in the product as a function of time. The corresponding value for the primary butyl compound (S_N2 reaction) is 1.0072.

In both reactions, therefore, the C—S bond is being broken in the transition state but this appears to be more advanced in the S_N1 reaction, as expected. Many studies of secondary (hydrogen) isotope effects on these reactions have been made. The change in hybridization of carbon from sp^3 to sp^2 is associated with changes in the vibrational frequencies of the carbonium ion (or transition state) particularly in the out-of-plane mode and hence the energy of the

$$\begin{array}{ccc}
\underset{\underset{\text{Me}}{\overset{\text{Me}}{\Big|}}}{\overset{\text{Me}}{\searrow}}\text{C}\!-\!\overset{+}{\text{S}}\text{Me}_2 & \longrightarrow & \left[\underset{\underset{\text{Me}}{\overset{\text{Me}}{\Big|}}}{\overset{\text{Me}}{\searrow}}\overset{\delta+}{\text{C}}\cdots\overset{\delta+}{\text{S}}\text{Me}_2\right] \longrightarrow \text{products}
\end{array}$$

$$\begin{array}{ccc}
\underset{\underset{\text{H}}{\overset{\text{C}_3\text{H}_7}{\Big|}}}{\overset{\text{C}_3\text{H}_7}{\searrow}}\text{C}\!-\!\overset{+}{\text{S}}\text{Me}_2 & \overset{\text{OH}^-}{\longrightarrow} & \left[\text{HO}\cdots\underset{\overset{\text{Me} \;\; \text{H}}{}}{\text{C}}\cdots\text{SMe}_2\right] \longrightarrow \text{products}
\end{array}$$

transition state is slightly dependent upon isotopic substitution at the α-carbon. Replacement of hydrogen by deuterium at the reaction centre retards the ionization rate by a factor amounting to 10–20% ($k_H/k_D = 1.1$–1.2) in typical S_N1 reactions, while in bimolecular displacements this factor is much smaller and may be less than unity ($k_H/k_D = 0.95$–1.00).

For similar reasons, though it is uncertain which vibrational frequencies are affected, isotopic substitution at the β-carbon also is associated with a secondary isotope effect, which if anything is slightly larger than the α-effect (1.15–1.25 per deuterium atom in unimolecular solvolyses). Typical values of isotope effects are given in Table 7.

Table 7. Isotope effects in solvolytic reactions

Reaction (Isotope position in bold type)	k_H/k_D (per deuterium atom)	
α-Isotope effects:		
CH_3—$Cl + H_2O \rightarrow CH_3OH + HCl$	0.97	S_N2
$Me_2CHOTos + H_2O \rightarrow Me_2CHOH + TosOH$	1.13	S_N1
$MeOCH_2Cl + H_2O \rightarrow MeOCH_2OH + HCl$	1.12	S_N1
β-Isotope effects:		
$Me_2C(Cl)CH_2CH_3 + EtOH \rightarrow Me_2C$—$CH_2CH_3$ $\qquad\qquad\qquad\qquad\qquad\qquad\quad \mid$ $\qquad\qquad\qquad\qquad\qquad\qquad\quad OEt$	1.12	S_N1
$(CH_3)_3C$—$Cl + MeOH \rightarrow (CH_3)_3C$—$OMe + HCl$	1.21	S_N1
OTos → OAc + AcOH → + TosOH	1.17 (*trans-d*) 1.22 (*cis-d*)	S_N1 S_N1

2.11   **STEREOCHEMISTRY**

A free carbonium ion has a symmetrical structure either as a result of possessing a plane of symmetry or the rapid inversion of a pyramidal structure, and therefore cannot exist in enantiomeric forms. It would be expected that solvolysis of an optically active substrate proceeding via intermediate carbonium ion should yield racemic product. This expectation is largely borne out but

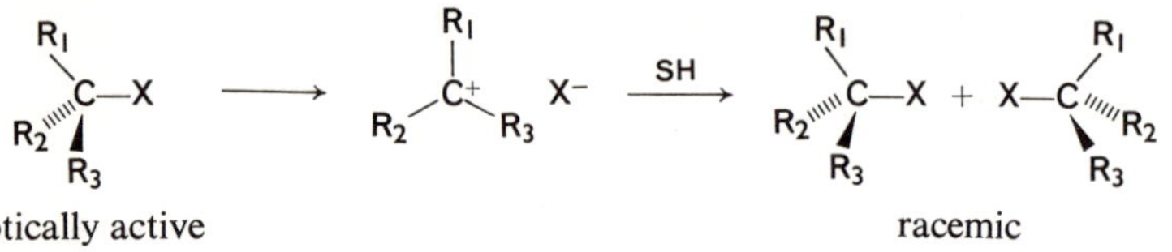

optically active   racemic

usually, in addition to the racemic product, there are greater or smaller amounts of inverted or even retained products. In any event, there is a sharp distinction between the stereochemistry of the S_N2 reaction which leads to inversion only.

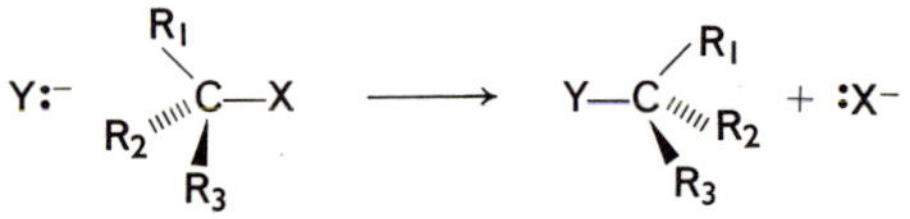

The excess of inverted product in a unimolecular reaction can result from a preference on the part of the solvent to attack the carbonium ion from the rearside due to shielding of the front face by the leaving group. However, the

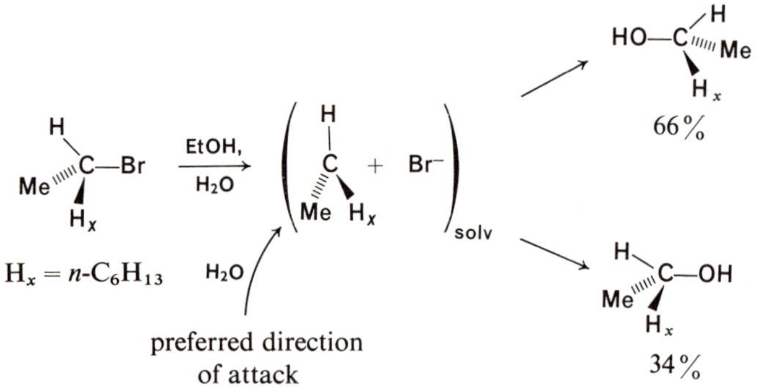

$H_x = n\text{-}C_6H_{13}$

preferred direction
of attack

same result could be observed if the reaction occurred by concurrent S_N1 and S_N2 processes.

Retention or partial retention of configuration can result, if the structure contains a 'configuration-holding' group, a nucleophilic group which can interact with the carbonium centre. 1,2,2-Triphenylethyl tosylate hydrolyses with about 30% retention; the reaction proceeds at least partly by the bridged ion,

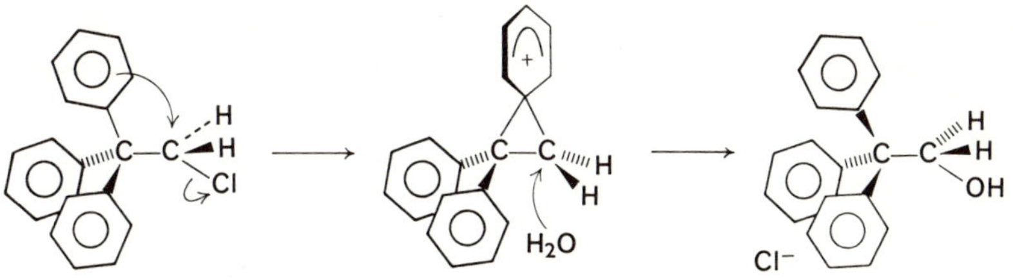

Reactions of this type are the subject of Chapter 3.

Yet another path to retention occurs in the reactions of secondary alcohols with thionyl chloride in dioxan (the solvent is important). The product is the chloride with retained stereochemistry formed via an intermediate chloro-

Table 8. Stereochemistry of unimolecular displacements

	Conditions	Inverted product (%) (remainder retained)
Me⟍ H᎐᎐᎐C—Br C₆H₁₃	60% Aq. EtOH 60% Aq. EtOH + Ag⁺	83 97
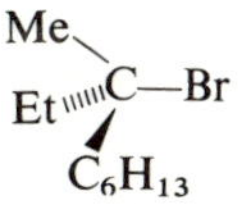 Me⟍ Et᎐᎐᎐C—Br C₆H₁₃	80% Aqueous acetone	60
Me⟍ H᎐᎐᎐C—Cl Ph	Water Aq. acetone	58 51
Me⟍ H᎐᎐᎐C—Cl COO⁻	Water Water + Ag⁺	0 68

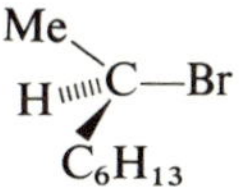

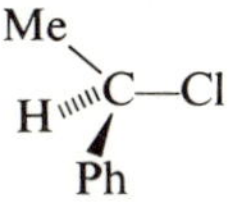

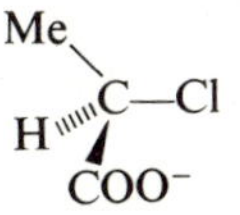

(Participation by
—COO⁻ Section 3.1)

sulphite, which decomposes at the reflux temperature of the dioxan; this is denoted an $S_N i$ (internal) reaction.

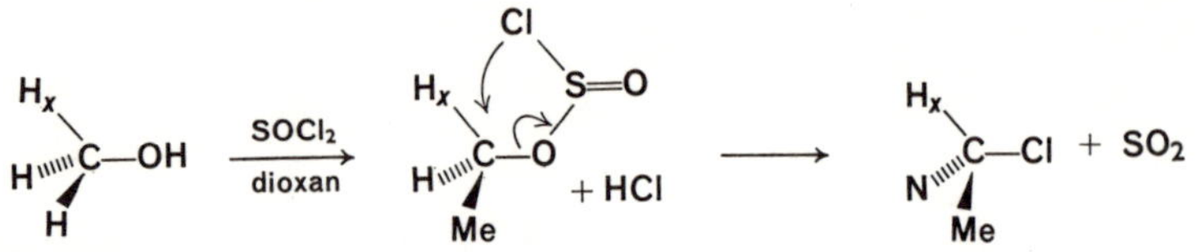

chlorosulphite

The amount of racemization depends to a considerable extent upon the lifetime and consequently the stability of the carbonium ion. Tertiary and benzylic carbonium ions, being longer lived, have a greater chance of achieving a symmetrical environment than secondary, as is seen in Table 8.

2.12   VINYL AND ARYL CATIONS

It is well known that vinyl and aryl halides are very resistant towards hydrolysis even in the presence of strong bases. Steric hindrance would make an $S_N 2$ attack difficult and it appears further that halide or sulphonate attached to sp^2-hybridized carbon is loth to undergo ionic dissociation. Vinyl cations can

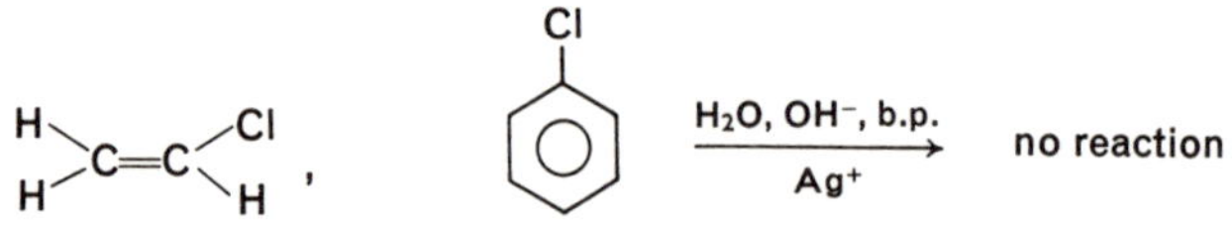

be generated, particularly if an α-aryl group is present and a highly ionizing solvent is employed. Bromostilbenes (25) undergo solvolysis in acetic acid although at rates much slower than the saturated analogues, to give mixtures of *cis* and *trans* acetates (26a, 26b). The Hammett reaction constant (for the

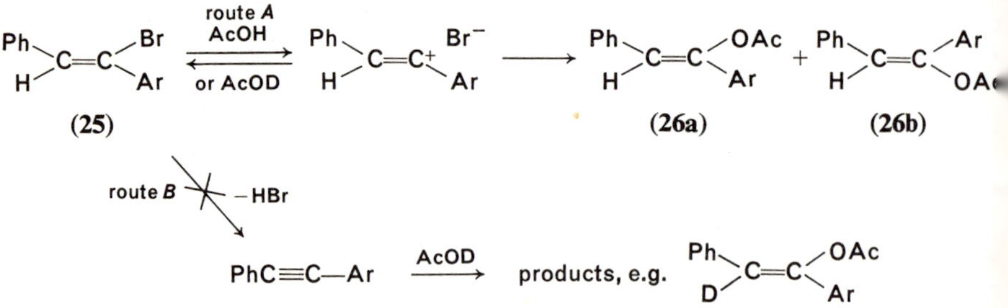

α-aryl substituted series, $\rho = -3.6$) is similar to that observed in other uni-molecular solvolyses, and the rates are sensitive to the nature of the solvent. Furthermore, in deuterated acetic acid, no deuterium is incorporated into the product, which rules out an elimination–addition mechanism via diphenyl-acetylene (route B).

Phenyl cations (27) are never formed in the solvolyses of aryl halides which, if they react at all, use either an addition–elimination route (Section 4.25) or an aryne mechanism (Chapter 7). Phenyl cations are generally supposed to be intermediates in the hydrolysis of aryldiazonium salts. The decomposition in

$$\text{Ph—NH}_2 \xrightarrow{\text{HONO}} \text{Ph—N}_2^+ \longrightarrow \text{Ph}^+ + \text{N}_2$$

(27)

water shows a positive salt effect and added nucleophiles will compete for the intermediate (e.g. Br^-, Cl^-, H_2O^-) without affecting the rate of decomposition. Furthermore, arylation of added nitrobenzene will occur to give 3-nitrobi-phenyl (28). A cation would be expected to react *meta* to the nitro group whereas a radical, Ph·, would substitute more randomly.

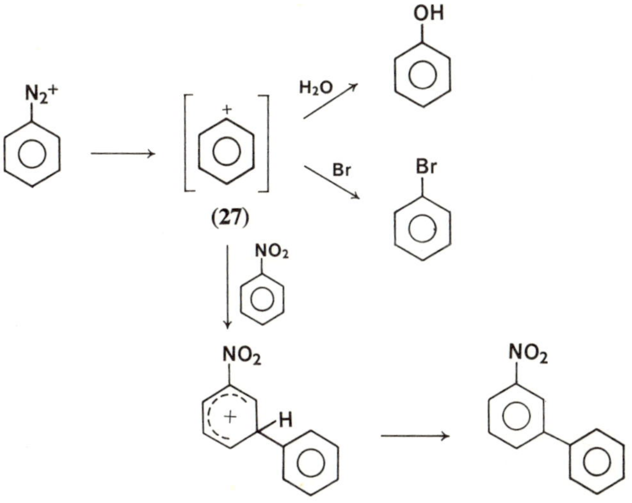

(27)

(28)

It is evident from these examples that vinyl and aryl cationic centres do not markedly interact with the associated π-systems and do not thereby derive stability due to delocalization.

2.13   CARBONIUM IONS FORMED FROM UNSATURATED SYSTEMS

A major route to carbonium ions is by the coordination of a cation, usually a proton, to an unsaturated system acting in the capacity of a nucleophile. Polar

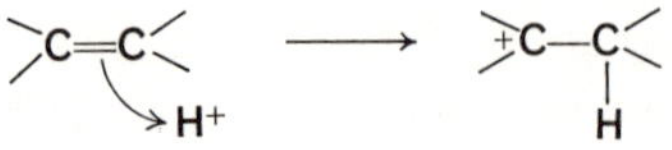

additions of HCl, HBr, HI, Br_2, ClBr, NOCl, H_2O, etc., to olefins and acetylenes proceed via intermediate carbonium ions produced in this way. A

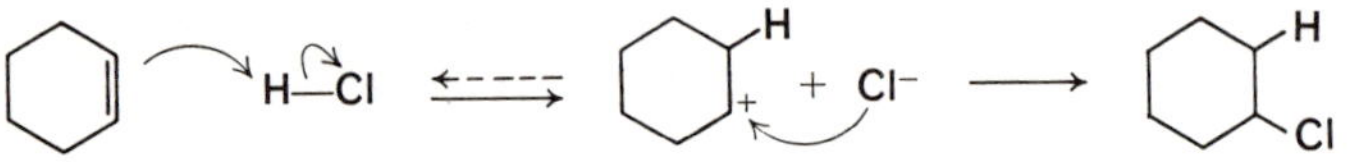

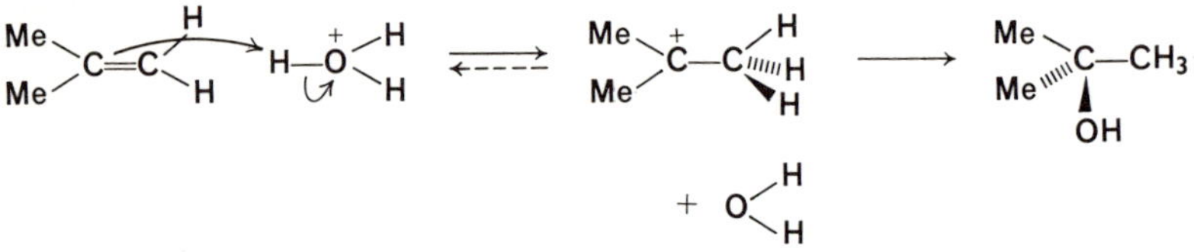

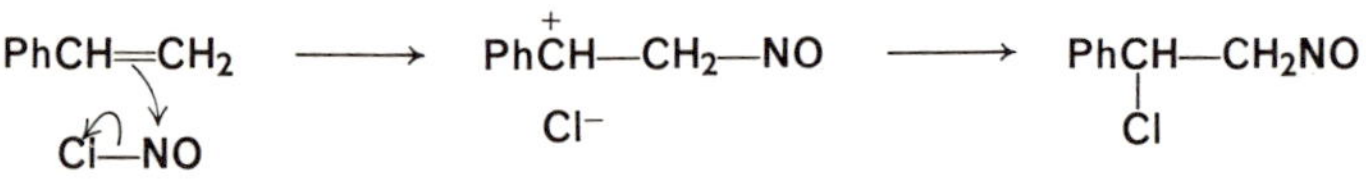

carbonium ion intermediate may be inferred from the effects of substituents on the bromination of styrenes, for example, for which rates follow σ^+. This indicates that donor substituents, Z, are capable of extensive conjugation

$$\log k_{rel} = \rho\sigma^+$$

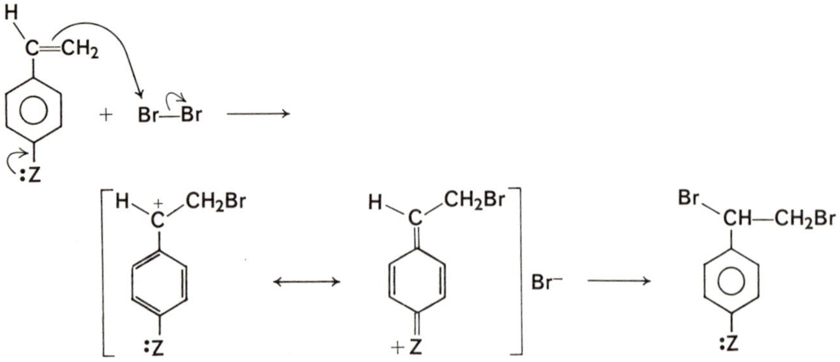

with the reaction centre in the transition state. Furthermore, the orientation of addition of an unsymmetrical reagent to an unsymmetrical olefin gives (at least as the major product) the adduct which may be explained as resulting from the more stable of the two possible carbonium ions. The initial coordination of electrophile to the olefin will give a tertiary carbonium ion in preference to a secondary, which in turn is preferred to a primary. This is a rational version of Markownikoff's rule which was originally stated to the effect that, when a species HX adds to an unsymmetrical olefin, the hydrogen coordinates to the carbon which bears the greater number of hydrogens. Olefins containing

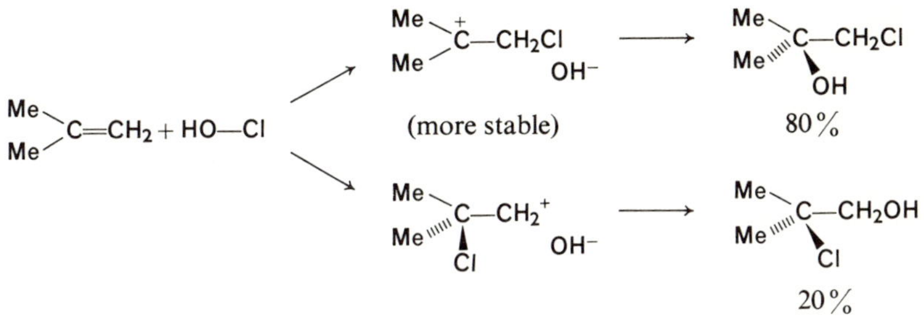

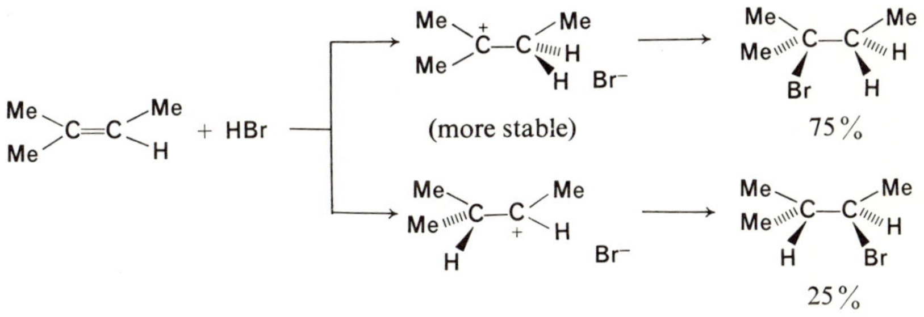

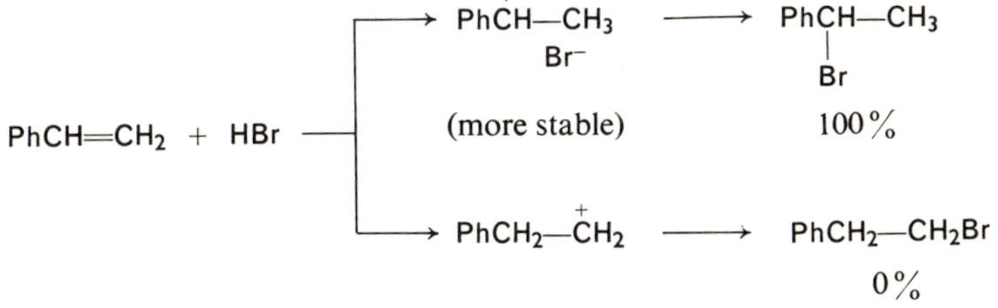

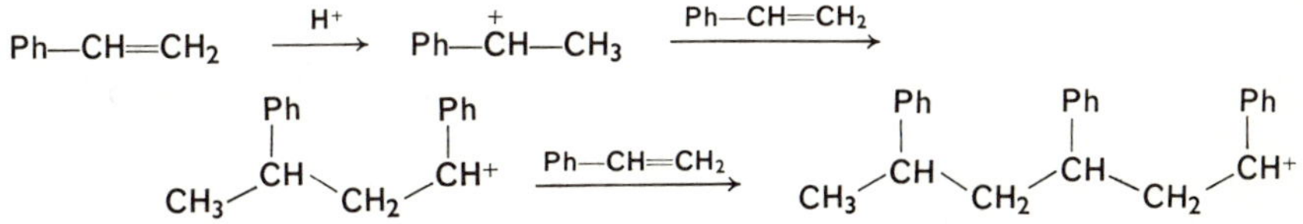

Cationic polymerization 'head-to-tail'

strong electron-withdrawing groups give products in the 'Anti-Markownikoff' sense but these examples are still explicable as being derived from the more stable carbonium ion. The juxtaposition of two positive charges is energetically highly disfavoured.

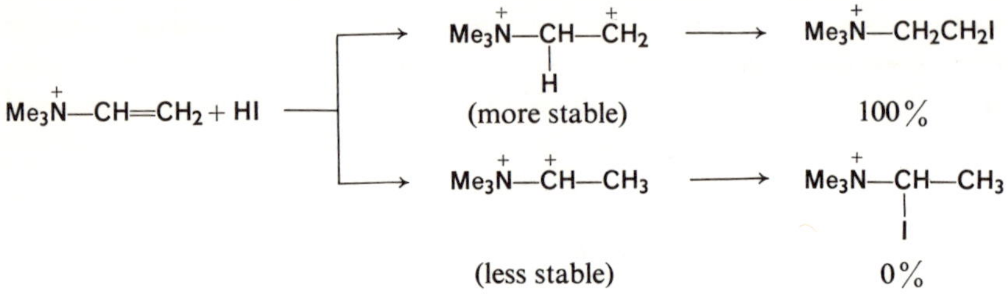

The formation of the carbonium ion is likely to be reversible at least where this involves a proton exchange, on account of the high mobility of the proton. It therefore appears that the formation of the carbonium ion is under thermo-dynamic control. This may not necessarily apply to the *products*, which are usually formed in ratios which do not reflect their respective stabilities. This is because the reactions are usually carried out under conditions such that the second step is not appreciably reversible and may not be rate determining. It is possible to find examples where the product from the first-formed carbonium ion is itself labile and is subsequently converted into a more stable product. The addition of HCl to isoprene (2-methylbutadiene, **29**) at −10° gives 2-chloro-2-methylbut-3-ene (**30**), which on standing is converted to the more stable 1-chloro-2-methylbut-2-ene (**31**), a non-terminal olefin. The inter-

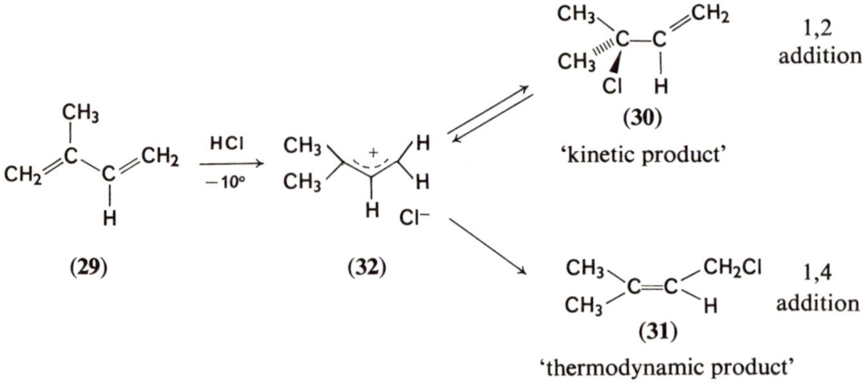

conversion requires the ionization of **30** back to the allylic cation **32**, the slow step in either an S_N1 or an $E1$ reaction. It is clear that these polar addition reactions are *E*1 reactions in reverse and consequently the intermediates and transition states should be the same for the two reactions by the principle of microscopic reversibility.

Addition and elimination phases of a reaction may be demonstrably shown to be interchangeable by relatively slight changes in conditions. Alcohols will frequently dehydrate in strong acid solution and olefins hydrate in a more dilute solution of the acid. There is a further point revealed by olefin hydration

$$\xrightarrow{E1}$$

$$Me_2CH\text{---}CH_3 \underset{-H^+}{\overset{H^+}{\rightleftarrows}} \underset{\overset{|}{+OH_2}}{Me_2CH\text{---}CH_3} \underset{H_2O}{\overset{-H_2O}{\rightleftarrows}} Me_2\overset{+}{C}\text{---}CH_3 \underset{H^+}{\overset{-H^+}{\rightleftarrows}} Me_2C\text{=}CH_2$$

$$\underset{\overset{|}{OH}}{}$$

$$\xleftarrow{\text{addition}}$$

studies, namely that hydration in D_2O/D_3O^+ leads to no deuterium incorporation nor rearrangement in the *unreacted* olefin. The former certainly, and probably also the latter, would be expected if protonation of the olefin were extensively reversible (i.e. a pre-equilibrium, the slow step being the reaction of water with the carbonium ion), e.g.

$$RCH_2\text{---}CH\text{=}CH_2 \underset{\xleftarrow{}}{\overset{D_3O^+}{\dashrightarrow}} RCH_2\text{---}\overset{+}{CH}\text{---}CH_2D \overset{\times}{\longrightarrow} RCH_2\text{---}CH\text{=}CHD \text{ etc.}$$

$$\downarrow D_2O$$

$$RCH_2CH(OD)CH_2D \qquad RCH\text{=}CH\text{---}CH_2D$$

This suggests that the carbonium ion, once formed, goes on to products exclusively. Yet this would mean that the kinetic rate law should be,

$$\text{rate} = k_2[\text{olefin}]\,[H_3O^+]$$

whereas the observed rate law takes the form,

$$\text{rate} = k_2[\text{olefin}]\,h_0$$

where h_0 is the Hammett acidity function (Section 1.20). If the Hammett–Zucker postulate* were to be trusted, it would have to be inferred that the slow

---

* Rates proportional to $[H_3O^+]$ imply a molecule of water and H^+ in the transition state; those proportional to h_0 (or H_0) imply only H^+ and substrate in the transition state. This postulate is now generally believed to be unreliable.

step involved a molecule of olefin and a *proton* only, a result which could be accommodated if a further intermediate, a π-complex of these two components (33) were the species which took part in the slow step. This mechanism may be correct, though more definite evidence for the π-complex is lacking.

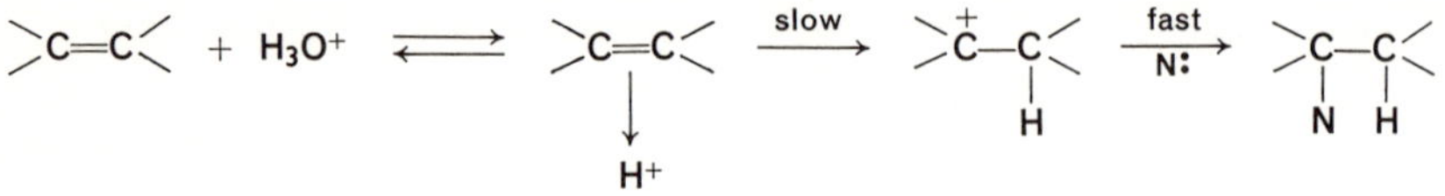

(33)

The addition of halogen such as bromine to the double bond is likewise a two-step reaction. It differs from the addition of hydrogen halide in being stereospecific leading to *trans* product. This is due to the intermediate being a bromonium ion (34) rather than an open carbonium ion, and is attacked by bromide from the rearside. This type of reaction is discussed more fully in Section 3.1. The addition of HBr by a radical chain process is discussed separately (Section 5.9).

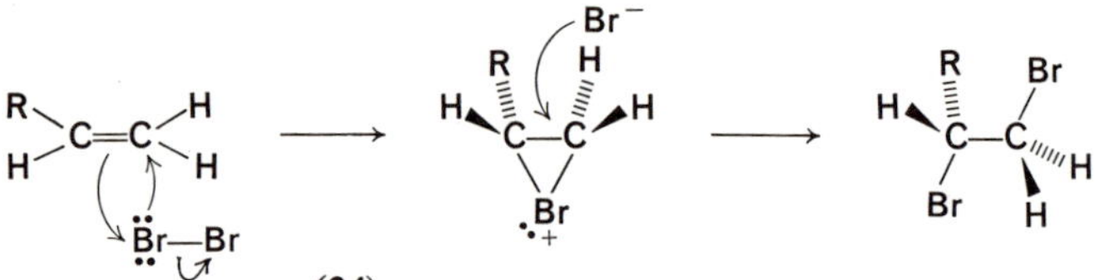

(34)

Decarboxylation of α,β-unsaturated acids can occur under acid catalysis. It is likely that this occurs by initial protonation of the double bond and loss of CO_2 from the resulting carbonium ion.

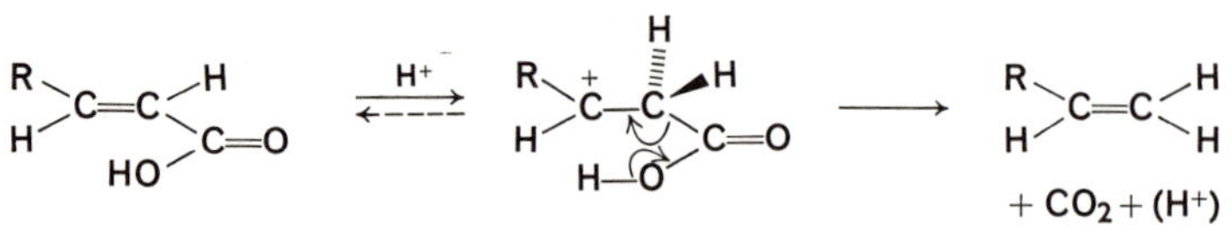

2.14   BENZENONIUM IONS AND ELECTROPHILIC AROMATIC SUBSTITUTION

The familiar substitution reactions at the aromatic nucleus—nitration, sulphonation, halogenation, Friedel–Crafts acylation and alkylation, for example—form a class of reactions with very similar characteristics and are therefore likely to occur by the same type of mechanism. Most such substitutions take place in acidic conditions or in the presence of Lewis acids (Table 9) which suggests that the attacking species is also an acid or electrophile. The leaving group is nearly always hydrogen which would then be displaced as a proton. There is now no reason to doubt that all these 'electrophilic substitutions' occur by an addition–elimination route rather than a concerted

Table 9. Electrophilic aromatic substitution reactions

	–E	Y	Reagents	Electrophilic species (E$^+$)
Nitration	$-NO_2$	H	HNO_3–H_2SO_4	NO_2^+
Sulphonation	$-SO_3H$	H	H_2SO_4 (SO_3)	SO_3
Friedel–Crafts				
alkylation	$-R$	H	Rhal–$AlCl_3$	$R^+AlCl_4^-$ (RCl + $AlCl_3$)
acylation	$-COR$	H	RCOhal–$AlCl_3$	$RCO^+AlCl_4^-$ (RCoCl + $AlCl_3$)
Proton exchange	H (D)	H (D, T)	H_2SO_4 (D_2SO_4) $HClO_4$ etc.	(H$^+$) (strong acids)
Halogenation	Cl, Br	H	Cl_2 (Br_2)–$AlCl_3$	Cl^+ ($AlCl_4^-$) Cl_2^+, $AlCl_3$ or Br_2 + $AlCl_3$ Br^+
Diazocoupling	$-N_2Ar$	H	ArN_2^+ Cl$^-$	ArN_2^+
Nitrosation	$-NO$	H	N_2O_4	NO^+
Mercuration	$-HgCl$ $-HgOAc$	H H	$HgCl_2$, $Hg(OAc)_2$	$\overset{+}{HgX}$
Thalliation	$-Tl$	H		Tl^{+++}, $Tl(OAc)_3$
Protodemercuration	$-H$	$-HgX$	$HClO_4$	(H$^+$)
Halodesililation	$-Cl$ $-Br$	$-SiR_3$ $-SiR_3$	Cl_2, Br_2	Cl^+, Br^+

displacement. This means that an intermediate is formed which comprises one molecule each of the electrophile and the aromatic compound. Good evidence exists that this species is a σ-complex of general structure **34a**, known as a 'benzenonium ion'. The generic name 'arenonium ion' is sometimes used, or the earlier term 'Wheland intermediate'; 'benzenium ion' is also used by some

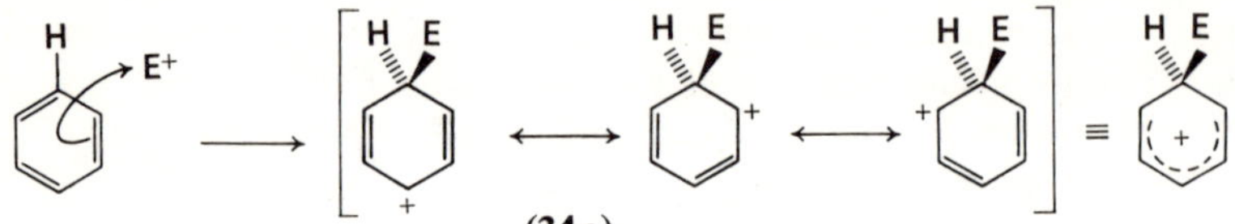

(34a)

authors. Benzenonium ions are a type of cyclic pentadiene cation. Nitration is one of the most thoroughly investigated of this class of reaction and will serve as an illustration of the evidence which has been used to deduce the presently accepted mechanism.

Mechanism of Aromatic Nitration

Nature of the reagent

The usual conditions for aromatic nitration are nitric–sulphuric acid mixtures ranging from 100% acid at e.g. 150°, for very unreactive substrates, to dilute aqueous nitric acid for the very reactive. Nitric acid in acetic acid, acetic anhydride or nitromethane, may also be used in appropriate circumstances. In concentrated sulphuric acid, nitric acid undergoes the following dissociation which lies far to the right.

$$O{=}\overset{+}{N}(OH)(\bar{O}) + H_2SO_4 \rightleftharpoons O{=}\overset{+}{N}(OH)(OH) \quad HSO_4^-$$

$$O{=}\overset{+}{N}(OH)(OH) \rightleftharpoons O{=}\overset{+}{N}{=}O + H_2O$$

$$H_2O + H_2SO_4 \rightleftharpoons H_3O^+ + HSO_4^-$$

Overall reaction:

$$HNO_3 + 2H_2SO_4 \rightleftharpoons NO_2^+ + H_3O^+ + 2HSO_4^-$$

Cryoscopic measurements show that each mole of nitric acid produces four moles of solute molecules or ions in sulphuric acid (van't Hoff *i* factor = 4). This is inferred since the depression of the freezing point of sulphuric acid by the addition of one mole of nitric acid is four times as great as that produced by one mole of an inert solute. This means that each nitric acid molecule is replaced by four ions or molecules, each of which separately contributes to the freezing point depression. The nitronium ion, NO_2^+, can be detected in this solution by its characteristic Raman and infrared spectrum ($\bar{v}_{max} = 1400\ cm^{-1}$) which may be deduced to arise from a linear triatomic species. Similar bands occur in the spectra of known crystalline nitronium salts such as the perchlorate, $NO_2^+ClO_4^-$ and nitrate $NO_2^+NO_3^-$ (nitric anhydride, N_2O_5). The equilibrium concentration of NO_2^+ diminishes with the addition of water to the solution, though reactive aromatic species may still be nitrated when NO_2^+ is no longer detectable. That the nitronium ion (rather than, say, nitric acid) is still the active nitrating agent even in dilute aqueous solution may be inferred from the following evidence. The kinetic form of the nitration of a reactive substrate such as phenol in a solution which contains a large excess of nitric acid is zero order, i.e.

$$\frac{-d[\text{phenol}]}{dt} = k_0$$

This arises since the rate-determining step is the formation of the nitrating agent. Furthermore, if the nitration is carried out in ^{18}O-labelled water the rate

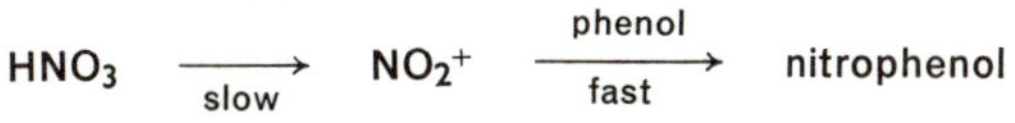

$$HNO_3 \xrightarrow[\text{slow}]{} NO_2^+ \xrightarrow[\text{fast}]{\text{phenol}} \text{nitrophenol}$$

of exchange of ^{18}O with the nitric acid is also zero order and the *same rate* as the nitration, which is consistent with the following scheme in which there is a common intermediate for both nitration and oxygen exchange. Preformed nitronium salts (e.g. in nitromethane solution) are excellent nitrating agents.

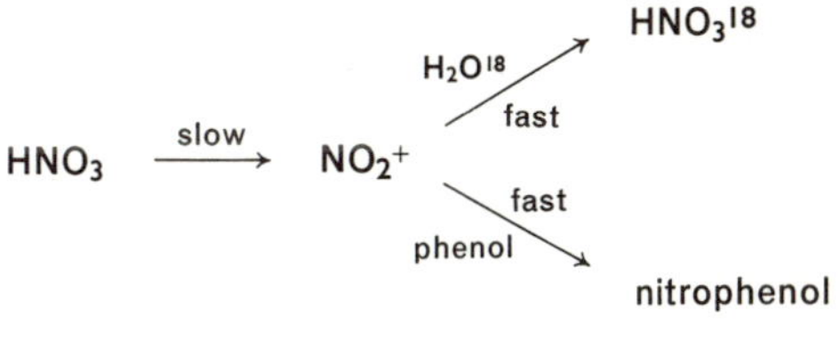

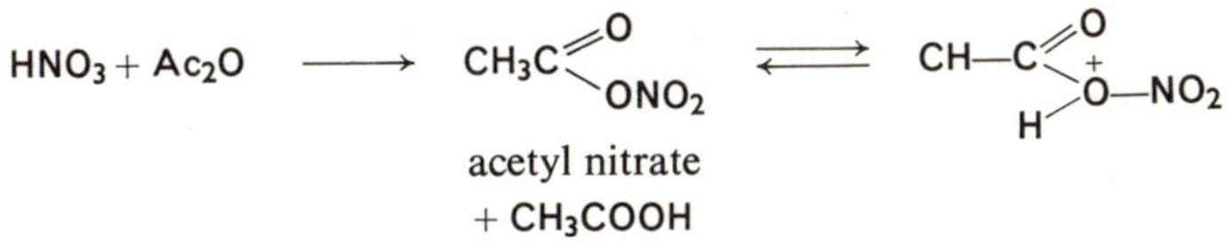

Slightly different results (such as changes in product ratio) are found when other systems are used, which suggests that other nitrating agents are possible. A mixture of nitric acid and acetic anhydride may owe its reactivity to acetyl nitrate or a protonated form. Westheimer and Kharasch observed that the rate of nitration in aqueous sulphuric acid showed a maximum at 90% sulphuric acid. This value, and indeed the whole plot of rate against composition, parallels the extent of protonation of tris(4-nitrophenyl)carbinol (35). The

$$\left(O_2N\text{—}\underset{\text{}}{\bigcirc}\text{—}\right)_3 C\text{—}OH + 2H_2SO_4 \rightleftharpoons \left(O_2N\text{—}\underset{\text{}}{\bigcirc}\text{—}\right)_3 C^+ + H_3O^+ + 2HSO_4^-$$

(35)

parallel between this dissociation and that of nitric acid is obvious and indicates that similar types of reaction are involved in the two processes.

2.15   CHARACTERISTICS OF THE NITRATION REACTION

Rates of nitration, like other electrophilic substitutions, are very sensitive to substituents in the ring. The following approximate data—it is difficult to measure accurately such a wide spread—are illustrative.

X	k_{rel}	Orientation of product
OMe	10^5	*ortho, para*
Me	600	*ortho, para*
1,3,5-triMe	10^8	2
H	1	—
Cl	0.15	*ortho, para*
CN	ca 0.01	*meta*
NO$_2$	ca 0.001	*meta*

Application of the Hammett equation (Section 1.10) leads to a value for $\rho = -5.9$ and, moreover, requires the use of 'modified' substituent constants, σ^+, for a linear correlation with $\log k_{rel}$. The interpretation of this is that the reaction is greatly facilitated by electron donation and this is accomplished with enhanced conjugation between a substituent bearing unshared electron pairs and the ring. Conjugation is much more pronounced when the substituent is located in the *ortho* or *para* positions than the *meta*. The initial attack of the nitronium ion on anisole may be written,

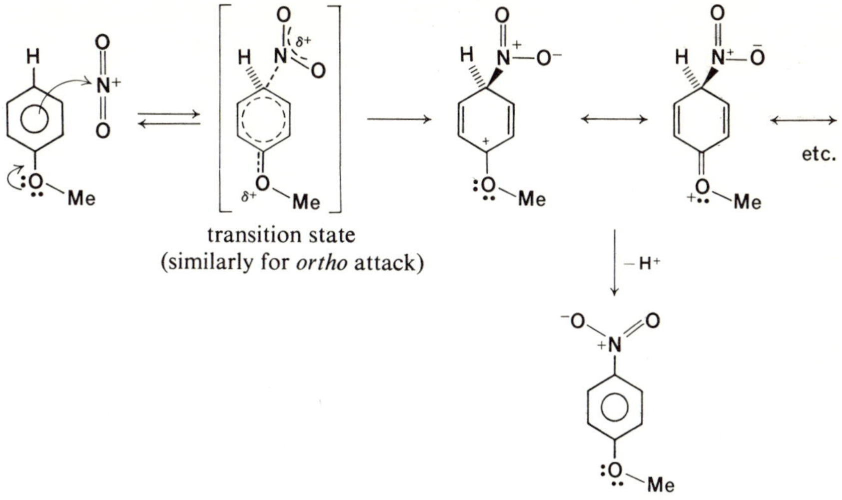

transition state
(similarly for *ortho* attack)

etc.

The orienting effects of substituents are best rationalized in terms of the stabilities of benzenonium ions. Those substituents which increase the rate of reaction (relative to benzene) do so by their capability to stabilize the benzenonium ion by conjugative interactions. These include $-NR_2$, $-OR$, $-SR$, and alkyl groups, all of which are well-known to direct substitution at the *ortho* and *para* positions. The same is true of halogen substituents though these deactivate due to an electron-withdrawing inductive effect which outweighs the conjugative stabilization.

The effects of substituents are often expressed as *Partial Rate Factors* defined as the rate of reaction (e.g. nitration) at a given position in a substituted benzene, relative to the rate at a single position in benzene itself. The following figures are Partial Rate Factors for nitration and are similar in magnitude to the values for other electrophilic substitutions:

(reference compound)

This indicates that toluene is greatly activated towards reaction in *ortho* and *para* positions, and less so at the *meta*; chlorobenzene is generally deactivated but less at *ortho* and *para* than at *meta*, while ethyl benzoate is strongly deactivated, but the *meta* position somewhat less than the others.

Stabilizing conjugation   Less stable

$X = N\langle, S\langle, O\langle$, hal

Deactivating substituents such as $-C\overset{\displaystyle O}{\underset{\displaystyle R}{\diagdown}}$, $-CN$, $-SO_2R$, $-NO_2$ all direct further substitution in the *meta* position where conjugation with the benzenonium ion is *least* pronounced, since now conjugation destabilizes the carbonium ion character of the ring as the substituents themselves are electron withdrawing,

Destabilizing conjugation   More stable

Since reaction rates depend upon the stabilities of the transition states, we may infer that these must resemble the benzenonium ions in character. This has been substantiated by finding that the calculated stabilities of a large number of benzenonium ions very accurately reflect the reactivities of the parent aromatic compounds (Figure 5, p. 159).

The kinetics of nitration reactions are often complex but two extreme cases may be recognized. Unreactive substrates such as nitrobenzene often exhibit second-order kinetics. In such cases the attack of the nitronium ion is the slow step. At the other extreme, highly reactive substrates such as anisole or phenol

$$HNO_3 \;\rightleftarrows\; NO_2^+ \xrightarrow[\text{slow}]{\text{ArH}} ArNO_2$$

rate $= k_2[HNO_3][ArH]$ or, if $[HNO_3]$ is in large excess, rate $= k_1{}'[ArH]$

show no rate dependence upon the substrate concentration. The attack of the nitronium ion is now not rate determining; the overall rate depends only upon the rate of generation of nitronium ion, and all reactive substrates nitrate at the same rate. Furthermore, the rate of incorporation of ^{18}O from water into

$$HNO_3 \underset{slow}{\overset{}{\rightleftarrows}} NO_2^+ \xrightarrow[fast]{Ar-H} ArNO_2$$

rate $= k_1[HNO_3]$ or if $[HNO_3]$ is in large
excess, rate $= k_0$, i.e. is zero order

nitric acid is equal to this zero-order rate. It would be a great coincidence if these two reactions did not involve the same slow step. Substrates whose reactivity falls between these two extremes may exhibit fractional order

$$HNO_3 \xrightarrow{slow} NO_2^+ \xrightarrow[fast]{H_3^{18}O} H^{18}NO_3$$

$$ArH \xrightarrow{fast} ArNO_2$$

dependence on nitric-acid concentration, (between 0 and 1) the value of which depends upon the conditions. Benzene and the halobenzenes are among those with intermediate reactivities. It should be noted that there is no reason to believe that there is a fundamental change of mechanism, despite the change in kinetic form. Physical evidence for the existence of benzenonium ions is given on p. 148.

The formation of product from the benzenonium ion requires the transfer of a proton to the strongest base present which may be water or bisulphate ion. This is clearly a fast process as indicated by the observation that, in general, no kinetic isotope effect is observed. For example, benzene and tritiated benzene are nitrated at the same rate. Absence of a primary isotope effect is good

$$k_H/k_T = 1$$

evidence that the reaction takes place in two steps since, though a CH bond must obviously be broken, it does not occur in the rate-determining step.

2.16   π-COMPLEX FORMATION

It has been suggested that the initial interaction between the nitronium ion and the aromatic substrate leads to the formation of a π-complex (charge-transfer complex, Section 10.9) in which the electrophile is bound weakly by overlap of a vacant orbital with the highest filled π-orbital in the aromatic molecule so that it is not specifically attached to any one carbon atom (36). One line of

$$O=\overset{+}{N}=O$$

(36)

evidence previously put forward in support of this was that the rate of nitration of toluene was much closer to that of benzene (ca 1.5 times) when the nitrating reagent was a nitronium salt (e.g. $NO_2^+ClO_4^-$) compared to the 'normal' value of ca 25 for nitration by nitric acid. It was proposed that in the former case the formation of the π-complex was rate determining as this agreed with the known behaviour of toluene towards other π-acids. However, it has since been shown

$$ArH + NO_2^+ \xrightarrow{\text{rate-determining}} ArH\ NO_2^+ \xrightarrow{\text{product-determining}} Ar\overset{H}{\underset{NO_2}{<}} \longrightarrow product$$

$$\pi\text{-complex} \qquad\qquad \sigma\text{-complex}$$

that nitration with nitronium salts is very rapid and much reaction has occurred before mixing of the reagents is complete. The evidence for this comes from the nitration of 1,2-diphenylethane (37) which forms far more dinitro compound than would statistically be expected. We can assume that the rings are non-interacting and that mononitration does not activate the other ring. This must

$$\langle O \rangle - CH_2 - CH_2 - \langle O \rangle \longrightarrow$$

(37)

$$O_2N\langle O \rangle CH_2CH_2 - \langle O \rangle \ +\ O_2N\langle O \rangle - CH_2CH_2\langle O \rangle NO_2$$

Theoretical ratio 2:1
Experimental ratio 0.205:1

mean that much reaction occurs when a given diphenylethane molecule finds itself in a high local concentration of nitronium salt in which case both rings are attacked, or else in a local environment where none is present and hence no reaction occurs, i.e. the solution is inhomogeneous during a significant part of the reaction. It is highly probable that the reaction rates of NO_2^+ with all 'activated' aromatic rings are extremely fast and probably at the encounter rate, being limited by the rate at which the two species can diffuse together.

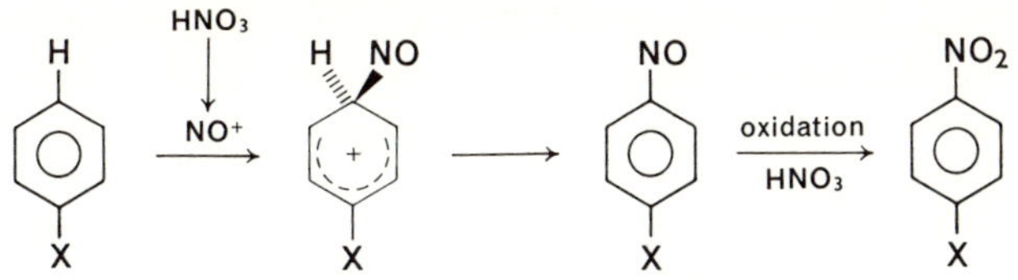

Furthermore, nitration via *nitrosation* may become important. The nitrosyl cation NO^+ is much less electrophilic than NO_2^+ but would react readily with an activated aromatic ring, e.g. $X = OMe$.

2.17   OTHER ANALOGOUS REACTIONS

In Table 9 is found a list of the more commonly met aromatic electrophilic substitutions which occur at a wide variety of aromatic and heteroaromatic nuclei. It will be noticed that the leaving group need not be the proton; many metallic substituents also can be displaced as electrophiles. It is generally considered that all these reactions occur, with minor modifications, by the same type of mechanism as nitration.

One difference which is apparent is the primary isotope effect shown by some reactions. This varies from quite small values for sulphonation ($k_H/k_D = 1.5$–1.75), to quite substantial ones for the diazo-coupling reaction,

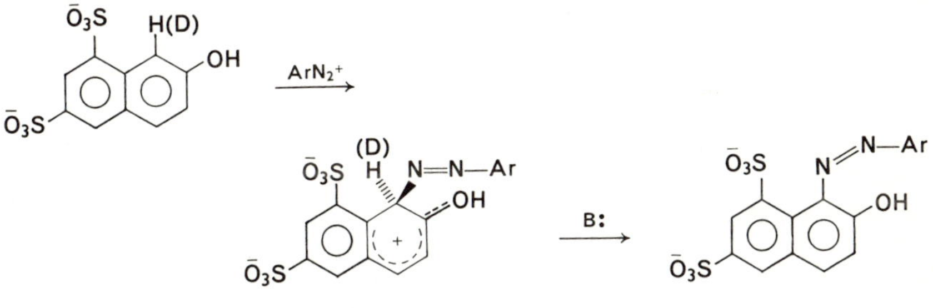

$$k_H/k_D \approx 3$$

and almost a maximum value for mercuration. It is clear that the loss of the

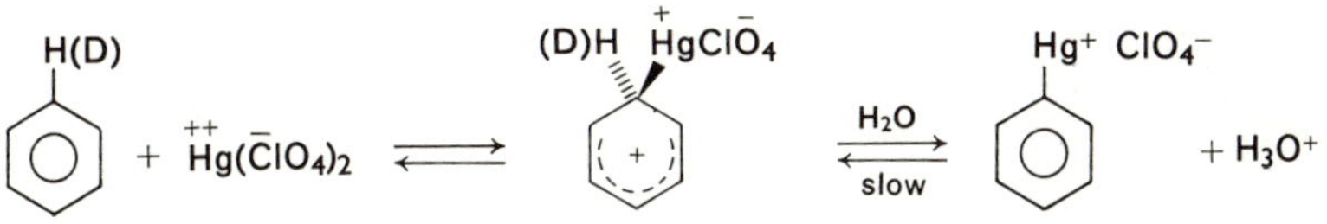

$$k_H/k_D = 5\text{–}7$$

proton from the ring is progressively becoming the rate-determining process for there is no reason to assume a basic change in mechanism. It is notable that these reactions are of rather weak electrophiles and that mercuration and sulphonation at least are reversible. The explanation of this shift in the timing of the reaction may lie in these facts. The formation of the benzenonium ion may then be treated as a pre-equilibrium but the rate depends upon the conversion of the intermediate to product.

Substituent effects for different reactions vary widely although all are assisted by electron donation. Halogenation and proton-exchange, for example, exhibit much larger (negative) ρ-values than nitration. This may be explained by supposing that the transition states for the former reactions lie closer to the benzenonium ion than is the case for nitration, so that the development of positive charge in the ring, and consequently conjugative interactions with the substituent, are more advanced.

2.18   CARBONIUM IONS IN OXIDATION REACTIONS

The conversion of a saturated hydrocarbon residue to a carbonium ion requires the removal of hydride ion, $H:^-$, a process which is conventionally regarded as an oxidation.

Chromium(VI) and some other high-valence metals (Section 9.2) are capable of this type of reaction, especially when a stabilized carbonium ion results.

$$Ph_3C-H + Cr^{VI} \longrightarrow Ph_3C^+ + Cr^{VI}$$

The stable triphenylmethyl cation can itself act as an oxidizing agent in a hydride transfer which yields an even more stable carbonium ion, such as the tropyllium ion from cycloheptatriene, (38) and the aromatization of dihydroanthracene (39). Alcohol oxidation frequently commences by the abstraction

tropyllium ion

(38)

(39)

of the α-hydrogen as hydride forming an oxocarbonium ion (protonated ketone, **40**) which subsequently loses a proton. Deuterium labelling reveals this pathway:

$$\begin{array}{c} Me \\ \diagdown \overset{+}{C}\!-\!OH \\ Me \diagup \end{array}$$

$$\underset{Me}{\overset{Me}{\diagdown}} D\!-\!\underset{Me}{\overset{|}{C}}\!-\!OH \quad \longrightarrow \quad Ph_3C\!-\!D \; + \; \underset{Me}{\overset{Me}{\diagdown}}\overset{+}{C}\!=\!\overset{+}{O}\!-\!H \quad \xrightarrow{-H^+} \quad \underset{Me}{\overset{Me}{\diagdown}}C\!=\!O$$

$$Ph_3C^+ \qquad\qquad\qquad\qquad\qquad\qquad (\mathbf{40})$$

The removal of one electron from an aromatic system would generate a carbonium ion with an unpaired electron—a radical cation. These are quite easily formed from polycyclic hydrocarbons such as anthracene and are further discussed in Section 5.5. Electrolytic oxidation provides a further

$$\xrightarrow[-1e]{H_2SO_4}$$

14 π-electrons   13 π-electrons $HSO_4^- + SO_2$ etc

route to certain carbonium ions. The Kolbe hydrocarbon synthesis, for example, occurs mainly by the generation and dimerization of alkyl radicals at an anode (Section 5.9). Some of these radicals are further oxidized to carbonium ions, leading to products of solvolysis, RS, as side reactions. The

$$R\!-\!COO^- \quad \xrightarrow[anode]{-e} \quad R\!-\!COO\cdot \quad \xrightarrow{-CO_2} \quad R\cdot \quad \xrightarrow{2\times} \quad R\!-\!R$$

$$\Big\downarrow anode$$

$$R^+ \quad \xrightarrow{SH} \quad R\!-\!S$$

yield of this side product increases the more stable is the carbonium ion (tertiary > secondary > primary) which also reflects the order of ease of oxidation of the radicals.

Anodic oxidation of olefins probably occurs via cation-radical and carbonium-ion intermediates. The anodic oxidation of butyrate ion (**41**) can

$$\underset{CH_3}{\overset{CH_3}{>}}C=C\underset{CH_3}{\overset{H}{<}} \xrightarrow[\text{anode}]{-e} \underset{CH_3}{\overset{CH_3}{>}}\overset{+}{C}-\overset{\cdot}{C}\underset{CH_3}{\overset{H}{<}} \xrightarrow{\text{AcOH}} \underset{CH_3}{\overset{CH_3}{>}}\overset{\cdot}{C}-C\underset{CH_3}{\overset{OAc}{\underset{/\!/\!/\!/H}{<}}}$$

$$\Big\downarrow -e$$

$$\underset{CH_3}{\overset{CH_2}{>}}C-C\underset{CH_3}{\overset{OAc}{\underset{/\!/\!/\!/H}{<}}} \underset{-H^+}{\overset{AcO^-}{\longleftarrow}} \underset{CH_3}{\overset{CH_3}{>}}\overset{+}{C}-C\underset{CH_3}{\overset{OAc}{\underset{/\!/\!/\!/H}{<}}}$$

give up to 20% cyclopropane (42) among the products. This is known to be formed from the 1-propyl cation (43), but *not* from the propyl radical (44).

$$CH_3\overset{CH_2}{\diagup}\diagdown CH_2-COO^- \longrightarrow CH_3\overset{CH_2}{\diagup}\diagdown \overset{\cdot}{CH_2} \xrightarrow{2\times} n\text{-}C_6H_{14}$$

(41)   (44)

$$\Big\downarrow e$$

$$CH_3\overset{CH_2}{\diagup}\diagdown CH_2^+ \longrightarrow \underset{CH_2-CH_2}{\overset{CH_2}{\triangle}}$$

(43)   (42)

2.19   LONG-LIVED CARBONIUM IONS

Compelling evidence for the intermediacy of carbonium ions in reactions comes from the direct observation and even isolation of examples which are formed in analogous processes and stabilized by virtue of their structure or their environment. In the latter connection it would be predicted, since carbonium ions are Lewis acids, that a stabilizing environment would require the absence of even weak bases, a polar solvent if in solution, and possibly the use of low temperatures to suppress rearrangements and other unimolecular reactions of the carbonium ions. Since an anion must be present, it must exhibit extremely low nucleophilic properties; ClO_4^-, BF_4^- and especially SbF_6^- best fulfil these requirements. Whereas prior to 1965 virtually the only carbonium ions which had been observed were the highly stable triarylmethyls, in the last few years new media have been developed in which many simple alkyl cations have a stable existence.

2.20   DI- AND TRI-ARYLMETHYL CATIONS

Triphenylmethyl (trityl) chloride (45), a colourless covalent compound, remains covalent in alcoholic solution or in less-polar solvents but in sulphuric

acid, trifluoroacetic acid, liquid SO_2 or acetonitrile, forms deep-yellow
solutions which contain the trityl cation (46). These facts were observed and

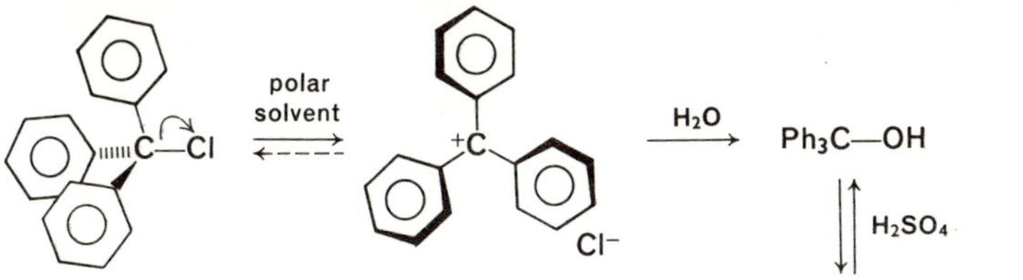

correctly interpreted by Gomberg, Hantzsch and others ca 1900. The corresponding perchlorate, nitrate, tetrafluoborate, etc., are ionic compounds in
the crystalline state; as are many substituted analogues and also the dicar-
bonium ion, 47,

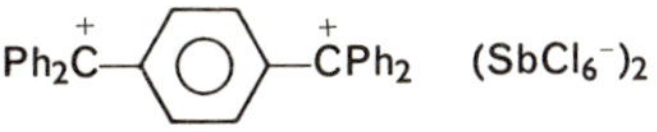

The trityl cation is not able to achieve total coplanarity due to steric interactions
between *ortho* hydrogens on the adjacent benzene rings. The result is a
propellor-shaped ion with the aromatic rings twisted 54° from the plane of the
trigonal carbon.

The stabilities of triarylmethyl cations are increased by electron-donating
substituents which can further delocalize the charge. This is evident from the
rates of reaction with water,

$$Ar_3C^+ + H_2O \longrightarrow Ar_3COH + H^+$$

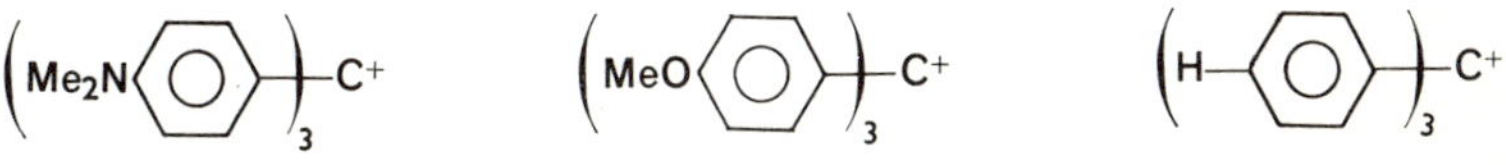

| Solvolysis rate (water, at 25°) | Stable | $0.3\ 1\ mol^{-1}\ s^{-1}$ | $10^3\ 1\ mol^{-1}\ s^{-1}$ |

Diarylmethyl cations are much less stable than triarylmethyl but can be
formed by solution of the halides in sulphuric acid. Protonation of 1,1-diaryl-

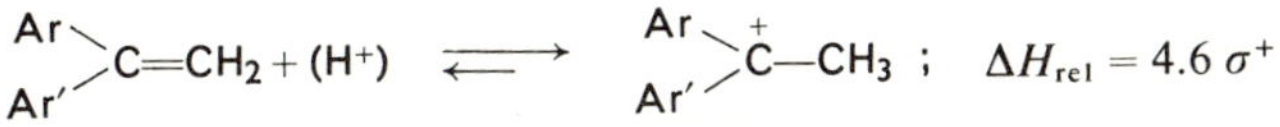

ethylenes produces stable tertiary diarylmethyl cations, (**48**). Heats of proto-nation of a series of diarylethylenes are found to be proportional to σ^+ for the substituent groups, pointing to the importance of through-conjugation.

2.21   **AROMATIC CATIONS**

By far the most stable of all purely carbon cations are a group of carbonium ions which owe their stability to the possession of aromatic character. According to Hückel's definition of aromaticity, planar, cyclic, completely conjugated systems containing k π-electrons are aromatic when $k = (4n + 2)$ and non- (anti-) aromatic when $k = 4n$, n being an integer including 0.

Thus aromatic π-systems contain 2, 6, 10, ..., electrons and will have the high stability associated with a totally symmetric ground state and a closed-shell system of π-molecular orbitals, Figure 3. The most familiar aromatic system (annulene) is benzene. Annulenes with $(4n + 3)$-membered rings and $(4n + 2)$ π-electrons are cationic aromatics, each carbon of the ring bearing an equal

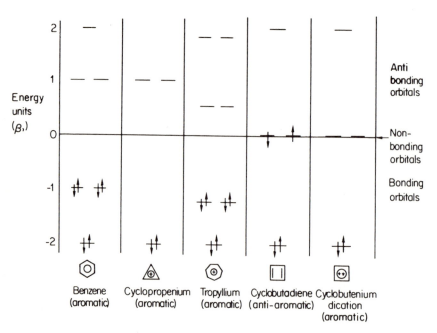

Figure 3. Molecular-orbital occupancy diagrams of some aromatic and antiaromatic systems.

share of the positive charge. The two most familiar systems are the cyclo-propenium ion (**49**) and the tropyllium ion (**50**), which possess two and six π-electron systems respectively ($n = 0$ and $n = 1$). The molecular orbitals of some aromatic and some anti-aromatic species are given in Figure 3.

(49)

(50)

The tropyllium ion was first characterized by Doering in 1954. Most preparations make use of a hydride removal from cycloheptatriene by strong Lewis acid (**38**). Tropyllium salts, including the halides, are ionic, water-stable compounds. All the carbon atoms of the ring are equivalent and bear $\frac{1}{7}$ of the positive charge resulting in a singlet n.m.r. spectrum at 0.8 τ. This very low value is due to a combination of positive charge and the diamagnetic ring current* which is diagnostic of aromatic rings.

Cyclopropenium salts (**49**) can be even more stable than tropyllium compounds. The parent ion (first prepared by Breslow in 1963) has a singlet proton-nuclear resonance at −1.1 τ, due to the diamagnetic ring current and the additional deshielding by $\frac{1}{3}$ of a positive charge on the ring carbon.

4, 8, 12, ..., -membered cyclic systems are antiaromatic if neutral, but in the form of dications or dianions will be transformed to aromatic compounds with

---

* The n.m.r. signal of benzene (2.63 τ) falls at a considerably lower field strength than that of an olefinic proton (ca 4 τ). This is due to a 'ring current' or movement of π-electrons in the plane of the ring producing an additional magnetic field, h, which supplements the applied field, H, outside the ring. Hence a lower than normal applied field is required for resonance. This is a diamagnetic ring current and is found in all aromatic molecules and ions. Anti-aromatic ($4n$) systems show a ring current in the opposite sense (paramagnetic), which

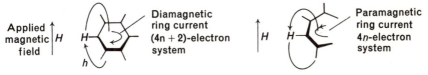

opposes the applied field outside the ring. Hence external protons resonate at an unusually high applied field.

2, 6, 10, ..., π-electrons. Though cyclobutadiene (**51**) has been shown to be a highly reactive, transient molecule, derivatives of the dication, e.g. **52**, are readily prepared and are two-electron systems analogous to the cyclopropenium ion. The dication from cyclooctatetraene (**53**), which is essentially an olefinic

(**51**)

(**52**)

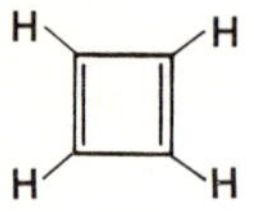

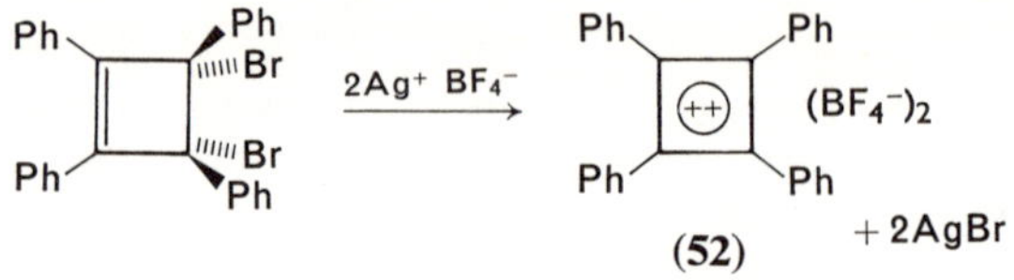

compound due to non-planarity, has also been recognized (**54**). For a discussion of the related homoaromatic cations, e.g. homotropyllium, see Chapter 3.

(**53**) (**54**)

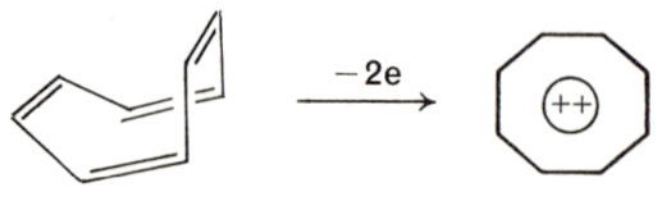

5-Membered annulenes are aromatic as the negative ion, **55** (Section 1.10), with six π-electrons, but as the cation, **56**, have only four π-electrons and are *destabilized* in contrast to the *acyclic* pentadiene cation **57**, which is highly

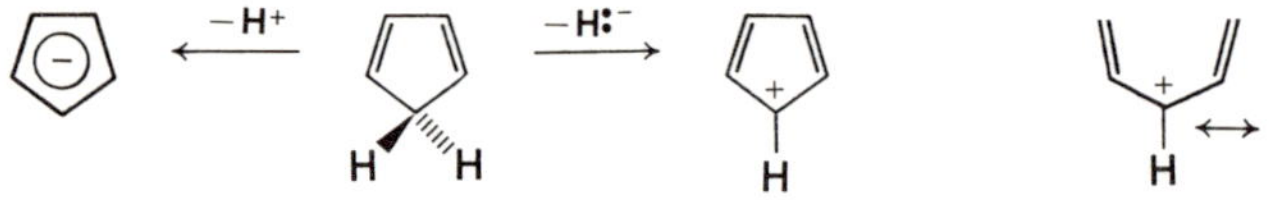

aromatic (6π) (**55**) | cyclopentadiene | antiaromatic (4π) (**56**) | pentadiene cation (stable) (**57**)

stable. For this reason, 9-methylfluorenol (**58**) is less prone to dissociation in strong acid than diphenylethanol (**59**) by a factor of 10^6 (although carbanions based on the fluorene system are common, Section 4.10).

(**58**)

$pk_{R+} = -16.6$

(**59**)

$pk_{R+} = -10.4$

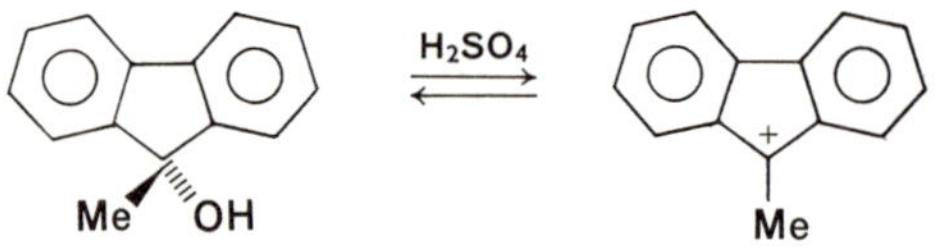

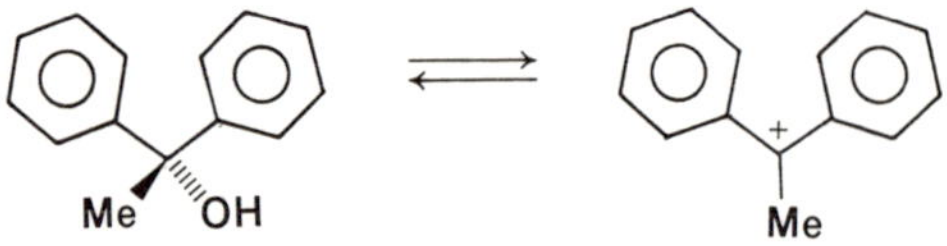

Cyclopentadienyl chloride, though doubly an allylic halide, refuses to solvolyse altogether.

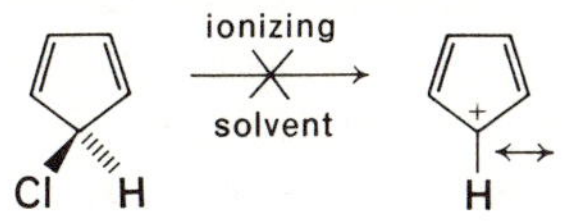

2.22   CARBONIUM IONS IN HIGHLY ACIDIC MEDIA

Sulphuric acid is an excellent medium for protonation of weak bases and solvation of the ions produced. Conjugated dienes and trienes are sufficiently strong bases to be extensively protonated in moderately concentrated acid with the formation of allylic cations. In the following examples, the concentrations of acid refer to conditions under which 50% protonation occurs ($K = 1$), and the figures to the proton-resonance chemical shifts. The cations **60–62** are quite

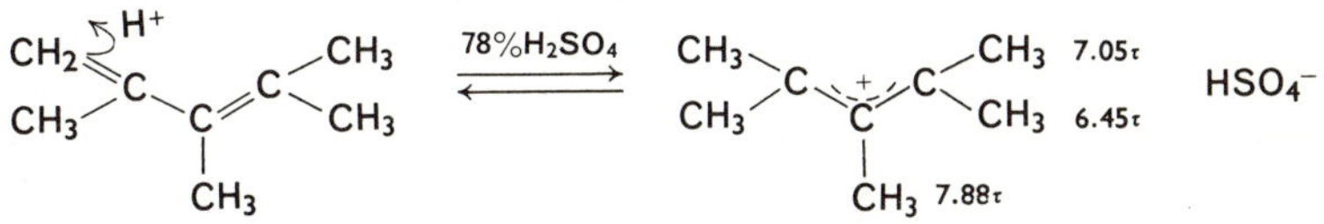

(60)

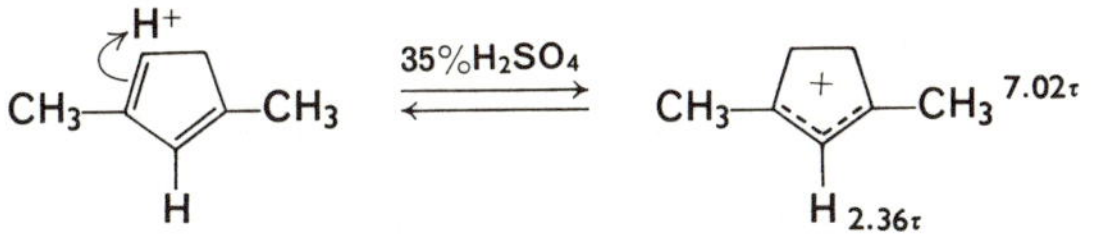

(61)

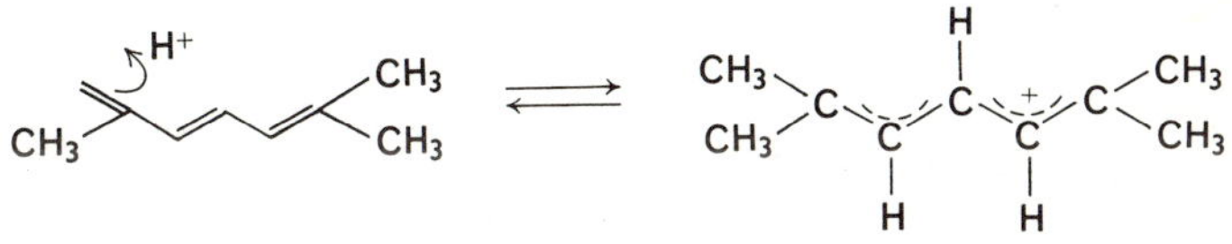

(62)

stable in solution and can be characterized by their electronic and n.m.r. spectra, which confirm their symmetry. Restricted rotation is apparent in the open-chain allylic cation **60**, since the two terminal methyl groups are non-equivalent in their n.m.r. spectra. The perchloropropenyl cation (**63**) is even more stable and can be isolated as the ionic tetrachloroaluminate. Weaker

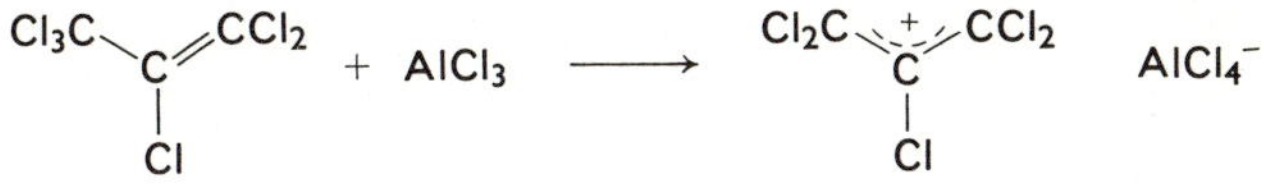

(63)

bases such as monoolefins are protonated in concentrated sulphuric acid solution. However, the more reactive alkyl cations produced polymerize or form covalent esters with the bisulphite ion present, which is therefore too nucleophilic to permit a high concentration of the carbonium ion to form. In 1962, G. A. Olah showed that an excellent medium for the stabilization of carbonium ions could be formed from a mixture of antimony pentachloride with liquid HF or, better, fluorosulphonic acid, FSO_2OH, and diluted if

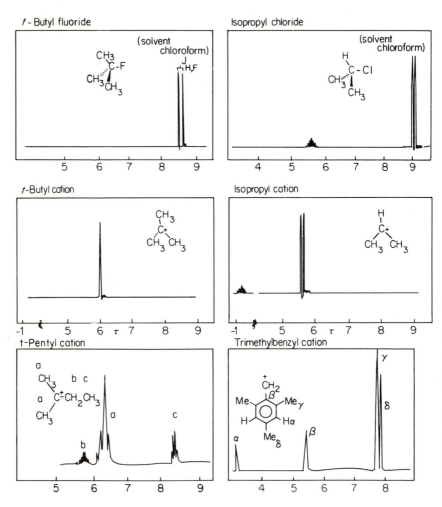

Figure 4. Proton n.m.r. spectra of some carbonium ions in superacid solution.

$$CH_2=C\overset{CH_3}{\underset{CH_3}{\diagup}} + H_2SO_4 \longrightarrow CH_3-\overset{+}{C}\overset{CH_3}{\underset{CH_3}{\diagup}} HSO_4^- \longrightarrow CH_3-\underset{\underset{OSO_2OH}{|}}{C(CH_3)_2}$$

$$\downarrow$$

polymerization

necessary with liquid SO_2 or sulphonyl chlorofluoride, SO_2ClF, which is preferred. These mixtures are more protonating than sulphuric acid and also less nucleophilic. They are, in fact the most highly acidic media known and may be considered to contain the acids, $HSbF_6$ or $H_2Sb_2F_{11}$; the term 'superacid' is often applied to these solutions.

It was soon found that t-butyl fluoride reacted with SbF_5 to give a stable t-butyl cation (**64**) in solution, the n.m.r. spectrum of the covalent fluoride (doublet due to coupling of the protons with fluorine) changing to a singlet spectrum with a considerable downfield shift (Figure 4). The same species was formed when t-butanol, or its chloroformate (**65**), other halides, isobutene and even 1- and 2-butyl halides, were dissolved in superacid solution. In the last two cases, rearrangement of the less stable 1- and 2-butyl cations, **66** and **67**,

$$(CH_3)_3C-OH \xrightarrow{SOCl_2} (CH_3)_3C-O\overset{\diagdown}{\underset{Cl}{\diagup}}C=O$$

(**65**) $(CH_3)_3C-hal$

$$\Big\downarrow -CO_2, Cl^-$$

$$CH_2=C\overset{CH_3}{\underset{CH_3}{\diagup}} \longrightarrow \underset{CH_3}{\overset{CH_3}{\underset{\diagdown}{|}}}\overset{|}{\underset{CH_3}{C_+}} \quad SbF_6^- \quad \overset{rearrangement}{\longleftarrow} \quad CH_3\overset{+}{C}HCH_2CH_3$$

(**64**)   (**67**)

$$\Big\uparrow \text{rearrangement} \qquad\qquad\qquad\qquad \Big\uparrow \quad hal$$

$$CH_3CH_2CH_2CH_2-hal \longrightarrow CH_3CH_2CH_2\overset{+}{C}H_2 \qquad CH_3-CH-CH_2CH_3$$

(**66**)

All reactions in SbF_5—HF or FSO_3H at low temperature

has occurred. Similarly, propyl and pentyl halides and the related olefins and alcohols lead to the isopropyl (**67**) and t-pentyl cations (**68**), respectively. The

$$
\begin{array}{ccc}
\begin{array}{c}CH_3\\ \diagdown\\ CH_3\end{array}\!\!\!CH\!-\!Cl & \longrightarrow & \begin{array}{c}CH_3\\ \diagdown\\ CH_3\end{array}\!\!\!\overset{+}{C}\!-\!H & \longleftarrow & H\!-\!C\!\!\begin{array}{c}\diagup CH_3\\ \diagdown CH_2\end{array}
\end{array}
$$

(67)

$$CH_3CH_2CH_2Cl$$

$$
\begin{array}{ccc}
C_5H_{11}Cl & \longrightarrow & \begin{array}{c}CH_3\\ \diagdown\\ CH_3\end{array}\!\!\!\overset{+}{C}\!-\!CH_2CH_3 & \longleftarrow & C_5H_{10}
\end{array}
$$

all isomers   **(68)**

disappearance of proton–fluorine coupling in the proton n.m.r. spectra of alkyl fluorides in superacid medium does not by itself prove that the free alkyl cation is present, although the downfield shift indicates the accumulation of positive charge. Similar spectra could be obtained from a rapidly exchanging system,

$$
\begin{array}{ccc}
\overset{\backslash}{\underset{\diagup}{C}}\!-\!\overset{*}{F}\cdots SbF_5 & \rightleftharpoons & \overset{\backslash}{\underset{\diagup}{C}}\!-\!F\cdots \begin{array}{c}SbF_4\\ |\\ *F\end{array}
\end{array}
$$

The ^{13}C n.m.r. spectra are less ambiguous since the chemical shift of each carbon atom present is directly proportional to the charge carried; a shift of 200 p.p.m. per electronic charge, downfield if positive and upfield if negative, is often assumed. The t-butyl cation has a central resonance at -135 p.p.m. (relative to $^{13}CS_2$) corresponding to a charge of $+0.7$ e, more positive than carbon in CS_2. Similarly the central carbons in the isopropyl **(67)** and t-pentyl **(68)** cations were found to resonate at -125 and -139 p.p.m., respectively. The former shows coupling to the 2-proton of 169 Hz which is characteristic of sp^2 hybridized carbon. The high positive charges carried by the central carbon atoms—essentially a whole electronic charge—and evidence from proton coupling constants are sufficient to confirm the carbonium ion nature of the species giving rise to these signals. Another tool which has been of much use in this connection is Raman spectroscopy. As predicted, simple alkyl cations do not show ultraviolet maxima above 210 nm, the shortest wavelength at which measurements can be made in superacid solution.

While dozens of tertiary cations have now been observed, primary cations (except benzylic types) are not observable, and secondary ones known to be stable are limited to the isopropyl cation, 2-butyl and secondary benzyl cations (Table 10). Attempted generation of primary or secondary cations usually results in rearrangement to the much more stable tertiary isomers (Section

Table 10. Some primary and secondary cations stable in superacid solution

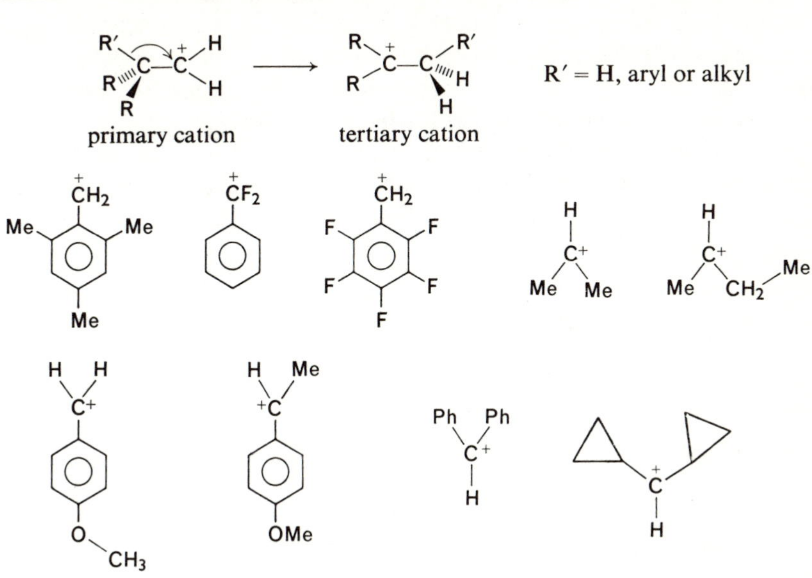

primary cation   tertiary cation $R' = H$, aryl or alkyl

2.29) by one or more 1,2-shifts of hydride or alkide ion (for example the formation of **69** and **70**).

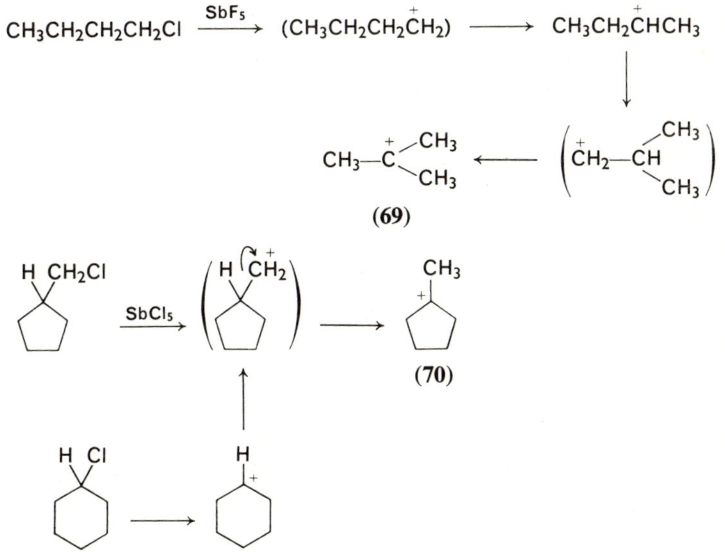

(69)

(70)

Aromatic hydrocarbons are protonated (i.e. act as bases) in superacid solution. Protonation equilibria lie farther to the right the more alkyl or other electron-donating groups are attached to the ring. The products are benzenonium ions (71, 72) many of which have now been studied by their n.m.r. spectra as stable species at low temperature. The parent ion, protonated benzene shows distinct resonances for the two methylene protons indicating that it is non-planar (71). The charge density around the ring (as shown in 71a is mainly concentrated at positions 2, 4 and 6. These species are inter-

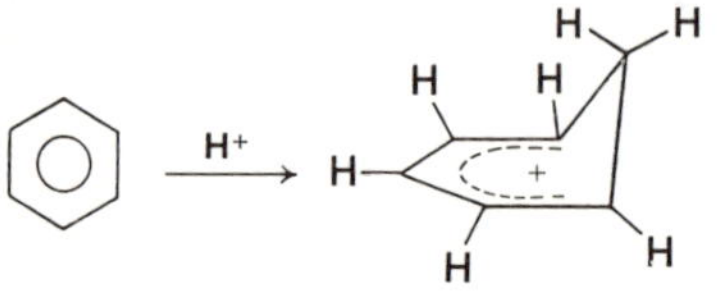

(71) 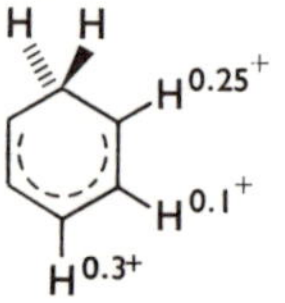   (71a)

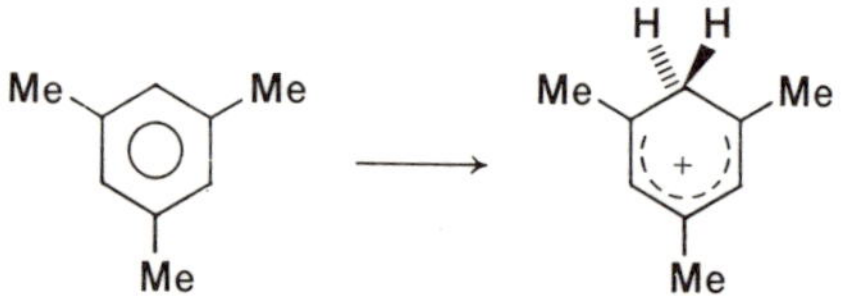

(72)

mediates in aromatic proton exchange (p. 127) but recently the nitro- and chlorobenzenonium ions (73 and 74) have been observed, previously postulated as intermediates in nitration and chlorination of benzene, respectively.

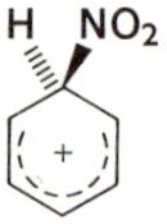

(73)    (74)

Very many examples of carbonium ions stabilized by adjacent heteroatoms are known to exist in SbF_5–HSO_3F solution; they are formed as before by dissociation or protonation of suitable precursors. Further examples of stabilized carbonium ions are discussed in Section 3.2.

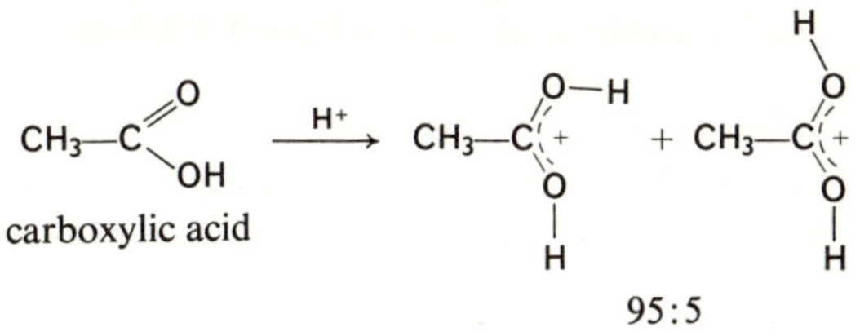

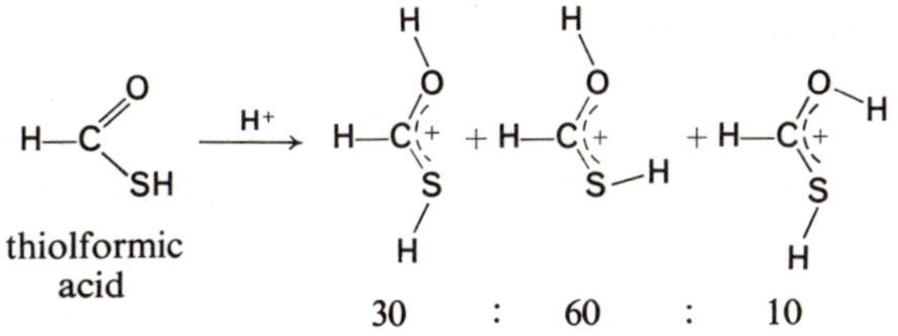

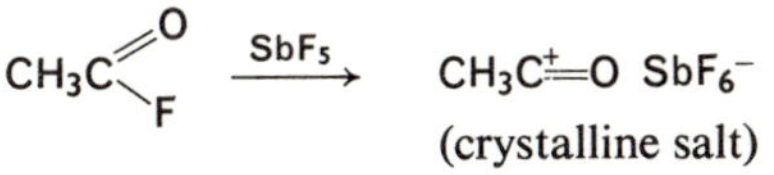

The reaction schemes are shown in the following figure:

$$CH_3-C\overset{O}{\underset{OH}{\diagdown}} \quad \xrightarrow{H^+} \quad CH_3-C\overset{O-H}{\underset{O}{\diagup}}+ \quad + \quad CH_3-C\overset{O}{\underset{O}{\diagup}}+$$

carboxylic acid

95:5

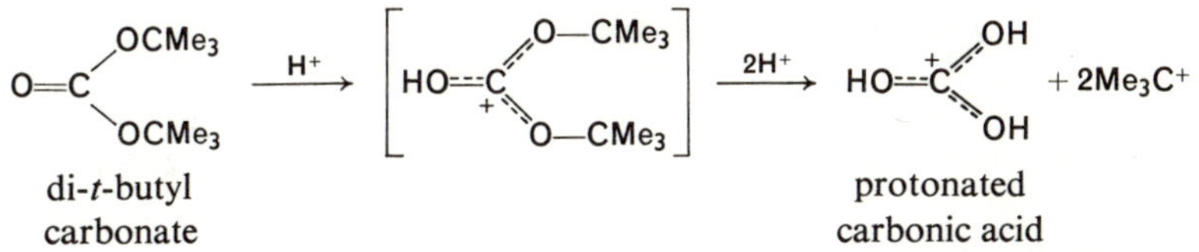

$$O=C\overset{OCMe_3}{\underset{OCMe_3}{\diagdown}} \quad \xrightarrow{H^+} \quad \left[HO=C\overset{O-CMe_3}{\underset{O-CMe_3}{\diagup}}\right] \quad \xrightarrow{2H^+} \quad HO=C\overset{OH}{\underset{OH}{\diagup}} + 2Me_3C^+$$

di-t-butyl   protonated
carbonate   carbonic acid

$$H-C\overset{O}{\underset{SH}{\diagdown}} \quad \xrightarrow{H^+} \quad H-C\overset{O}{\underset{S}{\diagup}}+ \quad + \quad H-C\overset{O}{\underset{S-H}{\diagup}}+ \quad + \quad H-C\overset{O-H}{\underset{S}{\diagup}}+$$

thiolformic
acid

30   :   60   :   10

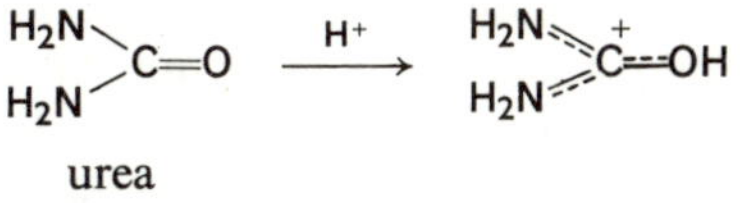

$$\overset{H_2N}{\underset{H_2N}{\diagup}}C=O \quad \xrightarrow{H^+} \quad \overset{H_2N}{\underset{H_2N}{\diagup}}C=OH$$

urea

$$CH_3C\overset{O}{\underset{F}{\diagdown}} \quad \xrightarrow{SbF_5} \quad CH_3C^+=O \ SbF_6^-$$

(crystalline salt)

2.23 THE MEASUREMENT OF CARBONIUM ION STABILITY

Almost any process which leads to the formation of a carbonium ion can be adapted to yield quantitative information on its stability, at least in a comparative sense. Carbonium ion stability is usually equated with ease of formation from a standard type of precursor under standard conditions and measured by a difference in free energy or enthalpy. In the case of isomeric carbonium ions, stabilities can be compared by free energies of formation, ΔH_f. Frequently these measurements may be practical only for a limited range of structural types. Little is known of absolute bond energies or heats of formation though there are now available many theoretical calculations of these quantities.

The Ionization Potentials (I.P.'s) of Radicals

Removal of an electron from an alkyl radical leads to a carbonium ion. This process may be accomplished in the gas phase by electron bombardment and the experiment is performed in the mass spectrometer. A stream of free

$$R{-}ONO \xrightarrow{\;\Delta\;} R{\cdot} + e^- \longrightarrow R^+ + 2e^-$$

$$\text{alkyl nitrite} \qquad (+NO_2) \qquad (+ \text{ kinetic energy})$$

radicals (prepared, for example, by pyrolysis of an alkyl nitrite in the inlet tube) may be introduced into the ionization chamber and bombarded by electrons (ideally mono-energetic electrons) whose kinetic energy is steadily increased. By extrapolation, the 'threshold' energy at which the ion, R^+, is first detected is equated with the enthalpy of ionization of the radical. This 'appearance potential'—the first ionization potential (i.p.) of the radical—is directly a measure of the stability of the carbonium ion relative to the corresponding radical (Table 11).

Structural features such as conjugation, chain branching and the presence of heteroatoms may affect the stabilities of both radical and carbonium ion in the same direction. Consequently, appearance potentials need to be interpreted with care. Since their heats of formation are an additive function of bond energies, alkanes are better standards for comparison than alkyl radicals and the appropriate quantities are better known. By applying the following thermal cycle,

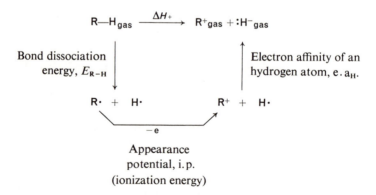

An estimate of the energy of heterolysis of the alkane can be made in terms of a C—H bond energy (homolysis), the appearance potential of the alkyl radical and the electron affinity for hydrogen,

$$R\text{---}H_{gas} \longrightarrow R^+{}_{gas} + H^-{}_{gas} + \Delta H_+$$

$$\Delta H_+ = (E_{R-H} + \text{i.p.}_{R\cdot} + \text{e.a}_{H\cdot})$$

ΔH_+ may then be equated with the stability of the carbonium ion. One difficulty which may arise when applying this method is the uncertainty in the structure of both radical and carbonium ion in the mass spectrometer. For instance, the n-propyl radical may be generated and subjected to ionization leading to the 2-propyl cation by a rearrangement, or it could in principle rearrange to the isopropyl radical prior to ionization. It is difficult to tell to what extent the

Table 11. Heats of formation of carbonium ions from appearance potentials

(a) By ionization of the radicals:		
R·	i.p.	ΔH_+ (cation) (kJ mol^{-1})
CH_3	950	1080
C_2H_5	835	943
n-C_3H_7	835	912
n-C_4H_9	835	885
iso-C_3H_7	760	810
t-C_4H_9	715	730
t-C_5H_{11}	685	678
NeoC$_5$H$_{11}$	805	815
Cyclopropyl		1000
Cyclobutyl		890
Cyclopentyl		815
Cyclohexyl		782
Allyl, CH_2=CH—CH_2		955
Propargyl CH≡C—CH_2		1105
Tropyllium		920
p-Cyanobenzyl	806	
p-Chlorobenzyl	766	
Benzyl	748	
p-Methoxybenzyl	660	
(b) By ionization of the hydrocarbon:		
$CH_4 \rightarrow CH_3{}^+$		1080
$CH_3CH_2CH_3 \rightarrow CH_3\overset{+}{C}H\cdot CH_3$		810
CH_2=$CH_2 \rightarrow CH_2$=CH^+		1190
CH_3CH=$CH_2 \rightarrow CH_2$—CH—$CH_2{}^+$		985
CH_2=C=$CH_2 \rightarrow CH_2$=C=CH^+		1180

$$CH_3CH_2CH_2 \cdot \longrightarrow (CH_3CH_2CH_2^+) \longrightarrow CH_3\overset{+}{C}HCH_3$$
$$\searrow CH_3\overset{\cdot}{C}HCH_3 \nearrow$$

rearrangement (an exothermic process) could have assisted the ionization of the radical whose measured i.p. would then be low. Similarly, the benzyl radical ionizes to the more stable tropyllium cation (75). However, the

(75)

reasonable values obtained for ionization potentials, and the differences shown by, for example, primary and tertiary isomers, indicates that rearrangements must in the main take place after ionization.

Kinetic Methods

One of the most commonly used criteria by which to judge stabilities of carbonium ions is their rate of formation in a solvolytic reaction. The rates which are to be compared should ideally be the 'pure ionization' component of the reaction, fk_c, though this is seldom known for secondary substrates, and the comparison should properly be made between free energies of activation rather than rates of reaction at one temperature. Many authors, nonetheless, do draw comparisons from total rates of solvolysis at one temperature. With the best data, values obtained are differences in free energy between the starting material and the transition state for heterolysis in which solvent is an important stabilizing factor. Table 12 lists a number of examples in which the kinetics of small series of compounds are compared.

Other kinetic data, such as from electrophilic additions to unsaturated compounds, may also be used for this purpose (Table 12).

2.24   EQUILIBRIUM STUDIES

Highly stable carbonium ions may be obtained in equilibrium with their products of reaction with a nucleophile. A series of such carbonium ions may be compared by their equilibrium constants under standard conditions.

Di- and tri-arylmethyl halides may be induced to ionize reversibly in polar media and the position of equilibrium arranged to lie sufficiently far to the right

Table 12. Unimolecular solvolytic rate constants and their relationship to carbonium ion stability

Substrate	Solvent, etc.	k_1 ($\times 10^5$ s^{-1})
PhCH$_2$Cl	Ethanol	0.0314
p-MeO-C$_6$H$_4$CHPh $\mid$ Cl	Ethanol	534
p-Br-C$_6$H$_4$CHPh $\mid$ Cl	Ethanol	1.90
Ph$_2$CHCl	Ethanol	5.75
PhCHMe $\mid$ Cl	Ethanol	3.75

p-MeO $\geqslant$ p-H > p-Br as stabilizing groups
Aryl > Me > H

Ph$_2$CHCl	80% aq. acetone, 25°	7
Ph$_2$CHBr	80% aq. acetone, 25°	153
Ph$_2$CCl$_2$	80% aq. acetone, 25°	9.9
Ph$_3$CCl	80% aq. acetone, 25°	approx 9 $\times$ 10^5

Ph $\geqslant$ H > Cl as stabilizing groups
Br > Cl (as leaving groups)

tBuCl	MeOH	0.018
tBuCl	EtOH	0.022
tBuCl	80% EtOH–H$_2$O	38.5
Me$_2$C$\diagup^{Et}_{\diagdown Cl}$	80% EtOH–H$_2$O	2.0
CH$_2$=CHCH$_2$Cl	50% EtOH–H$_2$O	0.016
CH$_2$=CH—CHCl $\mid$ Me	50% EtOH–H$_2$O	1.22

H$_2$O > MeOH > EtOH (as ionizing media)
Me > Et as stabilizing groups
t-alkyl ~ secondary allyl

Cl
$\mid$
X—CMe$_2$

X = Me	aqueous ethanol	0.59
Et	aqueous ethanol	1.00
nPr	aqueous ethanol	0.94
iPr	aqueous ethanol	0.53
t-Bu	aqueous ethanol	0.73

Table 12—*continued*

Substrate	Solvent, etc.	k_1 $(\times 10^5 \text{ s}^{-1})$
$(t\text{-Bu})_2\text{C}\underset{\text{Me}}{\overset{\text{Cl}}{\diagup}}$	aqueous ethanol	322

Electronic effects of alkyl groups are similar (slightly electron releasing) and show small differences due to hyperconjugation (q.v.) and to steric factors (such as release of strain upon ionization, which is especially marked in the last compound) and steric hindrance to solvation during ionization

Isopropyl chloride	AcOH, 60°	1.00
Cyclopropyl chloride	AcOH, 60°	2×10^{-5}
Cyclobutyl chloride	AcOH, 60°	12
Cyclopentyl chloride	AcOH, 60°	14
Cyclohexyl chloride	AcOH, 60°	0.88
Cycloheptyl chloride	AcOH, 60°	27
Cyclooctyl chloride	AcOH, 60°	251

Rings containing 4, 5, 7 or more, carbon atoms are more reactive than the model secondary compound (isopropyl) since strain relief occurs on ionization (eclipsing interactions, etc.). The cyclopropyl compound is unreactive for a similar reason to vinyl compounds, which it resembles, in possessing π-character.

Relative rates of electrophilic addition to olefins and acetylenes

$$\underset{R}{\overset{}{\diagdown}}\text{C}=\text{CH}_2 + \text{E}^+ \xrightarrow{\text{slow}} \underset{R}{\overset{}{\diagdown}}\text{C}^+-\text{CH}_2\text{E} \longrightarrow \text{products}$$

Reagent	(E$^+$ = Br$^+$, H$^+$) Bromine in			H_3O^+ in H_2O
	AcOH	MeOH	H_2O	
$\dfrac{k_{\text{PhCH}=\text{CH}_2}}{k_{\text{PhC}\equiv\text{CH}}}$	2590	100	0.63	0.67
$\dfrac{k_{\text{EtCH}=\text{CHEt}}}{k_{\text{EtC}\equiv\text{CEt}}}$	372×10^3			16.6

Alkyl cation is more easily formed than a vinyl cation but much depends upon the solvent. The solvents of higher dielectric constant (higher polarity) tend to favour reaction of the alkene over the alkyne.

$$Ar_3C\text{—}Cl \underset{\longleftarrow}{\overset{K}{\longrightarrow}} Ar_3C^+ + Cl^-$$

$$\lambda_{max} = 260 \text{ nm} \qquad\qquad 520 \text{ nm}$$

so that accurate determinations of K can be made. The very distinct spectra of the two species make the analysis possible by spectrophotometry.

Similar measurements may be made on solutions of the corresponding alcohols in strong acid solution

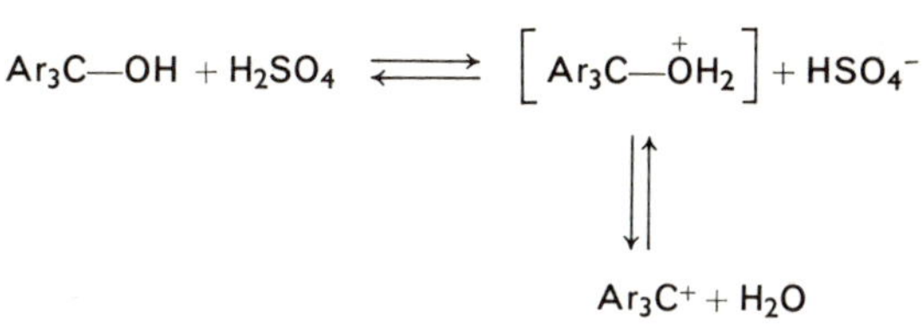

$$Ar_3C\text{—}OH + H_2SO_4 \underset{\longleftarrow}{\longrightarrow} \left[Ar_3C\text{—}\overset{+}{O}H_2 \right] + HSO_4^-$$

$$\updownarrow$$

$$Ar_3C^+ + H_2O$$

and upon cyclopropenols (**76**) in water:

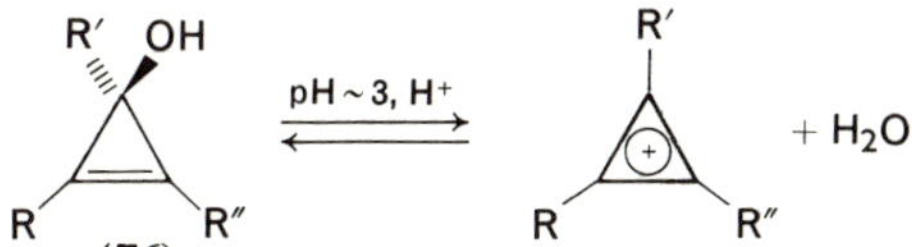

In principle, the more-reactive alkyl cations could be studied in equilibrium in superacid solution, but this has not so far been attempted. The dissociations of these alcohols follow the appropriate acidity function (Section 1.22), where

$$pK_{R^+} = H_R - \log\frac{[ROH]}{[R^+]}$$

Since H_R is a property of the medium, pK_{R^+} measures the dissociation tendency and the stability of the carbonium ion under the conditions used: the more positive pK_{R^+}, the more stable the cation.

Methods of measurement of the equilibrium constants include conductivity, potentiometry and proton n.m.r. The stabilities of benzenonium ions (protonated aromatic molecules) have been assessed by their partition coefficients between liquid HF and an organic solvent. Some results of equilibrium studies are given in Table 13.

Table 13. Dissociation constants of substituted triphenylmethyl chlorides in liquid SO_2

$$p\text{-}XC_6H_4\overset{\displaystyle Cl}{\underset{\displaystyle |}{C}}Ph_2 \overset{K}{\rightleftharpoons} p\text{-}XC_6H_4\overset{+}{C}Ph_2 + Cl^-$$

X	K_{rel}
Me	16
t-Bu	18
Ph	5.9
H	1
Cl	0.31

t-Bu ~ Me > Ph > H > Cl

Values of pK_{R+} for some alcohols

$$p\text{-}XC_6H_4\overset{\displaystyle OH}{\underset{\displaystyle |}{C}}Ph_2 \overset{K}{\rightleftharpoons} p\text{-}XC_6H_4\overset{+}{C}Ph_2$$

X	pK_R^+
OMe	−3.4
Me	−5.4
t Bu	−6.1
H	−6.6
NO$_2$	−9.1

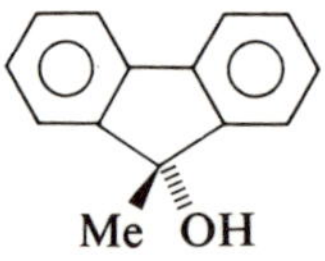

    −16.6

    +4.7

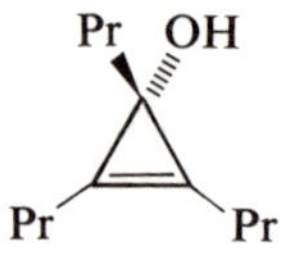

    +7.2

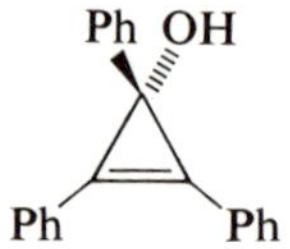

    +3.1

Table 13 *cont.*

X	pK_R^+

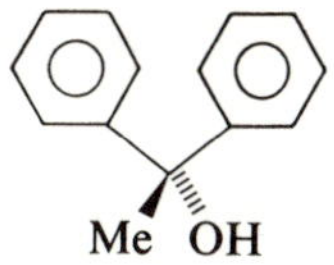

-10.4

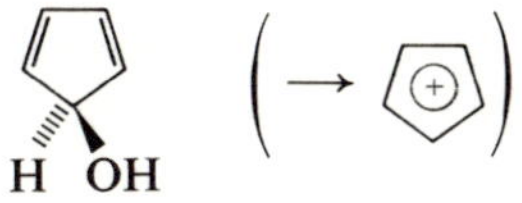

(→) -40

2.25   **HEATS OF FORMATION**

Arnett has recently measured some heats of formation of aliphatic carbonium ions and benzenonium ions at low temperatures in SbF_5–HSO_3F solution. The more exothermic the reaction, the more stable is the carbonium ion (Table 14).

Table 14. Enthalpies of formation of some carbonium ions

$$ROH + \text{'}HSbF_6\text{'} \rightarrow R^+ SbF_6^- (+H_2O)$$

R^+	$-\Delta H$ (kJ mol^{-1})
Me_2CH^+	67
Me_3C^+	97
Ph_2CH^+	162
Ph_3C^+	184

$ArH + \text{'}HSbF_6\text{'}$	$ArH_2^+ SbF_6^-$
m-xylene	27
Durene[a]	56
Hexamethylbenzene	65
Anthracene[b]	73
Azulene[c]	135

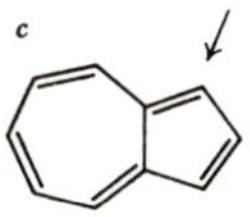

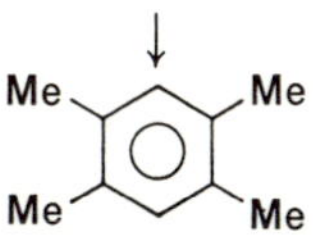

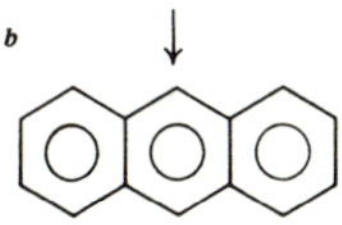

Positions of protonation as indicated by arrow

$$\left(X\!-\!\!\left\langle\bigcirc\right\rangle\!\!\right)_2\!\!-\!C\!=\!CH_2 \quad\xrightarrow{H^+}\quad \left(X\!-\!\!\left\langle\bigcirc\right\rangle\!\!\right)_2\!\!-\!\overset{+}{C}\!-\!CH_3 \quad X = H,\ 61\ \rho = 6.75$$

$$\left(X\!-\!\!\left\langle\bigcirc\right\rangle\!\!\right)_2\!\!-\!C\!=\!O \quad\xrightarrow{H^+}\quad \left(X\!-\!\!\left\langle\bigcirc\right\rangle\!\!\right)_2\!\!-\!\overset{+}{C}\!\cdots OH \quad X = H,\ 48\ \rho = 4.59$$

2.26   THEORETICAL APPROACHES

The availability of high speed computers has made it possible to apply approximate molecular-orbital theory to quite large molecules and to obtain theoretical values for heats of formation. Carbonium ions can be treated in this way and any desired geometry imposed on the model. However, even the most sophisticated methods of calculation contain serious approximations and the results are of little use as absolute values. *Relative* energies, charge densities and other properties are probably correctly predicted and so enable us to obtain at least a qualitative understanding of the electronic properties and factors which stabilize carbonium ions. The values obtained necessarily apply to the gas phase and cannot take account of solvation, Table 15.

Table 15. A comparison of theoretical and experimental heats of heterolysis
Ionizing hydrogen underlined. $R\!-\!H \rightarrow R^+ + H^-$

RH	R^+	ΔH_{expl}	ΔH_{calc} kJ mol^{-1}
CH_4	Methyl	1380	2110
$CH_3C\underline{H}_3$	Ethyl	1252	1840
$CH_3CH_2C\underline{H}_3$	1-Propyl	1248	1835
$CH_3C\underline{H}_2CH_3$	2-Propyl	1160	1655
$CH_3CH_2CH_2C\underline{H}_3$	1-Butyl	1246	1780
$CH_3CH_2C\underline{H}_2CH_3$	2-Butyl	1172	1710
$(C\underline{H}_3)_3CH$	Isobutyl	1186	1778
$(CH_3)_3C\underline{H}$	t-Butyl	1100	1500

Correlation between rates of aromatic substitution reactions and calculated stabilities of the benzenonium ion intermediates have been successful. The total heats of formation (from the atoms) of the aromatic compound and the corresponding benzenonium ion can be calculated and their difference, Λ, is

$$\Lambda = \Delta H_f(ArH_2^+) - \Delta H_f(ArH)$$

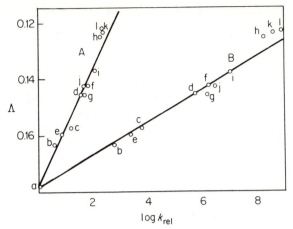

Figure 5. Relationship between relative reactivity, $\log k_{rel}$ and calculated energy of protonation, Λ, for a series of methylbenzenes. Curve A refers to mercuration, curve B to bromination.

related to the stability of the benzenonium ion. Although values of ΔH_f do not agree with experimental ones, values of Λ correlate well with reaction rates for the whole series of methylbenzenes (Figure 5). The calculations predict correctly the preferred positions of protonation of the substituted benzenes, i.e. the positions at which substitution preferentially occurs,

The successful correlation of Λ with reaction rates implies that the transition state for electrophilic substitution resembles the benzenonium ion and that orientation in substitution is primarily determined by the formation of the most stable benzenonium ion.

2.27   **CONCLUSIONS**

Experimental and theoretical studies agree that carbonium-ion stability increases, other factors being equal, in the order:

$$\text{Tertiary} > \text{Secondary} > \text{Primary} > CH_3^+$$

The effects of stabilizing groups attached to the positive carbon are:

$$R_2N- > RO- > Ar- > RCH{=}CH- > Alkyl > H$$

Among aryl groups, substituents in the *para* position stabilize an adjacent cationic centre inversely as their value of σ (or rather σ^+); i.e. $RO- > R > H >$ hal $> CN > NO_2$, etc. Among simple alkyl groups, methyl is best able to stabilize an adjacent positive charge, followed by secondary and then tertiary groups. This is the order of the number of α-hydrogen atoms on which much of the positive charge actually resides, a phenomenon which is described as 'hyperconjugation' and represented by the valence-bond structures **77** and **78** (though this is only a minor effect).

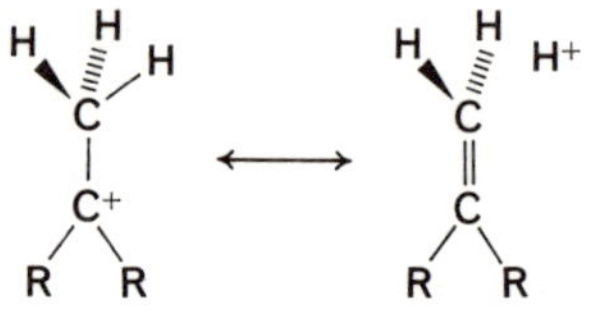

(77)   (78)

Stabilization by olefinic substitution increases with the length of the conjugated system in general and the charge becomes more widely delocalized, mainly on alternate carbon atoms. Cyclopropyl groups (which are known to

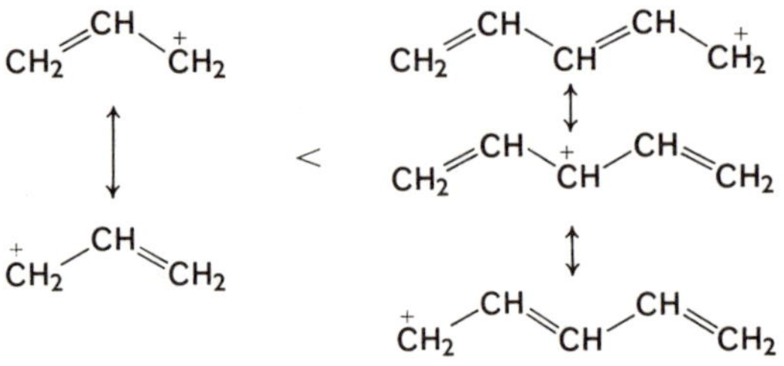

exhibit characteristics of π-systems), are also capable of considerable charge delocalization.

The spreading or delocalization of charge on to other atoms is probably the greatest single factor in the stabilization of simple cations. The exceptions are the aromatic and antiaromatic systems which by virtue of their molecular-orbital properties are, respectively, far more stable and far less stable than would be predicted from comparisons of acyclic systems.

2.28   THE STRUCTURES OF CARBONIUM IONS

The racemization which usually accompanies solvolyses via carbonium ions can be accommodated by either a planar structure, **79** (using sp^2 hybrid orbitals), or a rapidly inverting pyramidal one (using sp^3), **80**. The Raman

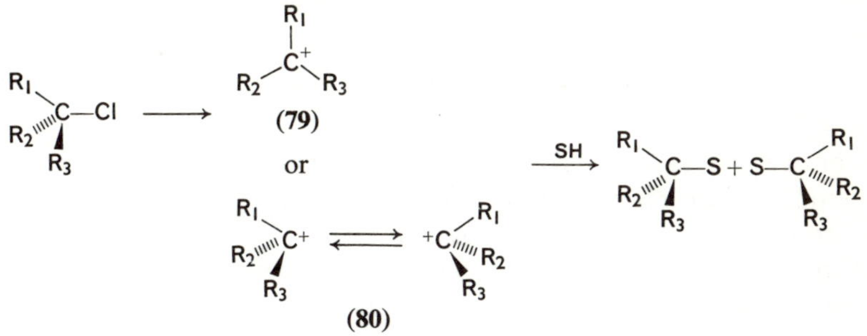

spectrum of the *t*-butyl cation (in superacid solution) is best interpreted as that of a species with C_{3v} symmetry, i.e. the planar symmetrical cation. Further-more, the large long-range coupling interactions (J) between protons on carbons adjacent to the cationic centre agree with values calculated for an sp^2-hybridized species, e.g. **81**. Coupling constants (^{13}C—H) in the isopropyl cation are also in accordance with predictions for a planar ion.

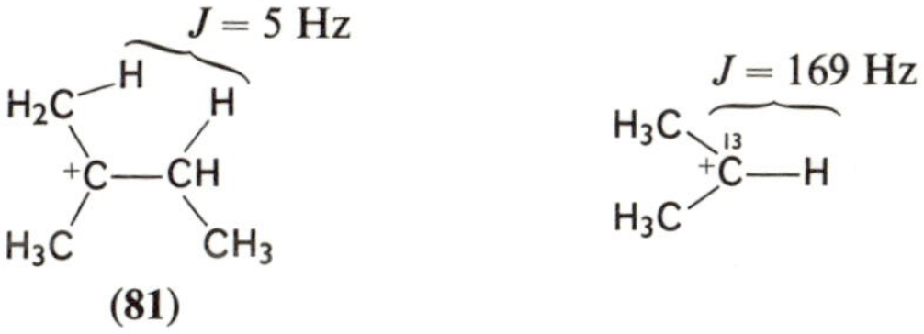

On chemical grounds, the lack of reactivity of bicyclic halides with halogen at the bridgehead has been explained in terms of the unfavourable non-planar carbonium ion which would be produced upon ionization, for example **82**. On

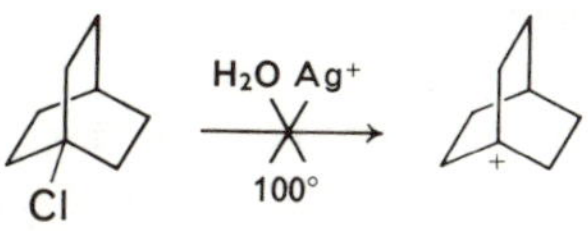

the other hand this inertness could be due to the inability of the solvent (water) to approach and solvate the rear side of the ion on account of its bicyclic structure. Bridgehead halides such as adamantyl chloride (**83**) ionize normally in superacid solution and indeed (p. 176) bicyclic cations often isomerize to

(83)

the bridgehead cations which are necessarily non-planar. Thus, at least in non-conjugated carbonium ions, a non-planar geometry does not seem to be too strongly inhibited, though is not preferred.

A planar configuration would appear to be important for efficient conjugation with π-systems. X-ray diffraction study of crystalline carbonium ions,

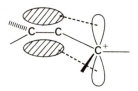

such as the triphenylmethyl cation, confirms the planar arrangement at the central carbon. X-ray studies have also been carried out on the heptamethylbenzenonium ion, a stable species. The carbon–carbon bond lengths confirm the partial double-bond character between the trigonal carbon atoms;

heptamethylbenzenonium ion   acetyllium ion

The acetyllium ion, CH_3CO^+, has been obtained in crystalline form and shown to be linear.

The distribution of charge in a carbonium ion depends markedly upon the attached substituent groups. ^{13}C resonance provides a direct measure of the charge density at carbon and Table 16 illustrates the abilities of various groups to accept positive charge.

Table 16. ^{13}C n.m.r. chemical shifts of carbonium ions (central carbon resonance), relative to $^{13}CS_2$

$R_1R_2R_3{}^{13}C^+$	Chemical shift (p.p.m.)	Calculated charge on central carbon relative to CS_2 (e)
Me Me Me	−135	+0.68
Me Me H	−125	+0.62
Me Me cyclopropyl	−87	+0.43
Me Me OH	−55	+0.27
Ph Ph Ph	−18	+0.09
Me OH OH	−1.6	−0.008
OH OH OH	+28	−0.14
OH OH OMe	+31	−0.15

^{13}C-Nuclear Magnetic Resonance Spectra

The carbon isotope, ^{13}C, which occurs to the extent of 1.1 % in natural carbon, has an isotopic spin of $\frac{1}{2}$ and is capable of giving nuclear resonance spectra analogous to proton n.m.r. The major carbon isotope, ^{12}C, has zero spin number and so is incapable of resonance. Rather sophisticated apparatus is necessary to obtain good ^{13}C n.m.r. spectra on account of the low natural abundance, but this is now routine and spectra are often simpler than the corresponding proton spectra since each individual carbon has very little probability of coupling with another ^{13}C nucleus, only with attached protons. Furthermore, the chemical shift of each carbon resonance is closely related to the electron density at the carbon atom. Thus ^{13}C n.m.r. is a versatile probe of the electronic distribution in stable carbonium ions and in other molecules.

Isopropyl chloride, when dissolved in superacid solution, produces a ^{13}C spectrum very much shifted downfield due to the positive charge in the isopropyl cation to which it is converted. The chemical shifts of a selection of carbonium ions (central carbons) is given in Table 16. It is clear that the positive charge on this carbon diminishes with increasing numbers of conjugating, and especially heteroatom, substituents and may even become negative.

^{13}C chemical shifts are of the order 200 p.p.m.* per electronic charge so that the isopropyl cation must bear approximately 65% of the charge on the central carbon, the remainder being on the hydrogens rather than the methyl carbons.

* p.p.m. = parts per million.

X-ray Photoelectron Spectroscopy of Carbonium Ions

Tertiary alkyl cations that can be obtained in superacid solution may be converted to the solid crystalline (hexafluoroantimonate or fluorosulphonate) salts by evaporation of the medium at low temperature. Photoelectron spectroscopy is an interesting spectroscopic tool which has been used on these salts. A monochromatic beam of X-rays of energy E_x causes the ejection of electrons from the substance under investigation and these include both valence and $1s$ electrons of carbon. The energies of the ejected 'photoelectrons', E_p are measured magnetically and hence the original binding energy E_b is given (with small corrections) by

$$E_b = E_x - E_p$$

The binding energies of these core electrons is naturally related to the charge on the atom: the greater the positive charge the more difficult it is to remove an electron. This technique permits the direct assessment of charge magnitude and distribution in a carbonium ion. Thus the t-butyl cation has a spectrum consisting of two peaks in the ratio $1:3$, and indicating an increase in C-$1s$ ionization energy of 4.5 eV for the central carbon and 1.1 eV for the methyl carbons relative to the standard form of carbon, graphite (Figure 7). Delocalized cations such as trityl show a small and very similar shift for all carbons.

Calculations suggest that electron donation from alkyl groups (hyperconjugation) is provided by the electrons of the C—H bonds and it is the hydrogens on which the charge largely resides (Figure 6a). The calculated charge densities of a benzenonium ion are shown in Figure 6b; the *ortho* and *para* positions, as expected, bear most of the positive charge.

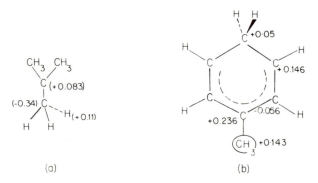

(a)   (b)

Figure 6. Calculated charge distribution in carbonium ions.

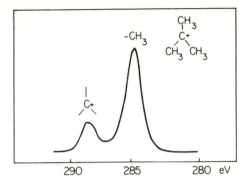

Figure 7. C-1s X-ray photoelectron spectrum
of the t-butyl cation.

2.29   **REACTIONS OF CARBONIUM IONS**

The reactions of carbonium ions in many instances are characteristic and serve
as additional evidence for their transitory formation.

Coordination to a Nucleophile

In its capacity as a Lewis acid, a carbonium ion is capable of coordination to
a nucleophile, the usual fate of this species when generated in solution. In a
solvolytic reaction, all nucleophiles present will compete for a carbonium ion to
give a multiplicity of products. The added azide ion, a fairly strong nucleophile,

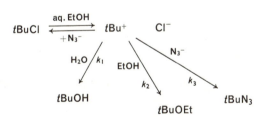

$$\text{BuOH}:\text{BuOEt}:\text{BuN}_3 = k_1\,[\text{H}_2\text{O}]: k_2\,[\text{EtOH}]: k_3\,[\text{N}_3^-]$$

produces alkyl azide among the products without materially altering the rate
of disappearance of butyl chloride. To species as reactive as alkyl cations,
almost any atom with a lone pair can act as a nucleophile.

Coordination to the π-system of an olefin can lead to polymerization in which the propagating species is another carbonium ion:

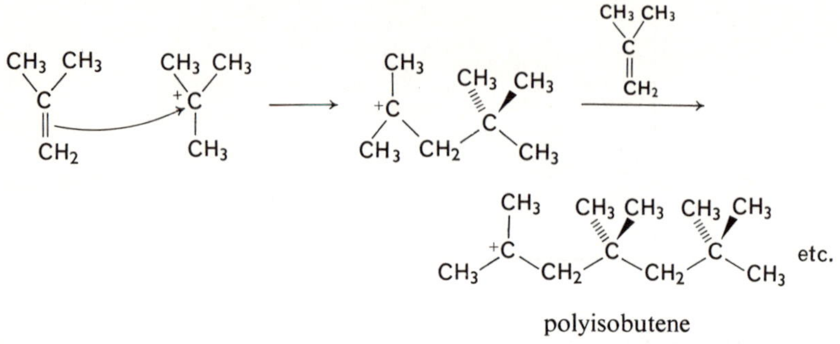

polyisobutene

Friedel–Crafts alkylation and acylation, on the other hand, can be regarded as examples of carbonium ions coordinating to aromatic moieties. In these

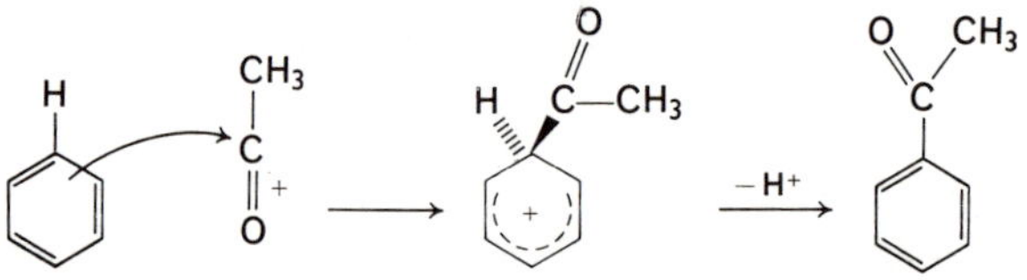

reactions the aromatic character is preserved by subsequent proton loss.

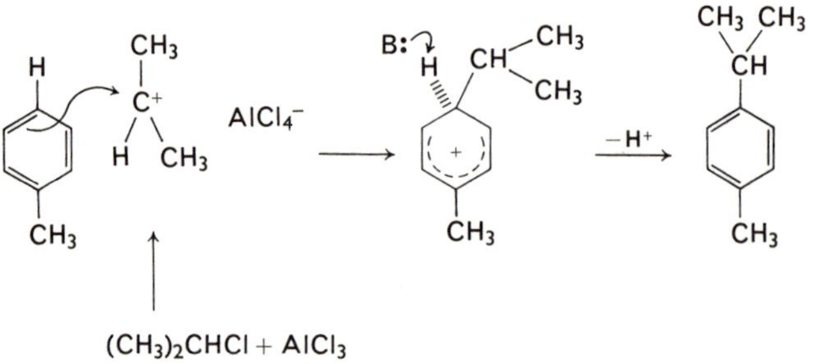

$(CH_3)_2CHCl + AlCl_3$

(B, the base, would likely be Cl^- here)

Loss of a β-proton

A nucleophile, acting as a base, may remove a β-proton if present as an alternative to attacking the carbonium centre. The product is an olefin or, in the case of a benzenonium ion, an aromatic molecule. Olefin formation is usually concurrent with nucleophilic substitution in unimolecular solvolyses, while proton loss from a benzenonium ion is the second (fast) step in aromatic substitution reactions (see above). Where two or more distinct β-hydrogens are available, that which is preferentially removed is the one leading to the most stable olefin, normally the most highly substituted ethylene (Saytzeff's rule). If the β–CH bond retained its integrity in the transition state there would be no kinetic isotope effect for deuterium substitution except a secondary effect (Sections 1.12, 2.10) amounting to $k_H/k_D \approx 1.2$. In fact, rather larger isotope effects are found for eliminations which undoubtedly proceed via carbonium ions, although the values are much less than a full primary effect (ca 7). This suggests that in the transition state, the solvent molecule which removes the β-proton is already coordinating to it weakly.

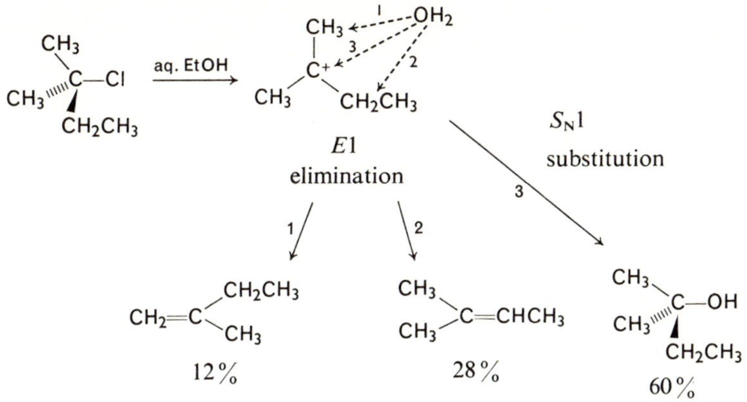

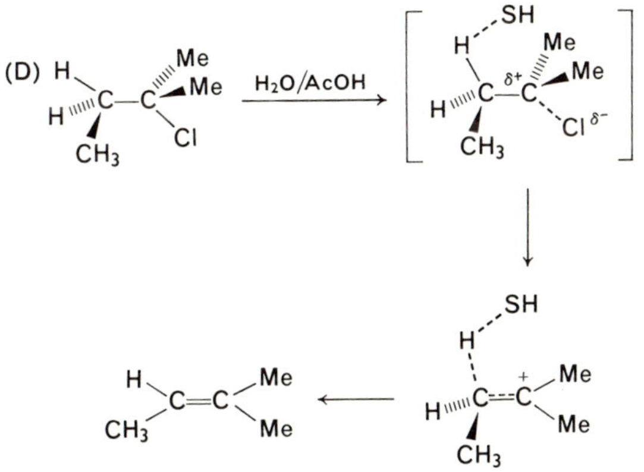

Rearrangement

A characteristic reaction of carbonium ions is their tendency to rearrange to a more stable isomer by the migration of a group to the positive carbon from an adjacent position. Aryl and alkyl groups or hydrogen will migrate with their bonding electrons so these processes may be described as 1,2-hydride or -alkyde shifts. The term Wagner–Meerwein rearrangement is often applied to these cationic isomerizations.

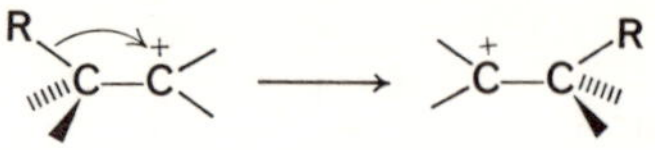

For example, reactions expected to yield the neopentyl cation, (84) usually lead to products which have the *t*-pentyl carbon skeleton. This is so typical that it may be used as a diagnostic test for a carbonium ion mechanism.

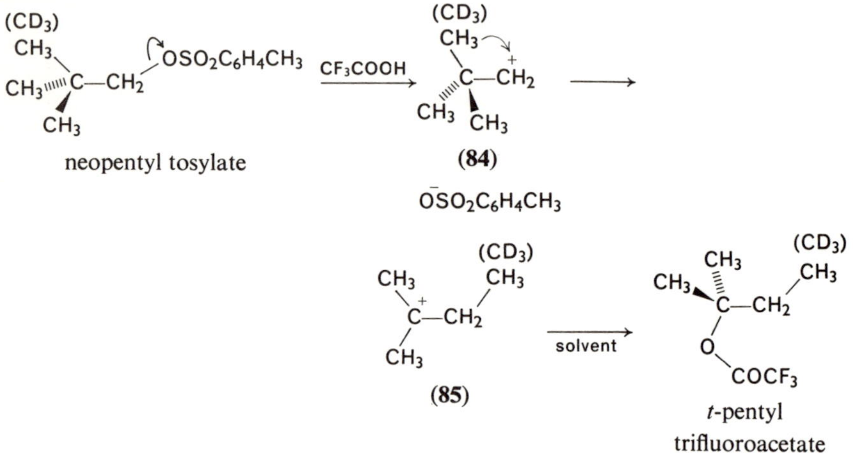

neopentyl tosylate

(84)

(85)

t-pentyl trifluoroacetate

In the above examples no effect on the rate is observed if deuterium is incorporated into the migrating methyl group. It is inferred that the rearrangement is not rate determining but that it occurs rapidly after the formation of an intermediate neopentyl cation to form initially the far more stable *t*-pentyl cation, (85). Similarly, the dehydration under acidic conditions of 'pinacolyl alcohol' (86) leads to olefins with a rearranged carbon skeleton. Reactions

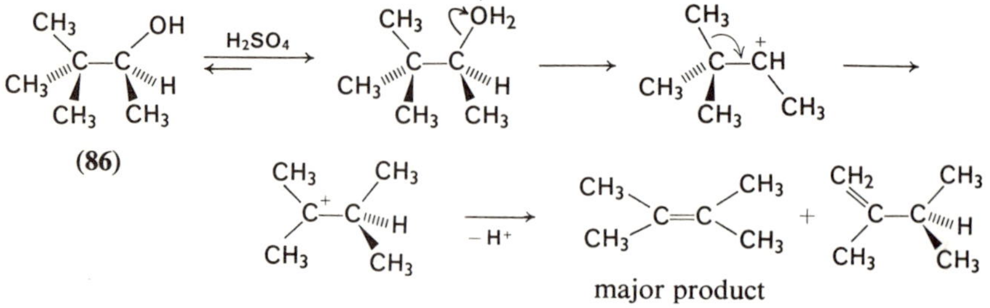

(86)

major product

involving the 1-propyl cation often give products with the isopropyl skeleton. Friedel–Crafts alkylation of benzene with 1-propyl chloride and aluminium chloride leads to the formation of isopropylbenzene. The reaction of a

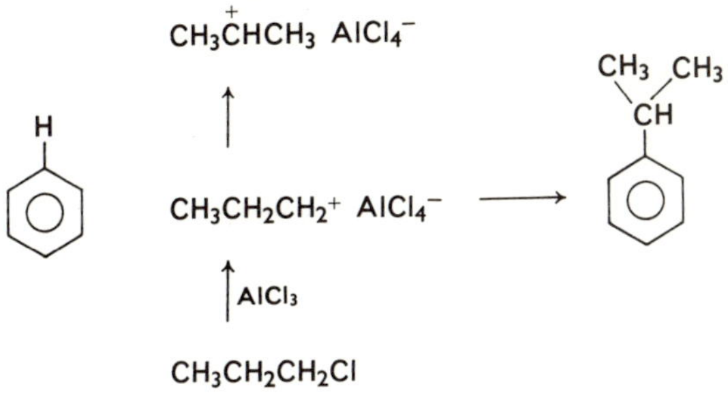

$$CH_3\overset{+}{C}HCH_3 \ AlCl_4^-$$

$$CH_3CH_2CH_2^+ \ AlCl_4^- \longrightarrow$$

$$\uparrow AlCl_3$$

$$CH_3CH_2CH_2Cl$$

1-propylaryltriazene with carboxylic acid gives variable amounts of 1- and iso-propyl esters in the products.

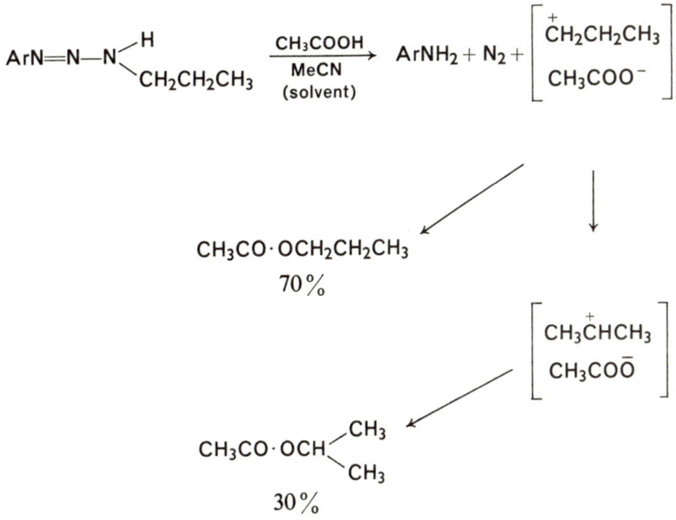

$$ArN{=}N{-}N\overset{H}{\underset{CH_2CH_2CH_3}{\diagdown}} \quad \xrightarrow[\substack{MeCN \\ (solvent)}]{CH_3COOH} \quad ArNH_2 + N_2 + \left[\begin{array}{c} \overset{+}{C}H_2CH_2CH_3 \\ CH_3COO^- \end{array}\right]$$

$$CH_3CO\cdot OCH_2CH_2CH_3$$
$$70\%$$

$$\left[\begin{array}{c} CH_3\overset{+}{C}HCH_3 \\ CH_3CO\bar{O} \end{array}\right]$$

$$CH_3CO\cdot OCH\overset{CH_3}{\underset{CH_3}{\diagup}}$$
$$30\%$$

    Bicyclic compounds including terpenes provide a rich variety of examples of the Wagner–Meerwein rearrangement; e.g. the acid-catalyzed dehydration of fenchol (**87**) produces four isomeric olefins (fenchenes). In all cases, final stabilization of the cation occurs by proton loss. The rearrangement of pinacols (tetra-substituted ethylene glycols), **88**, in acid solution and pinacolic de-amination of corresponding β-hydroxyamines, (**89**) are further closely related

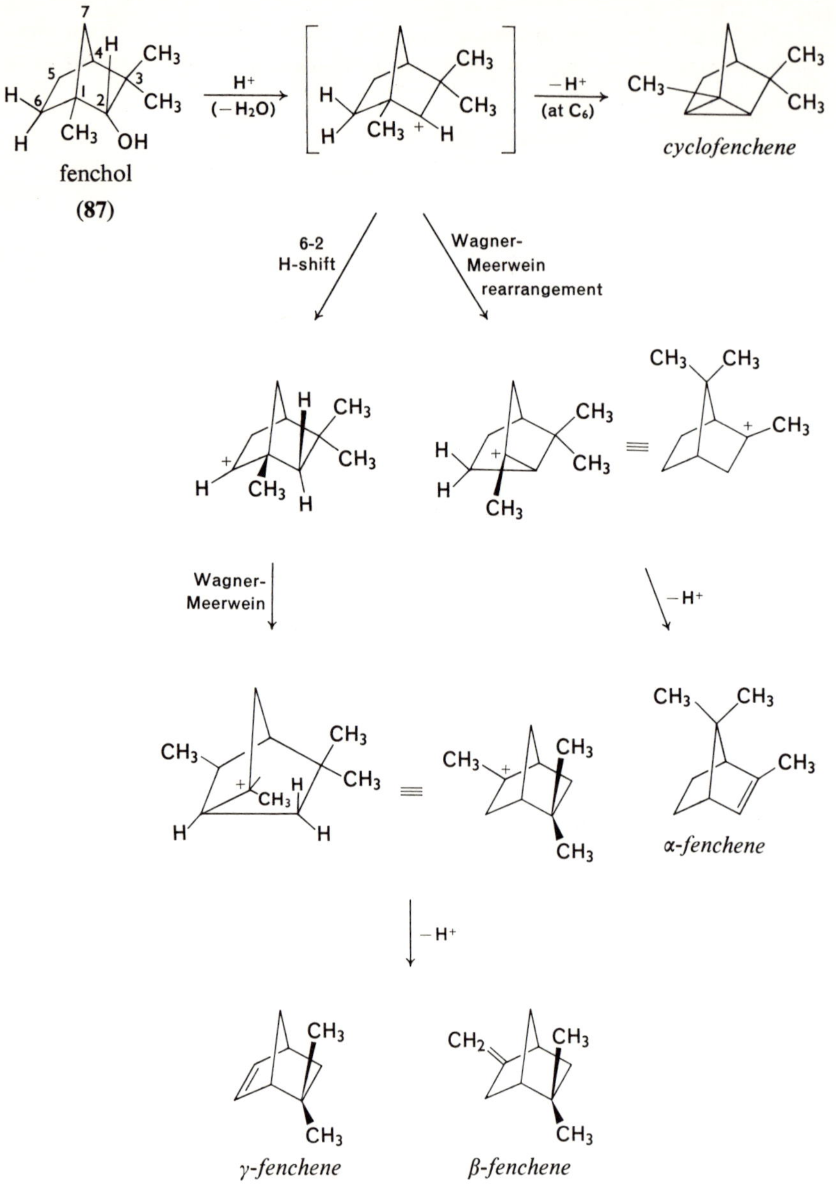
fenchol
(87)
H+
(−H2O)
−H+
(at C6)
cyclofenchene
6-2
H-shift
Wagner-
Meerwein
rearrangement
Wagner-
Meerwein
−H+
−H+
α-fenchene
γ-fenchene
β-fenchene

examples of the Wagner–Meerwein rearrangement. In the following case a tertiary cation rearranges to an oxocarbonium ion which is of greater stability.

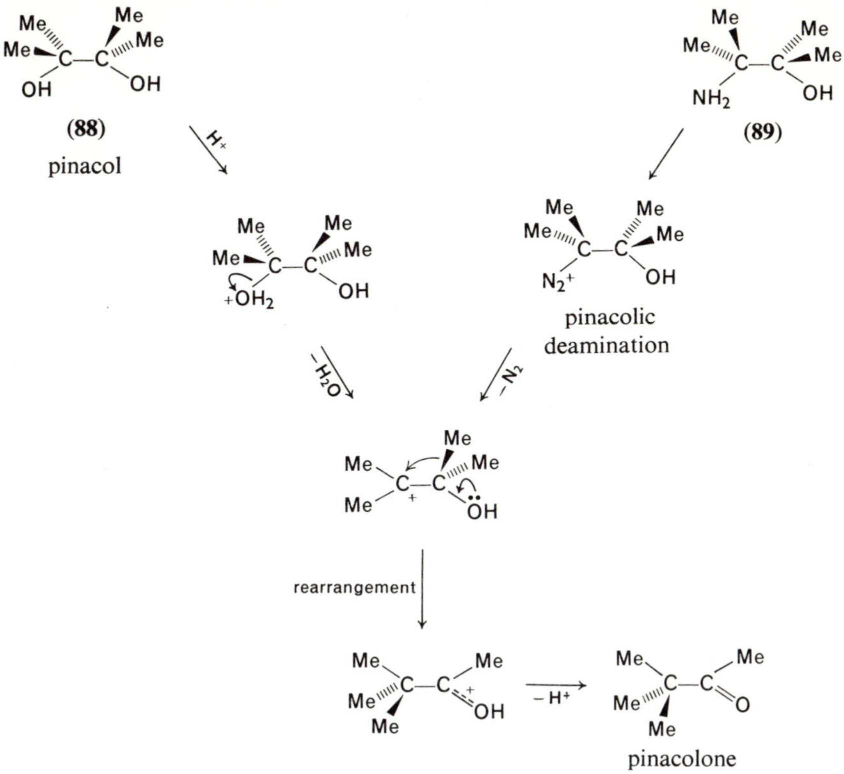

(88)

pinacol

(89)

pinacolic
deamination

rearrangement

pinacolone

Where a choice of migrating group is available, the order of preference ('migratory aptitude') is: aryl > alkyl > hydrogen.

Carbonium Ion Dynamics and Mechanisms of Rearrangement

The extraordinary stability of many simple carbonium ions in superacid media coupled with the use of both proton and ^{13}C n.m.r. has enabled a surprising picture of carbonium-ion structure to be built up. As has been indicated (p. 143) allylic cations show restricted rotation about the partial double bonds (e.g. **60** and **62**). Rotation will occur at a sufficiently high temperature and the barrier for conversion of *cis, trans* to *trans, trans*-dimethylallyl cations (**90–91**) has been shown to be 100 kJ mol^{-1}. The mechanism could be a direct twisting of the bond or it could involve coordination of an anion (e.g. F$^-$) followed by rotation and loss of the anion. This large activation barrier, however, represents a *minimum* value for coordination of an anion to

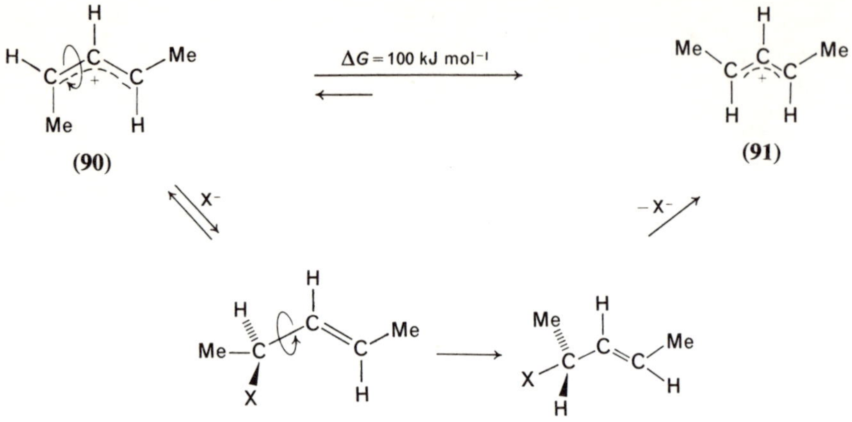

$\Delta G = 100 \text{ kJ mol}^{-1}$

(90)

(91)

X⁻

−X⁻

a carbonium ion in superacid solution. Such high barriers to coordination and to elimination of a β-proton provide ideal conditions for the study of rearrangements.

The isopropyl cation (92), the simplest carbonium ion which may be directly observed, shows an n.m.r. spectrum at 0–40° consisting of broad resonances indicative of rapid exchange between the two types of proton. One might explain this by a series of 1,2-hydride shifts, the 1-propyl cation being a transient intermediate (92–92a). If, however, one of the carbon atoms is

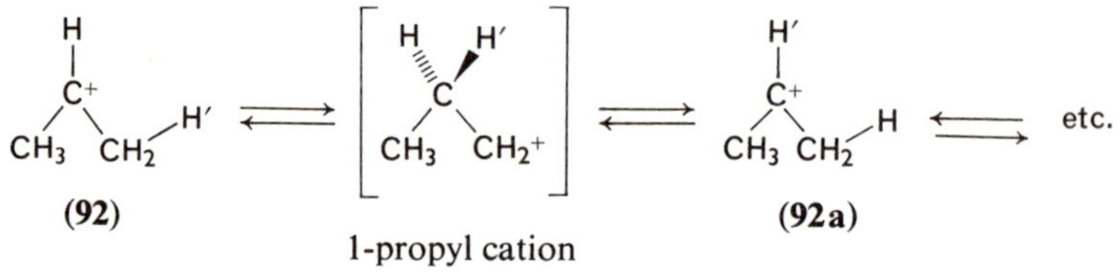

(92)

1-propyl cation

(92a)

etc.

labelled with ^{13}C, it is clear from the spectrum that the carbons also are freely interchanging. The above mechanism does not permit of an explanation of this result.

The introduction of a further intermediate (93), which may be described as 'protonated cyclopropane', can provide an explanation of both carbon and hydrogen interchange since the latter may exchange across the ring while the ring may open in any of three directions. Compelling evidence for the intermediacy of protonated cyclopropanes in the rearrangements of carbonium ions is accumulating though their precise structure is not settled; the species

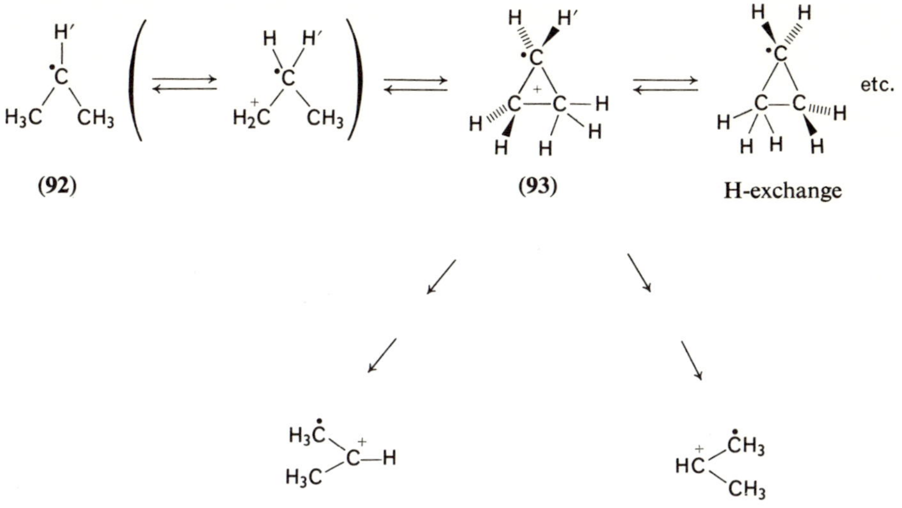

(92)   (93)

H-exchange

C-exchange

are variously represented as 'edge-protonated' (94) or 'corner-protonated', (95) structures. If these are truly intermediates, rather than transition states, they are related to the penta-coordinate carbon compounds described in Section 2.30 and should not be regarded as exhibiting carbon with five full covalencies.

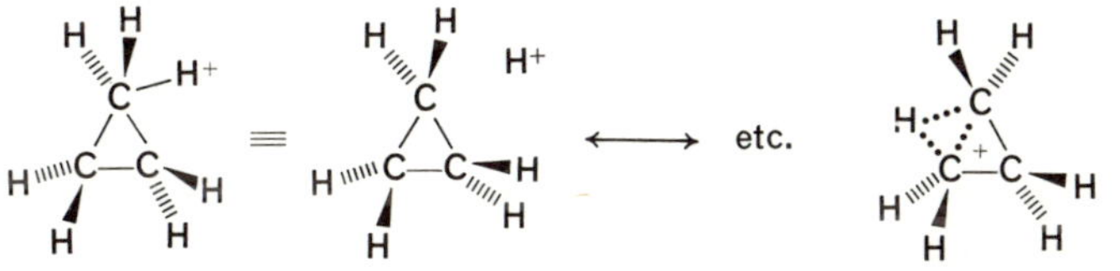

(95)   (94)

'corner-protonated'   'edge-protonated'

Careful addition of 2-butyl chloride to superacid at $-110°$ will yield the 2-butyl cation (96) stable for long periods. Several rearrangement processes have been revealed by n.m.r. spectroscopy. At all temperatures there is a rapid inter-change of protons at C_2 and C_3 (2,3 H-exchange) and there is a second route whereby all protons are exchanged within the ion with an activation energy of only 30 kJ mol^{-1}. Postulation of protonated methylcyclopropane (97) as intermediate explains these results. Finally, rearrangement to the t-butyl cation (98), a stable species, occurs.

$$CH_3\overset{\underset{\displaystyle |}{Cl}}{CH}CH_2CH_3 \xrightarrow[-110°]{SbF_5} CH_3\overset{+}{C}HCH_2CH_3 \underset{\longleftarrow}{\overset{2,3\ H\text{-exchange}}{\longrightarrow}} CH_3CH_2\overset{+}{C}HCH_3$$

$$(96)$$

$$\begin{array}{ccc}
\underset{\displaystyle \underset{H_3}{C}}{\overset{\displaystyle CH_3}{\underset{\diagdown}{CH}}\!\!-\!\!CH_2} & \underset{etc.}{\rightleftarrows} & CH_3\!\!-\!\!\underset{\displaystyle CH_2}{\overset{+}{CH}}\!\!-\!\!CH_3 & \longrightarrow & \overset{\displaystyle CH_3}{\diagdown}\underset{\displaystyle {}^+CH_2}{\overset{\displaystyle \diagup CH_3}{CH}}
\end{array}$$

$$(97)$$

$$\overset{\displaystyle CH_3\ \ CH_3}{\underset{\displaystyle CH_3}{\overset{\diagdown \diagup}{C^+}}}$$

$$(98)$$

In a similar way, the *t*-heptyl cation exchanges methyl groups at moderately high temperatures (**99–100**) and the three isomeric *t*-hexyl cations (**101–103**) coexist in equilibrium in superacid solution in equal proportions indicating their similar energies. The dynamic equilibria thus revealed may equally be explained in terms of protonated cyclopropane intermediates (**104**) and (**105**). Alternative mechanisms would require primary and secondary cations (**106–108**) to be formed intermediately and the 'protonated cyclopropanes' to be transition states. In the 2-butyl cation rearrangement it has been calculated that the intermediacy of primary cations would more than double the observed activation energy. The rapid degenerate rearrangement* of the *t*-heptyl cation (**99**) involving methyl-group migration occurs very rapidly and n.m.r. data set a maximum energy of activation of 25 kJ mol^{-1} whereas it is known that a corresponding secondary heptyl cation would be 45–60 kJ mol^{-1} higher in energy than the tertiary ion. It seems likely that protonated cyclopropanes are involved widely in Wagner–Meerwein rearrangements but perhaps not universally, at least in media other than superacids.

If the neopentyl cation, formed under solvolytic conditions, were to be transformed to *t*-pentyl via a protonated cyclopropane (**109**) one would expect hydrogen scrambling (revealed by deuterium labelling) to occur between the migrating group and the migration terminus and also that these two carbons

---

* A 'degenerate' rearrangement is one in which reactant and product are formally identical species, though individual atoms will have interchanged.

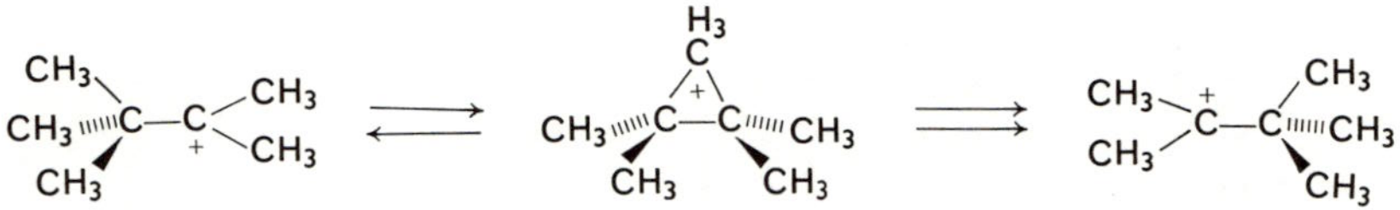

(99) (100)

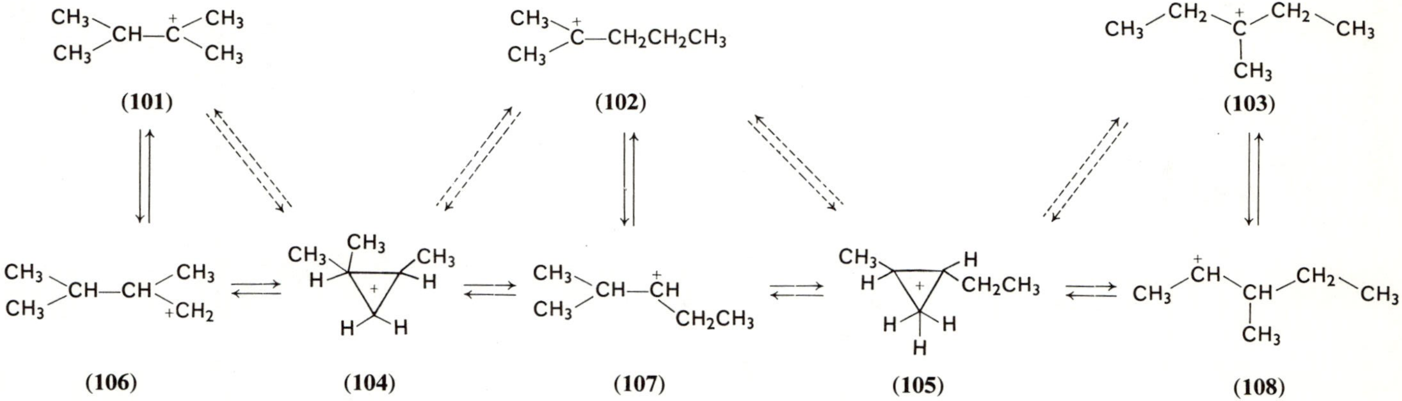

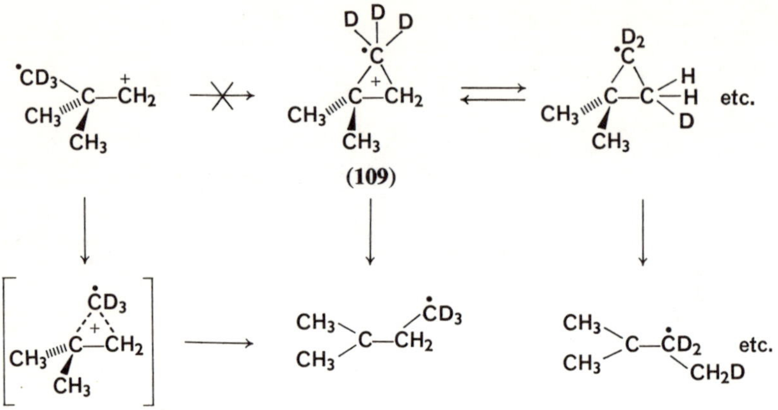

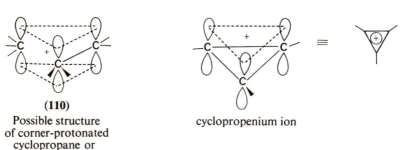

become equivalent. This does not appear to be the case. Under these conditions, either the protonated cyclopropane proceeds to product without ever becoming symmetrical and at a rate too fast for proton scrambling to occur, or else it should be regarded as a transition state. It should be pointed out that the corner-protonated species (**110**) has some analogy with a cyclopropenium ion (Section 2.19) and no doubt derives stability from the pseudoaromatic character it possesses by virtue of its two delocalized electrons.

(110)
Possible structure
of corner-protonated
cyclopropane or
transition state

cyclopropenium ion

Further Carbonium Ion Rearrangements

Bicyclo[2,2,2] and [3,2,1]octyl cations are converted into the bicyclo[3,3,0]octyl cation, **111**:

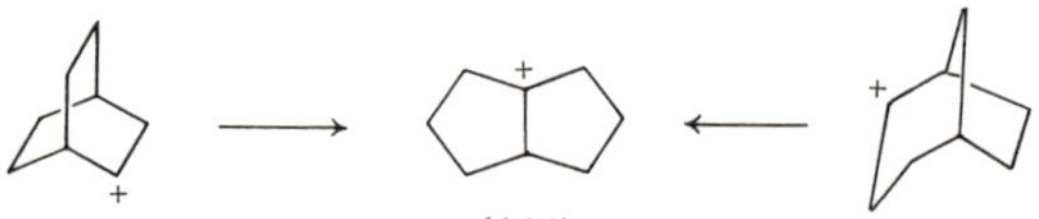

(111)

And similarly with the methyl homologue, **112**:

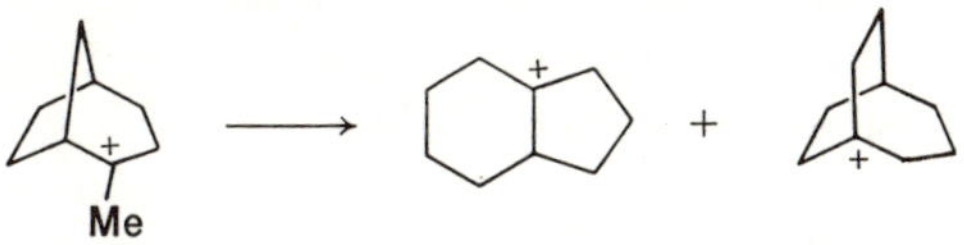

(112)

The interesting feature of these rearrangements is the apparent preference to form bridgehead cations (even from tertiary precursors) which are likely to be non-planar and formed with difficulty by solvolysis. The homocubyl cation, **113**, undergoes a rapid degenerate rearrangement so that in a short period of time all carbons have exchanged and become equivalent and a single proton-resonance signal is observed. A more complex example is the 9-barbaralyl

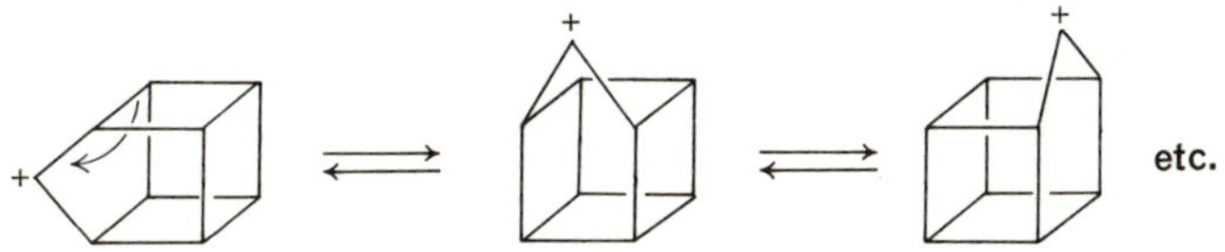

etc.

(113)

cation (**114**) in which all carbons rapidly interchange by a combination of Wagner–Meerwein and Cope rearrangements.†

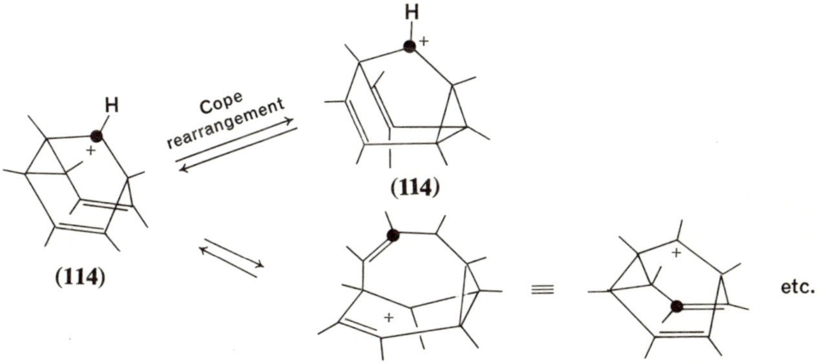

(114)

(114)

(114a)

etc.

† The Cope rearrangement is the concerted interconversion of two hexa-1,5-diene structures,

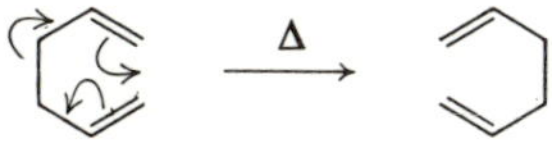

Rapid rearrangements can be revealed by a study of ^{13}C n.m.r. spectra. The
t-hexyl cation (115) has two magnetically distinct methyl groups (α and β)
giving two resonances. If by using a separate, tunable, radiofrequency source,
the frequency corresponding to one of these signals is strongly irradiated
('saturated'), it no longer absorbs since all the nuclei are promoted to the
upper energy state. Furthermore, the resonance due to the other methyl group
which was not irradiated also diminishes. This indicates that the two methyls
are rapidly interchanging; a protonated cyclopropane intermediate has been
suggested as intermediate in this rearrangement. The cyclopentyl cation shows

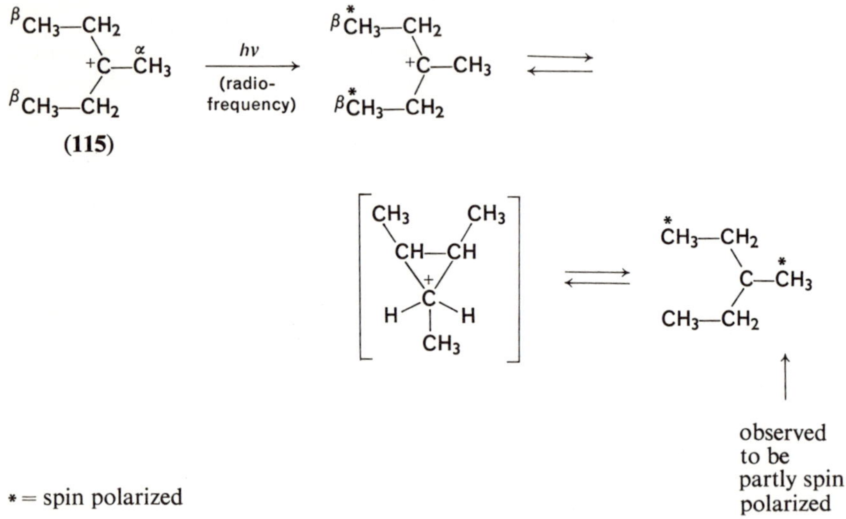

* = spin polarized

a single ^{13}C resonance but split into a 10-line multiplet due to coupling with all
nine protons to an equal extent. The degenerate rearrangement, 116 ⇌ etc.,
makes the protons equivalent. The acid-catalysed interchange of aldehydes

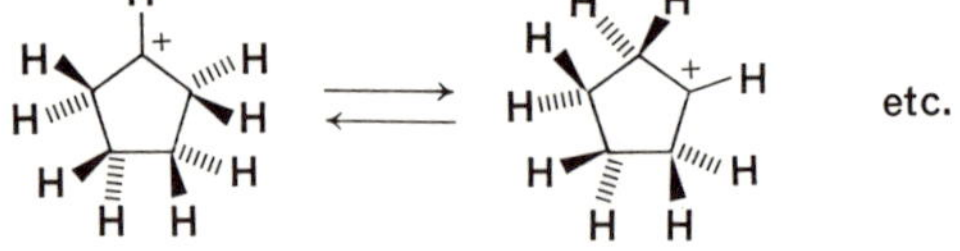

(116)   (116a)   etc.

and ketones proceeds via a series of 1,2-shifts within the protonated carbonyl
compound.

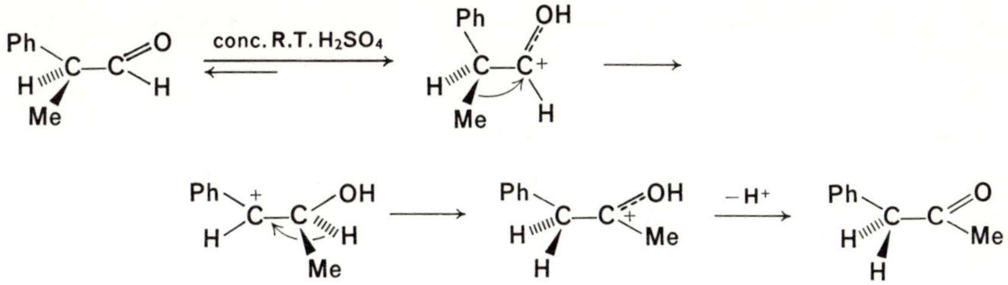

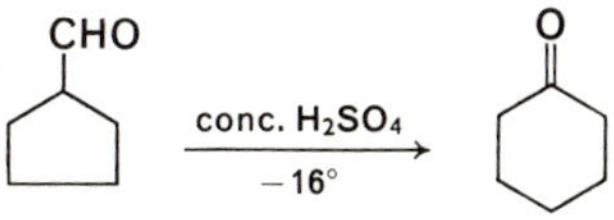

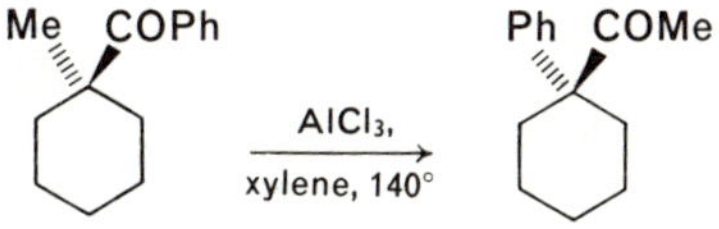

Electrocyclic Rearrangements

The cyclizations of linear conjugated carbonium ions to cyclic isomers with one less double bond are electrocyclic processes (Section 1.19) whose stereo-specificity is the result of orbital symmetry control. Thus, the interconversion of allyl and cyclopropyl cations is disrotatory in the ground state while the pentadienyl-cyclopentenyl rearrangement is conrotatory. The directions of rotation to effect ring closure are apparent from the symmetries of the highest filled orbitals of the conjugated systems;

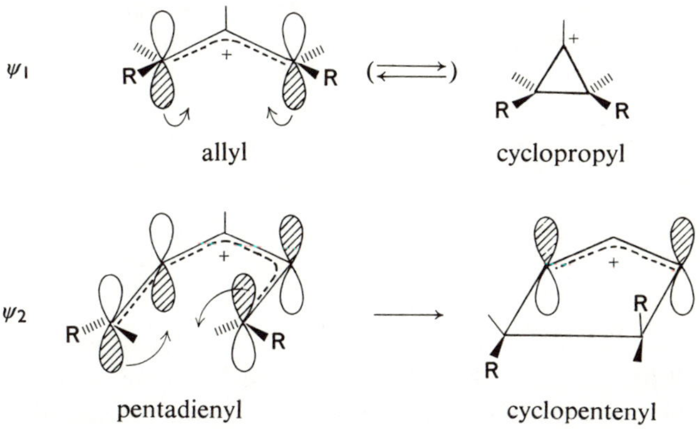

ψ_1 allyl   ($\rightleftharpoons$)   cyclopropyl

ψ_2 pentadienyl   cyclopentenyl

These relationships may be compared with analogous processes in conjugated carbanions (p. 277). The first process occurs spontaneously in the direction of the open system. The three stereo-isomers of 2,3-dimethyl-1-chlorocyclopropane (117a, b, c) ionize in superacid to give three distinct geometrical isomers of the dimethylallyl cation (118), stable under these conditions and distinguishable by their n.m.r. spectra. Ring opening must commence as

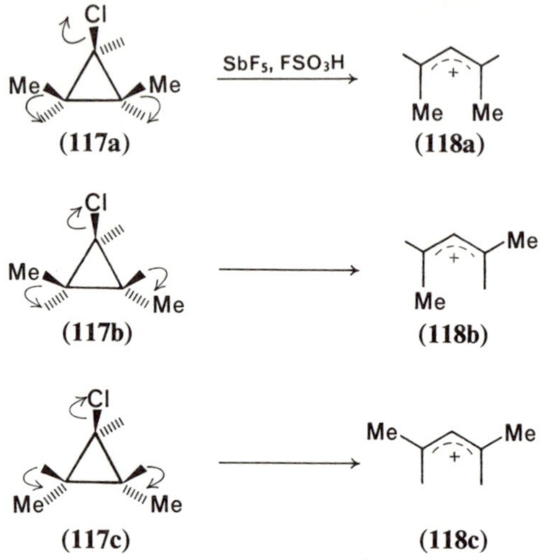

ionization proceeds, without an actual cyclopropyl cation being formed otherwise the same cyclopropyl cation would be formed from 117a and 117b leading then to a mixture of the corresponding allyl cations (118a and 118b) (since there are two disrotatory modes of rotation).

In the homologous series, cyclopentenyl cations form from the linear isomers. An early example is the formation of cyclopentenone (119) from divinyl ketone (120) under the influence of acid. The carbinol (121) ionizes in

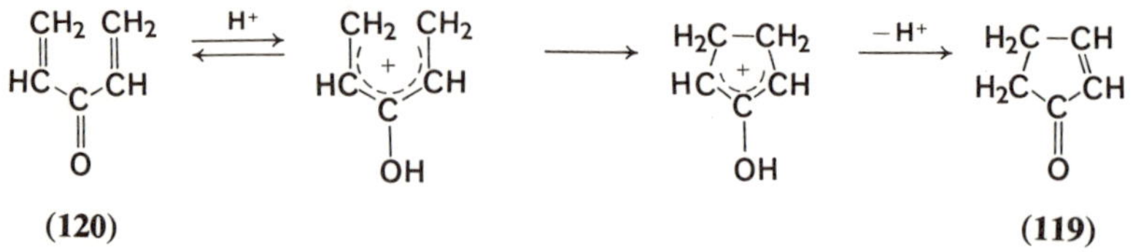

strong acid to the highly labile carbonium ion (122) which very rapidly cyclizes (half-life in superacid at −55° approximately 2 minutes), the ultimately stable product being (123). The opposite stereochemical restraints are enforced

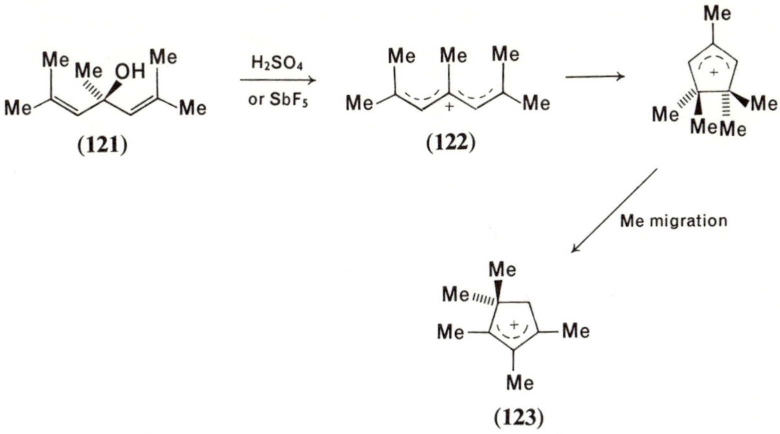

(121) (122) (123)

Me migration

upon these reactions when the reagents are in excited electronic states—i.e. under photochemical conditions. Benzenonium ions—cyclic pentadienyl cations—cannot, on account of the methylene bridge, twist in a disrotatory fashion and are stable in superacid. Upon irradiating with ultraviolet light they cyclize by the now permitted conrotatory mode as the highest occupied orbital is Ψ_3.

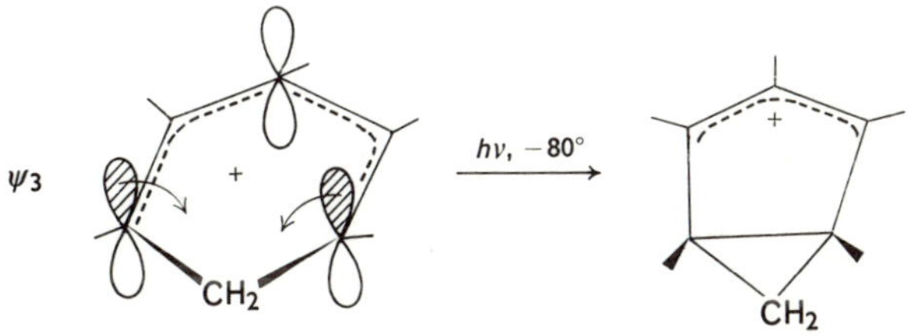

Sigmatropic Rearrangements

The migration of an atom or group to an unsaturated centre from a site adjacent is termed a 'sigmatropic' process of which the Wagner–Meerwein rearrangement is the simplest. These reactions, which are orbital symmetry controlled, are by no means limited to cationic rearrangements but a variety of sigmatropic rearrangements are found to occur in conjugated carbonium ions. The type of process is denoted by two numbers signifying the chain lengths of the migrating group and of the framework along which migration occurs, both between migration origin and migration terminus, thus;

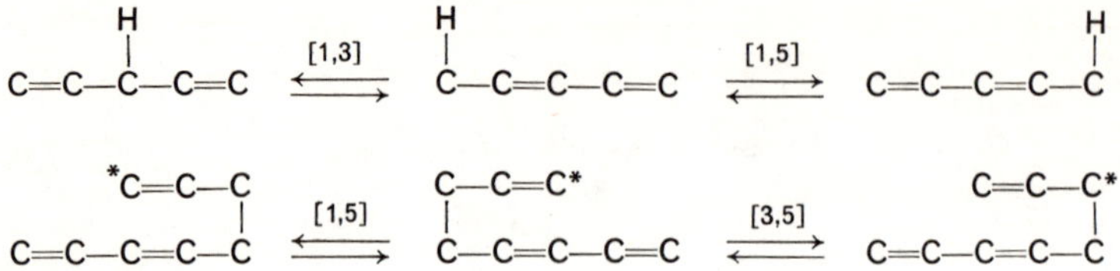

The following possibilities are realized in the rearrangements of the allylcyclohexadienol (**124**) which, in the presence of acid is converted to the allyltoluenes, (**125 a**, **b**, **c**) via the benzenonium ion (**126**).

(**124**) (**126**) (**126**)

(**125a**) (**125b**) (**125c**)

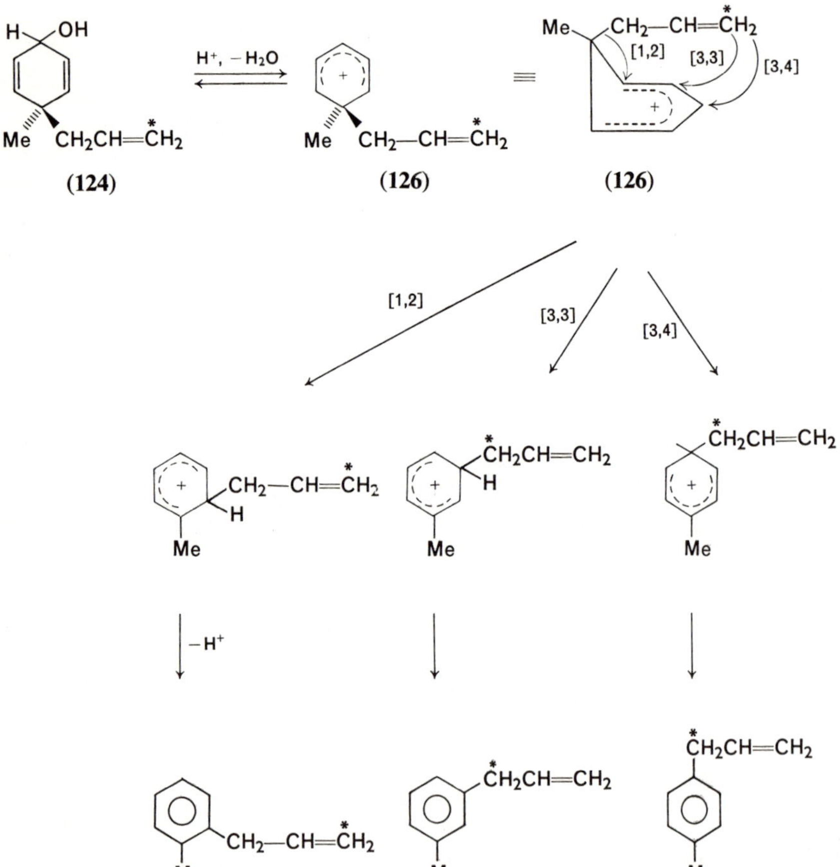

Benzenonium ions exhibit considerable mobility of the protons and substituent methyl groups. The parent ion (127) has four distinct proton resonances at −130° indicative of a static structure (on the n.m.r. time scale) while at −70° this has collapsed to a singlet signal due to the rapid interchange of the protons by [1,2] or other sigmatropic rearrangements. The most stable

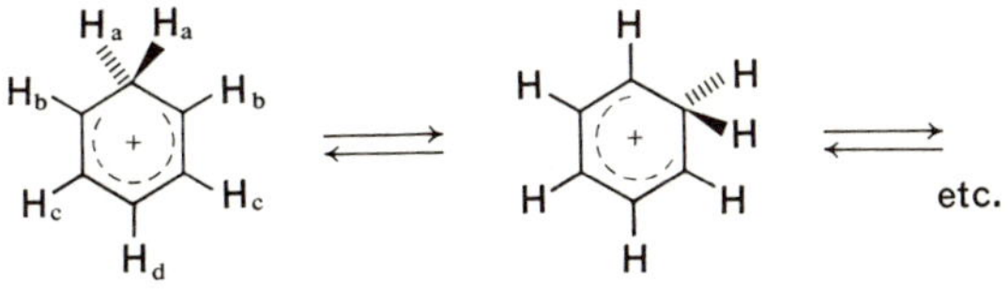

(127)

isomers of methylated benzenonium ions have the methyl groups in the 2, 4, or 6 positions where they are best able to stabilize the structure by their meso-meric effect. Methyl migrations frequently occur at observable rates and lead to more stable isomers. Presumably a series of 1,2-shifts of hydride and

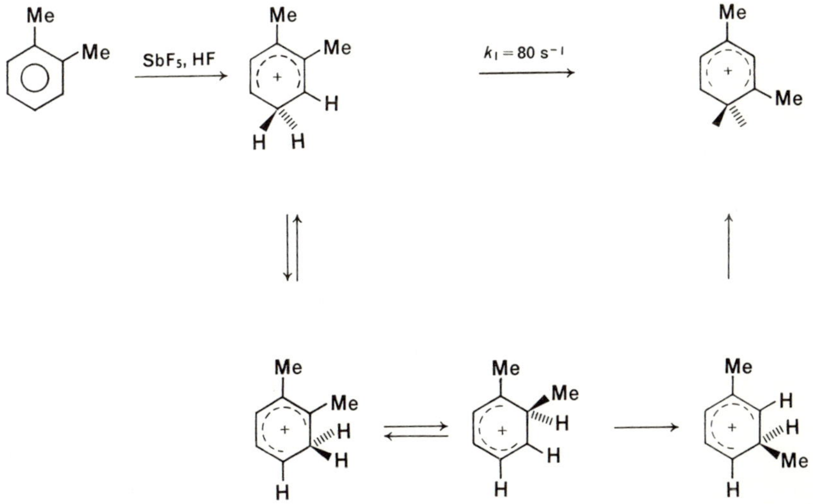

methide groups is required to cause this isomerization. Isomerization of the methylated benzene can thus occur and can be studied by n.m.r. spectroscopy.

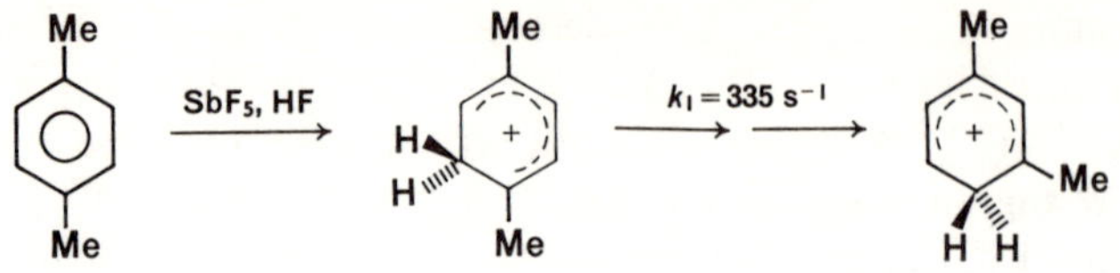

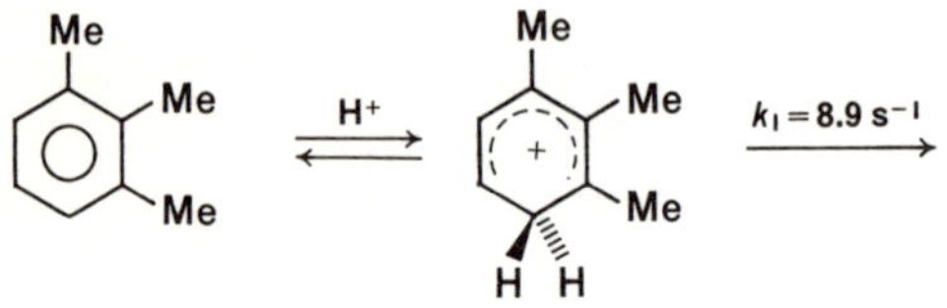

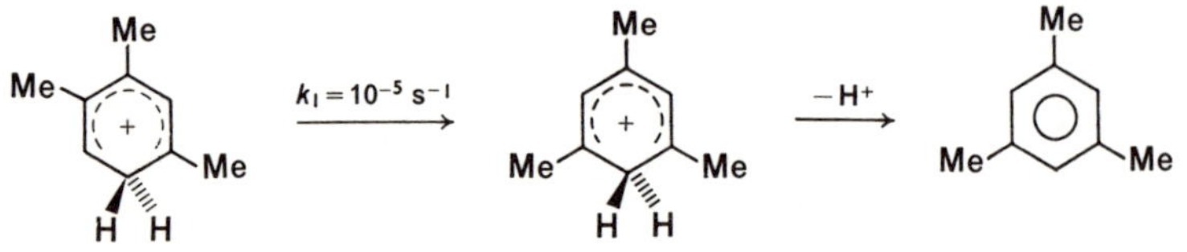

Similar rearrangements have long been known to occur during substitution of alkylated benzenes (the Jacobsen rearrangement),

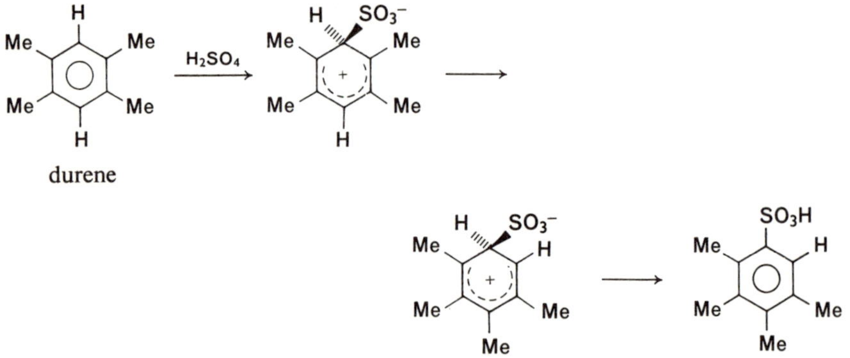

durene

and the isomerizations of methylbenzenes by BF_3 (e.g. 1,2,3- and 1,2,4-trimethyl to 1,3,5-, the most basic isomer). Benzenonium ions will isomerize to bicyclo[3,1,0]hexyl cations (**128–129**) when irradiated with ultraviolet light; the latter undergo a degenerate rearrangement whereby all carbons of the five-membered ring become equivalent.

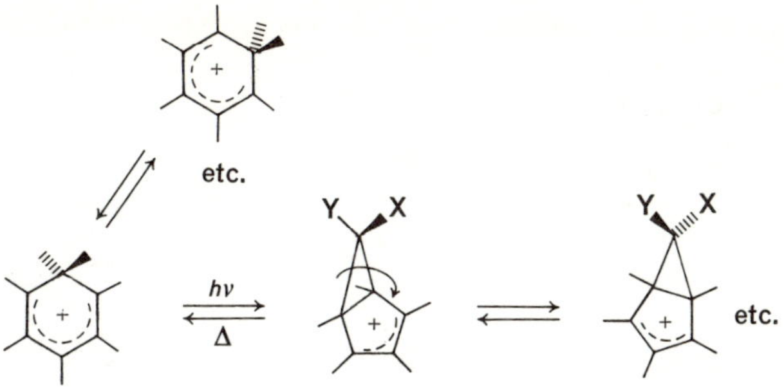

(128)   **(129)**   **(129a)**

The *Dienone–Phenol Rearrangement* is a sigmatropic rearrangement of a cyclohexadienone, e.g. (130) proceeds via an intermediate benzenonium ion. This undergoes a 1,2 shift of one of the groups at the tetrahedral carbon,

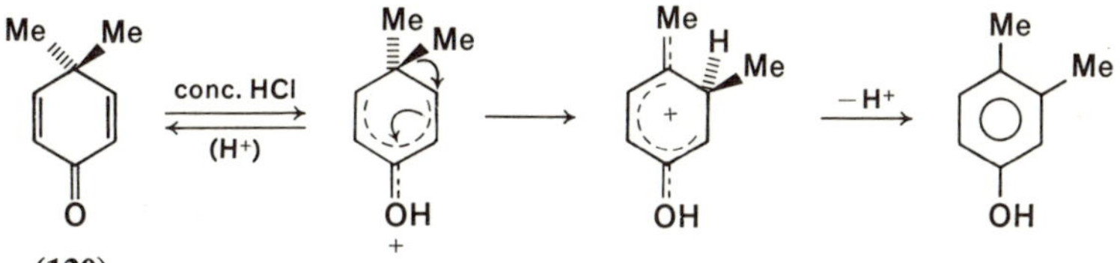

(130)

Quite startling skeletal rearrangements can result, for example, in the steroid field, (131).

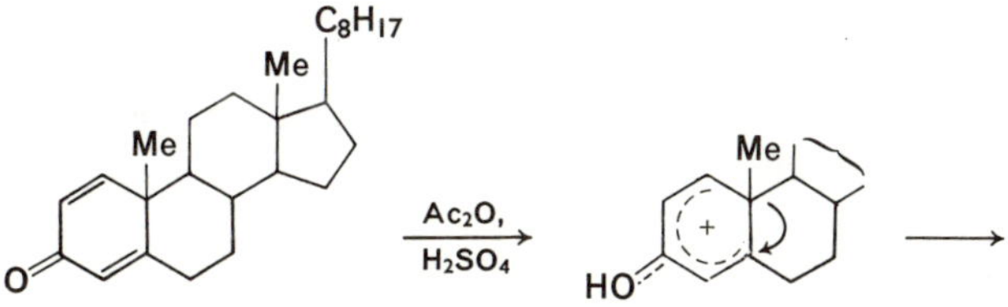

(131)

cholesta-1,4-dien-3-one

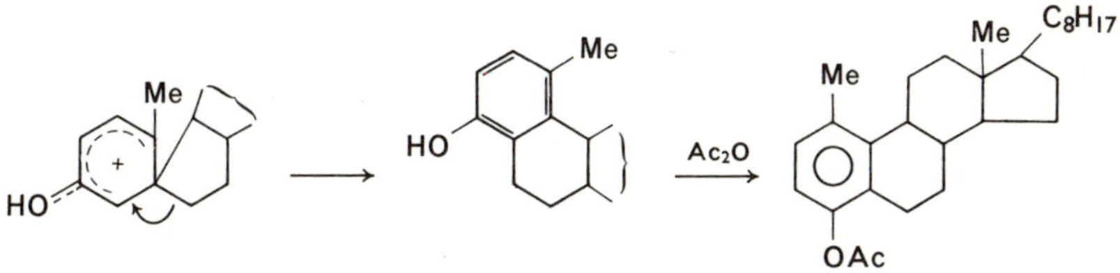

Vinyl cations may also rearrange by a 1,2-shift across the double bond,

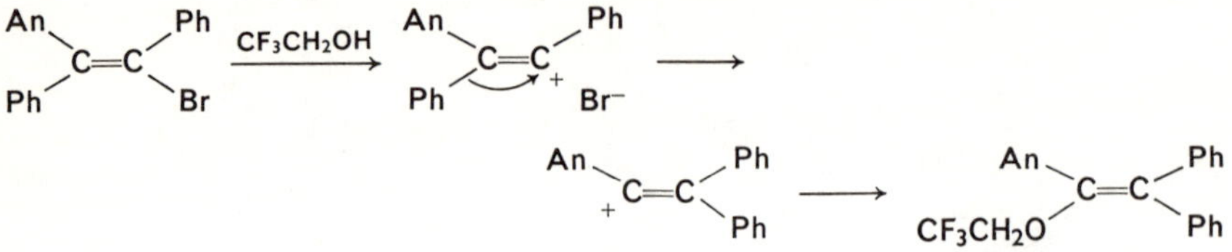

$$\begin{array}{c}\text{An}\\\end{array}\!\!\!\diagdown\!\!\!\!\begin{array}{c}\\\text{Ph}\end{array}$$

or from a neighbouring saturated centre if a more stable cation results:

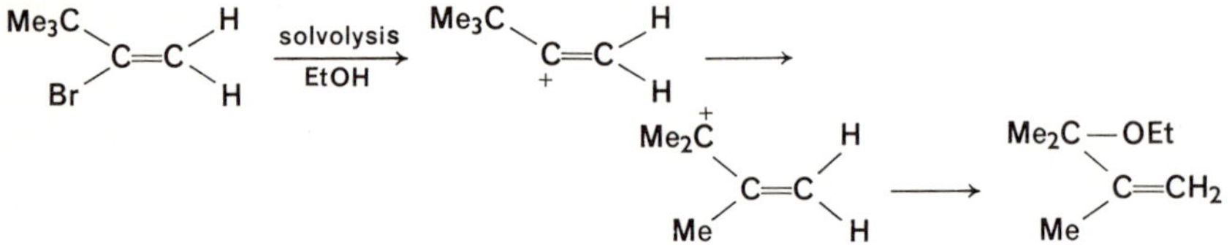

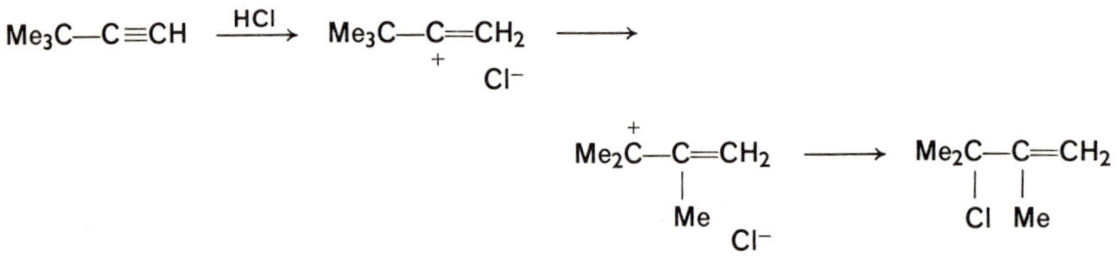

Anionotropy

Double-bond shifts are commonplace when allylic and polyenylic cations are involved (S_N1' process) since the attack of a nucleophile may occur at any of the alternate carbon atoms of the conjugated system which share the positive charge. Rearrangements of this kind, however, are not necessarily diagnostic

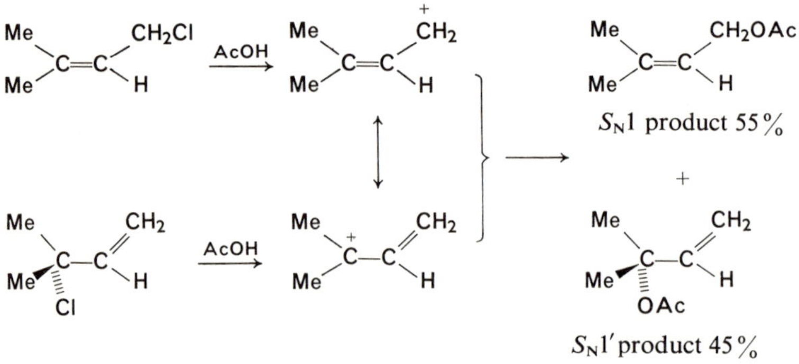

S_N1 product 55%

S_N1' product 45%

of carbonium ions and may also occur by a concerted bimolecular mechanism (designated S_N2').

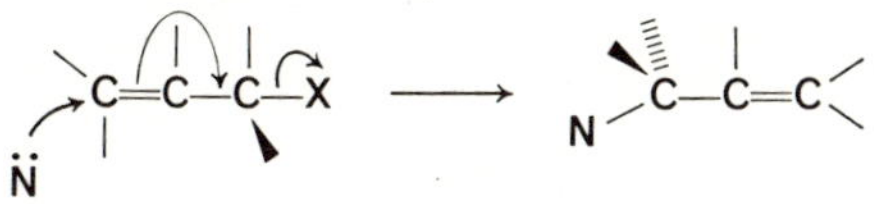

S_N2' mechanism

This is an example of an *anionotropic* rearrangement in which a nucleophilic group becomes detached from carbon and later rejoins another part of the conjugated carbonium ion. This type of rearrangement may occur in unsaturated alcohols (132) as to bring double bonds into conjugation.

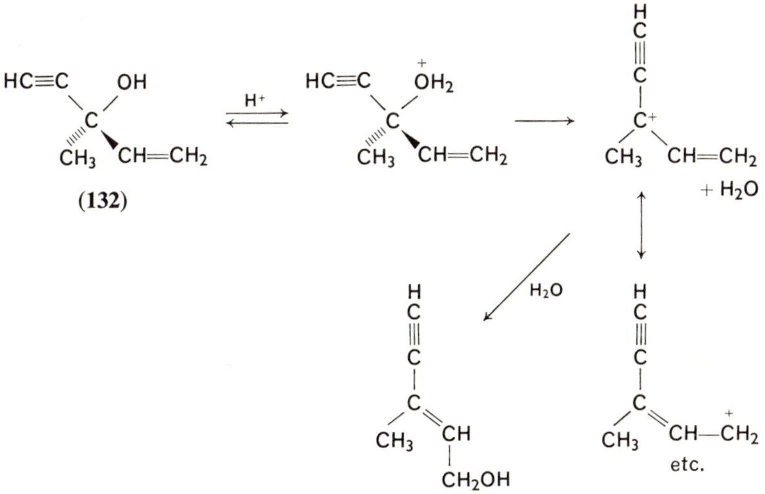

(132)

Loss of a Group other than Hydrogen at the β-Carbon

Fission of a carbon–carbon bond in a carbonium ion can occur to give an olefin and a second cation particularly when this is stabilized by heteroatoms with lone pairs located in the γ-position. This is the major decomposition

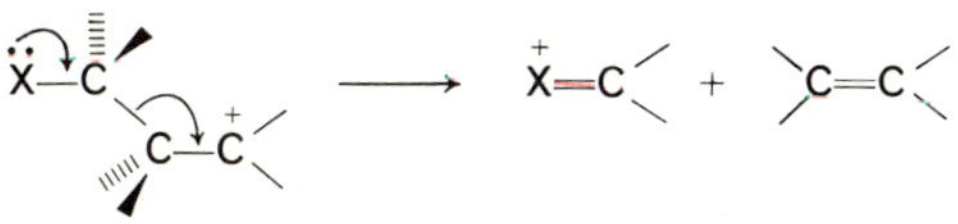

pathway of (133) formed from the corresponding chloride in aqueous solution at much the same rate as the model compound 134, which cannot fragment in

$$Me_2\overset{..}{N}-CH_2 \quad \overset{Me}{\underset{CH_2-C}{}} \overset{Me}{\underset{Cl}{}} \xrightarrow[\text{slow}]{\text{aq. EtOH}} Me_2\overset{..}{N}-CH_2 \quad \overset{Me}{\underset{CH_2-C\overset{+}{}}{}} \overset{Me}{} \longrightarrow$$

(133)   Cl⁻

fragmentation   solvolysis

$$Me_2CH-CH_2 \quad \overset{Me}{\underset{CH_2-C}{}}\overset{Me}{\underset{Cl}{}} \qquad Me_2\overset{+}{N}=CH_2 + H_2C=CMe_2 \qquad Me_2N-CH_2 \quad \overset{Me}{\underset{CH_2-C}{}}\overset{Me}{\underset{OH}{}}$$

(134)

$$Me_2NH + CH_2O$$

50%   50%

the same way but confirms that the γ-amino group only comes into action in the subsequent fast step. Ketones may undergo fission in strong acid media if a potentially stable carbonium ion is attached to the β-carbon (**135**).

$$\overset{Me}{\underset{Me}{}}C\underset{\overset{\|}{O}}{}-CH_2-C\overset{Me}{\underset{Me}{\overset{Me}{}}} \xrightarrow{HSO_3F} \overset{Me}{\underset{Me}{}}C\underset{\overset{\|}{+OH}}{}CH_2-C\overset{Me}{\underset{Me}{\overset{Me}{}}} \longrightarrow$$

(135)

$$\overset{Me}{\underset{OH}{}}C=CH_2 \quad + \quad \overset{Me}{\underset{Me}{}}\overset{+}{C}-Me$$

$$\overset{Me}{\underset{\overset{\|}{O}}{}}C\overset{Me}{}$$

Redox Reactions

Carbonium ions in oxidation reactions are treated in Chapter 9. In general, these reactions involve hydride-transfer steps either from or to a carbonium ion. A simple example is the conversion of cycloheptatriene to the tropyllium ion by a more reactive carbonium ion such as the triphenylmethyl cation (**46**). The trityl cation is reduced and the cycloheptatriene oxidized. Hydride abstraction by a triarylmethyl cation from a variety of donors has been shown to occur, particularly from primary and secondary alcohols:

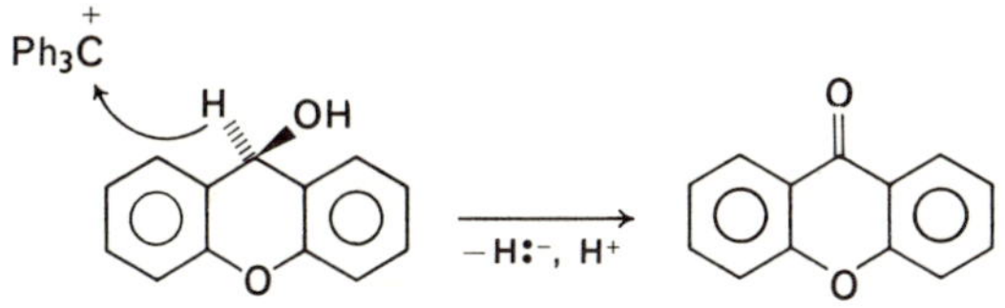

There is a formal similarity between these reactions and carbonyl disproportionations (Section 8.9) such as the Cannizzaro reaction in which an aldehyde or ketone may be considered to act as a carbonium oxide, **136** and accept an hydride ion:

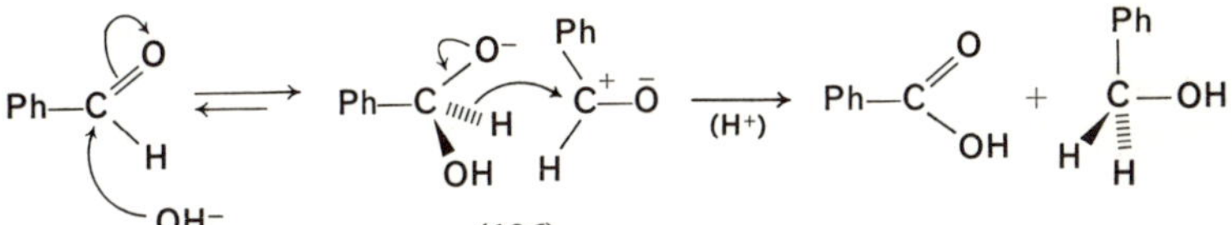

$$2PhCHO \xrightarrow{(OH^-)} PhCOOH + PhCH_2OH$$

In principle, this amounts to the coordination of a nucleophile to the carbonium ion, but a nucleophile (H^-) which is still bound in a covalent bond. Good hydride donors such as borohydride, BH_4^-, and aluminohydride, AlH_4^-, readily reduce carbonium ions and serve as 'trapping' reagents for these intermediates if they can be generated in the neutral media which the reagents require.

2.30   PENTACOORDINATE CARBONIUM IONS

In superacid solution, electrophilic reactions can occur even at saturated C—H bonds as in paraffin hydrocarbons. For instance, hydrogen–deuterium exchange will occur and also protolysis, the breaking of carbon–carbon bonds. The work of Olah and of Hogeveen have indicated that an intermediate must be involved in which five groups (hydrogen or alkyl) are coordinated to one carbon. The deuterium exchange of methane, it is suggested, takes place as follows:

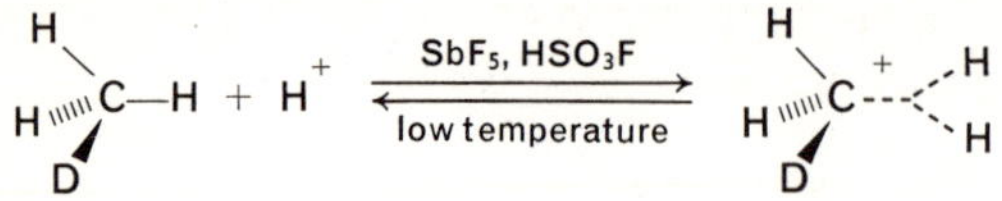

(137)

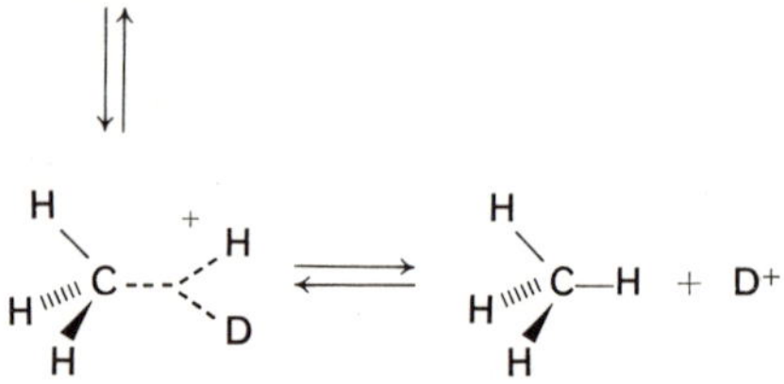

(139)   (138)

The bonding in the pentacoordinate carbonium ion (137) is likely to require a three-centre bond (broken lines)—a two-electron bond which interacts with three nuclei as in the boron hydrides (139). The alternative structures (138) etc. may interchange rapidly so that all hydrogens become equivalent. Intermediates of this type are considered to be common in superacid chemistry of hydrocarbons.

The exchange reaction of paraffins with molecular hydrogen has been realized and is likely to be analogous:

$$\Delta H = 75 \text{ kJ mol}^{-1} \longrightarrow \Delta S = -38 \text{ J K}^{-1}$$

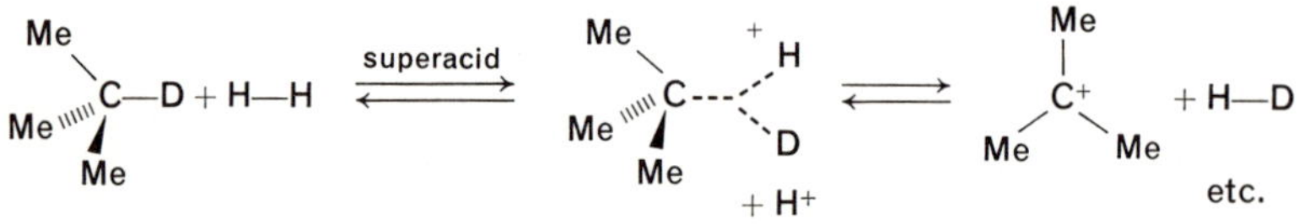

$$\Delta H = 53 \text{ kJ mol}^{-1} \qquad \Delta S = -105 \text{ J K}^{-1}$$

Reduction of alkyl cations is quite general on treatment with hydrogen in superacid solution; secondary cations are reduced 10^5–10^6 times more rapidly than tertiary, e.g.

$$Me_3C^+ + H\text{—}H \longrightarrow Me_3C\text{—}H + H^+ \quad k_2 = 3 \times 10^{-3} \text{ l mol}^{-1} \text{ s}^{-1}$$

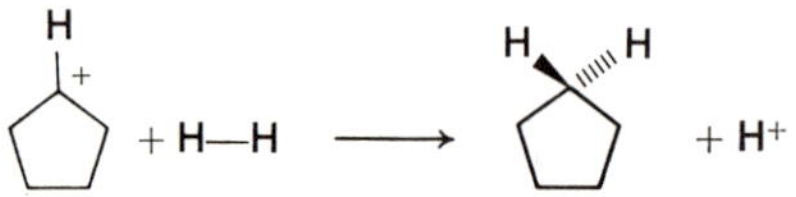

$$+ H^+ \quad k_2 = 900 \text{ l mol}^{-1} \text{ s}^{-1}$$

Protolysis of neopentane (**140**) leads to the formation of the *t*-butyl cation via a pentacoordinate intermediate, **141**. The reverse of this, alkylation of

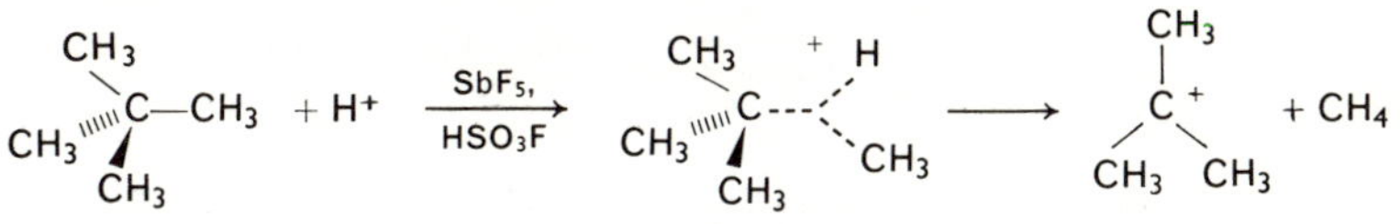

(**140**)   (**141**)

$$\Delta H^{\ddagger} = 89 \text{ kJ mol}^{-1}, \; \Delta S^{\ddagger} = -25 \text{ J K}^{-1}$$

paraffins by alkyl cations will occur though it is not a favourable reaction; thus, the preformed *t*-butyl cation will insert into isobutane via **142** to give about 2% 2,2,3,3-tetramethylbutane (**143**). Alkylation by the less sterically-hindered

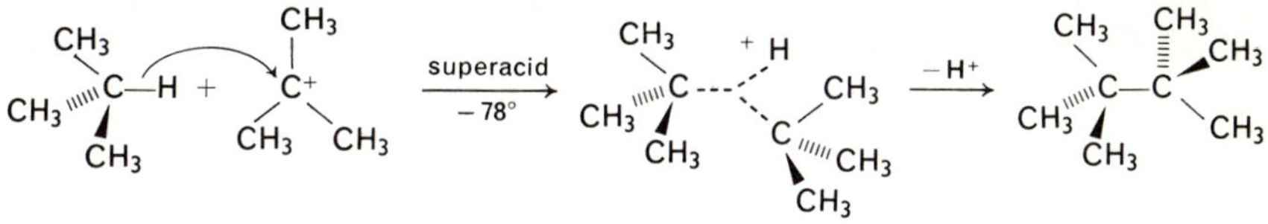

(**142**)   (**143**)

2%

+ other products

and more electrophilic isopropyl cation gave 12% of the alkylated paraffin, **144**. Other electrophilic reactions at alkanes such as nitration (by $NO_2{}^+PF_6{}^-$) and

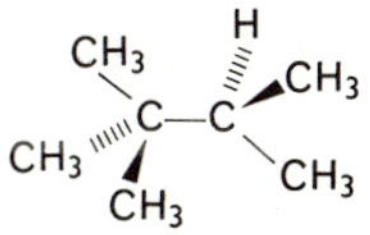

(**144**)

chlorination (by Cl_2–$AlCl_3$ in the dark) have also been achieved in low yields and may in future become important. The characteristics of the reactions, and presumably the intermediates, appear to be similar to those of the previous examples. At the time of writing, no pentacoordinate carbonium ions have been detected spectroscopically in solution but the ions, e.g. $CH_5{}^+$, are well known in the mass spectrometer.

Evidence is available that these electrophilic substitutions take place by front-face attack with retention of configuration. Consider the following rates of hydride (:H^-) and methide (:$CH_3{}^-$) abstraction by the proton and the

$$\begin{array}{ll}
\text{Me}\\
\;\;\;\;\;\;\diagdown\\
\text{Me}\,\text{\tiny{||||}}\text{C}\!-\!\text{CH}_3\\
\;\;\;\;\;\diagup\\
\text{Me}
\end{array}
\begin{array}{l}
a \xrightarrow[\text{superacid}]{\text{H}^+} \overset{+}{\text{C}}\text{Me}_3 + \text{CH}_4 \qquad \overset{k_{rel}}{10^4}\\[2ex]
b \xrightarrow[\text{superacid}]{\overset{+}{\text{C}}\text{Me}_3} \overset{+}{\text{C}}\text{Me}_3 + \text{CH}_3\!-\!\text{CMe}_3 \;\; 1
\end{array}$$

$$\begin{array}{ll}
\text{Me}\\
\;\;\;\;\;\;\diagdown\\
\text{Me}\,\text{\tiny{||||}}\text{C}\!-\!\text{H}\\
\;\;\;\;\;\diagup\\
\text{Me}
\end{array}
\begin{array}{l}
c \xrightarrow[\text{superacid}]{\text{H}^+} \overset{+}{\text{C}}\text{Me}_3 + \text{H}_2 \qquad\qquad 1\\[2ex]
d \xrightarrow[\text{superacid}]{\overset{+}{\text{C}}\text{Me}_3} \overset{+}{\text{C}}\text{Me}_3 + t\text{Bu}\!-\!\text{H} \quad 10^6
\end{array}$$

t-butyl cation. The point to note here is the low rate of reaction *b* which is probably due to steric hindrance since the *t*-butyl cation is evidently no weak electrophile (as shown by reactions *c* and *d*). Steric hindrance would be expected to be severe in a front-face attack (**145**) but much less in the alternative rear-face attack which would lead to inversion (**146**). It is no doubt significant

$$\begin{array}{c}
\text{Me}\;\;\;\;\;\;\;\;\;\overset{+}{}\\
\;\;\;\;\diagdown\;\;\;\;_{\text{.-}}\text{CMe}_3\\
\text{Me}\,\text{\tiny{||||}}\text{C} \text{-----}\\
\;\;\;\;\diagup\;\;\;\;^{\text{..}}\text{Me}\\
\text{Me}
\end{array}
\qquad\qquad
\begin{array}{c}
\text{H}\\
|\;\;\;\;_+\\
\text{Me}_3\text{C}\text{---}\text{C}\text{---}\text{CMe}_3\\
\;\;\;\;\;\diagup\!\!\diagdown\\
\text{H}\;\;\;\text{H}
\end{array}$$

$$\qquad\qquad\textbf{(145)}\qquad\qquad\qquad\qquad\qquad\textbf{(146)}$$

that other electrophilic reactions at saturated carbon (e.g. metal–metal displacements, Section 4.22) occur by a front-face attack of the electrophile.

2.31   CARBONIUM IONS IN THE GAS PHASE

Thermal reactions in the gas phase do not in general produce carbonium ions since, in the absence of solvent to stabilize ionic fragments, homolytic processes are more favourable energetically. It is possible both to produce carbonium

$$\text{CH}_3{}^+ \;\; :\!\text{CH}_3{}^- \xleftarrow[\text{heterolysis}]{\;\;\;\times\;\;\;} \text{CH}_3\!-\!\text{CH}_3 \xrightarrow[\text{homolysis}]{\text{ca } 700°} \text{CH}_3\!\cdot\; + \;\cdot\text{CH}_3$$

ions in the gas phase and to study their reactions by means of the mass spectrometer. A stream of vapour of an organic compound is admitted at about 10^{-6} Torr to an ionizing chamber where interaction of the molecules with a stream of electrons occurs. A significant fraction of the organic molecules are ionized by loss of an electron and the cation radicals formed are accelerated away by an electric field. These may undergo subsequent decomposition to give smaller fragments including carbonium ions which are detected by an electrometer after mass sorting by a variable electromagnet and characterized by their ratio of mass to charge, m/e (e is usually +1, sometimes +2). A few examples

$$X\!-\!Y \xrightarrow{\;e^-\;} (X\!-\!Y)^+ \longrightarrow X^+ + Y\cdot$$

$$+\,2e \qquad\qquad \text{detected} \quad \text{not detected}$$

will give an idea of the scope of this technique. A long-chain hydrocarbon after ionization tends to lose a succession of ethylene fragments, also terminal methyl groups, giving rise to a series of ions 14 mass units apart. With branched-

$$CH_3CH_2CH_2CH_2CH_2CH_3 \xrightarrow{-CH_3} CH_3CH_2CH_2CH_2CH_3{}^+ \longrightarrow CH_3CH_2CH_3{}^+$$

$$\xrightarrow{-C_2H_4} CH_3CH_2CH_2CH_3{}^+ \longrightarrow CH_3CH_3{}^+ \qquad \text{etc.}$$

chain examples, decomposition to a tertiary carbonium ion is more favoured than that leading to secondary or primary. Substituted benzophenones (**147**) form a molecular ion which decomposes to a benzyl cation and thence to a phenyl cation:

(**147**)

m/e 91

m/e 77

The relative favourabilities of paths a and b (as shown by the abundance of the ions detected) depends upon the substituent, electron-donating (+M) substituents favouring a. In fact the relative abundance of the phenyl cation ($m/e = 77$) and of the benzoyl cation ($m/e = 91$), though see below, increase linearly with the Hammett σ-value for the substituent ($\rho = 1$), showing that at least in some instances, concepts from solution chemistry are applicable.

The decomposition of molecular ions from aromatic molecules which contain carbon side-chain, lead to a large fragment at $m/e = 91$. This undergoes subsequent decomposition with loss of C_2H_2. If hydrogen is replaced specifically in the molecule by deuterium or carbon by ^{13}C, the C_2H_2 fragment appears to split out randomly indicating that all hydrogens and all carbons are equivalent. This indicated that $m/e = 91$ is probably the tropyllium ion rather than the benzyl cation, although this rearrangement is not known in solution. The structures of ions in the mass spectrometer must not be assumed to be

$$C_5H_4D^+ + C_2H_2 \quad 5$$
$$\overline{}$$
$$C_5H_5^+ + C_2HD \quad 2$$

those directly derived from the precursor. Another aromatic structure, the cyclopropenium ion **148** has been proposed for the fragment at $m/e = 39$ which occurs in the mass spectra of many compounds.

Cyclic decompositions are common, e.g. the loss of an olefin fragment from an ester recalls the thermal decomposition of the neutral molecule:

(148)

2.32   ION-CYCLOTRON RESONANCE (i.c.r.)

An ion, produced in the gas phase by electron bombardment and electro-statically accelerated, follows a circular trajectory in a perpendicular magnetic field. If a complete circular pathway is allowed (compared with just a sector as in the mass spectrometer), the instrument is, in effect, a cyclotron and the motions of ions are described by the usual equations, namely;

$$\omega_c = ne/m$$

$$r = V/\omega_c$$

where ω_c is the frequency of rotation of the ion in the circular path of radius r; V is the velocity of the ion, ne its charge and m its mass.

Interaction of radiation (frequency v) with the rotating ion results in absorption of energy at the resonance condition,

$$\omega_c = v$$

It will be noticed that ω_c and hence v is independent of V; on absorbing further quanta of energy, the ion merely moves into a larger orbit. The ions present in the cyclotron may be detected in terms of their mass/charge ratio by detecting the resonances with applied radiofrequency radiation, so that i.c.r. spectra give essentially the same information as the mass spectrometer. However, since the primary ions are able to spend a far longer time in the cyclotron than in the mass spectrometer, the probability of collisions with neutral molecules in the atmosphere of the instrument is greatly enhanced. Reactions between molecules and ions may be studied by injecting ions into the cyclotron which contains a small pressure of the reagent molecules. New ions are formed and may be detected by their characteristic m/e values. The following examples illustrate the use of this technique.

Ions derived from Methyl Cyanide

The following scheme of reactions has been proposed for reactions of the methyl cyanide ion with the neutral molecule and other species. Numbers refer to m/e for the species.

$$CH_3CNH^+ \xleftarrow{(H^+)} CH_3CN \longrightarrow CH_3\overset{+}{C}N \xrightarrow{CH_3I} CH_3CNCH_3 + I\cdot$$

$m/e = 42$ 41   56

$\downarrow Me_2C=O$ $\downarrow (-H)$

Me—C—Me + MeCN $CH_2\overset{+}{C}N \xrightarrow{CH_3CN} C_3H_4N^+ + HCN$

+ OH

59   40   54

$\downarrow (-H)$

$$CH\overset{+}{C}N \xrightarrow{CH_3CN} C_3H_3N^+ + HCN$$

39   53

Isomeric Cations

It is sometimes possible to differentiate between isomeric ions by their characteristic reactions. Among the isomers of $C_2H_5O^+$, the oxo-carbonium ion, $CH_3\overset{+}{O}{=}CH_2$ (formed by electron bombardment of CH_3OCH_2R) typically undergoes Me^+ and H^- transfers to neutral molecules,

$$CH_3OCH_2CH_3 \xrightarrow{\ e^-\ } \underset{m/e\,=\,45}{CH_3\overset{+}{O}{=}CH_2} \xrightarrow{\ CH_3OCH_2CH_3\ } \underset{59}{CH_3OCH_3 + CH_3OCHCH_3}\ \ 94\%$$

$$\underset{61}{CH_3\overset{+}{O}\underset{CH_3}{\overset{CH_3}{\diagdown\!\!\diagup}}}\qquad 6\%$$

The isomers, $CH_3\overset{+}{\underset{H}{C}}{=}\overset{+}{O}H$ (protonated acetaldehyde) and $CH_2{-}CH_2$ with $\overset{+}{O}H$ bridging

(protonated ethylene oxide), undergo exclusively proton-transfer reactions.

$$CH_3\overset{+}{\underset{H}{C}}{=}\overset{+}{O}H + ROH \longrightarrow CH_3CHO + R\overset{+}{O}H_2$$

Negative as well as positive ions may be studied by i.c.r. The gas-phase acid dissociations of alcohols have been measured in this way.

Gas-Phase Acidities

Electron bombardment of a mixture of two alcohols in the presence of a small amount of water vapour (all at ca 10^{-5} mm pressure) initiates the following sequence of ion–molecule reactions:

$$H_2O + e^- \longrightarrow H^- + OH^{\bullet}$$
$$H^- + H_2O \longrightarrow OH^- + H_2$$
$$ROH + OH^- \longrightarrow RO^- + H_2O$$
$$ROH + H^- \longrightarrow RO^- + H_2$$

The intensities of the different ions, RO^-, produced in this way give a measure of the gas-phase acidity properties, which may be markedly different from those in a condensed phase where solvation masks the properties of an ion. The i.c.r. technique gives the acidity order, tBuOH > iPrOH > EtOH >

MeOH > H_2O; in solution, water is a much stronger acid than any of the alcohols.

2.33 NITRENIUM IONS

The nitrogen analogues of carbonium ions would contain divalent nitrogen with a positive charge. Such species, known as nitrenium ions are relatively unimportant in chemistry but nonetheless have been shown to occur as reactive intermediates. The expected pathway, analogous to carbonium-ion formation, would involve the heterolysis of a weakly nucleophilic leaving group from nitrogen:

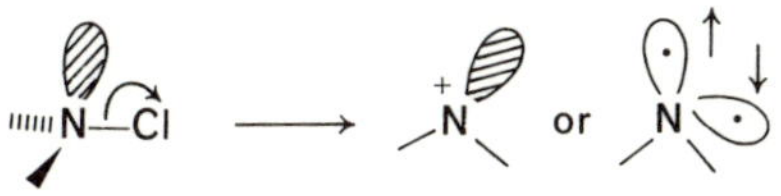
N
Cl
N
or
N

Thus, the compound **149** appears to form a nitrenium ion which undergoes a Wagner–Meerwein rearrangement and a bicyclo[3,2,1]octyl derivative (**150**) is formed. Similar rearrangements occur in the azanorbornane series (**151–152**).

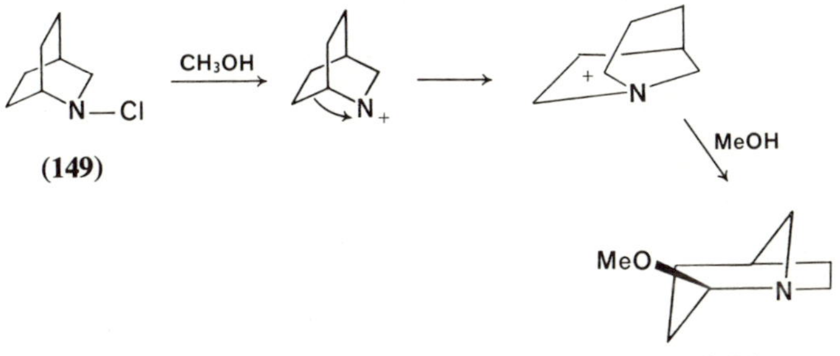
N—Cl
CH₃OH
N
N
MeOH
MeO
N
(149)

(150)

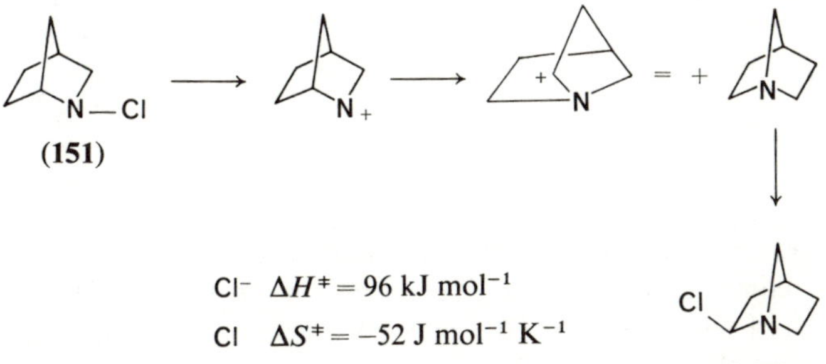
N—Cl
(151)
N
N
=
N
Cl⁻ ΔH‡ = 96 kJ mol⁻¹
Cl ΔS‡ = −52 J mol⁻¹ K⁻¹
Cl
N

(152)

The activation parameters are typical for an S_N1 process in which the slow step is the separation of oppositely charged ions.

N-chloroaziridines (153) have also been postulated to solvolyse via nitrenium ions (154). In this case, the intermediate is analogous to a cyclopropyl cation (155), which is known to undergo symmetry-allowed disrotatory ring opening to the allyl cation (156) (p. 179). The same process occurs in the cyclic nitrenium ion and an aza-allyl cation (157), which when formed hydrolyses to

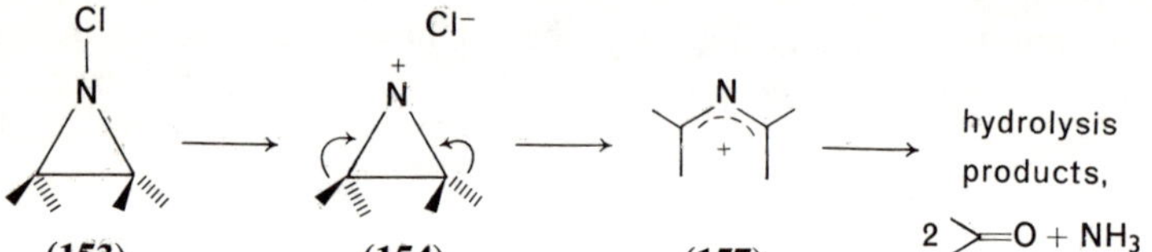

two carbonyl fragments and ammonia. At least one addition of a nitrenium ion to a double bond has been observed:

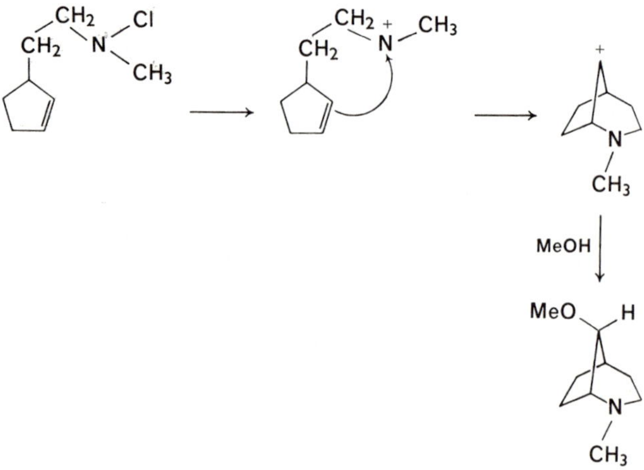

This method offers an interesting and valuable route to the formation of azabicyclic compounds. Whereas the nitrenium-ion interpretation of these reactions has been given here, the possibility remains that they are radical processes: homolytic fission of N-hal bonds is very facile.

SUGGESTIONS FOR FURTHER READING

General texts, see Chapter 1

D. Bethell and V. Gold, *Carbonium Ions*, Academic Press, New York, 1967.

D. M. Brouwer and H. Hogeveen, 'Electrophilic substitutions at alkanes and in alkylcarbonium ions', *Progr. Phys. Org. Chem.*, **9**, 179 (1972).

R. J. Gillespie and T. E. Peel, 'Superacid systems', *Adv. Phys. Org. Chem.*, **9**, 1 (1971).

L. M. Jackman and S. Sternhall, *Applications of NMR Spectroscopy in Organic Chemistry*, Vol. 5, Chap. 3–9, Pergamon, London, 1969.

G. A. Olah, 'Stable carbonium ions in solution', *Science*, **168**, 1298 (1970).

G. A. Olah, 'Mechanism of Electrophilic Aromatic Substitutions' *Accounts Chem. Res.* **4**, 240 (1971).

G. A. Olah, 'The electron donor single bond in organic chemistry', *Chem. Brit.*, **8**, 282 (1972).

G. A. Olah and P. von R. Schleyer, *Carbonium Ions*, 5 volumes, Wiley–Interscience, New York, 1970.

A. Streitwieser, *Solvolytic Displacement Reactions*, McGraw-Hill, New York, 1962.

Chapter 3

Neighbouring Group Participation and Non-Classical Ions

3.1 PARTICIPATION BY UNSHARED ELECTRON PAIRS

Substituent groups may affect reaction rates and equilibria by perturbing the electronic environment of the reaction centre—inductive and mesomeric effects. The former particularly are known to fall off rapidly with distance as illustrated by the dissociation constants of the cyano acids:

	K_A	K_A (relative)
CH_3COOH	1.7×10^{-5}	1
$NCCH_2COOH$	3.55×10^{-3}	209
$NCCH_2CH_2COOH$	1.01×10^{-4}	6
$NCCH_2CH_2CH_2COOH$	3.66×10^{-5}	2.1

While this will be the pattern for inductive effects, many examples are known in which the substituent causes large rate increases from a remote position. Consider the rates of the following solvolysis reactions in formic acid (Table 1):

Table 1. Solvolytic rates of ω-methoxyalkyl sulphonates

$$R\text{—OBs} + HCOOH \xrightarrow{k} R\text{—OCHO} + HOBs \ (Bs = \underset{\displaystyle \underset{Br}{\bigcirc}}{SO_2}$$

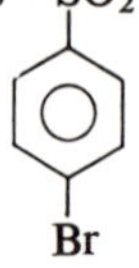

Compound No.	R—OBs	k_{rel} (75°)
(1)	$Me(CH_2)_2CH_2OBs$	1.00
(2)	$MeO(CH_2)_1CH_2OBs$	0.10
(3)	$MeO(CH_2)_2CH_2OBs$	0.33
(4)	$MeO(CH_2)_3CH_2OBs$	461.0
(5)	$MeO(CH_2)_4CH_2OBs$	32.6
(6)	$MeO(CH_2)_5CH_2OBs$	1.13

1-Butyl bromobenzenesulphonate (brosylate) (1) is included for comparison with the series of methoxybrosylates (2–6). It undoubtedly reacts by an S_N2 attack of formic acid (direct pathway, k_s). 3-Methoxypropyl brosylate (2) is slower than 1—evidently due to the inductive withdrawal of electrons by the methoxyl group which makes the departure of the leaving group more difficult,

The same effect is apparent in 3, although attenuated on account of the extra carbon between the methoxyl group and the reaction centre. A sudden dramatic increase in rate is found to occur when the methoxyl group is removed still further (4); this clearly cannot be an inductive effect nor, since the carbon chain is saturated, can resonance effects be operating. The rate increase is due to the appearance of an alternative and more favourable pathway involving the methoxyl group. This acts as an internal nucleophile and displaces brosylate via a favourable five-membered transition state (7) to form an intermediate oxonium ion (8). The intermediate is rapidly attacked in turn by formic acid and the expected product, the methoxy formate (9) is

obtained. The internal-displacement pathway, k_Δ, occurs concurrently with (and is considerably faster than) the normal route by attack of formic acid on the brosylate, k_s. A similar interpretation holds for the 5-methoxypentyl homologue (5), although we must infer that the assisted route is less favourable than in the case of the lower homologue and would require the formation of a six-membered ring (10). Due to the increasing flexibility of the system this is less readily formed than the five-membered ring. Larger rings form with even

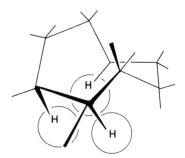

greater difficulty not only on account of greater flexibility and the decreasing chance of the methoxyl group coming into contact with the reaction centre but also since these medium rings are themselves strained due to internal repulsions between hydrogen atoms (Figure 1). Compound 6 and higher homologues therefore solvolyse at approximately the unassisted rate, k_s, that of 1.

Figure 1. Internal non-bended repulsions in an eight-membered ring.

The involvement of a functional group (almost always a nucleophile) with a reaction centre in the same molecule leading to reaction via a cyclic inter-mediate, is known as 'Neighbouring group participation'. If this participation is evident as an increase in rate, the phenomenon is described as *Anchimeric Assistance* (by some, *Synartetic Acceleration*). The great majority of examples of neighbouring group participation are nucleophilic displacements at saturated carbon. The internal nucleophile may be any atom with an unshared pair (divalent oxygen or sulphur, trivalent nitrogen or phosphorus and the halogens) or which can supply an electron pair from a π-system or even in some cases from a σ-bond. The most favourable intermediates are those in which a three, five or six-membered ring is formed. The following are the criteria which may be applied to determine whether neighbouring group participation is occurring. They need not necessarily all be satisfied, but frequently all are.

(a) Enhancement of the reaction rate may occur. Retardation of rate is never the result of participation since the unassisted pathway is always available. Anchimeric assistance is common but not inevitable; its measurement depends upon the availability of a suitable 'model' compound with which to predict the unassisted rate. It is convenient to express the anchimeric assistance, A, as:

$$A = \frac{k \text{ (substrate)}}{k \text{ (model compound)}} = \frac{(k_s + k_c + k_\Delta)^*}{(k_s + k_c)}$$

Any change in structure is likely to have some effect upon the rate and so no model compound can be said truly to react at the unassisted rate of the substrate under investigation. The credibility of this criterion depends upon obtaining a fairly large value for A, say above 10, which cannot be accounted for by other effects. The values of k_{rel} in Table 1 are the values of A (butyl brosylate as model compound). Their magnitudes are such that we can say with confidence that anchimeric assistance is occurring in **4** and **5**, but in **6** the small increase in rate is inconclusive as to whether any assistance at all is present.

The solvent is very important in determining the relative contributions to the overall rate from assisted and unassisted pathways. In the above examples, the assisted pathway has the more polar transition state and becomes progressively more important as the polarity of the medium is increased, thus:

	EtOH	AcOH	HCOOH
A for MeO(CH$_2$)$_4$OBS (relative to CH$_3$(CH$_2$)$_3$OBs)	22.2	425.0	610.0

At the same time, the nucleophilic power of the solvent decreases, hence k_s diminishes further enhancing the factor A. For predicting solvent effects in general, one would have to consider the specific case in connection with the principles set out in Section 1.17.

(b) The stereochemistry of an assisted displacement is very characteristic, *retention of configuration* being the rule. More precisely, this is the result of a double inversion of configuration consequent upon two S_N2 processes. In

* To summarize the terms (cf. Chapter 2): k_s refers to direct (S_N2) displacement by solvent. k_c to ionization (S_N1) assisted by solvation. k_Δ to internal (neighbouring group) displacement.

order to observe this effect, the reaction centre must be asymmetric. For example, deamination of α-aminoacids by nitrous acid may lead to α-hydroxy-acids of the same configuration. L-alanine (**11**) undergoes carboxyl-assisted deamination via an α-lactone (**12**), which rapidly ring-opens to L-lactic acid.

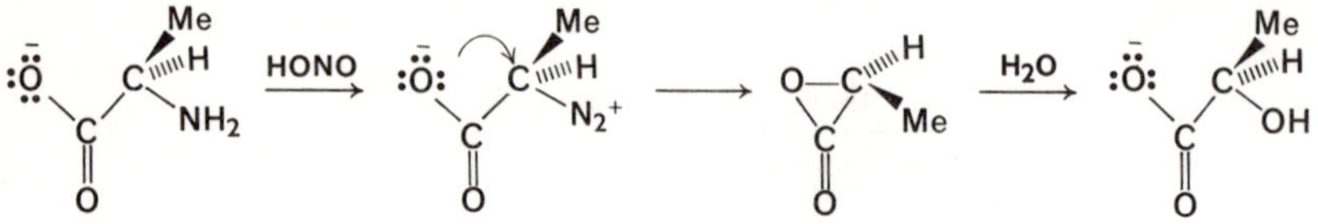

| (**11**) | | (**12**) | |
| L-alanine | diazonium salt | α-lactone intermediate | L-lactic acid |

(c) It may be possible to isolate the cyclic intermediate or to divert it to recognizable products from which its transient existence may be inferred. The hydrolysis of the chlorhydrin (**13**), with retention, may be arrested at the intermediate epoxide stage (**14**) by adding only one equivalent of alkali. The epoxide is quite stable and may be isolated or the hydrolysis continued to the glycol by more alkali. The assisted hydrolysis of 4-hydroxybutyl halides (**15**)

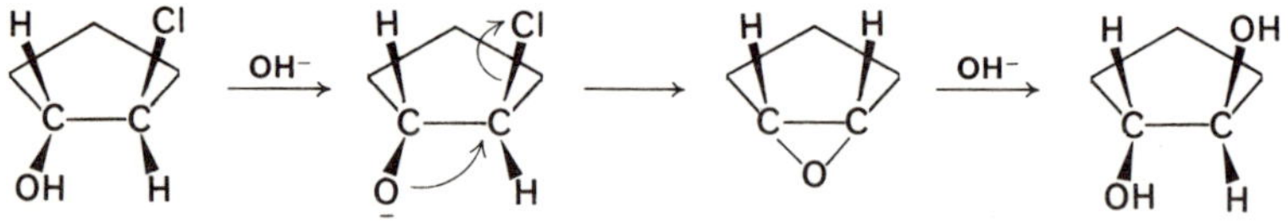

(**13**)   (**14**)

or esters produces tetrahydrofuran (**16**) as a product. The intermediate oxonium ion **17** more readily loses a proton from oxygen than undergoes ring opening.

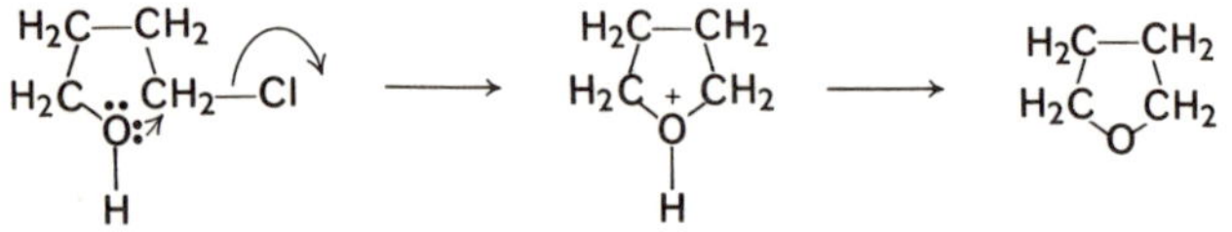

(**15**)   (**17**)   (**16**)

(d) The intermediate may be capable of ring opening in two directions giving rise to the possibility of the apparent migration of the neighbouring nucleophilic group. Isotopic labelling may be necessary to observe this,

$$\begin{array}{ccc}
\text{H}_2\text{C}-\text{CH}_2 & \text{H}_2\text{C}-\text{CH}_2 & \text{H}_2\text{C}-\text{CH}_2 \\
\text{H}_2\text{C} \cdots {}^{14}\text{CH}_2-\text{OBs} & \text{H}_2\text{C} \quad {}^{14}\text{CH}_2 & \text{H}_2\text{C} \quad {}^{14}\text{CH}_2\text{OCHO} \\
\text{O:} & \text{O}^+ & \text{OMe} \\
\text{Me} & \text{N:} \quad \text{Me} \quad \text{N} &
\end{array}$$

with HCOOH giving the products shown.

$$+$$

$$\begin{array}{c}
\text{H}_2\text{C}-\text{CH}_2 \\
\text{H}_2\text{C} \quad {}^{14}\text{CH}_2\text{OMe} \\
\text{O}-\text{C}=\text{O} \\
\textbf{M}
\end{array}$$

(e) The criterion of plausibility should not be neglected. If neighbouring group participation is inferred, there should be a suitable physical model to explain the observations. As a counterexample, the increase in the rate of hydrolysis of the toluenesulphonate (**18**) upon the introduction of a 17-keto group cannot be a neighbouring group effect of the type discussed, since the steroid ring system is rigid and the positions 3 and 17 are separated by ca 0.8 nm (nine bonds). Evidence in favour of some transient intermediates is available

$$\frac{k_1(X = C{=}O)}{k_1(X = CH_2)} \sim 4$$

from their isolation or identification under other circumstances. Oxonium ions are well known (H_3O^+ in solution) and may be isolable, e.g. trimethyl-oxonium salts, $Me_3O^+ ClO_4^-$. Recently an α-lactone was isolated.* It was very reactive towards nucleophilic ring opening. Halonium ions containing divalent positive chlorine, bromine and iodine, are likely intermediates in halogen participation. Crystalline examples have been isolated, e.g. **18**, and cyclic examples are stable in superacid solution, e.g. **19**. They have previously been mentioned in connection with halogen addition to olefins (p. 122).

$$\begin{array}{cc}
\overset{+}{\text{hal}} \quad SbF_6^- & \text{Me} \quad \overset{H}{\underset{}{\bigwedge}} \quad \text{Me} \\
\text{Me} \quad \text{Me} & C-C \quad H \\
\textbf{(18)} & \overset{}{\underset{+}{\text{Br}}} \\
\text{hal} = \text{Cl, Br, I} & \textbf{(19)}
\end{array}$$

$$* \quad \begin{array}{c} O-O \\ O{=}C \quad C{=}O \\ C \\ Me \quad Me \end{array} \xrightarrow{h\nu} \begin{array}{c} O-C{=}O \\ C \\ Me \quad Me \end{array} + CO_2$$

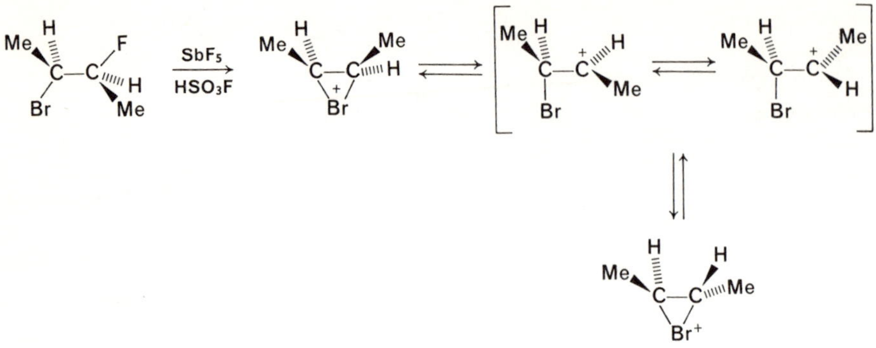

The *trans* addition of bromine to an olefin is best explained by an intermediate bromonium ion decomposed by S_N2 attack of Br^-:

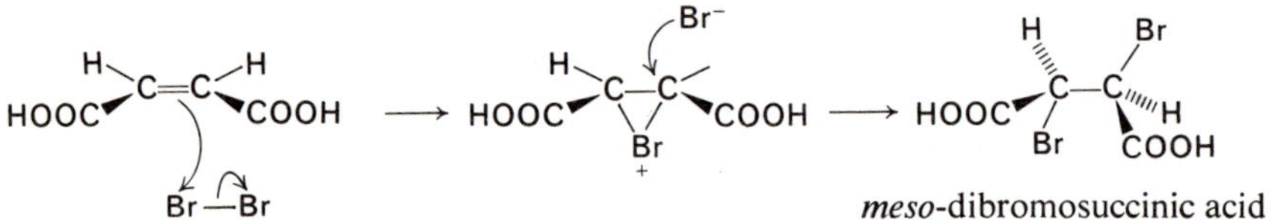

meso-dibromosuccinic acid

Neighbouring group participation is a widespread phenomenon in the field of nucleophilic displacement reactions. The following examples serve to amplify the principles discussed.

The chlorine atoms of mustard gas, 2,2'-bis(chloroethyl)-sulphide (**20**) are very labile towards hydrolysis or displacement by nucleophiles in general. It has been established that the reactions take place by participation of sulphur via an intermediate episulphonium ion (**21**). The molecule is capable of reacting

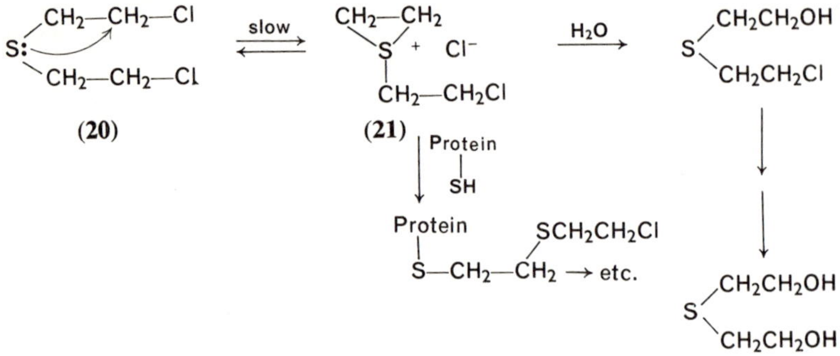

at two nucleophilic sites and of crosslinking protein or other biologically active molecules by alkylation, which accounts for its intensely poisonous nature. Hydolysis is independent of hydroxide concentration but is retarded by added chloride ion because of the reversibility of the slow step. In fact the sulphonium ion (21) is competed for by all nucleophiles present and the reaction has the kinetic characteristics of an S_N1 reaction (Section 2.3). Similar reaction mechanisms are used by β-chloroamines; compounds with two such functional groups are known as 'nitrogen mustards', e.g. 22, and have some therapeutic use in the treatment of leukemia. A rearrangement

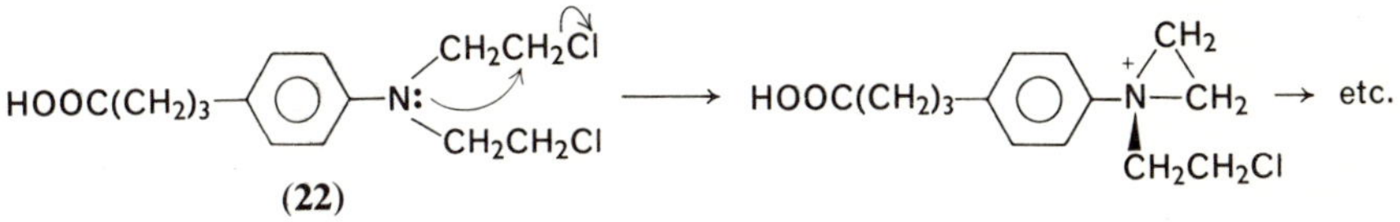

(22)

'Chlorambucil'

occurs in the related reaction, assisted by sulphur. *Trans*-2-hydroxycyclohexyl

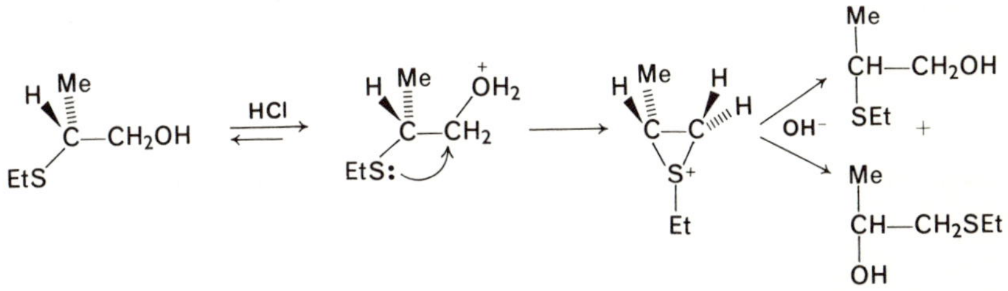

chloride (23) reacts in aqueous alkaline solution to give the *trans* glycol (24) at a rate 100 times greater than the *cis* isomer (25) which also gives *trans* glycol. The *trans* chlorohydrin alone is suitably substituted for oxygen participation; the intermediate epoxide (26) is isolable. The hydrolysis of aspirin sodium salt

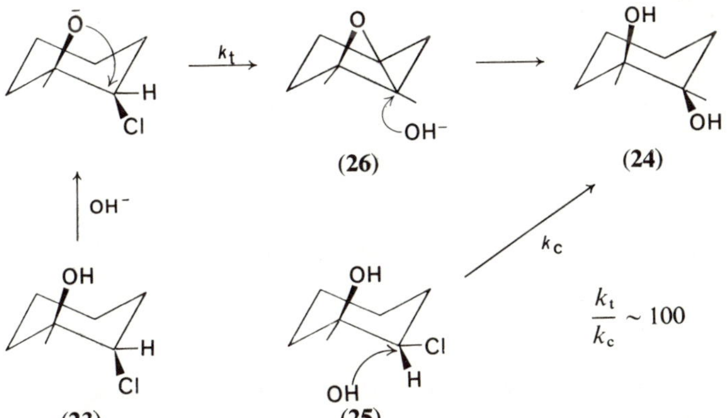

(26)

(24)

(23)

trans

(25)

cis

$$\frac{k_t}{k_c} \sim 100$$

is greatly assisted by the neighbouring carboxylate ion acting as a general base. Similarly, phenylsalicyl phosphate reacts some 10^8 times faster than the

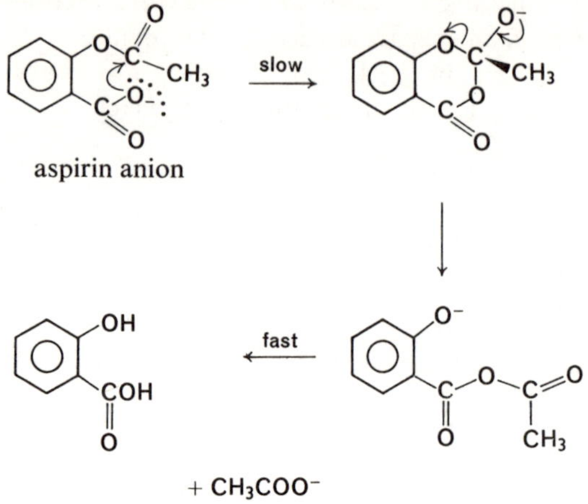

aspirin anion

$+ CH_3COO^-$

para-substituted isomer. Retention of configuration and rearrangement often

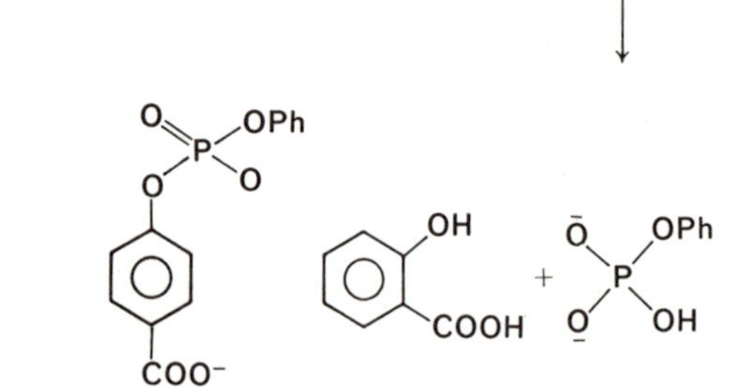

phenylsalicyl phosphate

occurs in displacements from β-halogeno compounds; their readiness to take part is in the order I > Br > Cl, the order of their ease in expanding the valence shell. This explanation is consistent with the observation that only one of the

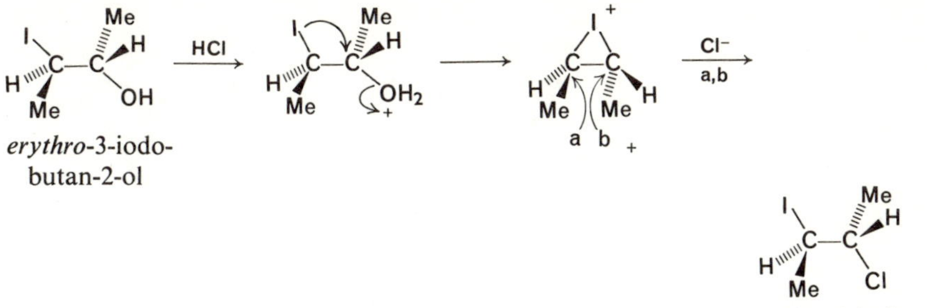

erythro-3-iodo-
butan-2-ol

erythro-3-iodo-
2-chlorobutane

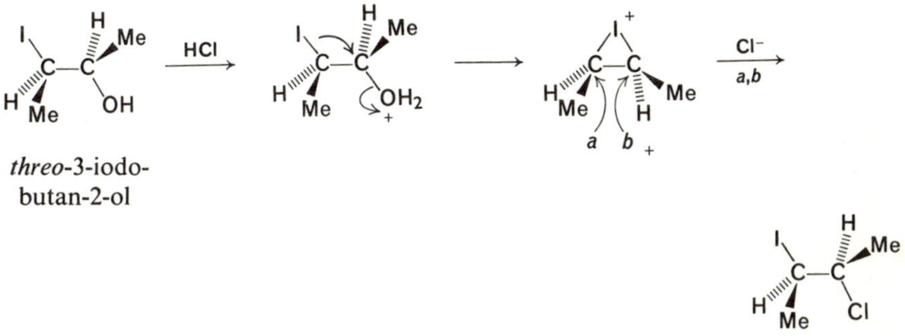

threo-3-iodo-
butan-2-ol

threo-3-iodo-
2-chlorobutane

two identical iodines in 1,4-diiodobutane (**27**) may be replaced in reaction with mercuric chloride. Chlorine in the iodochloride, **28**, is not able to assist the

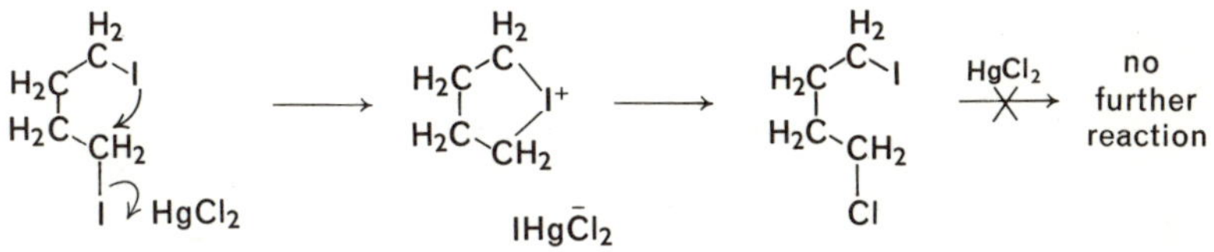

(**27**) $IHg\bar{C}l_2$ (**28**) $HgCl_2$ no further reaction

displacement of a second iodide. β,γ-Unsaturated acids add bromine and iodine in ether to form a bromolactone or iodolactone via the halonium ion: for example, the addition of iodine to give first a β-lactone and on standing, a γ-lactone—an example of kinetic and thermodynamic control of the product.

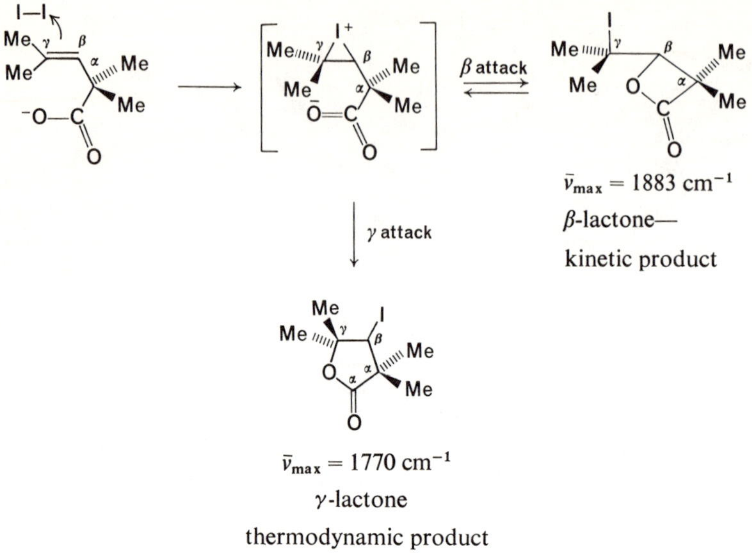
β attack
γ attack
$\bar{\nu}_{max} = 1883\ cm^{-1}$
β-lactone—
kinetic product
$\bar{\nu}_{max} = 1770\ cm^{-1}$
γ-lactone
thermodynamic product

Rearrangement is the evidence for halogen assistance in the cleavage of the ether (**29**) by trimethyloxonium trifluoroacetate; the more reactive tertiary position of the intermediate bromonium ion (**30**) is preferentially attacked, which indicates an S_N1-like mode of reaction. Considerable rearrangement is

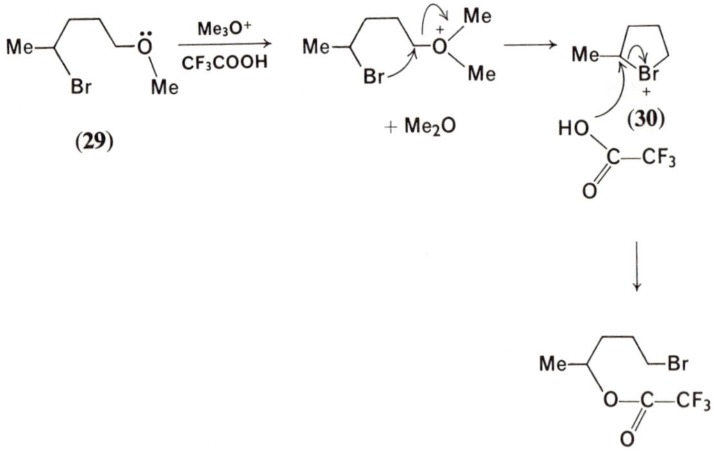
Me_3O^+
CF_3COOH
$+ Me_2O$
(29)
(30)

observed in the addition of hypobromous acid to allyl bromide (**31**), followed here by isotopic labelling. Carbonyl participation is relatively unfavourable

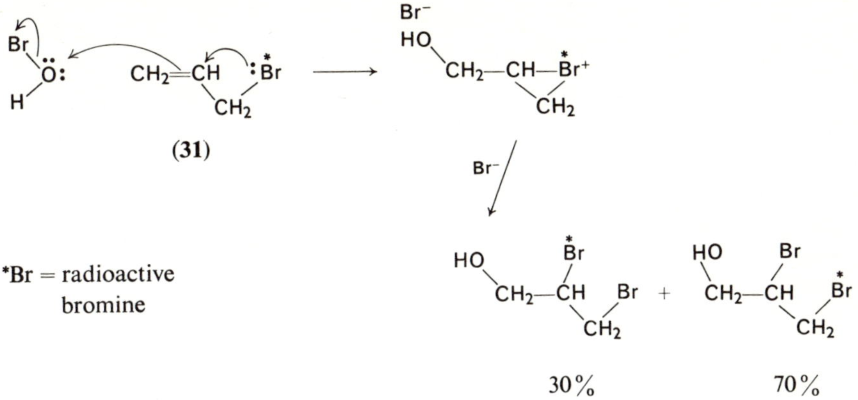

(31)

*Br = radioactive
bromine

30%   70%

since the nucleophilic character of the carbonyl oxygen is rather low. One
example is the base-catalysed cyclization of a series of acetoacetate derivatives
(32). Normally, cyclization to carbon occurs (33), but when n = 3, and a
cyclobutane would be expected, cyclization to oxygen is observed and a
six-membered ring produced (34). In this instance, the internal nucleophile
may be the enol 32a. Another example of a different type is the 10^5-fold

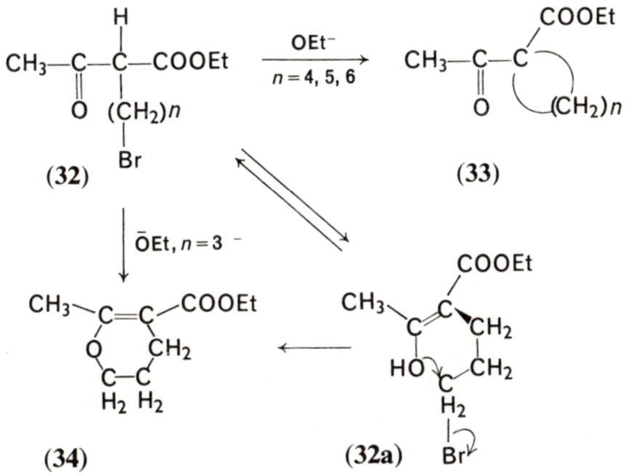

acceleration of the hydrolysis of methyl benzoate produced by an *ortho* formyl
group (35). It is likely that the neighbouring group is actually an oxide ion
(36) formed by nucleophilic attack on the formyl group leading to a cyclic
product (37).

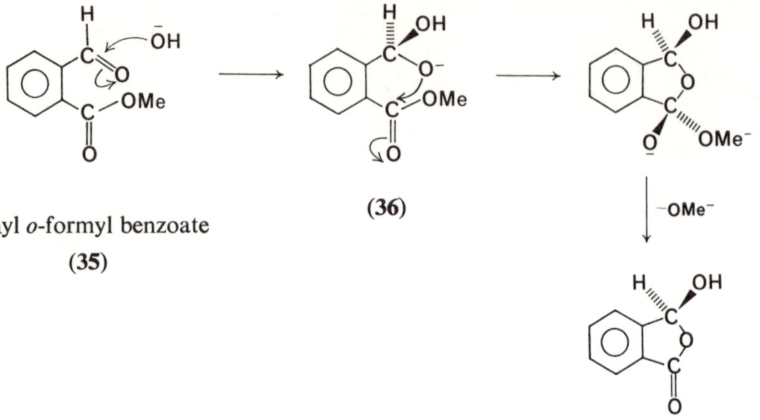

Methyl *o*-formyl benzoate

(35)

(36)

(37)

3.2 PARTICIPATION BY π-ELECTRONS

The criteria established for the confirmation of neighbouring group participation are frequently satisfied when the nucleophilic groups have only π-electrons. The rate of acetolysis of the pent-3-enyl ester (38) is very fast compared to a saturated analogue. It is clear that the π-electrons are playing a

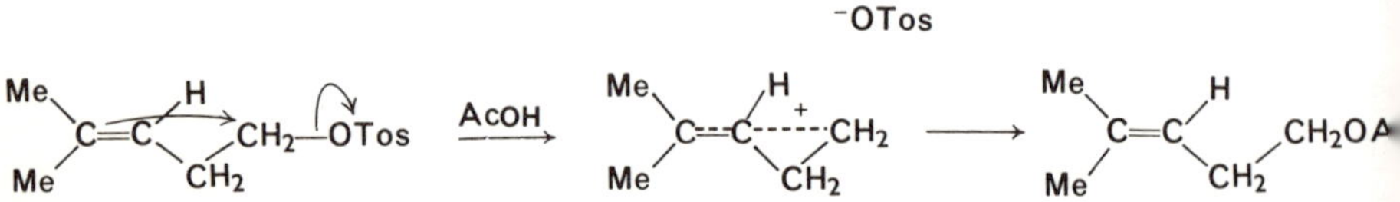

(38)   (39)

an 'homoallylic cation'

part in the reaction and formal conjugation with the reaction is not possible because of the intervening saturated carbon. The problem is then how to formulate the cyclic intermediate which results from the donation of π-electrons to the reaction site, non-commitally written as **39**. Some possibilities are indicated by formulae **40 a–d**. **40a** expresses no interaction between the π-bond and the reaction site and can be disregarded; **40b** expresses the full donation of an electron pair from the olefinic bond. An ion of this type should also be produced from **41** but this compound leads to different products and therefore the intermediates must be different. The structures **40c** and **40d**

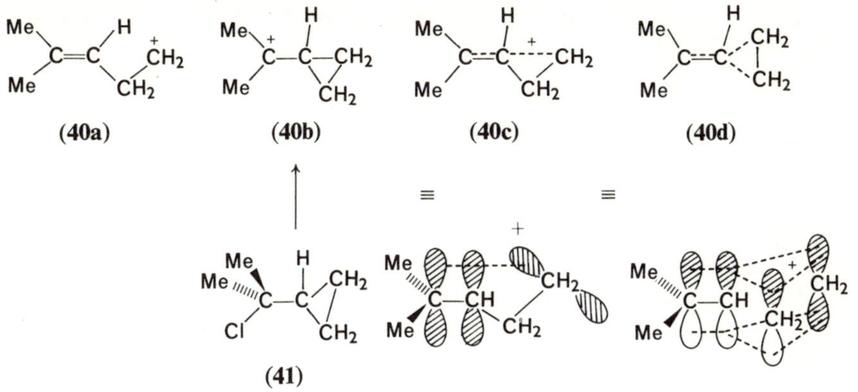

(40a)   (40b)   (40c)   (40d)

(41)

require the overlap of carbon-$2p$ orbitals on three and four atoms respectively to form a new set of π-molecular orbitals. The unusual feature in these structures is that overlap is postulated between atoms which are not also formally connected by σ-bonds—'non-classical' structures. The positive charge is thereby spread over these atoms and constitutes a *homoallylic* system (homo ≡ one more carbon than); the analogy with the allyl cation **42** may be seen.

Solvolysis of the deuterium-labelled compound **43**, results in 'scrambling' of

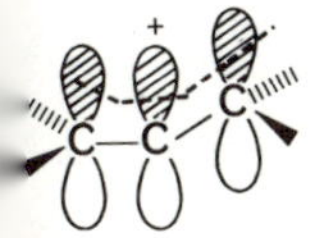

(42) 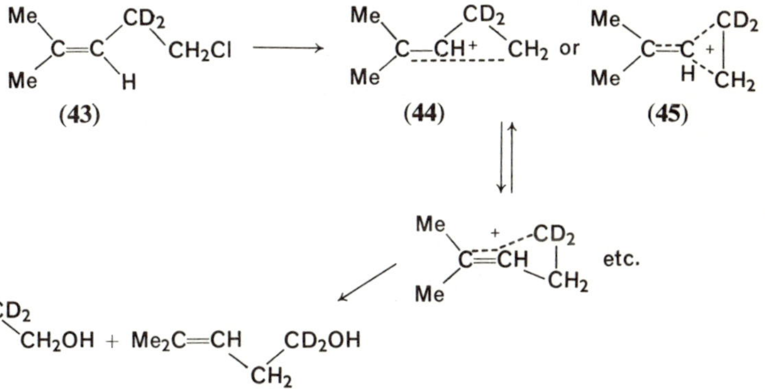   (43)   (44)   (45)

the label indicating that C-1 and C-2 become equivalent. Whether this indicates an intermediate of the type **40d** or of **40c** which can rearrange (i.e. **45** or **44**) may be debated, but perhaps the latter is to be preferred since other examples of homoallylic participation occur in which anchimeric assistance is observed but no rearrangement. For instance, cholesteryl sulphonates (**46**) solvolyse in acetic acid some 500 times faster than the saturated cholestanyl compounds (**47**). The homoallyl cation is an example of a *non-classical ion*, a species which exhibits conjugation between atoms which are not formally connected by σ-bonds.

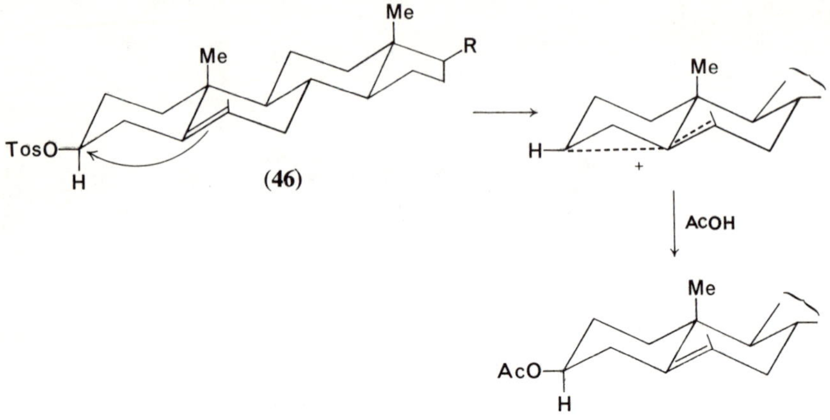

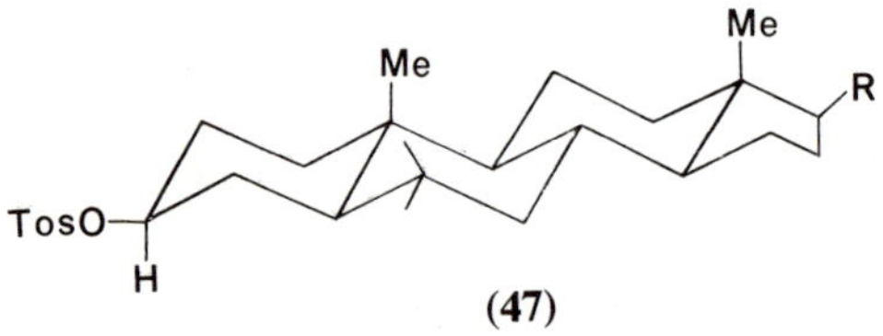

(46)

AcOH

(47)

Homoaromatic Cations

The tropyllium cation has been discussed (Section 2.21) as being of great stability. Seven interacting C-2p orbitals are required with six electrons populating the resulting molecular orbitals. Cyclic conjugation may still occur if the seven-carbon ring is interrupted by a saturated carbon forming a *homotropyllium ion.* Such a structure is assigned to the stable cation which

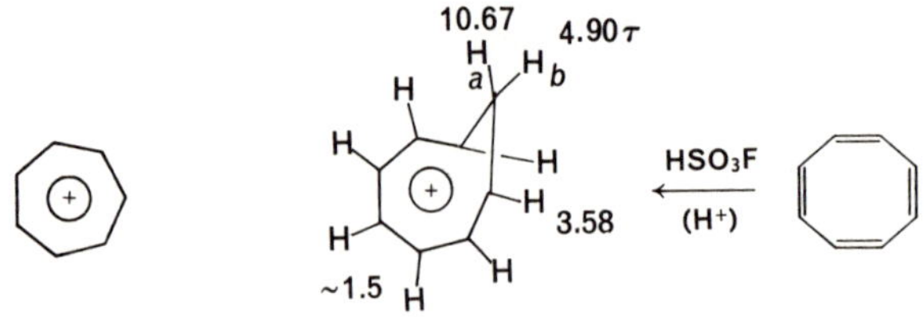

tropyllium   homotropyllium

results from the protonation of cyclooctatetraene in superacid solution. The n.m.r. chemical shifts of the peripheral protons are characteristic of a delocalized and aromatic cation, while the methylene protons *a* and *b* are shielded and deshielded respectively by the paramagnetic ring current. A number of related structures are known and also some stable bis-homo-tropyllium ions, in which the aromatic structure is twice interrupted by a saturated carbon, for example:

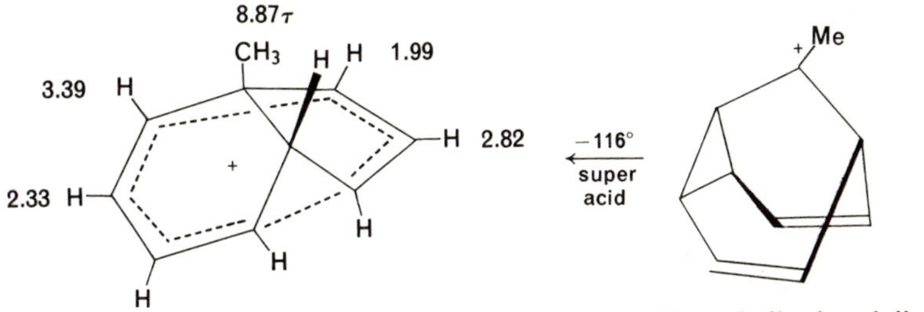

'9-methylbarbaralyl'
cation

One of the most dramatic examples of π-participation is to be found in the bicyclo[2,2,1]heptyl (norbornyl) series. The relative rates of acetolysis of 7-norbornyl tosylate (**48**), and the unsaturated analogues *syn*-7-norbornenyl (**49**), *anti*-7-norbornenyl (**50**) and norbornadienyl tosylates (**51**), are shown below.

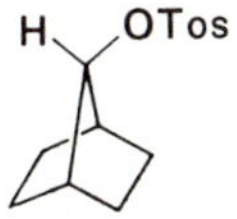

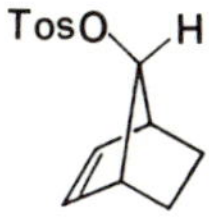

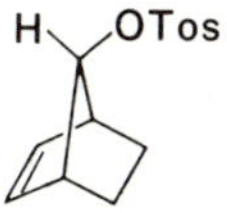

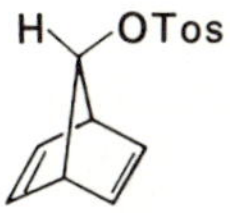

(48)	(49)	(50)	(51)
	syn	anti	
$k_{rel}(AcOH)$ 1	$10^{3.72}$	$10^{11.1}$	10^{14}

All react by the S_N1 route. The system is highly strained and is sensitive to relief of bond-angle strain, which is probably involved in the considerable rate difference between **48** and **49**. The interesting feature is the great difference in rate (more than 10 millionfold) between the rates of the *syn* and *anti* isomers (**49** and **50**). This is undoubtedly due to participation by the π-electrons in the latter case where they are suitably located for a rearside attack (**50a**). The system is a homoallylic one but the symmetry and puckering of the ring makes participation particularly favourable, and retention of configuration is observed whereas the *syn* isomer reacts with inversion (**49a**). Even more

effective participation occurs in the solvolysis of norbornadienyl compounds
(51).

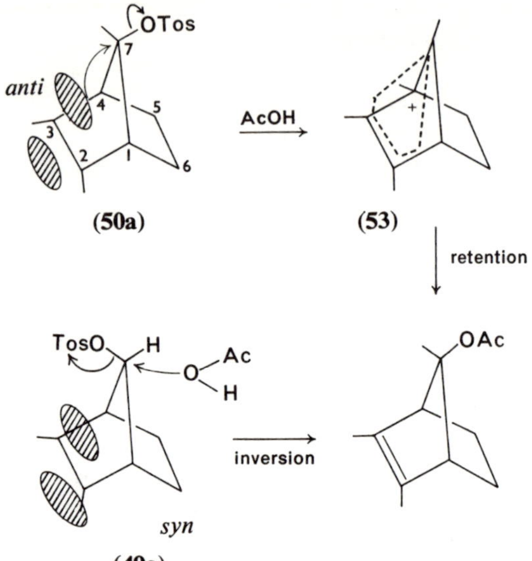

(50a)   (53)

retention

(49a)
syn

inversion

The structure of the intermediate cation 53 formed from the *anti*-norbornenyl
compound has been the subject of a major controversy. Chemical evidence
showed that C-2 and C-3 remained equivalent and this was interpreted by
H. C. Brown as indicating two tricyclic cations (52a and 52b) in rapid equilibrium, whereas S. Winstein (1969) described the carbonium ion in terms of a
single structure (53) formed by the overlap of three carbon $2p$ orbitals at C-2,
C-3 and C-7. As with the homoallylic ion (40) this is a *non-classical* structure.
It is assumed that the overlap of these orbitals (containing jointly two electrons)
gives rise to a set of molecular orbitals which are analogous to the cyclo-

(50) 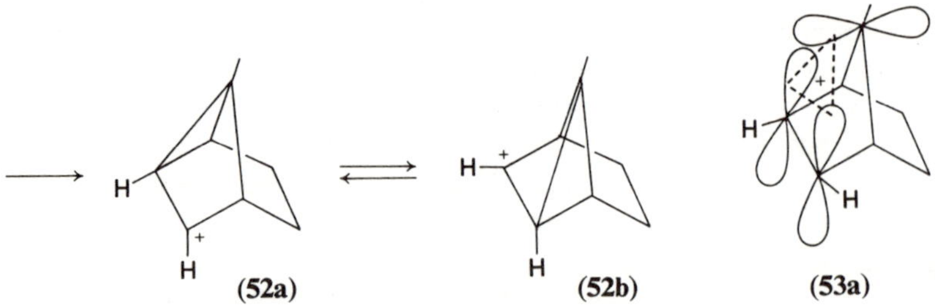   (52a)   (52b)   (53a)

propenium ion. That is, it is a *bis-homocyclopropenium ion*,* the three atoms of
the quasi-aromatic system being interrupted by two saturated carbons (53a).
The case for the non-classical representation has been considerably strength-
ened by the direct observation of the ion as a stable species in superacid
solution. Extraction of either 54 or 55 into SbF_5–HSO_3F solution at −50°
produces a species whose n.m.r. spectrum is consistent with 53 (i.e. 53b), the
chemical shifts for all the protons as shown. H-2 and H-3 appear at low field
indicating that they are in an environment of considerable positive charge. This
could still be consistent with a structure 52. However, the spectrum of the
2-methylnorbornenyl cation (56) shows resonances of H-3 and H-7 almost

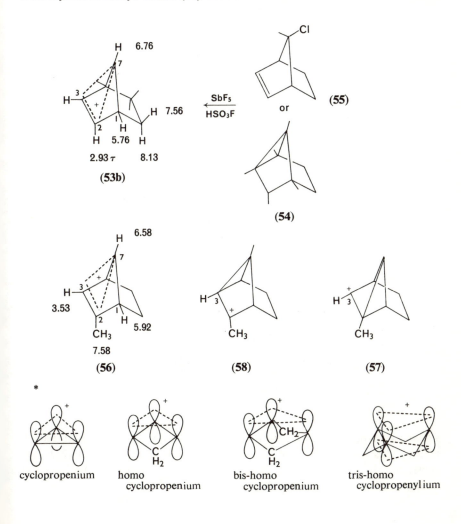

unchanged. Now in this case, the two tricyclic ions **57** and **58** are not identical and the isomer **58** would be greatly preferred to **57** since the former is tertiary. Hence one would expect H-3 to bear little or no positive charge and its resonance to be above 6τ, if that structure were correct. Another line of evidence which supports the non-classical formulation comes from photo-electron spectroscopy (Section 2.28). The X-ray p.e. spectrum of the norbornyl cation can be obtained at low temperature, Figure 2, and, since the time scale of electron ejection is $\sim 10^{-16}$ s, represents the instantaneous structure. This

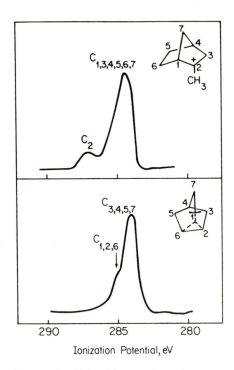

Figure 2. C-1s X-ray photoelectron spectrum of the norbornyl and 2-methyl-norbornyl cations.

shows two types of C-1s ionization potentials in the ratio 2:5 separated by 1.7 eV. This suggests that the species is non-classical and does not contain a centre at which the charge is concentrated such as occurs in the 2-methyl analogue (**56**).

3.3 THE CALCULATION OF UNASSISTED RATES

The identification of anchimeric assistance depends upon being able to make an estimate of the expected rate in the absence of such assistance. Comparisons with model compounds may be satisfactory in some cases (e.g. compounds **1–6**) but it becomes more difficult to select a suitable model when dealing with strained ring compounds such as the norbornyl series. The removal of the double bond, though preventing π-participation, may significantly affect the rates in other ways. Thus we might expect the rates of **48** and **49** to be the same as neither derives assistance from the π-electrons. A useful and quantitative approach to this problem is possible by a method due primarily to Halford, Schleyer and Foote. It was observed that the rates of acetolysis (**25**) of a series of *secondary* tosylates correlated with the carbonyl stretching frequencies (from infrared spectra) of the corresponding ketones. This relationship is

$$\log_{10} k_{rel} = 0.125 \, (1715 \; \bar{v}_{co})$$

improved if terms for torsional strain, non-bonding interactions and inductive effects in the tosylate, are added. The torsional term measures strain energy associated with deviations from a perfectly staggered conformation ($\phi = 60°$). If relieved, this can assist the ionization process. The non-bonding and inductive terms are usually small and somewhat arbitrary and can often be neglected. The equation becomes,

$$\log k_{rel} = 0.125(1715 - \bar{v}_{co}) + 1.32\Sigma_i(1 - \cos 3\phi_i)$$

One can define the amount of anchimeric assistance, A, as before by

$$A = \frac{k_{rel}(\text{observed})}{k_{rel}(\text{calculated})}$$

and some values of these observed and calculated rates are given in Table 2. A sharp differentiation can be made between systems which react with, and those which react without, π-participation.

Norbornadienyl (7-bicyclo[2,2,1]heptadienyl) tosylate (**51**) is even more reactive than *anti*-norbornenyl (**50**); evidently even more effective π-assistance

Table 2. Calculated and observed rates of acetolysis of secondary
toluenesulphonates, R—OTos $\rightarrow$ R$^+$ $\rightarrow$ R—OAc

R	k_{rel}(obs)	k_{rel}(calc)	A
Cyclohexyl	1.0	1.2	1
Cyclopentyl	7	7	1
Cyclooctyl	900	880	1
7-Norbornyl, **48**	10^{-7}	10^{-7}	1
exo-2-Norbornyl[a]	155	0.078	$10^{3.3}$
anti-7-Norbornenyl, **50**	13000	7.6×10^{-7}	$10^{12.9}$
Cholesteryl, **46**	100	0.02	$10^{3.7}$
anti-7-Benzonorbornadienyl[b]	0.075	2×10^{-10}	$10^{9.1}$

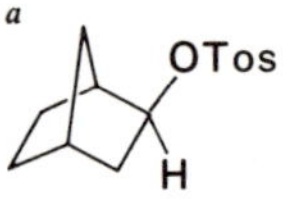

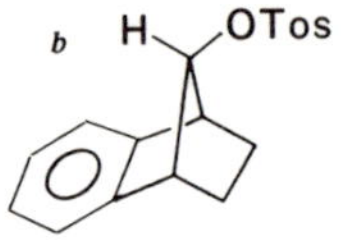

is occurring. The norbornadienyl cation can be generated and observed as a
stable species in superacid and shows separate resonances for the two pairs of
vinyl protons, only one being conjugated with C-7 and sharing the positive
charge (**51a**). The 7-methyl homologue (**53a**) shows a similar spectrum at $-50°$

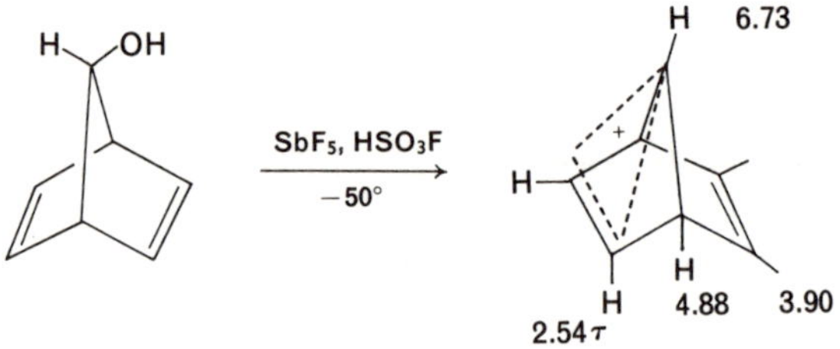

while at $-14°$ the vinyl protons merge to give a single signal. Evidently a
'bridge-flipping' equilibrium (**54a**) occurs and is slowed down at low temperatures so that the individual structures give a characteristic resonance spectrum.
The availability of this type of equilibrium most likely produces additional
stabilization in the dienyl cation, hence greater reactivity.

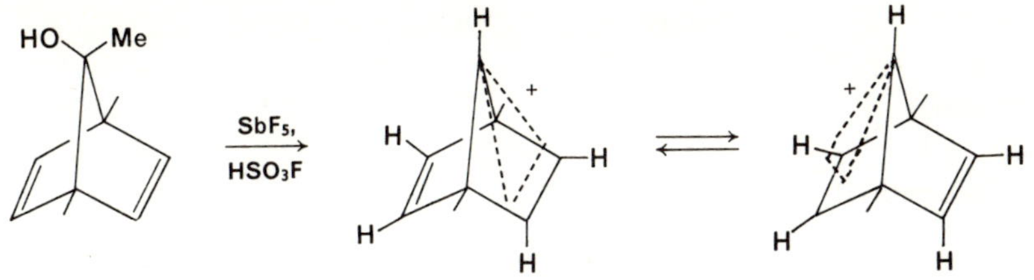

(53a)   **(54a)**

Phenyl participation is well established in solvolytic reactions. A β-aryl group will frequently impose retention of configuration on a reaction; the following sequence requires, in addition, the phenyl group to migrate to the adjacent carbon in exactly 50% of the product.

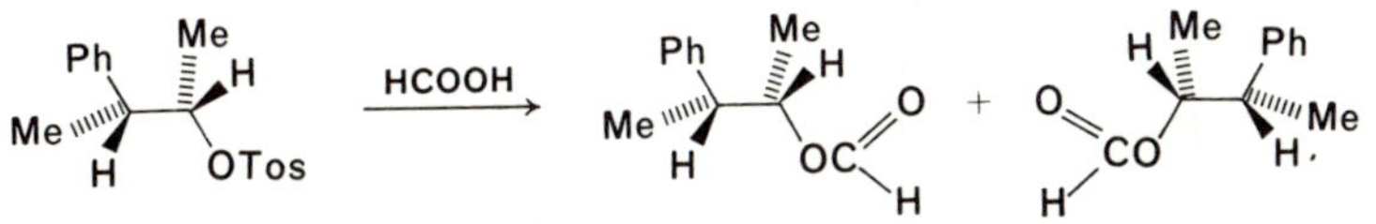

L-*threo**   HCOOH   DL-*threo* (>99%)

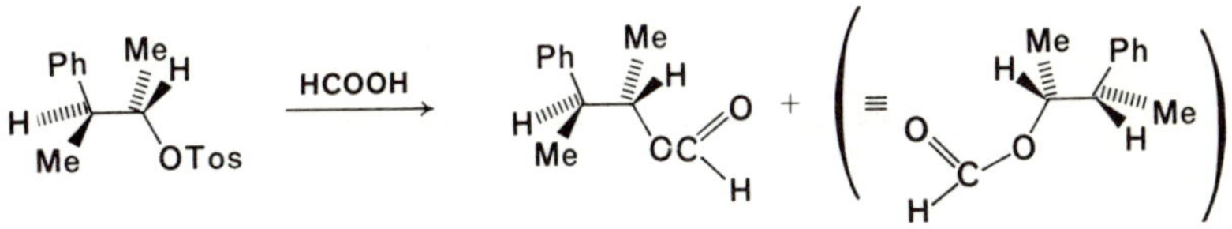

D-*erythro**   HCOOH   D-*erythro* (>99%)

A β-phenyl group can undoubtedly cause anchimeric assistance, but the effect is only pronounced in the most limiting solvents. In ethanol and acetic acid, β-phenylethyl sulphonates (**55a**) react slower than ethyl (**56a**), but in formic and especially in trifluoroacetic acids, more ionizing solvents, the rate acceleration becomes very marked. Concurrently, entropies of activation become less negative, indicating a change towards a less-crowded transition state.

(55a)   **(56a)**

* For the stereochemical terminology, see e.g. K. Mislow, *Introduction to Stereochemistry*, Benjamin, New York, 1965.

	EtOH	CH$_3$COOH	HCOOH	CF$_3$COOH
k (phenylethyl)				
k (ethyl)	0.24	0.37	2.1	3040
$\Delta S^{\ddagger}$ J K^{-1} (phenethyl)	−85	−72	−40	−35

The intermediate cation in these phenyl-assisted solvolyses is likely to be an ethylenephenonium ion with the structure **57a**; it is in fact a type of benzenonium ion (Section 2.22) and further evidence for this comes from the observation of the n.m.r. spectrum of the parent ion (Figure 3) and of the *p*-methoxy derivative **58a** in superacid, and the isolation of an alcohol with the structure **59**

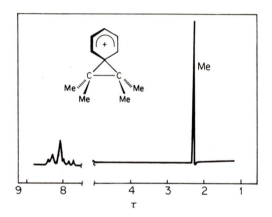

Figure 3. Proton n.m.r. spectrum of the ethylene-phenonium cation.

from the hydrolysis of β-phenylethyl compounds **55**. The ion **57a** can clearly be attacked at either carbon giving a product in half of which (attack at C$_\beta$) phenyl migration has occurred. In an isotopic labelling experiment, the deutero ester **60** gives rise to a rearranged product **61**, but unreacted ester shows no rearrangement. The two carbons only become equivalent during the irreversible slow step of the reaction. The secondary analogue (**62**) undergoes

$$PhCH_2CD_2OTos \xrightarrow{CF_3COOH} PhCH_2CD_2OCOCF_3 + PhCD_2CH_2OCOCF_3$$

$$\text{(60)} \hspace{6cm} \text{(61)}$$

solvolysis by both assisted (k_A) and unassisted (k_s) routes. The relative amounts of the two can be estimated from the amount of inverted product (k_s) to that of

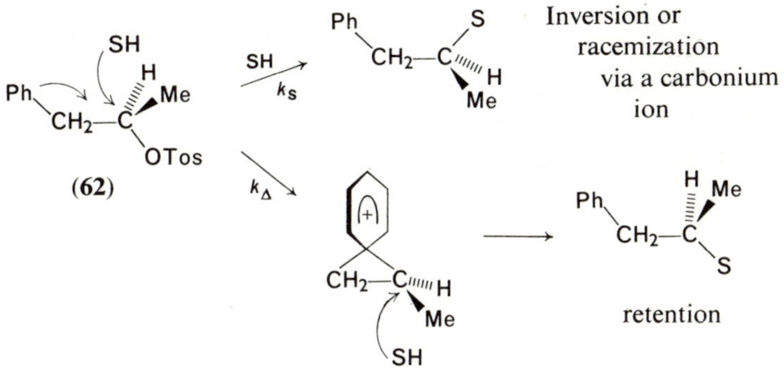

retained stereochemistry (k_A). By this type of analysis, the importance of the anchimerically assisted route is found to increase with the polarity (limiting character) of the solvent:

Solvent	k_s	k_A
Ethanol	1.31	0.099
Acetic acid	0.38	1.11
Formic acid	46.0	258
Trifluoroacetic acid	6.6	36800

Similarly it has been found that the importance of anchimeric assistance is greatly dependent upon the availability of electron-releasing groups in the benzene ring:

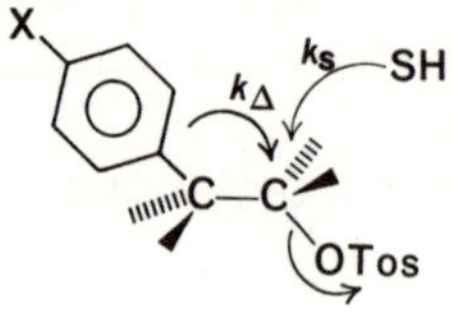

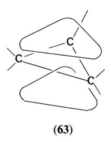

X	$10^7 k_s$	$10^7 k_c$
Cl	1.97	0.39
H	2.07	2.39
Me	2.10	19.0
MeO	3.2	228
O⁻*		ca10⁸

Electron-releasing ability

Donor groups will plausibly stabilize ethylenephenonium ions, e.g. **58**.

3.4   PARTICIPATION BY CYCLOPROPANE RINGS

The conjugating properties of a double bond and a cyclopropane ring bear many analogies. The electrons which form the skeletal framework of the three-membered ring occupy orbitals which possess π-character (**63**):

(63)

It has been found that a cyclopropane ring can participate at the rearside of an ionization reaction when suitably situated, in much the same way as a double bond. The geometry of the system is crucial as may be seen from the following examples (**64–67**) in the norborane series:

* The product in this case is

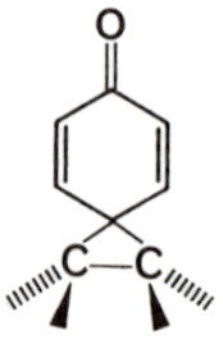

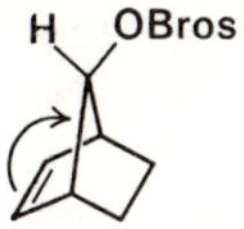

H OBros
(64)

H OBros
exo, anti
(65)

H OBros
endo, anti
(66)

BrosO H
endo, syn
(67)

k_{rel} 10^9
(acetolysis)
π-assisted

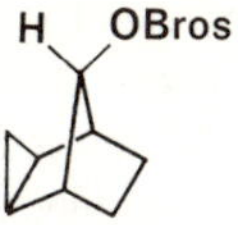

1

unassisted

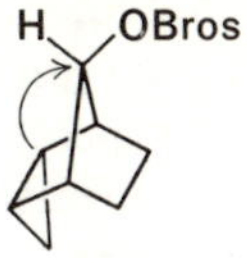

10^{12}
cyclopropyl
assisted

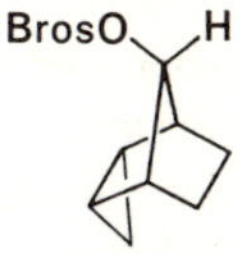

10

unassisted

$\left(\begin{matrix} PNB = \\ p\text{-nitrobenzoate} \end{matrix} \right)$

H H H
H H OPNB

H OPNB

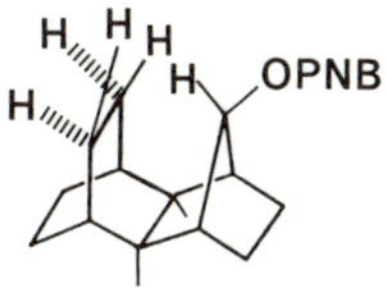

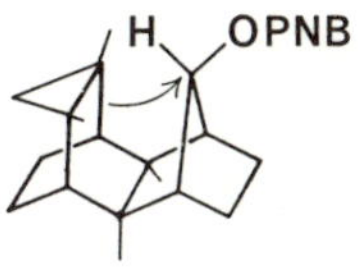

k_{rel}

1

$10^{8\cdot2}$

Only the *endo, anti* isomer (66) shows an assisted ionization, the degree of assistance being very large, i.e. the remote cyclopropane ring must be presented to the rearside of the displacement centre 'edge on'. The intermediate formed after ionization of 66 may be expressed in terms of a non-classical carbonium ion 68 (a tris-homo-cyclopropenium ion, footnote p. 217) although the alternative pair of classical ions (70) is not yet unequivocally excluded. The product of acetolysis is largely rearranged (69). The cyclopropylmethyl

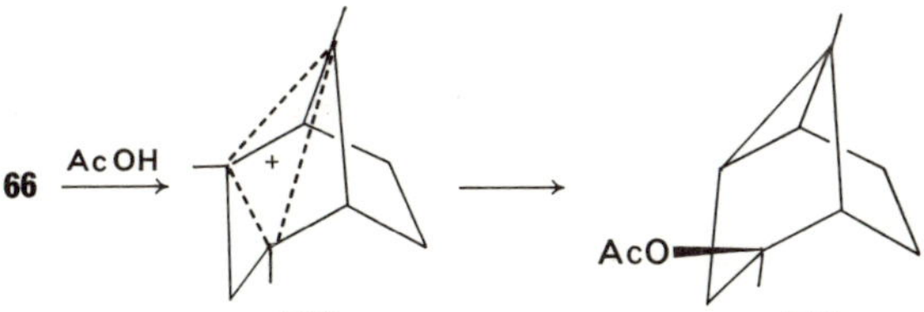

66 →AcOH→ (68) → AcO (69)

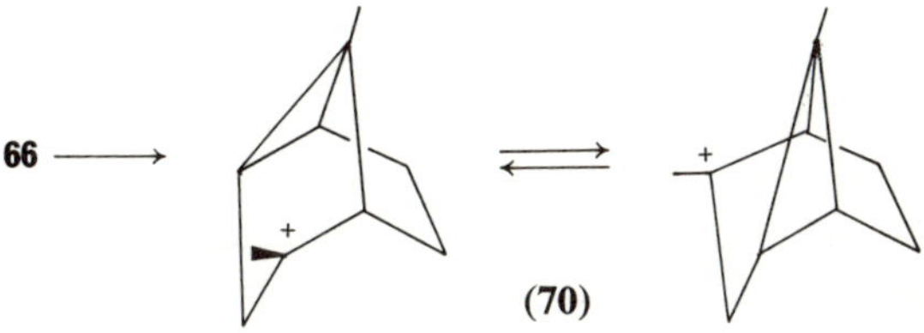

66 → (70)

system (71) has been studied intensively in recent years. The n.m.r. spectrum of a simple derivative, the dimethylcyclopropyl carbonium ion (72) has been observed directly in superacid solution. At temperatures above about −50°, the resonance due to the methyl groups is averaged due to their rapid rotation. Below that temperature the methyl resonances appear separate which indicates that the conformation 73 is the stable one, the energy barrier between that and the 'bisected' conformer 72 being 57.3 kJ mol^{-1}. This conclusion is supported by calculations of the stabilities of the two conformers. It is easy to see that

(71)

(72)   (73)

stabilization of an adjacent cationic carbon by the three-membered ring will only occur when it is able to assume the type of conformation with respect to the reaction centre as in 73. In the rigid adamantane derivative 74, the orientation of the cyclopropane ring (as in 72) is unfavourable and no rate acceleration relative to 75 occurs. The origin of the retardation is obscure. Very

(74)	(75)	
k_{rel} 1.6×10^{-3}	1	0.6

efficient cyclopropyl participation is able to occur in 76 and 77, which have correct orientations of the cyclopropylmethyl fragment (outlined boldly) for enhanced stability. The structure of the cyclopropylmethyl cation has been the subject of much speculation. The situation is complicated by the fact that

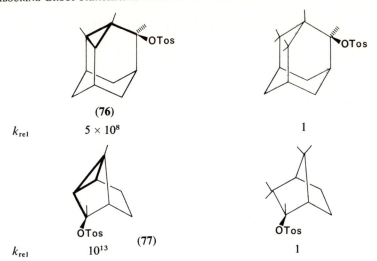

k_{rel} **(76)** 5×10^8 1

k_{rel} 10^{13} **(77)**   1

extensive rearrangements take place resulting in more or less complete carbon scrambling when starting from cyclopropylmethyl, cyclobutyl or allylcarbinyl substrates, each giving similar mixtures of products, showing that the intermediate carbonium ion is very labile. The parent ion has been observed in

superacid solution at $-80°$ by Olah, and its n.m.r. spectrum found to be consistent with either of the non-classical structures, each of which are undergoing a rapid degenerate rearrangement (**78, 79**).

or   **(78)**

(79)   etc.

3.5   **PARTICIPATION BY σ-ELECTRONS**

The S_N2 reactions of neopentyl halides (**80**) are very sluggish on account of the
steric hindrance to attack caused by the three β-methyl groups. The ionization
route to solvolysis (S_N1) is not attractive in solvents such as ethanol or acetic
acid as these compounds are primary halides. In very ionizing media however,
the ionization process is much more favourable and comes into play. The
result is a transition between neopentyl, being very much *less* reactive than
ethyl, to becoming very much *more* reactive, and moreover, undergoing
simultaneous rearrangement to the *t*-pentyl cation.

Solvent	EtOH	CF$_3$COOH	FSO$_3$H
$\dfrac{k_{\text{ethyl}}}{k_{\text{neopentyl}}}$	2.5×10^5	2.1×10^{-3}	1.2×10^{-4}

(80)   (81)

intermediate or

transition state?

Not only does the neopentyl compound lose its inert character when forced
into ionization, but it becomes exceptionally reactive. This suggests that the
ionization is assisted by electrons released by the σ-bond of the methyl group
as it migrates. The transition state (or possibly intermediate) in this rearrangement resembles the pentavalent species described in Section 2.30 (**81**). This
process is a Wagner–Meerwein rearrangement and it is possible that other such
changes are accelerated by σ-participation. However, against this explanation
it must be mentioned that in trifluoroacetic acid, neopentyl ester solvolysis
occurs without any secondary isotope effect at the migrating methyl group i.e.
the rates of solvolysis of **82** and **83** are the same. If the transition state was such
that partial migration of the methyl group had occurred (**84**) one would expect
a slightly lower rate for the deutero compound as is found in the pinacol
rearrangement of Ph$_2$C(OH)—C(CH$_3$)$_2$ [(CD$_3$)$_2$]—see p. 171.

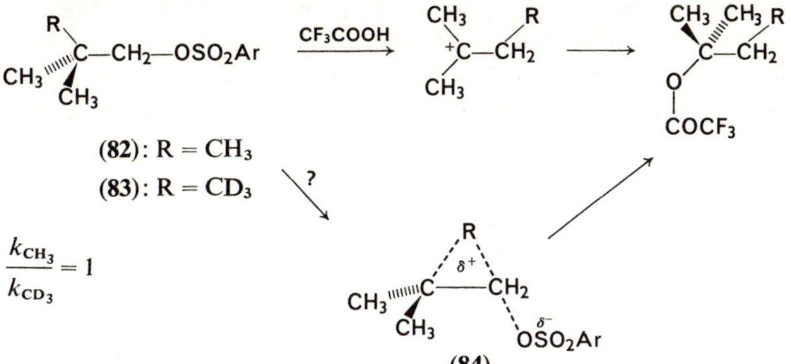

(82): R = CH₃

(83): R = CD₃

$$\frac{k_{CH_3}}{k_{CD_3}} = 1$$

(84)

The most widely discussed example of σ-participation is in solvolytic reactions of exo-norbornyl esters. The exo isomers, e.g. **85a** are much more reactive than the endo (**85b**) and skeletal rearrangement evidently occurs as C-2 eventually becomes distributed at positions 2 and 4 of the product. The rearrangement is due to a Wagner-Meerwein 1,2-shift which regenerates the norbornane skeleton.*

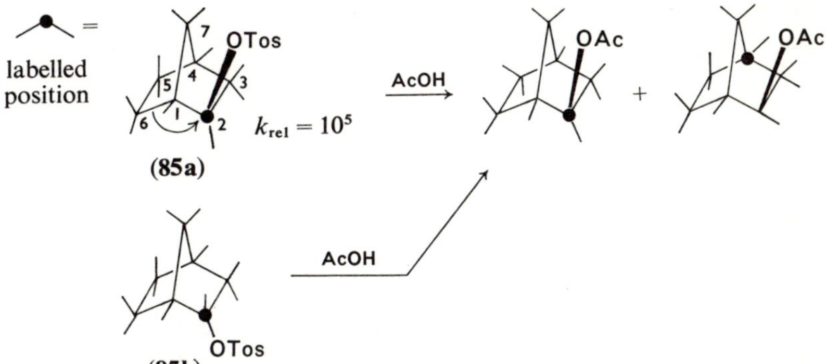

= labelled position

$k_{rel} = 10^5$

(85a)

(85b)

$k_{rel} = 1$

*

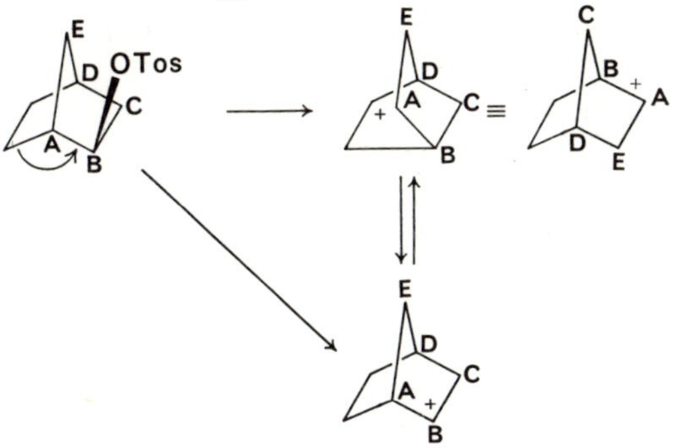

It has been plausibly argued that the high *exo/endo* rate ratio, the retention of configuration and rearrangement are best accommodated by a mechanism in which rearside σ-participation leads to an intermediate norbornyl cation described as the non-classical species (86). However it is becoming apparent

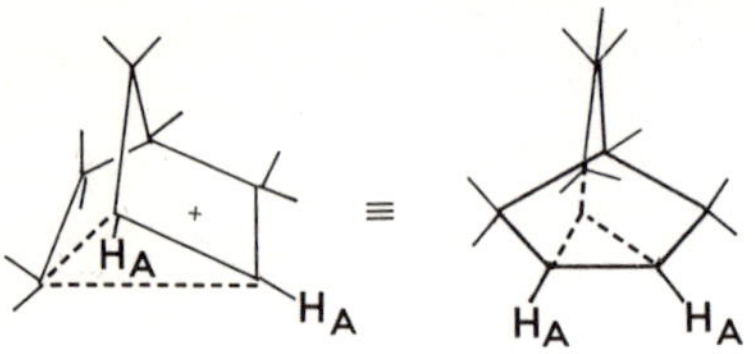

(86) 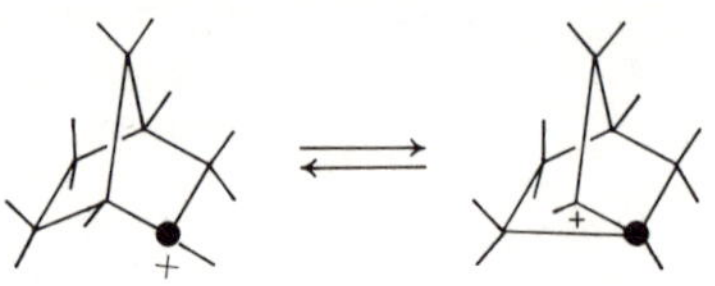   (87a)   (86)

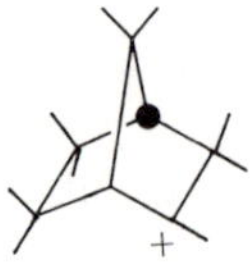

that the *exo* norbornyl esters are not unduly reactive; it is the *endo* isomers which are very *unreactive* and the reason for the latter is clear. Ionization forces the leaving group to approach the 6-*endo* hydrogen with a consequent increase in the energy of activation compared to the exo compound. The high

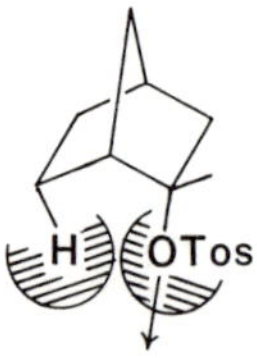

exo specificity of the products from both *exo* and *endo* esters is a consequence of this steric factor; the *exo* side is far more accessible to attack by a reagent. This factor may be illustrated by the product ratios obtained by addition to the olefins (88–90); 89 and 90 are even more hindered than norbornene (88) to

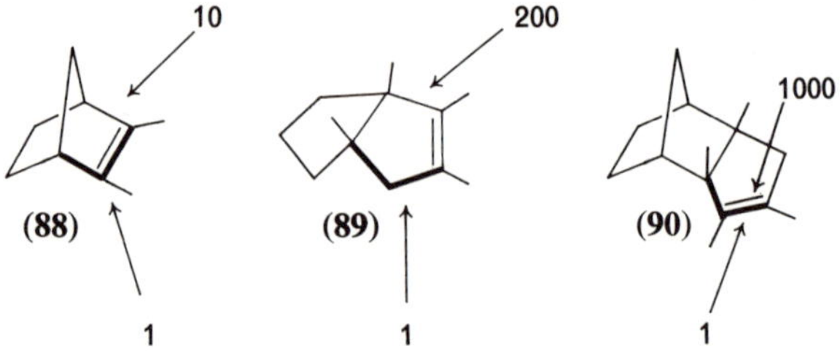

(88)   (89)   (90)

1   1   1

attack at their *endo* sides. One should therefore be cautious before assigning a non-classical structure to the norbornyl cation; the equilibrating classical structures **87a** ⇌ **87b** can accommodate the kinetic and product evidence adequately.

The norbornyl cation may be obtained as a stable species in superacid solution at low temperature and evidence presented that this species is the non-classical ion (**85**). Thus the n.m.r. spectrum shows sharp resonances assigned to protons A and B (Figure 4). The Roman spectrum has been interpreted as due to a species with a threefold rotation axis (which is possessed only by the non-classical ion) and the photoelectron spectrum (Section 2.28) in a solid matrix at $-180°$ has been reported to show carbon $1s$ ionization from positive carbon to neutral carbon in the ratio $2:5$. However these interpretations have not met with universal agreement and indeed the last-named

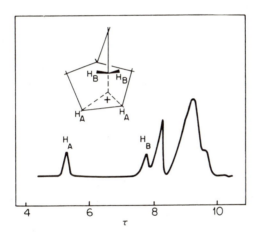

Figure 4. Proton n.m.r. spectrum of the 2-norbornyl cation.

ratio has been claimed by other workers as $1:6$—as expected from a classical ion. It seems fair to say that even the species in superacid may also be a classical ion and that the subject is still controversial.

3.6   ELECTROPHILIC PARTICIPATION

An electrophilic group, usually a proton donor, may facilitate an acid-catalysed reaction by an intramolecular proton transfer. Examples of this type of neighbouring group participation are less common than the nucleophilic type, possibly because proton transfer from an external acid is so fast in any case. An example is the hydrolysis of the salicylic acid acetal,

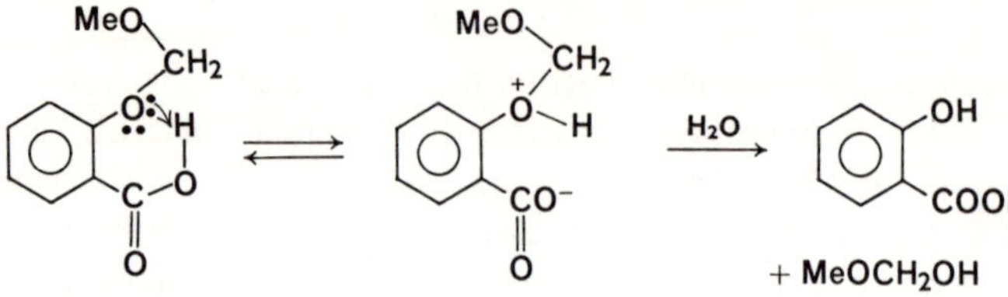

Acetal hydrolysis is general acid catalysed and normally the rate would vary inversely with pH; in this case, throughout the acidic region the rate of hydrolysis is almost independent of the pH of the medium.

The hydration of the acetylene (91) may occur by a 'push–pull' mechanism in which both intramolecular nucleophilic and electrophilic catalysis is

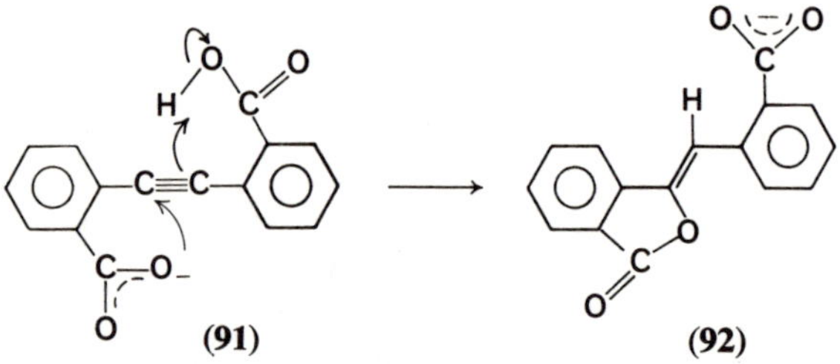

(91)   (92)

occurring. The rate is proportional to the concentration of *mono* anion and the product is the enol lactone 92. Similarly the hydrolysis of the acetal 93 is 10^5 times faster than that of the corresponding *para* isomer 94. Further examples of participation by carboxyl groups are discussed in Chapter 8.

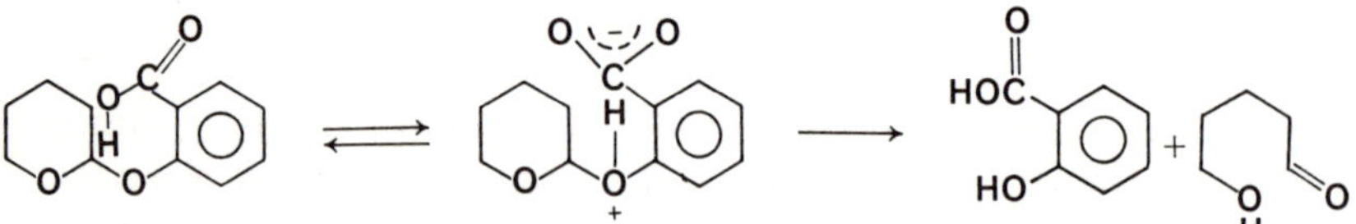

(93)

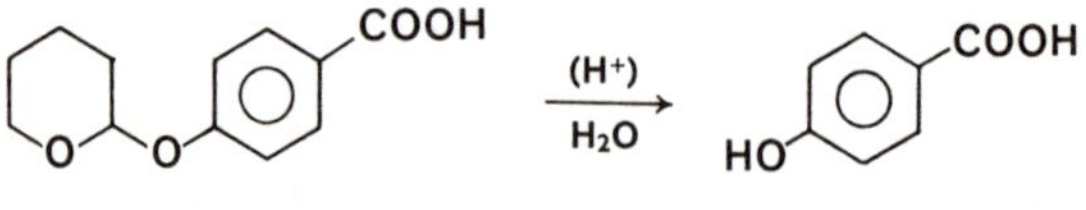

(94)

SUGGESTIONS FOR FURTHER READING

General texts, see Chapter 1

H. C. Brown, 'The Norbornyl Cation—Classical or Non-classical?' *Chem. Brit.*, **2**, 199 (1966).

B. C. Capon, 'Neighbouring group participation', *Quart. Rev.*, **1964**, 45.

G. Dann Sargent, 'Bridged, non-classical carbonium ions', *Quart. Rev.*, **1966**, 301.

S. Winstein, 'Non-classical ions and homoaromaticity', *Quart. Rev.*, **1969**, 141.

Chapter 4

Carbanions

INTRODUCTION

Many organic reactions appear to depend upon the ability of a C—H bond to break heterolytically under appropriate conditions and to give up a proton to a base. What remains is a species containing a trivalent carbon atom with lone

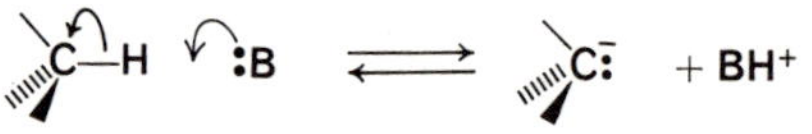

pair, isoelectronic with an amine, and known as a carbanion (**1**). The geometry

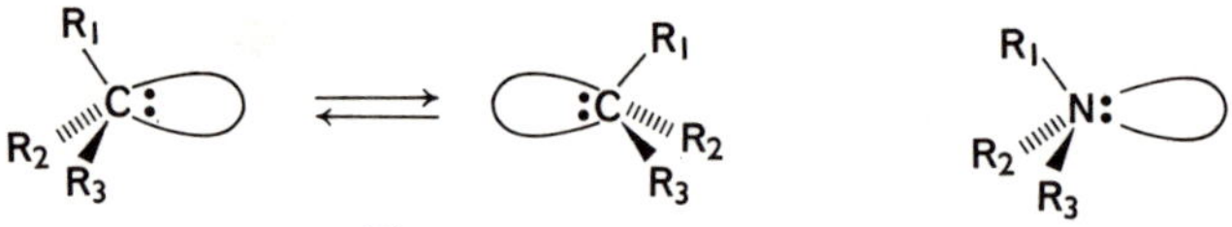

(**1**)

of a carbanion may be described in terms of a pyramidal model in which carbon uses sp^3-hybridized orbitals and readily undergoes inversion like an amine or a planar model with sp^2-hybridization and the lone pair in a p orbital which is more suitably disposed towards conjugation with a neighbouring unsaturated group (**2**). Which structure is more favourable depends upon the individual carbanion.

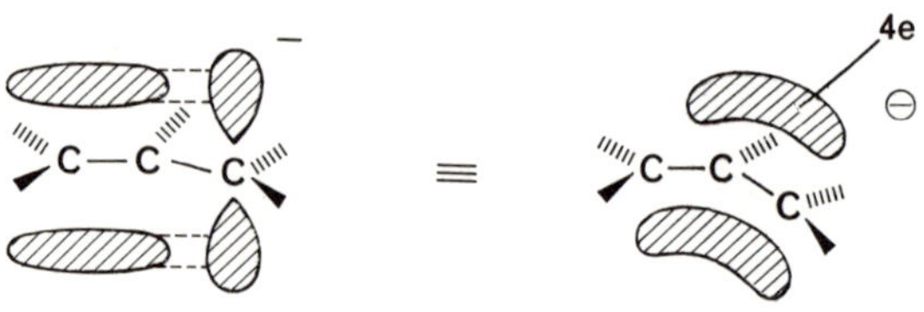

(**2**)

A carbanion is itself a base and a nucleophile generally and is the conjugate base of a *carbon acid* whose strength may be described in the usual way in terms of a dissociation equilibrium K is a measure of the strength of the carbon

$$\text{\textbackslash}C\text{—}H + :B \underset{k_{-1}}{\overset{k_1}{\rightleftharpoons}} \text{\textbackslash}C:^- + BH^+ \qquad K_a = k_1/k_{-1}$$

acid in the particular basic system used, the dissociation constant. The stability of a carbanion therefore may be equated with its basicity and hence the acidity of the corresponding carbon acid. This may be expressed either as the equilibrium constant, K_a, or the ionization rate, k_1.

In principle, any C—H bond may be considered to be capable of acting as a carbon acid, yet the acidities of C—H bonds span a range of more than 50 orders of magnitude, ranging from compounds comparable with the mineral acids to paraffin hydrocarbons whose acidity is barely discernible. Indeed, free, simple alkyl carbanions may be said not to exist as such; however, metallic derivatives, though essentially covalent, may have a good deal of carbanion character. Species such as **3**, **4**, and **5**, act as if they were simple carbanions in many reactions. In order to understand the chemical features

$$CH_3CH_2CH_2\overset{\delta^-}{C}H_2\text{—}\overset{\delta^+}{Li} \qquad \overset{\delta^-}{C}H_3\text{—}\overset{\delta^+}{MgBr} \qquad \overset{\delta^-}{Ph}\text{—}\overset{\delta^+}{Na}$$

$$(3) \qquad\qquad (4) \qquad\qquad (5)$$

which facilitate carbanion formation, an account of carbon-acidity measurements is first given before considering reactions which proceed by way of carbanionic intermediates.

4.2   EQUILIBRIUM FORMATION OF CARBANIONS— THERMODYNAMIC STABILITY

The value of the equilibrium constant, K, for the acid–base exchange,

$$R\text{—}H + :B \overset{K}{\rightleftharpoons} R^- + BH^+$$

is a measure of the strength of the carbon acid, RH, and of the stability of the carbanion, R^-. This is given by

$$K = \frac{[R^-][BH^+]}{[RH][B]} \log \left(\frac{f_{R^-} f_{BH^+}}{f_{RH} f_B} \right)$$

The activity coefficient terms, f, are usually omitted for lack of information of their values. This leaves an *apparent* dissociation constant, K' which is experimentally accessible. To obtain values of K', the concentrations of the four species at equilibrium need to be measured. Analytical methods usually require that these concentrations should not be too dissimilar or inaccuracies result. It follows that for reasonably reliable estimates of K, the basic strengths of B and R$^-$ should be roughly of the same order of magnitude. Hence, to cover the large range of carbon acidities which exist, a wide range of bases and a series of different media are required, each appropriate to a relatively small range of acidity.

'Superstrong' Carbon Acids

A number of cyanocarbons are known which are completely dissociated in water and whose strengths are comparable with the strong mineral acids. Tricyanomethane is one of the simplest examples; three cyano groups each exert a powerful $(-M)$ effect weakening the C—H bond and stabilizing the carbanion by conjugation. Acids such as tricyanomethane are too acidic to be

measured in water because, like sulphuric acid, they are completely dissociated. The ionization may be partially suppressed by added strong acid (common-ion effect) and under these conditions an equilibrium constant is measured. Table 1 gives relative acidity values for some 'superstrong' carbon acids in aqueous perchloric acid.

'Strong' Carbon Acids (pK ca 1–18)

Many carbon acids are appreciably ionized in water or at least in the presence of OH$^-$. In almost all cases the acidic proton is flanked by $(-M)$-substituent groups and the charge in the carbanion mainly resident upon a heteroatom. For example,

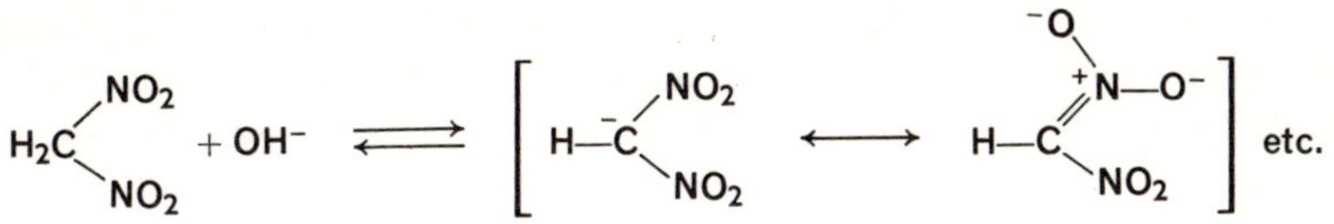

$$H_2C \overset{NO_2}{\underset{NO_2}{\diagdown}} + OH^- \ \rightleftharpoons \ \left[H-\overset{-}{C} \overset{NO_2}{\underset{NO_2}{\diagdown}} \longleftrightarrow H-C \overset{\overset{+}{N}\diagdown_{O^-}^{O^-}}{\underset{NO_2}{\diagdown}} \right] etc.$$

$$pK = 3.57$$

$$H_2C \overset{COCH_3}{\underset{COCH_3}{\diagdown}} + OH^- \ \rightleftharpoons \ H-\overset{-}{C} \overset{O\diagdown C - CH_3}{\underset{COCH_3}{\diagdown}} \longleftrightarrow H-C \overset{O^- \diagdown C-CH_3}{\underset{COCH_3}{\diagdown}} etc.$$

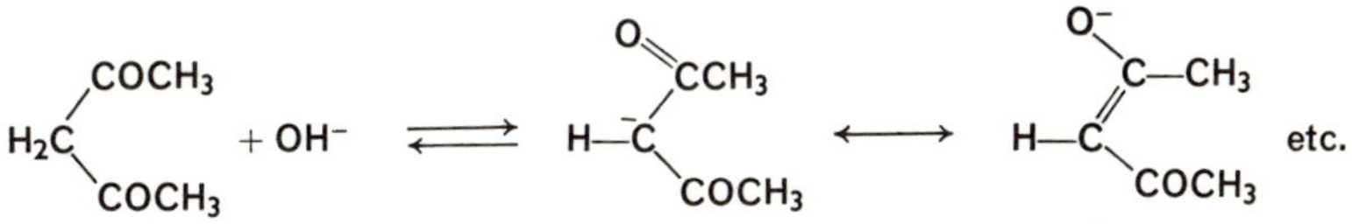

$$pK = 5.85$$

The values of pK for these compounds may be measured by normal techniques appropriate to weak acids in water (e.g. carboxylic acids and phenols), such as by measuring the pH of the solution at half neutralization when

$$pK = pH_{\frac{1}{2}} \quad \text{since} \quad pK = \frac{[A^-]}{[HA]}[H^+] \text{ and } [A^-] = [HA]$$

Table 1. Relative acidities of cyanocarbon acids (in aq. HClO$_4$)

Cyanocarbon acid (Acidic protons underlined)	pK_{rel}
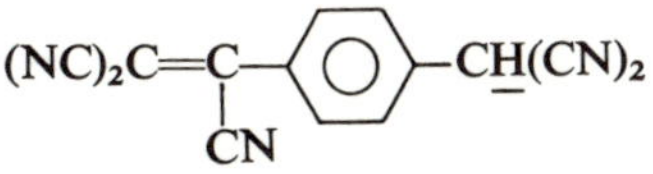 (NC)$_2$C=C(CN)—C$_6$H$_4$—CH(CN)$_2$	0 (reference)
(NC)$_2$CH.COOMe	−1.2
(NC)$_3$CH	−4.5
(NC)$_2$CH.C(CN)=C(CN)$_2$	−8.5
(NC)$_2$CH—C(=C(CN)$_2$)—CH(CN)$_2$	−8.5

Table 2 contains representative carbon acids in this range. The strongest purely *hydrocarbon* acid known at present is tris-(9-fluorenylidenemethyl)-methane (**13**), $pK = 6.2$, the reasons for the acidity of which are discussed in Section 4.10.

Table 2. Strengths of 'strong' carbon acids

	Carbon acid (Acidic proton underlined)	pK
(**6**)	$C\underline{H}(NO_2)_3$	0
(**7**)	$C\underline{H}(CN)_3$	0
(**8**)	$C\underline{H}(SO_2Me)_3$	0
(**9**)	$C\underline{H}_2(NO_2)_2$	3.57
(**10**)	$C\underline{H}_2(CHO)_2$	5.0
(**11**)	$CH_3COC\underline{H}_2COCF_3$	4.7
(**12**)	$C\underline{H}(COCH_3)_3$	5.85
(**13**)	$C\underline{H}(Fm)_3{}^a$	6.2
(**14**)	$C\underline{H}_2(Fm)_2{}^a$	8.8
(**15**)	$CH_3COC\underline{H}BrCOCH_3$	7.0
(**16**)	$CH_3COC\underline{H}_2COCH_3$	9.0
(**17**)	$C\underline{H}_2(CN)_2$	11.2
(**18**)	Fluoradene[b]	11
(**19**)	$CH_3COC\underline{H}EtCOCH_3$	12.7
(**20**)	9-Cyanofluorene 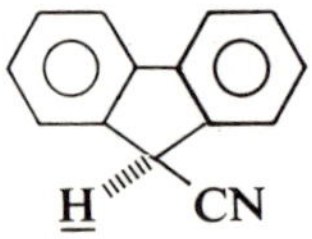	11.4
(**21**)	Cyclopentadiene 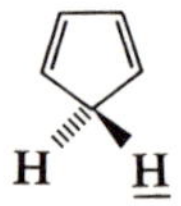	14

[a] Fm = 9–Fluorenylidenemethyl   [b]

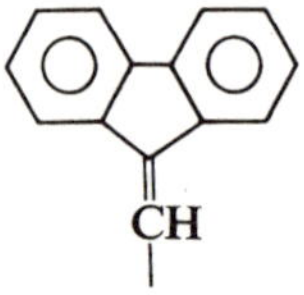

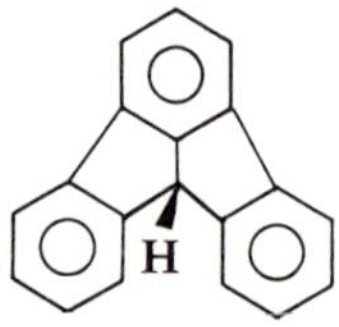

'Weak' Carbon Acids (pK 16–30)

Water, with the strongest base available OH⁻, is too acidic a solvent to support appreciable dissociation of acids with p$K < \sim16$. Useful dissociation equilibria may however be observed in more basic media such as sodium ethoxide in ethanol. Another, which has been introduced by Streitwieser, is cyclohexylamine, containing as base its caesium salt, caesium cyclohexylamide. This will

dene, p$K = 21$

bring about sufficient dissociation of acids of pK up to about 30, the limit being due to the acidity of the solvent. Cyclohexylamine is a protic, fairly polar solvent, hence ion pairing and other complications are likely to be unimportant. This is also minimized by the use of caesium, a very large cation, as the counter-ion.

The usual pH scale expressing hydrogen-ion activities cannot be used in non-aqueous media. Instead, a comparative scale of acidities is obtained by allowing a pair of weak acids to come to equilibrium,

$$RH + R'^- \rightleftharpoons R^- + R'H$$

If RH represents a weak acid of known pK, and R'H an acid slightly weaker than RH and of unknown pK,

$$pK_{RH} - pK_{R'H} = \Delta pK = -\log\frac{[R^-]}{[RH]} + \log\frac{[R'^-]}{[R'H]} \qquad (4.1)$$

If the extent of dissociation of each acid (i.e. the concentrations of the four species present) can be measured, the only unknown in equation 4.1 is p$K_{RH'}$. Spectrophotometric means may often be employed to determine the extent of dissociation of each acid since in many cases, conjugated hydrocarbons and their corresponding carbanions have quite different absorption spectra, as illustrated in Figure 1. It is not, of course, necessary to be able to measure separately the concentrations of both conjugate acid and base; if one is known, the other follows from the known amount of material added to the solution.

If $\Delta pK > $ ca 2, [R⁻]/[RH] may be very large or [R']/[R'H] very small which will lead to low precision in the estimation of these ratios. In order to

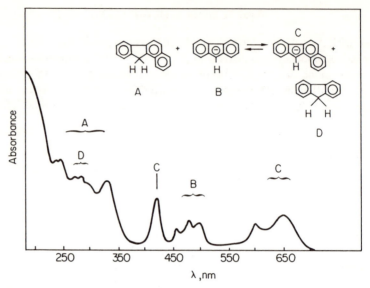

Figure 1. Typical ultraviolet spectrum of an equilibrium mixture of carbanions and their conjugate acids.

extend the acidity scale to weaker acids, R′ H is now used as the acid of known strength and compared with a slightly weaker carbon acid and the process continued until the ionization of yet weaker acids is limited by the ability of the medium itself to cause sufficient dissociation. In this way, values of pK have been obtained for many weak acids including hydrocarbons, whose special features render them more than usually acidic (pK 18–30). The scale is ultimately referred to an acid whose dissociation constant may be measured independently, e.g. 9-phenylfluorene (**22**), pK = 18.5 (Table 3). This type of approach was pioneered by McEwan who equilibrated the sodium or potassium salts of pairs of acids in ether:

$$R^-Na^+ + R'H \underset{}{\overset{Et_2O}{\rightleftharpoons}} RH + R'^-Na^+$$

The spectra observed are now known to be due to ion pairs rather than to free carbanions which result in slight differences in the acidities as measured by the McEwan and the Streitweiser methods.

   Media at present available are not sufficiently basic to support the dissociation of ultra-weak carbon acids—paraffins, simple olefins or benzene derivatives—which lie 10–15 pK units beyond triphenylmethane (**31**), (pK = 31.5).

Table 3. Dissociation constants of 'weak' carbon acids

Carbon acid (Acidic proton underlined)		pK
(22)	9-Phenylfluorene	18.5 (reference)
(23)	3,4-Benzfluorene	19.4
(24)	1,2-Benzfluorene	20.0
(25)	Fluorene	22.9
(26)	2,3-Benzfluorene	23.2
(27)	Indene	21.0
(28)	Acetylene	25.0
(29)	1,1,3-Triphenylpropene	26.4

For (28): HC≡CH

For (29): Ph—CH—C=C with Ph, Ph groups

Table 3. —*continued*

	Carbon acid (Acidic proton underlined)		p*K*
(30)	4-Phenyldibenzpyran	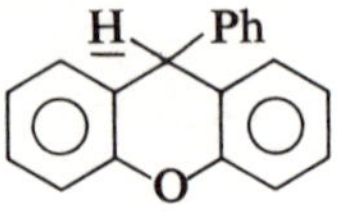	29.0
(31)	Triphenylmethane	$Ph_3C\underline{H}$	31.5
(32)	Diphenylmethane	$Ph_2C\underline{H}_2$	33.0
(33)	Toluene	$PhC\underline{H}_3$	ca 35
(34)	Tropilidene (cycloheptatriene)	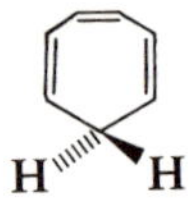	ca 36
(35)	Benzene	Ph—$\underline{H}$	ca 37
(36)	Cyclopropane	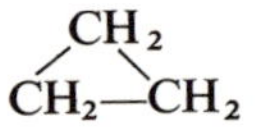	ca 39
(37)	Cyclohexane		ca 45

The best estimates of thermodynamic acidities for such compounds has been obtained from the positions of organometallic equilibria. For example, the system

$$R^-\text{—}Li^+ + Ph\text{—}I \underset{70^\circ}{\overset{Et_2O}{\rightleftharpoons}} R\text{—}I + Ph^-\text{—}Li^+$$

would be expected to lie further to the left as the stability of R^- increased, the equilibrium constant being a measure of the acidity of RH relative to PhH (benzene). Similarly the equilibrium

$$R_2Hg + R'_2Mg \underset{25^\circ}{\overset{THF}{\rightleftharpoons}} R_2Mg + R'_2Hg$$

would lie further to the left the more stable were R'^- compared to R^-, since the organomagnesium compounds are more ionic—'carbanion-like'—than the organomercury ones. Table 4 sets out some results obtained using these equilibria.

Table 4. Relative acidities of 'ultra-weak' carbon acids from the equilibrium, $R—Li + Ph—I \underset{}{\overset{k}{\rightleftharpoons}} R—I + Ph—Li$

	R—H	pK_{rel}
(38)	$CH_2{=}CH_2$	−2.41
(35)	Ph—H	0 (reference)
(36)	$\overset{CH_2}{\underset{CH_2{-}CH_2}{\diagup \diagdown}}$	0.98
(39)	$CH_3—CH_3$	3.50
(40)	Neopentane, $C(CH_3)_4$	5.46
(41)	Cyclobutane	6.14
(42)	Cyclopentane	6.90

4.3   RATES OF FORMATION OF CARBANIONS: KINETIC ACIDITY

As an alternative approach to the measurement of carbanion stability, the rate of a reaction in which carbanion formation is rate determining may be measured. Isotopic hydrogen exchange is the reaction which has been used almost exclusively for this purpose—a high rate of exchange with a protic solvent under standard conditions being equated with an high carbon acidity

$$R—H + S^- \underset{k_{-1}}{\overset{k_1 \text{ (slow)}}{\rightleftharpoons}} R^- + S—H$$

$$R^- + S—D \rightleftharpoons R—D + S^-$$

$$S—H = \text{protic solvent}$$

Either deuterium or tritium exchange may be used but, because of the essentially reversible nature of the exchange, an accurate measure of the rate of the forward reaction, k_1, can only be obtained if it is carried out to very low conversion—i.e. before any appreciable amount of the reverse reaction has occurred.

In principle, the kinetic method is applicable to the whole range of acidities. Very fast exchange rates could be followed by stopped-flow or temperature-jump methods, techniques appropriate to very fast reactions, while slower rates may conveniently be studied by n.m.r. spectroscopy for example, either in the spectrometer or by observation of a reaction mixture over a period of weeks if need be. This facility of extending reaction times indefinitely has led to the kinetic method being applied primarily to the study of very low acidities.

Deuterium exchange with very weak carbon acids has been extensively investigated by Shatenshtein using deuteroammonia, ND_3, as solvent both in the presence and absence of $K^+ND_2^-$. The reactions are carried out in steel or glass ampoules and temperatures up to 150° have been used with the most intransigent compounds. Under these conditions almost any proton will exchange to some extent. One may obtain individual rate constants for different

$$R-H + ND_2^- \; \rightleftharpoons \; R^- + ND_2H \; \xrightarrow{\;\;ND_3\;\;} \; R-D + ND_2^-$$

protons in the same compound, e.g. anisole (43), for which rates of exchange of *ortho*, *meta* and *para* protons may be obtained. Tables 5 and 6 set out some

Table 5. Proton exchange rates (kinetic acidities) of some carbon acids in water

Carbon acid (Acidic protons underlined)	$k_1(s^{-1})$
$C\underline{H}_2(NO_2)_2$	0.83
$C\underline{H}_2(CN)_2$	1.5×10^{-2}
$C\underline{H}_2(COCH_3)_2$	1.7×10^{-2}
$C\underline{H}_2(COCEt)_2$	2.5×10^{-5}
$C\underline{H}_3NO_2$	4.3×10^{-8}
$C\underline{H}_3COCH_3$	4.7×10^{-10}

Table 6. Proton exchange rates of some weak carbon acids in ammonia-d_3

	Carbon acid (Acidic protons underlined)	$k_1(s^{-1})$	Temperature (°C)	$\overset{+}{K}\overset{-}{ND_2}$ (M)
(27)	Indene	0.4	120	0
(25)	Fluorene	0.02	120	0
(44)	Acetophenone, $PhCOC\underline{H}_3$	4×10^{-3}	120	0
(45)	2-Acetylnaphthalene	1.5×10^{-5}	120	0
(31)	Triphenylmethane	2×10^{-7}	120	0
(32)	Diphenylmethane	7×10^{-9}	120	0
(46)	Hex-1-ene	10^{-4}	50	0.06
(47)	Isopentane, $Me_2CHCH_2CH_3$	10^{-7}	25	0.05
(48)	Ethylcyclopropane, $\begin{array}{c} CH_2 \\ | \quad{>}CHEt \\ CH_2 \end{array}$	10^{-4}	25	0.05

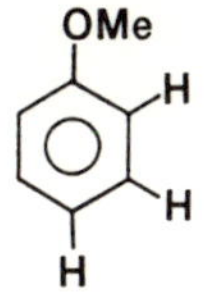

(43)

	$k_{exch\ (rel)}$
o	500
m	1
p	0.5

hydrogen-exchange rates determined in this way. Direct comparison of acidities is only possible where measurements have been conducted under the same conditions.

Characteristics of Proton-Exchange Reactions

Exchange rates are typically found to be accelerated by added salts indicating a transition state more ionic than the starting materials. For instance, the exchange rate of acetophenone (CH_3 group) in water increases by a factor of four in the presence of 3 M calcium nitrate. There is also a large primary kinetic-isotope effect; the relative rates of exchange for toluene, $C_6H_5CH_3$, and deuterotoluene, $C_6H_5CD_3$, are of the order $k_H/k_D = 11$–12 (per D atom) and must indicate a proton almost equally shared with the attacking base in the transition state.

4.4 THE FORMATION OF CARBANIONS IN NON-AQUEOUS MEDIA

Deprotonation of carbon acids to form carbanions will occur in solutions of sufficient basicity. The following are some of the systems more commonly used and the approximate minimum pK which can lead to observable dissociation:

(a) Water; p$K = 9$
(b) Water/OH^-; p$K = 16$
(c) Methanol/OMe^-; p$K = 22$
(d) Cyclohexylamine/alkali metal cyclohexylamide; p$K = 32$
(e) Butyl lithium in ether; possibly up to p$K = 40$
(f) Dipolar aprotic solvents are especially appropriate in the search for more and more basic media.

As has been mentioned (Section 1.18) solvents such as dimethylsulphoxide (49) (DMSO) are only poorly able to solvate anions. Basic anions such as alkoxide and the conjugate base of DMSO (50) behave more like 'free' ions in this type of medium and are able to exert their full powers as proton acceptors unhindered by solvating molecules. Strong bases such as these in DMSO,

$$R—H \qquad :\overset{..}{\underset{..}{O}}—CH_3 \qquad\qquad R—H \qquad :\overset{..}{\underset{..}{O}}—CH_3$$

unsolvated base   solvated base

dimethylacetamide (51), hexamethylphosphorictriamide (HMPT), (52), are capable of dissociating acids of $pK = 30$ to an appreciable extent and rank as some of the most powerfully basic solutions available. The basic properties of

(49)   Na(NaH) →   (50)   etc.   (51)   (52)

these solutions may be measured by indicator methods and expressed on the H_- scale (Section 1.22).

4.5   SPECTRA OF CARBANIONS

Using media of appropriate basicity, solutions of carbanions may be obtained from all but the weakest carbon acids. In this field it is relatively easy to characterize the species by their spectroscopic properties and provide additional evidence for their existence transitory or otherwise.

Electronic Spectra

Simple alkyl anions would not be expected to show ultraviolet absorption in the accessible range, but conjugation with aromatic rings, olefinic groups, etc., brings together the highest filled and lowest vacant orbitals and reduces the excitation energy. As a result, many conjugated carbanions absorb in the visible region (Table 7). A basic reason for the long-wavelength absorption lies in the fact that many conjugated carbanions (like carbonium ions) are odd, alternant

Table 7. Excitation energies of some carbanions

	Hydrocarbon, R—H λ_{max} (nm)	Carbanion, R:$^-$ λ_{max} (nm)
Fluorene	254, 263, 288, 300	452, 477, 510
2,3-Benzofluorene	262, 305, 317, 340	420, 600, 650
9-Phenylfluorene	265, 293, 304	487, 520
Triphenylmethane	256, 262, 269	488

systems and consequently possess a non-bonding molecular orbital, in this case doubly filled. The presence of this orbital, the highest filled, effectively halves the energy gap for the carbanion compared to the parent olefin (Figure 2). Molecular-orbital theory may be used to justify the observed excitation

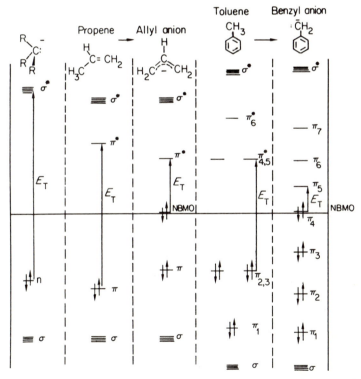

Figure 2. Energy level diagram for some carbanions showing the effect of non-bonding molecular orbitals in electronic transitions.

energies of carbanions. It may be noted that simple theory predicts the same excitation energies for carbanions as for the corresponding carbonium ions and radicals.

N.m.r. Spectra

Valuable evidence of carbanion structure is provided by n.m.r. spectroscopy. The negative charge or high electron density on carbon is shown by the resonance of the attached protons moving to a higher τ-value (i.e. indicating shielding). Furthermore, the alternation of negative charge which is predicted for conjugated carbanions (e.g. canonical structures **53a**) is clearly revealed; for example, the cyclic pentadienide ions (**53** and **54**) produced by the dissociation of 1,4-dienes,

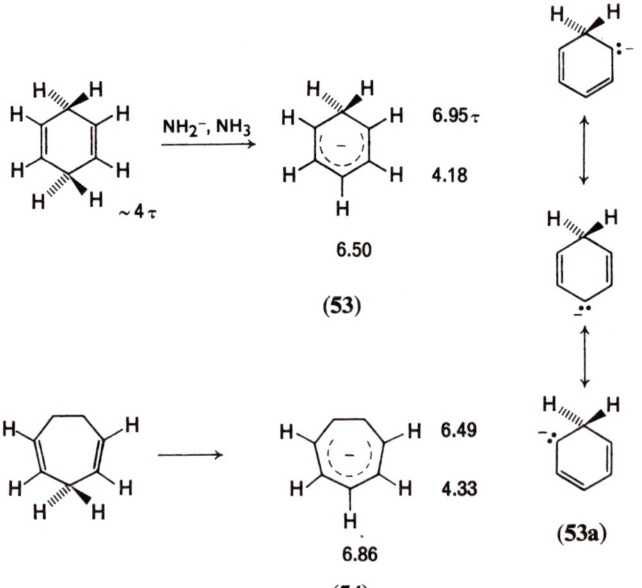

The cyclopentadienide ion (**55**) exhibits a single resonance at 4.39τ due to its symmetry and the equivalence of all five protons. The position of this resonance is the result of an interplay between the shielding effect of the (1/5)-negative charge and the diamagnetic ring-current tending to de-shield the protons; (cf. benzene, with the ring-current but no negative charge, shows a single resonance at lower field, 2.73τ, and tropyllium cation (Section 2.22) with ring current and positive charge, at 0.73τ).

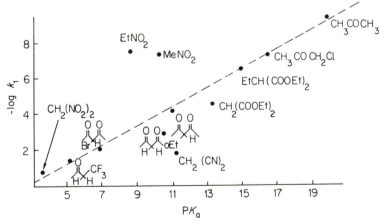

cyclopentadiene

cyclopentadienide
ion

4.6 THE RELATIONSHIP BETWEEN THERMODYNAMIC AND KINETIC ACIDITIES

Brønsted pointed out that a linear relationship should exist between kinetic and thermodynamic acidities, the stronger the acid, the faster it should ionize as expressed by the linear free-energy equation (Section 1.10):

$$\log K = \alpha \log k + \log c \qquad (\alpha, c \text{ are constants})$$

In practice, this relationship holds moderately well only for series of similar types of carbon acid such as cyanocarbons or substituted triarylmethanes. A general plot of log K against log k for all types of carbon acids shows many deviations from linearity (Figure 3); for instance, nitromethane exchanges its proton at a rate much lower than its equilibrium acidity would predict. It is

Figure 3. Plot of thermodynamic acidity (pK_a) against kinetic acidity (log K) for some carbon acids.

suggested that the extensive reorganization of charge in the anion **56** results in relatively little charge residing on carbon; therefore the carbon atom is less likely to abstract a proton from an acid than a base of comparable strength but less delocalized charge. In the acid form (**57**), no such reorganization of charge has occurred and the proton is relatively slow to depart.

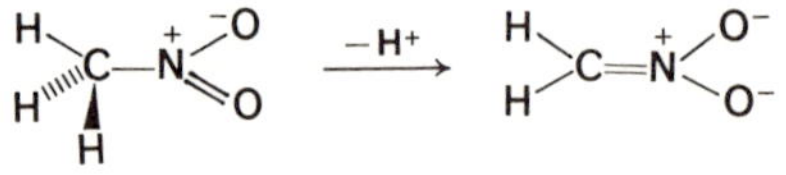

(**57**)   (**56**)

4.7   **ENOLIZATION**

The phenomenon of enolization is closely bound up with carbanion chemistry. A carbonyl compound (**58**), (aldehyde or ketone, ester or even a carboxylic anhydride) which has at least one hydrogen atom on an α-carbon may undergo an internal proton transfer which converts it into a tautomer, the *enol* form **59**. The equilibrium between keto and enol tautomers usually lies heavily on the side of the keto, but the reactivity of the enol is often far greater than that of the keto form and its formation as an intermediate is frequently necessary for reaction to occur. The conversion of keto to enol is both general-acid and general-base catalysed. Intermediates in the tautomerization are, respectively, the protonated carbonyl compound **60**, and the enolate ion **61**. The enolate

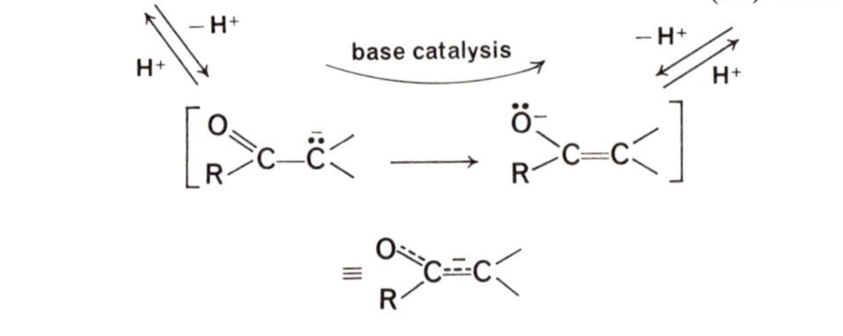

(**60**)

keto (**58**)   acid catalysis   (**59**) enol

base catalysis

(**61**) enolate

ion is a resonance-stabilized carbanion. Many of the moderately strong carbon acids (Table 2) dissociate to form enolate ions, e.g. dimethylmalonate (62).

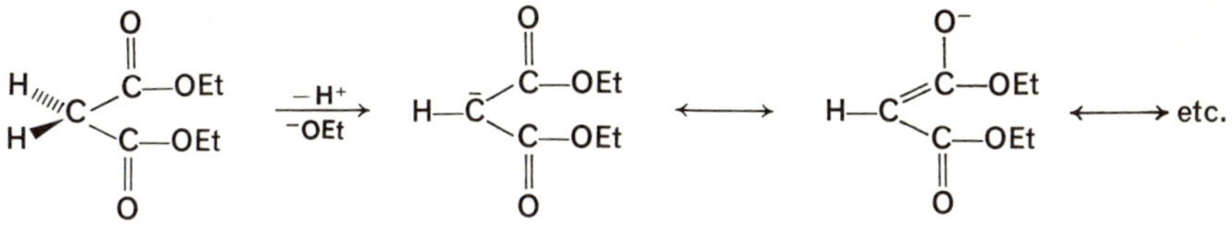

(62)

The equilibrium proportion of enol in pure simple aldehydes or ketones or in neutral solution is very small (and that in esters even smaller). 1,3-Dicarbonyl compounds on the other hand frequently contain a high proportion of enol which is stabilized by internal hydrogen bonding, e.g. in acetylacetone (63). Some typical keto–enol ratios are given in Table 8.

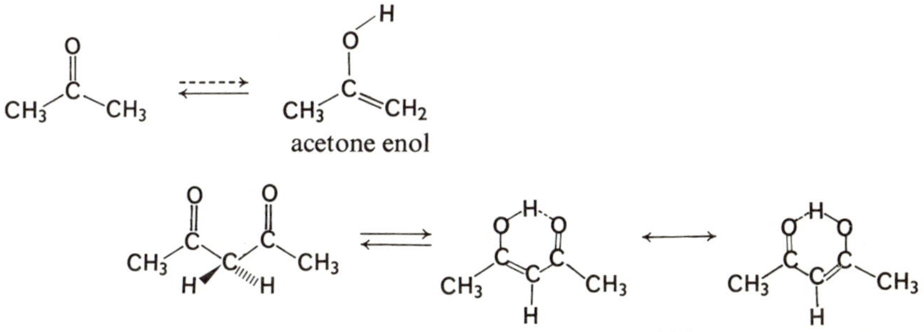

acetone enol

(63)

The determination of these ratios may make use of infrared spectroscopy if the two tautomers are present in comparable amounts. Distinct carbonyl-stretching frequencies are shown by the two forms. If the enol is present in very low concentration it may be determined chemically by a rapid bromine titration at low temperature. Only the enol form reacts and the titration may be carried out before a significant shift of the equilibrium has occurred. It was

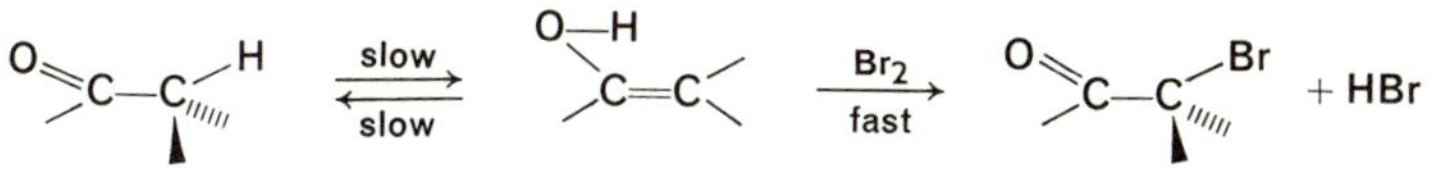

first demonstrated by Lapworth (around 1912) that substitution reactions at carbon α to a carbonyl group proceeded via the enol (or enolate). One of the

Table 8. Keto–Enol equilibria in pure compounds

Compound (Acidic proton underlined)		% enol, 25°
Acetone		2.5×10^{-4}
Cyclopentanone		4.8×10^{-3}
Ethyl cyclopentanonecarboxylate	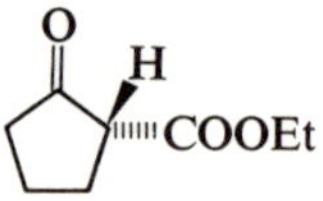	5.0
Ethyl cyclohexanonecarboxylate	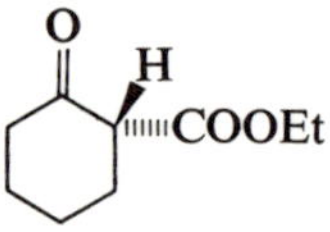	18.0
Ethyl acetoacetate, $CH_3COC\underline{H}_2COOEt$		7.5
Acetylacetone, $CH_3COC\underline{H}_2COCH_3$		78
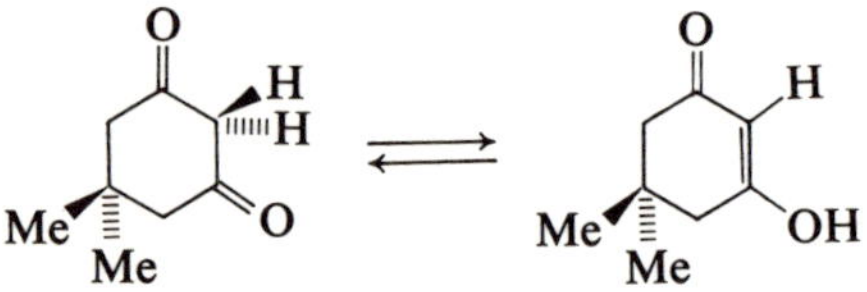 'Dimedone'		>95

telling pieces of evidence was the identical rates measured for different substitution reactions such as bromination, deuterium exchange, iodination,
racemization; it was plausibly postulated that all these reactions required the

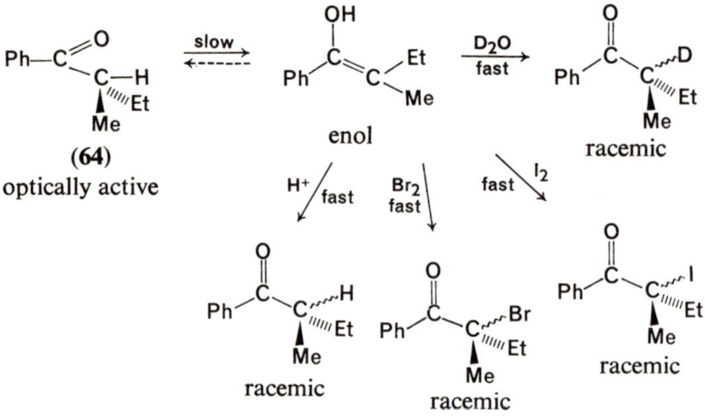

same slow step, the conversion of keto to enol. The reactions of optically active isobutyrophenone (64) illustrate these points. The intermediate is likely to be the enol in acid-catalysed reactions and the enolate ion under base-catalysed conditions, both of which have a plane of symmetry. The rate-determining loss of the α-proton is confirmed by the large primary isotope effects observed. These are also found in substitutions of aliphatic nitro compounds for which a similar enolization mechanism applies (Table 9).

nitro form   aci form
                  ('enol')

Table 9. Deuterium isotope effects for reactions occuring via enolization

Reaction (Labelled position underlined)	k_H/k_D
$C\underline{H}_3COCH_3 + Br_2 \xrightarrow{H_2O} CH_2BrCOCH_3$	7.7
$C\underline{H}_3{-}NO_2 + Br_2 \longrightarrow BrCH_2NO_2$	4.5–6.5
Racemization of	4.6

4.8   METAL-ORGANIC COMPOUNDS

A metal-carbon bond is invariably polarized with carbon the negative end, but the bond may range from a fully covalent to a fully ionic one, passing through various ion pairs, depending on the stability of the carbanion (which may become free), the electropositive character of the metal and the solvent.

polarized   intimate   solvent-   free
covalent   ion pair   separated   carbanion
                            ion pair

It is frequently difficult to know the exact nature of a carbanion, especially when formed transiently. The alkali metals, especially caesium, tend to favour the ionic state while compounds of the Group II, Group III and transition metals, are increasingly covalent. Some, such as the lithium alkyls are associated (**65**). The metal salt of a highly stable carbanion such as sodium cyclopentadienide, especially in dimethylsulphoxide, is likely to exist as free ions while sodium butyl (**66**), for instance, is entirely covalent.

$$CH_3-CH_2-CH_2-\overset{\delta^-}{CH_2}-\overset{\delta^+}{Na}$$

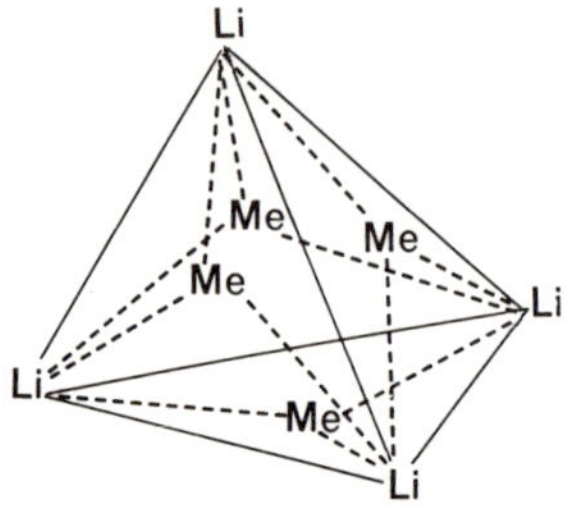

(**66**)

$(LiMe)_4 \equiv$

(**65**)

Alkali-metal alkyls and aryls behave in reactions as sources of carbanions; i.e. they behave as alkide donors. It is probably true to say that the salts of such

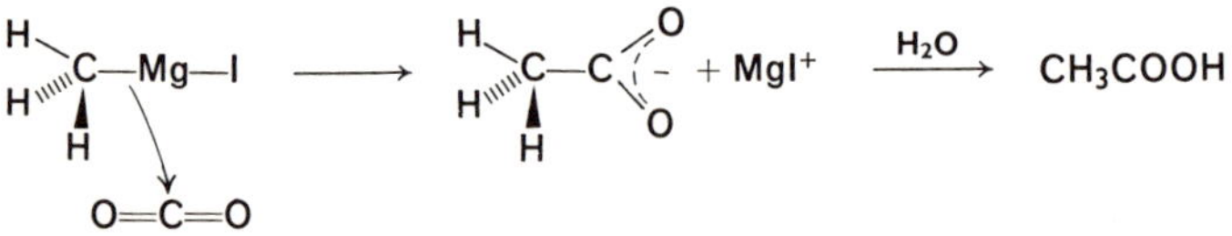

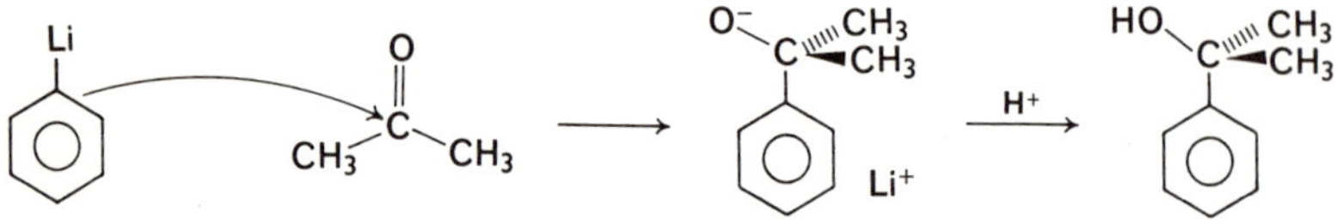

weak acids as paraffin hydrocarbons, benzene, etc., do not exist as ionic species but are always covalently bound to the metallic counter-ion.

It has recently been discovered that certain cyclic polyethers known as 'crown ethers' are capable of complexing alkali-metal cations, e.g. **67**. Addition of crown ether can remove the cation from the ion pair and leave a virtually free anion. This technique has been used to obtain spectra and equilibrium data of carbanions which in their absence, even in cyclohexylamine solution, show differences which must be due to the proximity of the counter-ion.

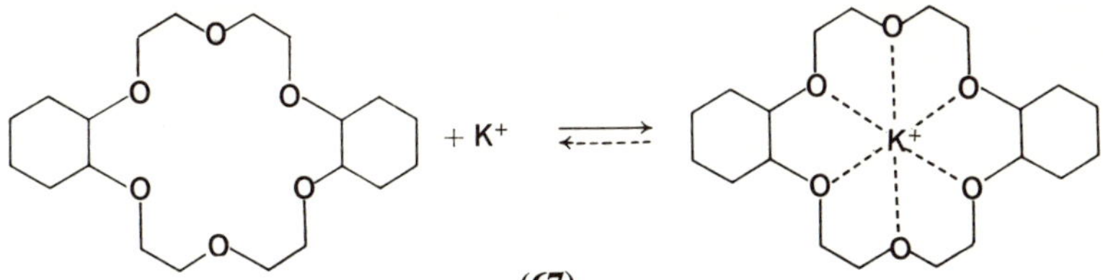

(67)

Dicyclohexano-18-crown-6'(2, 3, 11, 12-dicyclohexano-1, 4, 7, 10, 13, 16-hexa-oxacyclooctadecane)

The addition of crown ether to a reaction is becoming a useful diagnostic test for the involvement of ion pairs bringing about changes in rates and products by 'freeing' an anionic reagent from intimate association with a cation.

4.9   **THE STEREOCHEMISTRY OF CARBANIONS**

Since free simple aliphatic carbanions do not seem to exist it is a matter of speculation as to whether the pyramidal (sp^3) or planar (sp^2) geometries would be preferred. The former would be expected on theoretical grounds and by analogy with amines which are isoelectronic

Conjugated carbanions appear to favour a planar geometry by which π-overlap is more efficient. The X-ray crystal structure of the dinitrobutenamide ion (**68**) shows it to be a completely planar species. The phenylbutenyl

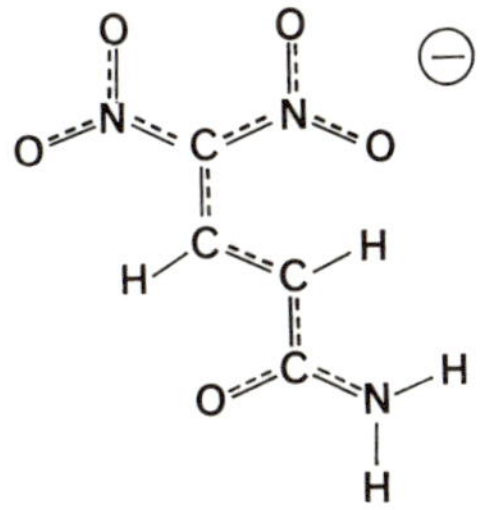

(68)

anion, an allylic carbanion, exists in two geometrically isomeric forms, **69** and **70**, interconverted by rotation about the partial double bonds. The rotation barrier is quite high (83 kJ mol^{-1}) but much less than that for rotation about a full double bond (about 250 kJ mol^{-1}). The benzene ring also shows restricted rotation ($\Delta H = 50$ kJ mol^{-1}) being most stable when it is coplanar with the allyl system. These findings are derived from the temperature-dependent n.m.r.

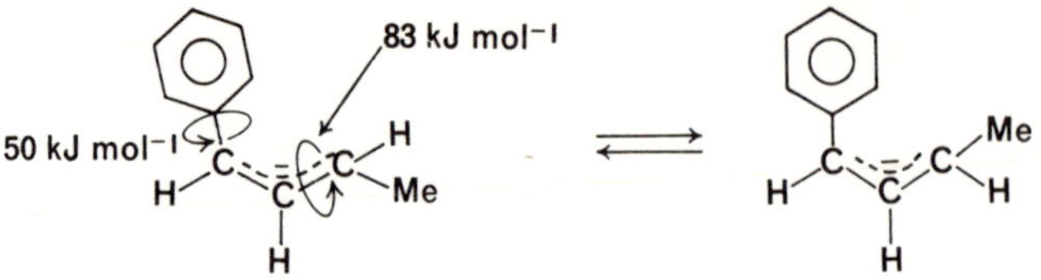

(69)   (70)

spectra of the ion. If a conjugated carbanion is constrained to a non-planar configuration its stability is usually reduced (the corresponding carbon acid is weaker). It is probably for this reason that tryptophane (71) is a weaker carbon acid than triphenylmethane (31). Similarly, the bicyclic dione (72) shows no

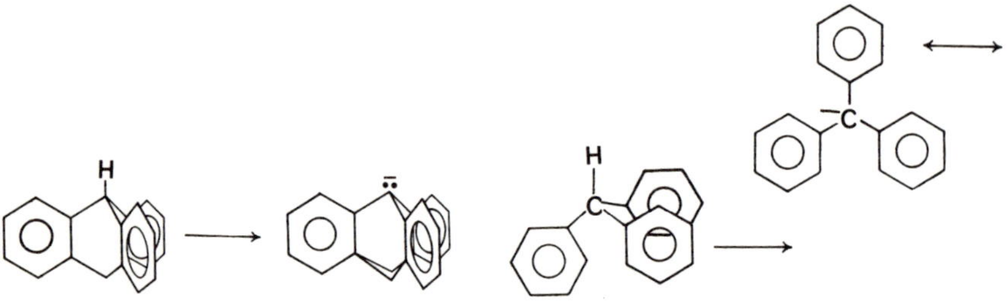

(71)   (31)

pK = 38   pK = 31·5

tendency to enolize. Not only is a planar structure impossible but additional strain energy is introduced by the bridgehead double bond which makes 72a unfavourable (Bredt's rule). c.f. Acetylacetone (16) Table 2; pK = 9. Whether

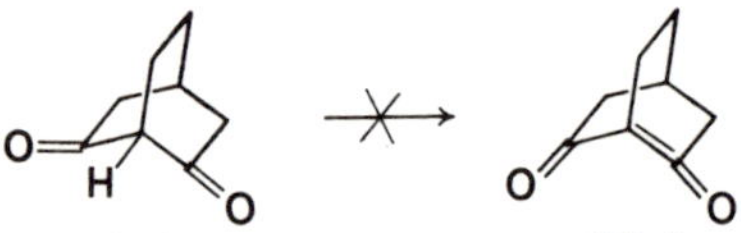

(72)   (72a)

an unconjugated carbanion were to adopt a planar or a pyramidal geometry, it would not show optical activity, since inversion would cause rapid racemization of an asymmetric carbanion. For this reason, it is difficult to prepare asymmetric metal alkyls, unless they are altogether covalent. Frequently, spontaneous racemization occurs by way of even a small amount of ionization. Asymmetric organomercury compounds are highly covalent and often quite stable, e.g. 73. On conversion to the corresponding lithium alkyl, 2-octyl lithium (74), racemization is much more rapid as this is a more ionic compound.

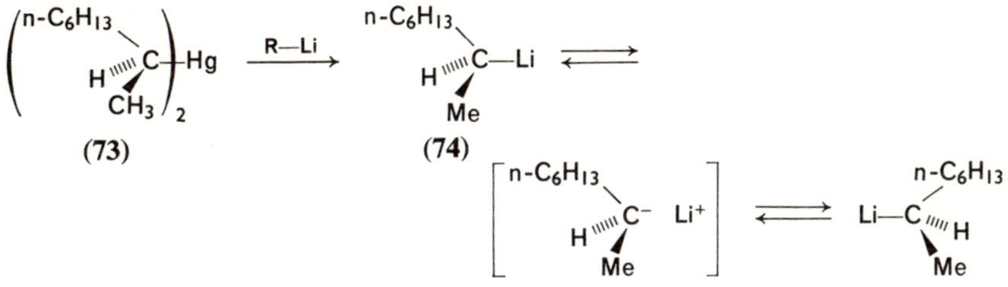

4.10   FACTORS WHICH AFFECT CARBANION STABILITY

We are now in a position to review the data of this section with a view to determining the factors which confer stability on a carbanion, and consequently
increase the strength of a carbon acid.

Delocalization of Charge

The ability of a carbanionic centre to conjugate with a neighbouring π-system,
particularly if the latter contains electronegative elements such as oxygen,
results in the spreading of negative charge on to alternate atoms and stabilization. The following examples of relatively stable delocalized carbanions
illustrate this factor.

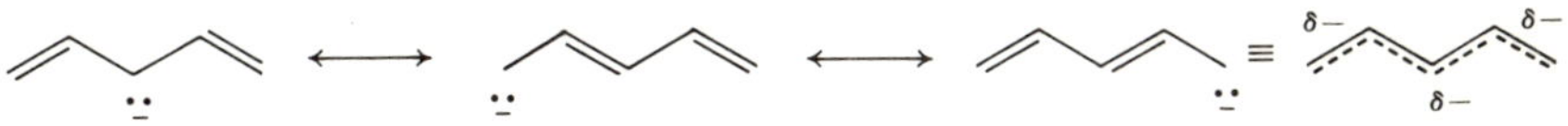

(Note alternation
of charge in odd-carbon
conjugated ion)

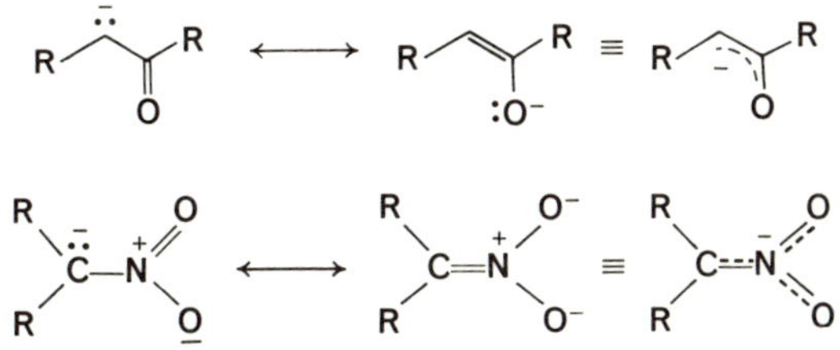

Generation of a carbanionic centre may link together isolated π-systems into a
conjugated whole. This is the major contributing factor to the acidity of triphenylmethane (31) (pK = 31.5) compared, for example, to isobutane (pK ca

40). Substituents which render an adjacent proton especially acidic are: nitro, $-NO_2$, cyano, $-CN$, sulphones, $-SO_2R$ and sulphonic esters, $-SO_2OR$, various carbonyl functions such as ketone, $-COR$ and carboxylic ester, $-COOR$; these act by a combination of mesomeric and inductive effects. Clearly, the more such groups adjacent to the proton, the more acidic it becomes (Table 2).

Inductive Effects

The proximity of a highly electronegative element of a cationic centre, even in the absence of conjugation, will render a C—H bond more acidic by polarization of the σ-bond framework. Thus, α-hydrogen exchange occurs more readily in quaternary ammonium ions than in the corresponding tertiary amines. Adjacent fluorine groups (e.g. CF_3) act primarily inductively as an

$$\overset{+}{R_3N}-CH_3 \xrightarrow{\text{base}} \overset{+}{R_3N}-\overset{-}{CH_2} \xrightarrow{D_2O} \overset{+}{R_3N}-CH_2D$$

acid-strengthening group. The juxtaposition of atoms with lone pairs may destabilize the carbanion because of electron–electron repulsion.

Effect of Hybridization

In general, the more s-orbital character contributed by carbon to a C—H bond, the more acidic it is. This may be considered due to the less directional character and consequently poorer overlap exhibited by a hybrid atomic orbital with a high degree of s-content, and results in the acidity order, paraffin < olefin < acetylene. Furthermore, an s-orbital has a non-zero probability at the nucleus and consequently s-electrons experience more of the effect of the positive charge, thus:

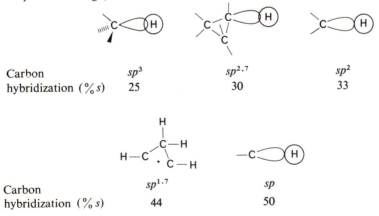

Carbon hybridization (% s)	sp^3	$sp^{2.7}$	sp^2
	25	30	33

Carbon hybridization (% s)	$sp^{1.7}$	sp
	44	50

Acetylenes are well known to be rather acidic as hydrocarbons go (pK for acetylene $= 25$) and readily form metallic compounds of an ionic character.

$$HC\equiv CH + Na^+NH_2 \longrightarrow HC\equiv C:^-Na^+ + NH_3$$

Cyclopropane rings occupy a position between paraffins and olefins with approximately 30% s-character in the C—H bonds and $pK \sim 39$. Cyclopropenes, with about 44% s-character are of comparable acidity to acetylenes. The acidities of cyclic ketones also increase as the ring size decreases and the s-character of the α-C—H bonds increase.

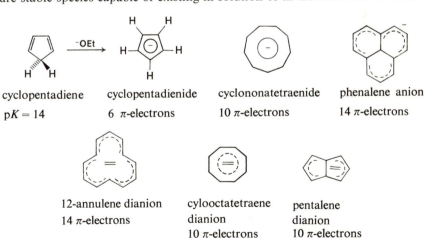

k_{rel} 290   85   12
(exchange)

Aromatic Character

Completely conjugated, planar, cyclic systems with $4n + 2$ (i.e. 2, 6, 10, 14, ...) delocalized π-electrons are aromatic according to the definition given by Hückel and are especially stable. The series of compounds with $(4n + 2)$ π-electrons in a $(4n + 1)$-membered ring are singly-charged aromatic carbanions and those with $(4n + 2)$ electrons in a $4n$-membered ring are doubly-charged carbanions. The following are some examples of aromatic carbanions. All are stable species capable of existing in solution or in the solid state as salts.

cyclopentadiene   cyclopentadienide   cyclononatetraenide   phenalene anion

$pK = 14$ 6 π-electrons   10 π-electrons   14 π-electrons

12-annulene dianion   cylooctatetraene   pentalene
14 π-electrons   dianion   dianion
                         10 π-electrons   10 π-electrons

Cyclopentadiene ($pK = 14$) is far more acidic than cyclohexadiene ($pK \sim 30$) and the benzo derivatives indene (27), fluorene (25) and their derivatives, are also fairly acidic (pK 21 and 23 respectively). The fluorenylidenemethyl (Fm) group owes its acid-strengthening properties to its ability to absorb an electron pair and form an aromatic anion; (Fm)$_3$CH (Table 2) is the most acidic hydrocarbon known ($pK_A = 6.2$).

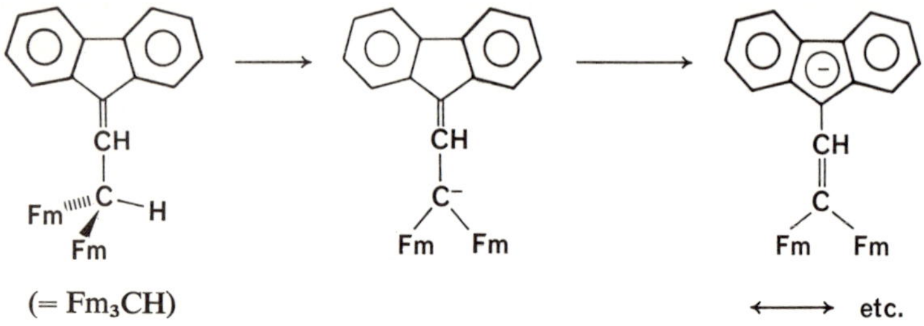

(= Fm$_3$CH)   ⟷   etc.

Homoconjugation and Homoaromaticity

π-Orbital overlap may occur even between non-adjacent carbons leading to increased stabilization by *homoconjugation*. Thus camphenilone (75) exchanges the *endo* β-protons at position 6 under basic conditions via an homoenolate (76). The acidity of 77 is explained in terms of the anion 78, stabilized by virtue

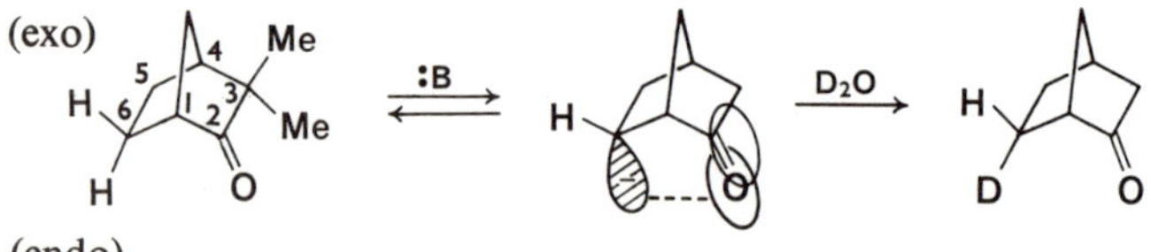

(75)   (76)

of its π-overlap with both the adjacent and the remote double bonds, has six π-electrons and is analogous to the cyclopentadienide ion—a *bis-homo-cyclopentadienide ion* ('bis-homo', indicating that the aromatic system is interrupted by *two* saturated carbons), is stable in solution; its n.m.r. spectrum confirms a symmetrical ion and agrees with the delocalization of charge at C_6 and C_7.

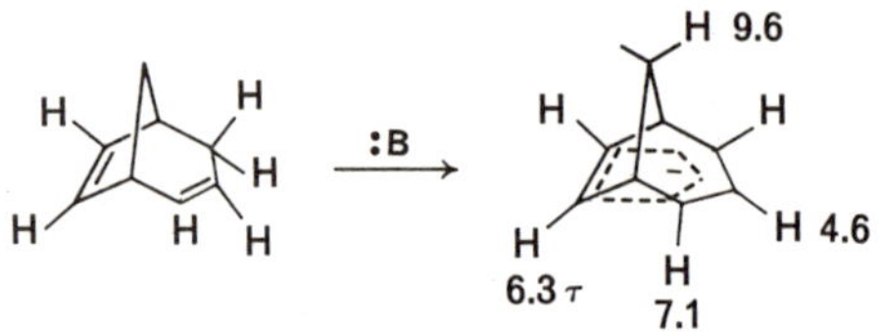

(77)   (78)

Another homoaromatic anion, related to the cyclooctatetraene dianion and stable in solution, is **78a** (cf. the corresponding cation, p. 214).

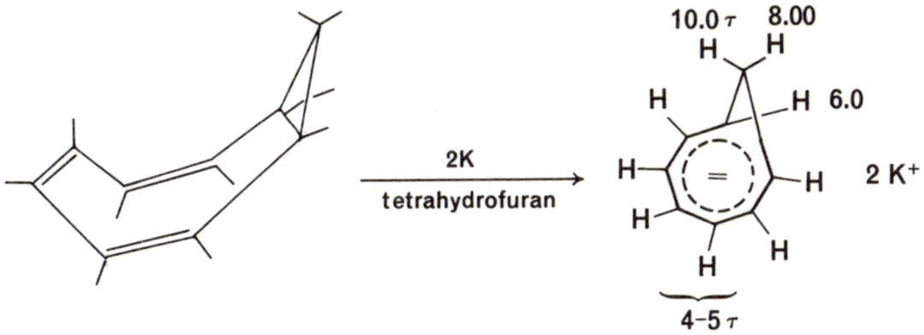

(78a)

d-Orbital Conjugation

Whereas an adjacent quaternary ammonium ion stabilizes a carbanion by its inductive effect, a quaternary phosphonium or tertiary sulphonium ion is far more effective. This is attributable to the overlap of the filled carbon $2p$-orbital with a $3d$-atomic orbital on phosphorus or sulphur which leads to an expansion of the valence shell, permissible only for the second-row elements. The result is an *ylide*, in which the charges are partially neutralized internally (Section 4.25). The methyl protons in **79** and **80** exchange some $10^6–10^7$ times

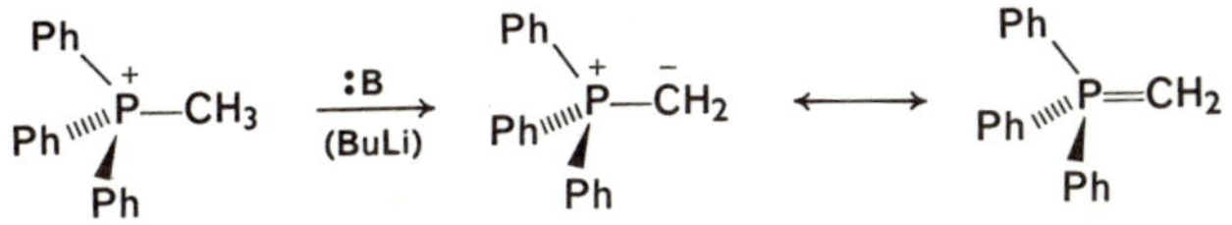

(79)   triphenylphosphonium methylide

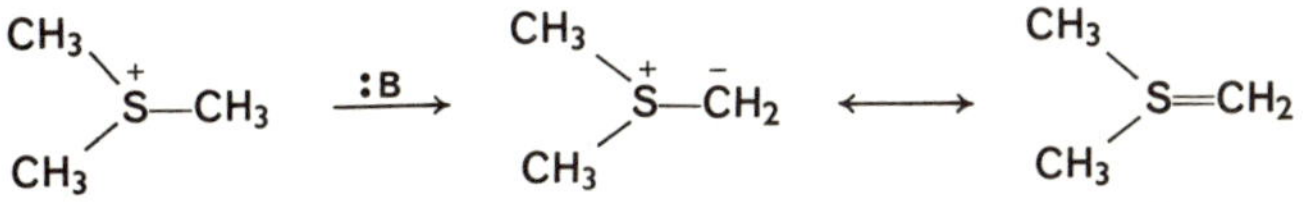

(80)   dimethylsulphonium methylide

faster than those in the nitrogen analogue Me_4N^+. For the same reason, chloroform ionizes much faster than fluoroform, CHF_3, despite the much greater inductive effects in the latter.

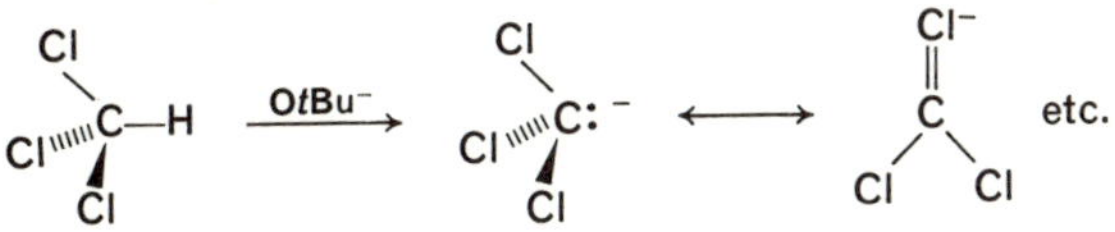

Steric Inhibition of Resonance

The above considerations apply only if a suitable geometry exists for favourable interactions with neighbouring groups. The β-diketone (**72**) does not enolize or exchange the bridgehead hydrogen under even strong basic conditions; the formation of a double bond at the bridgehead is sterically unfavourable as it introduces a large amount of angle strain (Bredt's rule).

4.11   CARBANIONS AS NUCLEOPHILES

The nucleophilicity of a carbanion towards saturated carbon roughly parallels its strength as a base (nucleophile towards hydrogen). Thus for the most part carbanions are powerful nucleophiles and engage in the usual reactions of this class of reagent. The carbanion may be formed transiently or as a long-lived intermediate or may be employed in the form of the organometallic compound.

4.12   NUCLEOPHILIC SUBSTITUTION AT
SATURATED CARBON

The alkylations of diethyl malonate, ethyl acetoacetate and similar β-dicarbonyl compounds, which are highly important as synthetic methods, are S_N2 reactions (Section 2.3) of the carbanions on saturated carbon in an alkyl halide or similar compound. The following examples are typical.

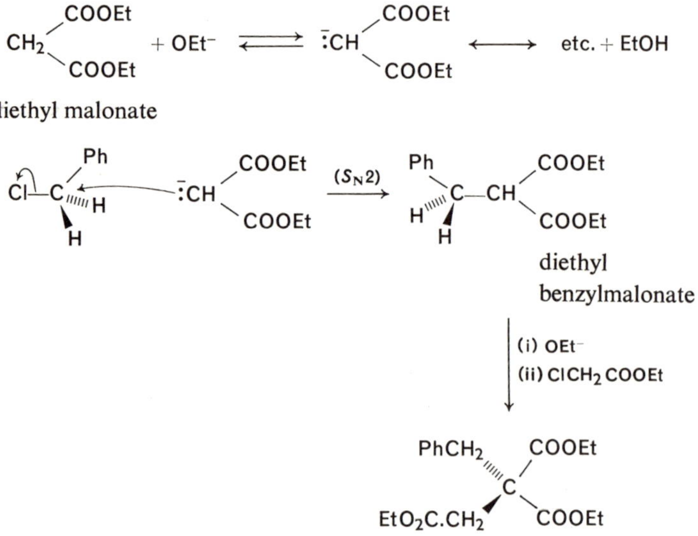

diethyl malonate

diethyl
benzylmalonate

(i) OEt⁻
(ii) ClCH₂COOEt

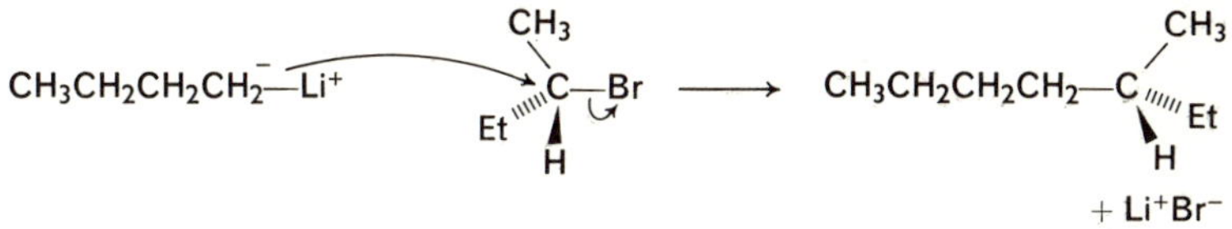

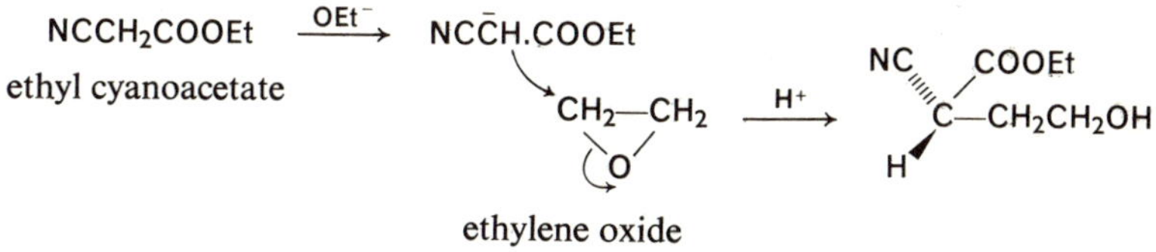

$$CH_3CH_2CH_2CH_2{-}Li^+ \quad + \quad \underset{\underset{H}{|}}{\overset{CH_3}{\overset{|}{\underset{Et}{C}}}}{-}Br \quad \longrightarrow \quad CH_3CH_2CH_2CH_2{-}\underset{\underset{H}{|}}{\overset{CH_3}{\overset{|}{C}}}{\cdots}Et$$

$$+ \; Li^+Br^-$$

$$NCCH_2COOEt \quad \xrightarrow{OEt^-} \quad NC\overset{-}{C}H.COOEt$$

ethyl cyanoacetate

$$\underset{\underset{O}{\diagdown}}{CH_2{-}CH_2} \quad \xrightarrow{H^+} \quad \underset{\underset{H}{}}{\overset{NC}{}}\overset{COOEt}{\underset{}{C}}{-}CH_2CH_2OH$$

ethylene oxide

4.13   CARBONYL α-SUBSTITUTIONS

Base-catalysed α-substitutions at enolizable carbonyl compounds (α-halogenation, isotopic exchange and racemization) readily occur under basic conditions via an enolate ion. In acidic solution the reactive intermediate is the

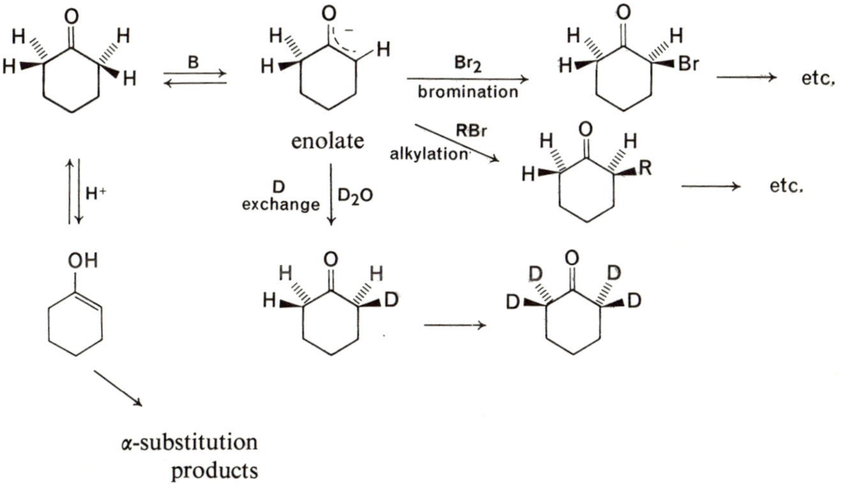

enolate

bromination

alkylation

D exchange

α-substitution
products

enol. Nucleophilic displacement at saturated carbon by the enolate ion could in principle lead to alkylation at the α-positions. This is not important as a preparative method on account of side reactions which occur (aldol condensations, Section 4.16, in particular) but the reaction may be achieved by first converting the ketone to an *enamine* (81)—a nitrogen analogue of an enol.

This then undergoes alkylation cleanly (Stork reaction); finally, the alkylated ketone is regenerated.

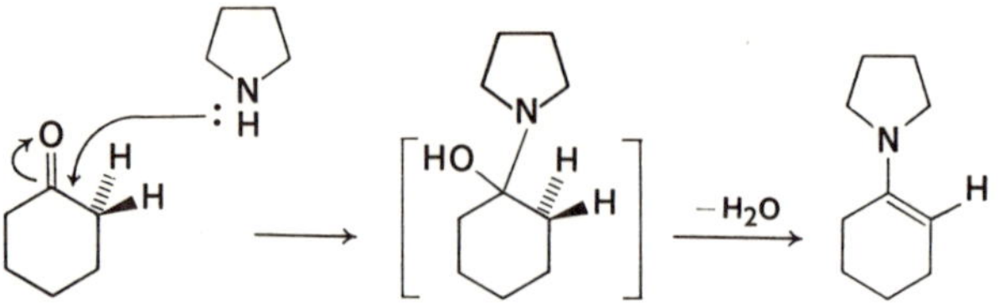

cyclohexanone
pyrrolidine enamine
(isolated)

(81)

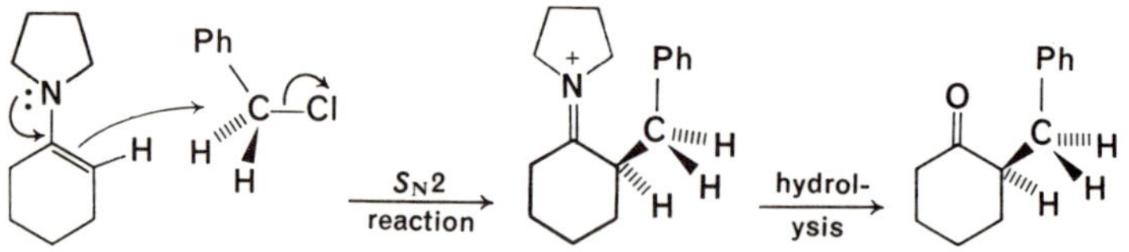

2-benzylcyclohexanone

Alkylation (Stork reaction)

4.14   **THE FAVORSKII REARRANGEMENT**

The Favorskii rearrangement is the conversion of an α-haloketone (possessing at least one α'-hydrogen) to a branched-chain ester under the action of alkoxide ion:

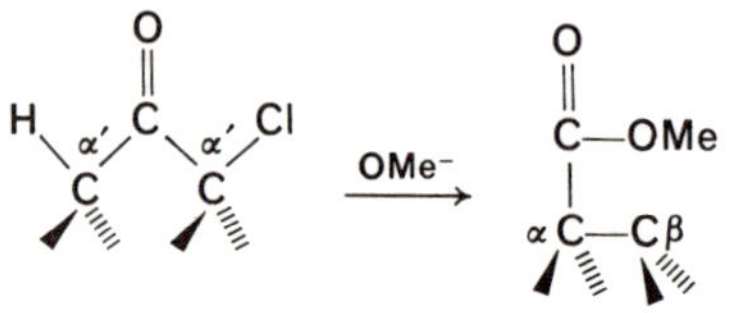

In some cases, isotopic exchange of the α-hydrogen occurs *prior* to the re-arrangement suggesting that carbanion formation occurs in a rapid initial step. Furthermore a carbon label at C_α in the starting material becomes equally distributed between C_α and C_β in the product which means that these two car-bons become equivalent in a symmetrical intermediate. A plausible mechanism of this rearrangement is as follows: the carbanion **82** initiates an internal S_N2 attack displacing halide ion to give a cyclopropanone (**83**). Such species

are known to be ring-opened by strong base and, obviously, both carbons in
the ring are equally likely to be attacked.

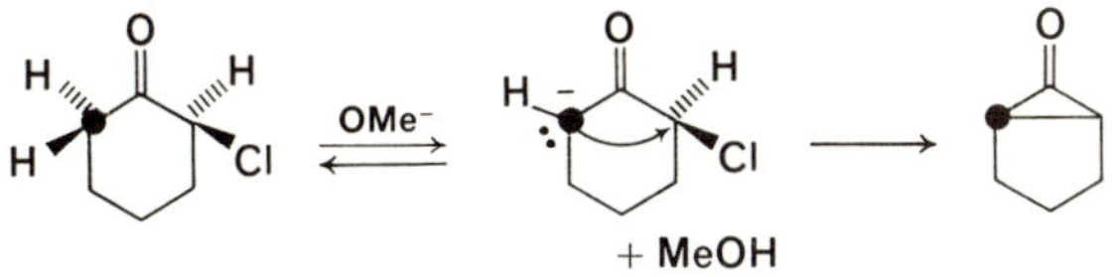

(82)   (83)

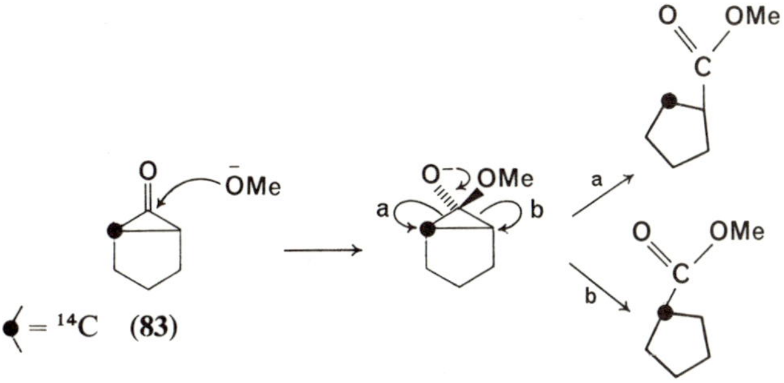

◄ = ¹⁴C (83)

4.15   MICHAEL ADDITION TO
                   ELECTROPHILIC OLEFINS

Olefins which bear electron-withdrawing substituents are subject to attack
by nucleophiles including carbanions. Thus, a compound with a fairly acidic
proton may be added to such a double bond in the presence of a base, a useful
synthetic procedure. The orientation of addition is such as to form a resonance-
stabilized carbanion as intermediate (84). It will be observed that this reaction
is a *vinylogous aldol* reaction (next section).

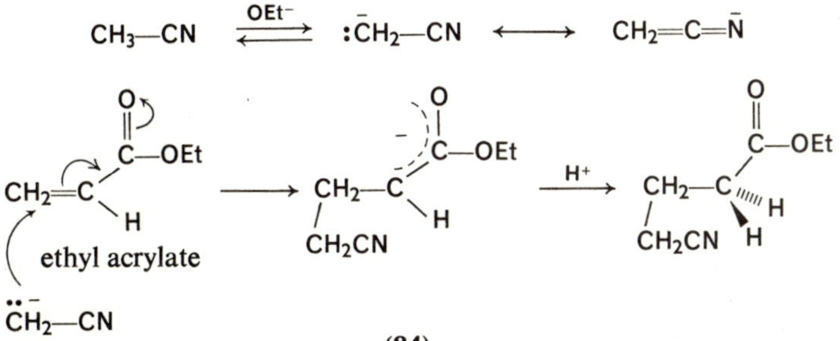

(84)

Aldol reaction   Michael addition

Anionic Polymerization

Anionic polymerization is essentially an extended Michael addition. Electrophilic olefins will add a strong base to give a stabilized carbanion, usually an enolate ion, which reacts in turn with more olefin leading to polymerization. The propagating species is a carbanion, otherwise the reaction resembles radical (Section 5.9) and cationic (Section 2.13) polymerization. A typical example is the polymerization of acrylonitrile initiated by amide ion:

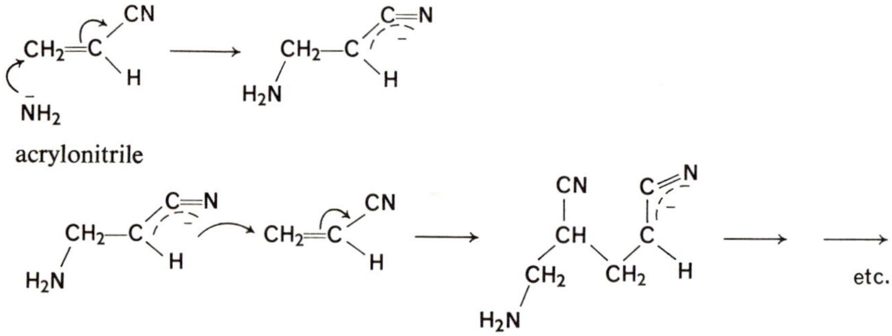

acrylonitrile

Only olefins which can form a fairly stable carbanion are polymerized in this way, and include acrylic esters, CH_2=CH—COOR, and styrene, PhCH=CH_2, which can form a benzylic anion. Initiators must be strong bases such as alkali-metal amides, butyl lithium and phenyl sodium.

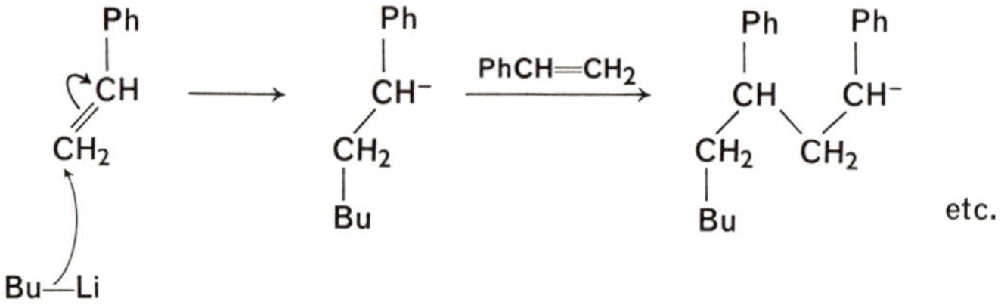

4.16 THE ALDOL FAMILY OF CONDENSATION REACTIONS

Carbanions readily add to a carbonyl function (85), (Section 8.7). The aldol reaction occurs when the enolate ion derived from an aldehyde or ketone adds to the carbonyl group of the same or a similar species. The base-catalysed dimerization of acetaldehyde is the prototype reaction. The condensation is, in principle, reversible (Section 4.17).

$$CH_3C\overset{O}{\underset{H}{\diagdown}} + OH^- \longrightarrow \ :^-CH_2-C\overset{O}{\underset{H}{\diagdown}} \longleftrightarrow CH_2{=}C\overset{O^-}{\underset{H}{\diagdown}}$$

enolate ion

$$\overset{H}{\underset{CH_3}{\diagdown}}C{=}O \quad :\bar{C}H_2-CHO \longrightarrow H^{\text{\tiny IIII}}\overset{O^-}{\underset{CH_3}{C}}CH_2-CHO$$

(85)

$\bigg\downarrow H^+$

$$H^{\text{\tiny IIII}}\overset{OH}{\underset{CH_3}{C}}CH_2CHO$$

β-hydroxybutyraldehyde (aldol)

A mixture of two aldehydes or ketones may lead to four products unless one of them has no α-hydrogen and cannot form an enolate. In these circumstances, 'crossed' aldol reactions become preparatively useful (the *Claisen–Schmidt Condensation*), for example acetone and benzaldehyde, when mixed in the presence of a little base, from dibenzalacetone **(86)**. In this example two successive aldol condensations occur, but the intermediate hydroxyketones dehydrate spontaneously to the α, β-unsaturated product, a common occurrence in these

$$\underset{CH_3\ \ CH_3}{\overset{O}{\|}}\underset{C}{\underset{}{}} \underset{\overset{OH^-}{\rightleftharpoons}}{} \underset{CH_3\ \ CH_2}{\overset{O}{\|}}\underset{C}{} \xrightarrow{PhC\overset{\diagup O}{\diagdown}_H} \underset{CH_3\ \ CH_2}{\overset{O}{\|}}\underset{C}{}\overset{O^-}{\underset{H}{C}}{-}Ph \longrightarrow CH_3{-}\overset{O}{\overset{\|}{C}}\underset{\overset{CH_2-CH}{\underset{OH}{|}}}{}{Ph}$$

$\bigg|{-}H_2O$

spontaneous

$\downarrow$

$$\overset{O}{\overset{\|}{C}}\underset{\overset{Ph}{\underset{H\ H}{}}}{}\text{dibenzalacetone} \quad \xleftarrow[\text{3) }-H_2O]{\text{1) }OH^-\ \ \ \text{2) }PhCHO} \xleftarrow[\text{sequence}]{\text{repeat}} \underset{CH_3}{\overset{O}{\overset{\|}{C}}}\overset{Ph}{\underset{H}{C}{=}C}$$

dibenzalacetone

(86)

reactions when this results in the formation of a conjugated system, and one which tends to force any equilibrium in favour of products (a 'condensation reaction', Section 8.7).

Analogous to this reaction is the *Knoevenagel Condensation* between an aromatic aldehyde and a β-ketoester or malonic ester, catalysed by a secondary amine such as piperidine. The function of the amine may be more than a base to cause enolization; it is suggested that it also reacts with the aldehyde to form the more reactive iminium ion (87).

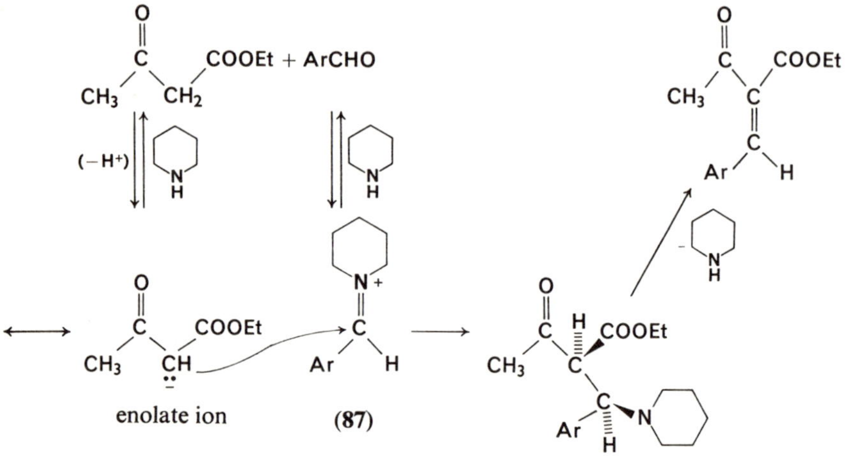

enolate ion   (87)

The *Stobbe Condensation* occurs between diethyl succinate (88) and an aldehyde or ketone. A very strong base is required to cause enolate formation and reversibility may be prevented by the formation of a cyclic intermediate.

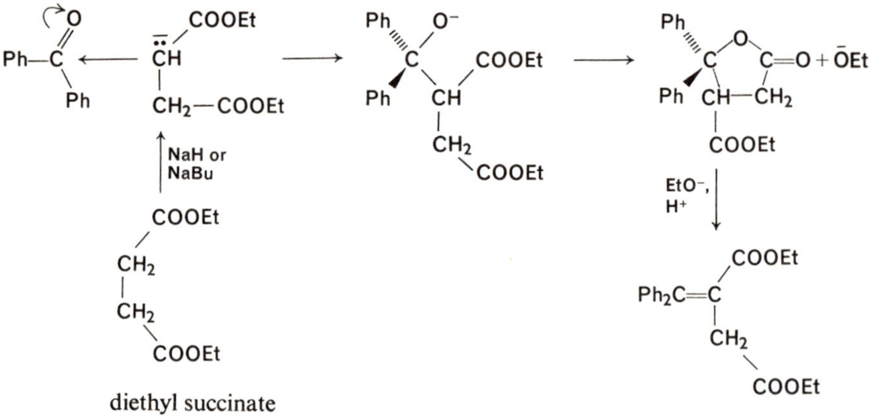

diethyl succinate

(88)

In the *Darzens Condensation,* an α-chloroester provides the carbanion. This reacts with a ketone to form an epoxy ester (glycidic ester).

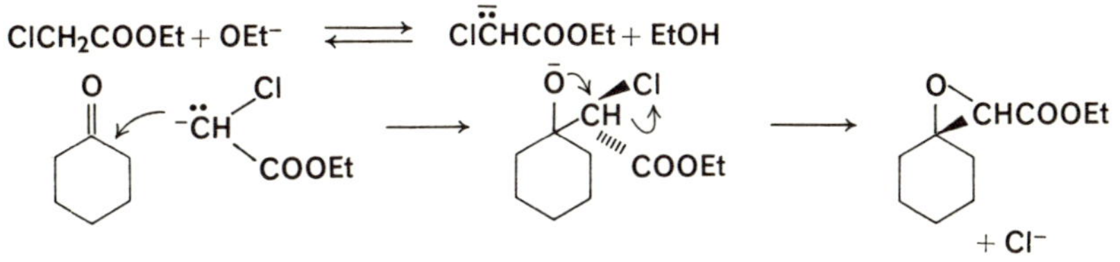

$$ClCH_2COOEt + OEt^- \rightleftharpoons Cl\overset{\cdot\cdot}{C}HCOOEt + EtOH$$

$$+ Cl^-$$

The *Claisen Condensation* occurs between the enolate ion from an ester and another molecule of the same ester; it is an adaptable method for the preparation of β-ketoesters such as ethyl acetoacetate (**89**).

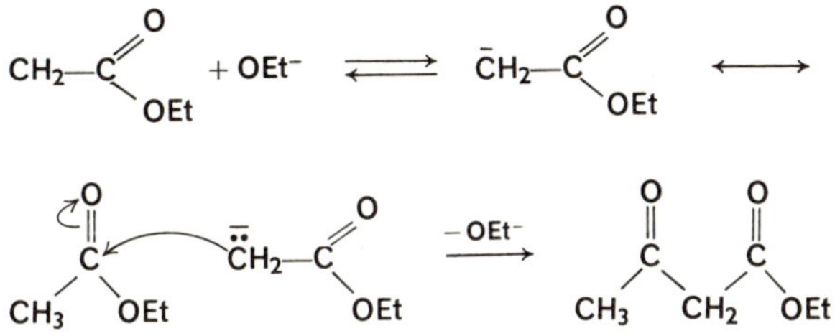

ethyl acetoacetate

(**89**)

The *Dieckmann Reaction* is an internal version of the Claisen condensation starting with a dicarboxylic ester.

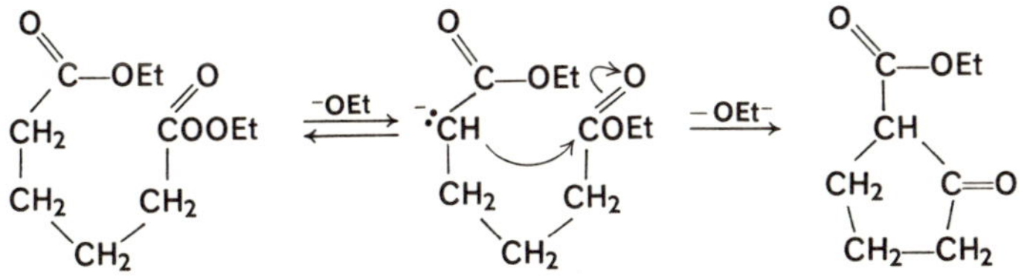

diethyl
adipate

ethyl cyclopentanonecarboxylate

The synthesis of cinnamic acids (**90**) by the *Perkin Reaction* requires carbanion formation from a carboxylic anhydride brought about by a carboxylate ion as base. The carbanion then reacts with an aromatic aldehyde as in previous examples. High temperatures are required since the base used is rather weak, but yields are good.

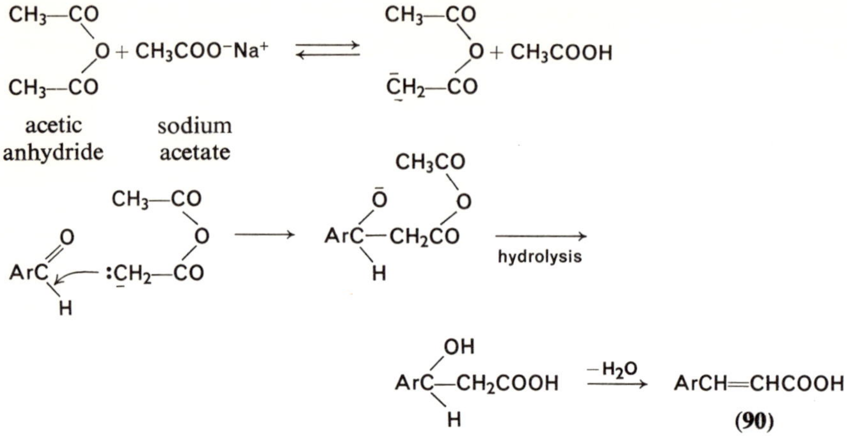

$$CH_3-CO\diagdown O + CH_3COO^-Na^+ \;\rightleftarrows\; CH_3-CO\diagdown O + CH_3COOH$$
$$CH_3-CO\diagup \qquad\qquad\qquad\qquad \bar{C}H_2-CO\diagup$$

acetic   sodium
anhydride   acetate

$$ArC\diagup\diagdown O \;-\; :CH_2-CO \longrightarrow ArC-CH_2CO \xrightarrow{\text{hydrolysis}}$$

$$ArC-CH_2COOH \xrightarrow{-H_2O} ArCH=CHCOOH$$

(90)

a cinnamic acid

The *Mannich Reaction* also belongs to this family; an enolate or other carbanion reacts with an aldehyde in the presence of ammonia or an amine as follows; an intermediate iminium ion is suggested (cf. **87**).

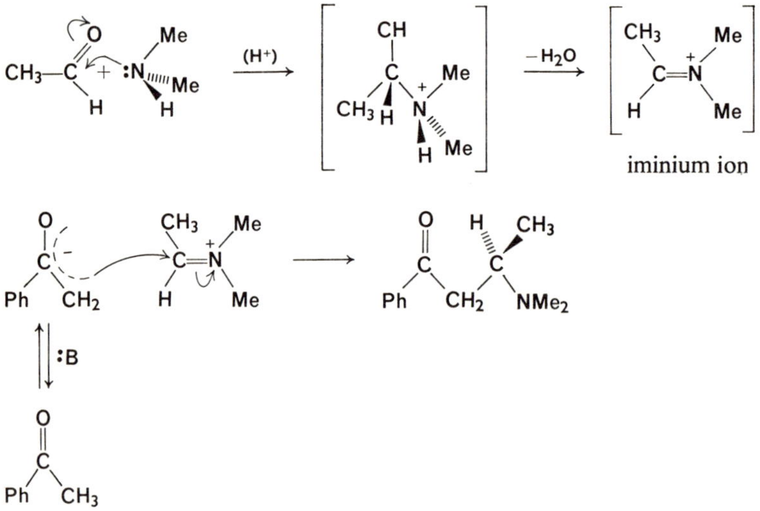

$$CH_3-C\diagdown{}_{H}^{O} + :N\diagdown{}_{H}^{Me}\diagdown Me \xrightarrow{(H^+)} \left[\; CH_3\;\diagup_{H}^{CH}\diagdown C \diagdown{}_{Me}^{+}N\diagup{}_{H}^{Me}Me \;\right] \xrightarrow{-H_2O} \left[\; CH_3\diagdown{}_{H}C={}^{+}N\diagup{}_{Me}^{Me} \;\right]$$

iminium ion

$$Ph\diagup{}_{CH_2}^{O}C \;\longrightarrow\; C={}^{+}N\diagdown{}_{Me}^{CH_3\;Me} \;\longrightarrow\; Ph\diagup{}_{CH_2}^{O}C \diagdown C{}_{NMe_2}^{H\;CH_3}$$

$$\updownarrow\; :B$$

$$Ph\diagup{}_{CH_3}^{O}C$$

4.17   **THE RETRO-ALDOL REACTION**

As mentioned above, aldol condensations are potentially reversible. An alkoxide ion may eject a carbanion and revert to a carbonyl compound. The ease of the reaction depends upon the stability of the carbanion being ejected and is commonly an enolate. An example which has been investigated in detail is the following:

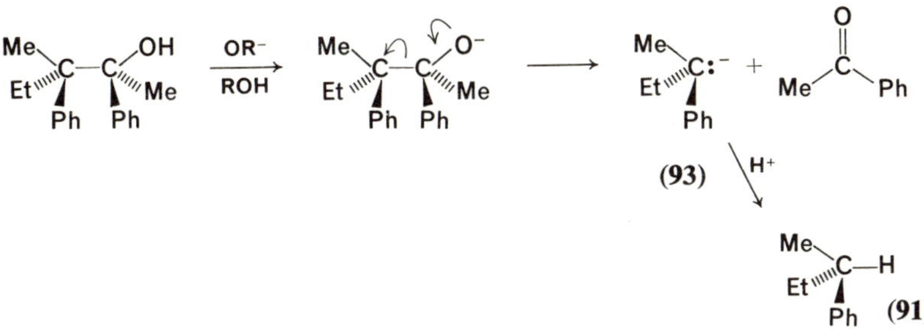

(93)

(91)

The stereochemistry of the 2-phenylbutane (91) produced depends markedly upon the solvent; in dioxan or other media of low ionizing power, almost complete retention of configuration occurs via a transition state such as 92. The cation also plays an important role, since replacement of potassium by a quaternary ammonium ion leads to racemization. Racemization and an enormous rate increase occurs in dimethylsulphoxide due evidently to the transient

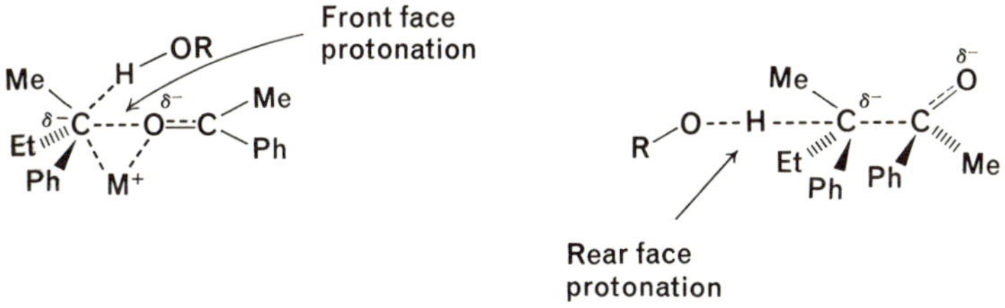

(92)   (94)

production of essentially free carbanions (93). Protic solvents such as ethylene glycol lead to racemization with excess inversion, via 94.

4.18   ELIMINATIONS VIA CARBANIONS

β-Nitroethyl phenyl ether (95) undergoes base-catalysed elimination of phenol to form nitroethylene (96), which adds to solvent to give the observed product, 97. The exchange of the β-proton occurs prior to elimination and the rate shows a large β-hydrogen isotope effect ($k_H/k_D = 7$). The rate-determining step must involve the loss of the β-proton but *not* the expulsion of phenoxide ion which occurs in a subsequent fast step. This type of elimination, via a

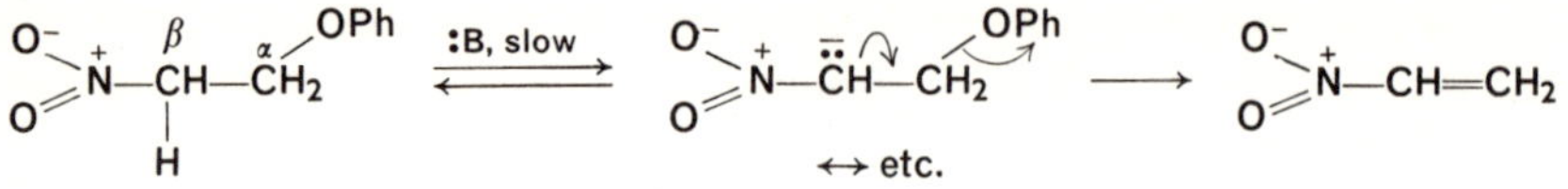

(95)   **(96)**

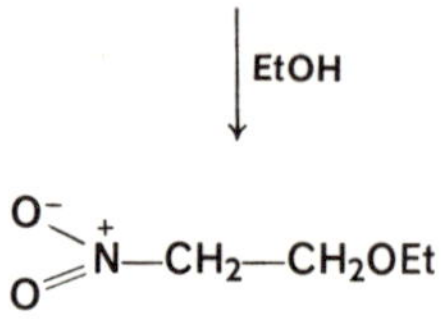

EtOH

(97)

carbanion, is designated $E1cB$ (first-order elimination via the *conjugate base*). Typically this mechanism is preferred for a substrate with a *rather acidic β-proton* and a *poor leaving group*. While the nitro group is one of the best activating groups, trialkylphosphonium, R_3P^+-, and dialkylsulphonium, R_2S^+-, are almost as good, although quaternary ammonium, R_3N^+-, is very poor on account of the lack of available *d*-orbitals (Section 4.10). Another example of the $E1cB$ reaction is the base-catalysed dehydration of 9-(hydroxymethyl)-fluorene (**98**):

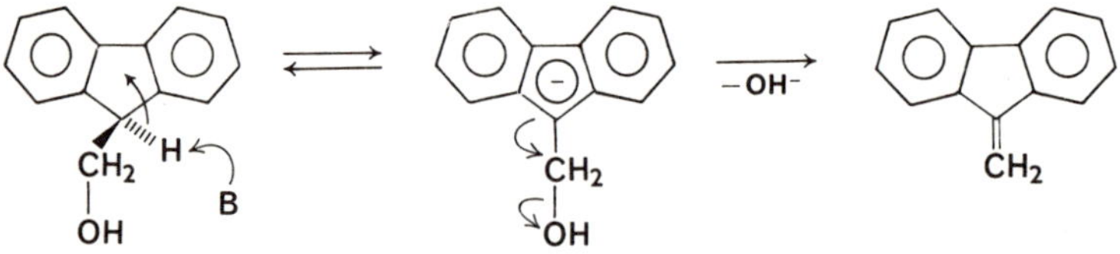

(98)

4.19   CARBOXYLATION AND DECARBOXYLATION

Strongly nucleophilic carbanions (usually in the form of a metal alkyl) react with carbon dioxide to form a carboxylate ion,

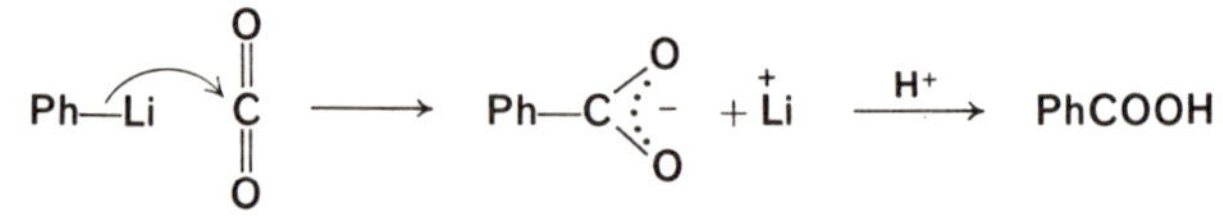

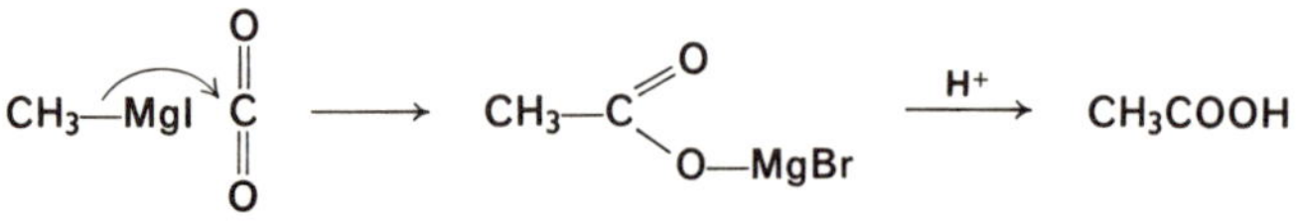

The reverse reaction can occur if a rather stable carbanion is available to be ejected, i.e. one of comparatively low nucleophilicity. β-Ketoacids readily decompose in aqueous solution; acetoacetic acid (99) yields acetone and CO_2; notice the similarity to the Retro-aldol reaction. Similarly, α-nitro, α-halo

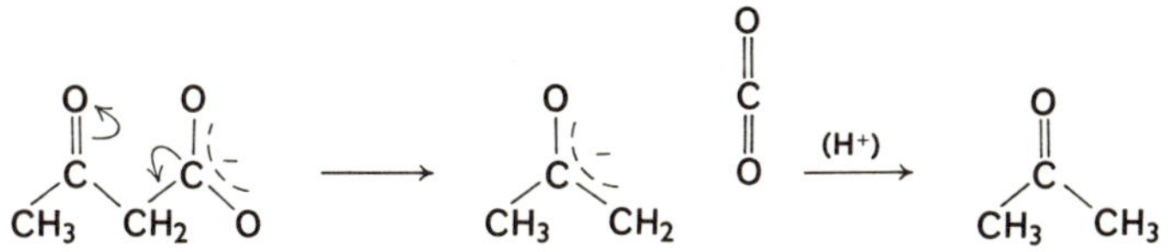

(99) acetone enolate ion acetone

and acetylenic acids are readily decarboxylated in basic solution. The latter reaction provides a convenient synthetic route to deuterochloroform (a

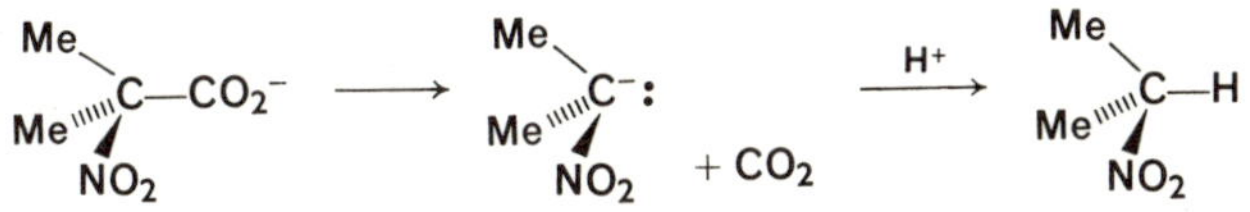

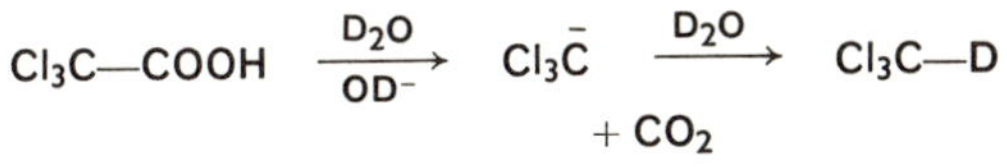

$$Cl_3C{-}COOH \xrightarrow[OD^-]{D_2O} Cl_3\bar{C} \xrightarrow{D_2O} Cl_3C{-}D$$
$$+ CO_2$$

useful solvent for n.m.r. spectroscopy). The intermediate carbanion in decarboxylation may be diverted to form an alkyl bromide in the presence of bromine:

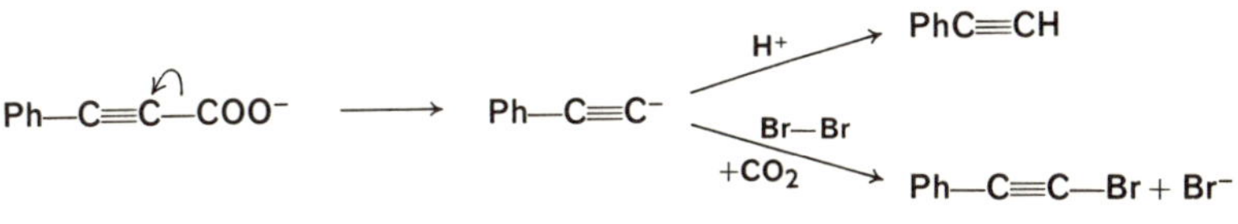

The formation of paraffin hydrocarbons by heating carboxylate salts with soda-lime no doubt proceeds by a similar type of mechanism, the high temperatures reflecting the need to force the reaction when unstabilized carbanions

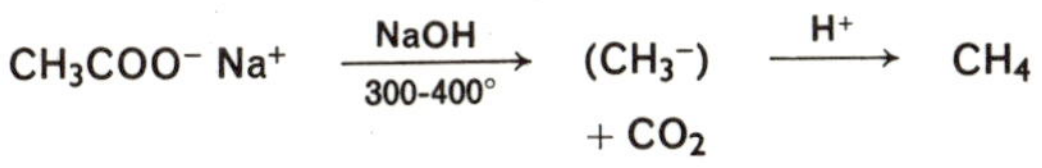

$$CH_3COO^- Na^+ \xrightarrow[300-400°]{NaOH} (CH_3^-) \xrightarrow{H^+} CH_4$$
$$+ CO_2$$

are involved. Substantial primary isotope effects are found for decarboxylation of (carbonyl-) labelled acids:

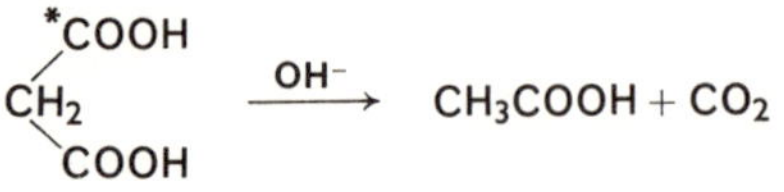

$$\begin{array}{c} \overset{*}{C}OOH \\ | \\ CH_2 \\ | \\ COOH \end{array} \xrightarrow{OH^-} CH_3COOH + CO_2$$

$$*^{13}C \qquad\qquad\qquad k_{^{12}C}/k_{^{13}C} = 1.04\text{-}1.06$$

$$C = {}^{12}C,$$

Other mechanisms of decarboxylation are available under the appropriate circumstances; in strongly acid conditions, carbonium ions may be intermediate (Section 2.13) while a cyclic transition state is favoured by β,γ-unsaturated acids (100):

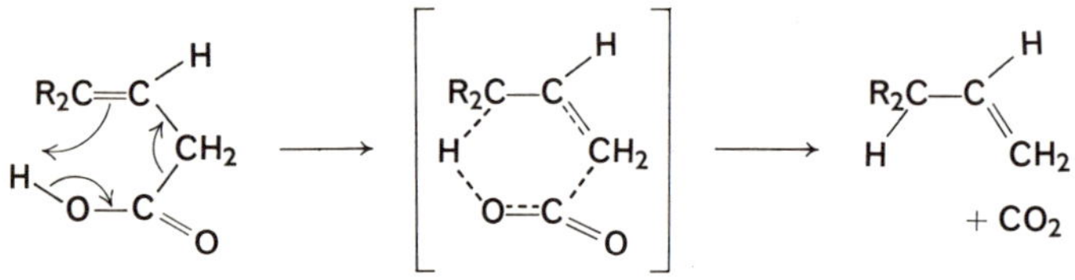

(100)

Similar cyclic transition states probably occur in decarboxylations of the undissociated β-ketoacids (101) in neutral or acid conditions. Radical mechanisms of decarboxylation are discussed later (Section 5.9).

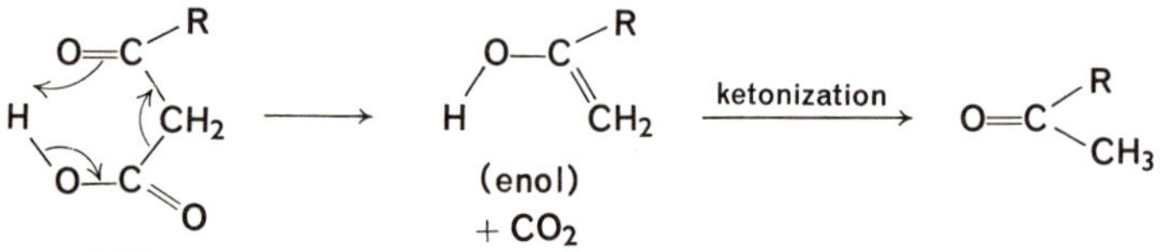

(101)

4.20   **ESTER HYDROLYSIS VIA A CARBANION**

The common mechanisms for hydrolysis of esters are discussed in Chapter 8. A rather rare hydrolytic mechanism via a carbanion has been recognized in certain cases such as β-ketoesters (102). These undergo initially an $E1cB$ elimination as shown by isotopic exchange of the acidic β-hydrogen atom prior to reaction, to give an intermediate keten derivative (103) which rapidly reacts with water in a normal polar addition process. An exactly analogous

$$
\underset{(102)}{R-\overset{\overset{\displaystyle O}{\|}}{C}-CH_2-\overset{\overset{\displaystyle O}{\|}}{C}-OPh} \;\underset{\xleftarrow{\hspace{1em}}}{\overset{:B}{\xrightarrow{\hspace{1em}}}}\; R-\overset{\overset{\displaystyle O}{\|}}{C}-\overset{..}{\underset{..}{C}}H-\overset{\overset{\displaystyle O}{\|}}{C}-OPh \;\longleftrightarrow\; R-\overset{\overset{\displaystyle O}{\|}}{C}-CH=\overset{\overset{\displaystyle O^-}{|}}{C}-OPh \text{ etc.}
$$

$$\longleftrightarrow$$

$$
\underset{\beta\text{-ketoacid}}{R-\overset{\overset{\displaystyle O}{\|}}{C}-\underset{\underset{\displaystyle H \quad OH}{|\quad\;|}}{CH}-C\overset{\displaystyle O}{\underset{}{}}} \;\xleftarrow{\;H_2O\;}\; \underset{\substack{(103)\\ \text{acyl keten}}}{R-\overset{\overset{\displaystyle O}{\|}}{C}-CH=C=O + OPh^-}
$$

mechanism has been identified in the hydrolysis of carbamate esters (urethanes)

$$
\overset{R}{\underset{H}{>}}N-C\overset{\displaystyle \nearrow O}{\underset{\searrow OR'}{}} .
$$

The intermediate inferred is an isocyanate.

$$
\underset{B:}{\overset{R}{\underset{H}{>}}N-C\overset{\displaystyle \nearrow O}{\underset{\searrow OR'}{}}} \;\longrightarrow\; \overset{R}{\underset{\;\;..}{>}}N-C\overset{\displaystyle \nearrow O}{\underset{\searrow OR'}{}} \;\longrightarrow\; \underset{\text{isocyanate}}{R-N=C=O + OR'^-}
$$

$$\downarrow H_2O$$

$$
R-NH_2 + CO_2 \;\xleftarrow{\hspace{2em}}\; \underset{\substack{\text{decarboxylation} \qquad\quad \text{carbamic}\\ \text{(spontaneous)} \qquad\qquad \text{acid}}}{R-\underset{\underset{\displaystyle H}{|}}{N}-C\overset{\displaystyle \nearrow O}{\underset{\searrow OH}{}}}
$$

4.21   **CARBANION REARRANGEMENTS**

Prototropic Changes

The re-protonation of a conjugated carbanion may occur at a site other than that from which the proton was initially abstracted. The result is the migration of a double bond and will be a favourable process if the product is more stable than the starting material. This process is known as prototropy. The β,γ-unsaturated cyanide **104**, for example, rearranges to the more stable conjugated system **105** under the action of the base. The rate of a prototropic change

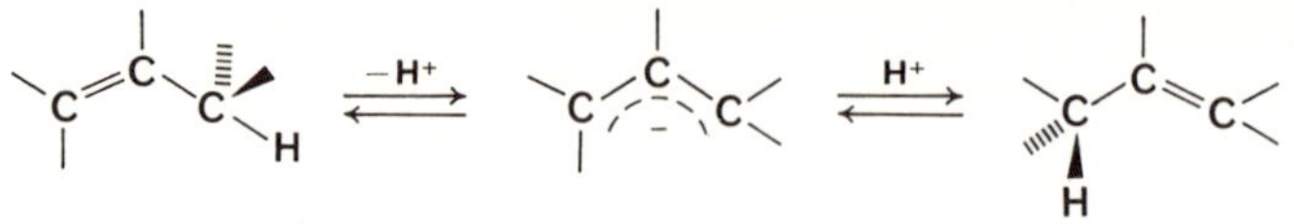

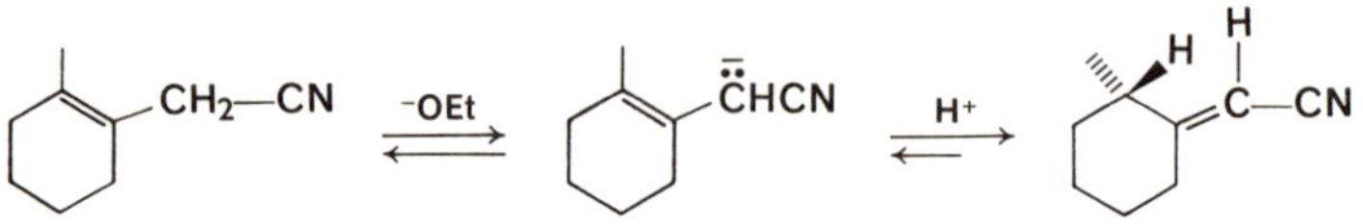

          (104)   **(105)**

can be used as a measure of the basicity of the medium. The following series have been studied,

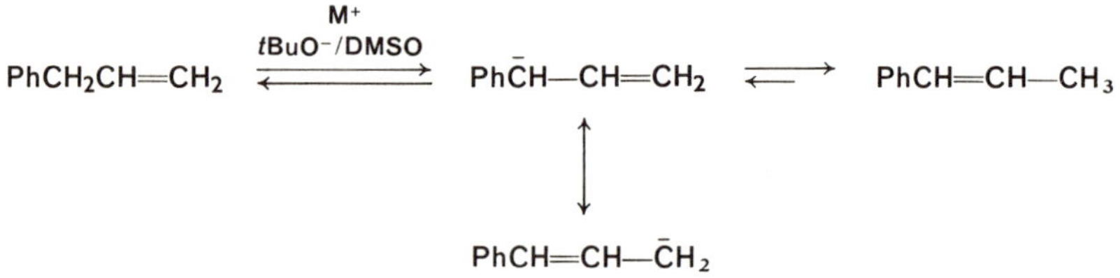

$$PhCH_2CH{=}CH_2 \underset{tBuO^-/DMSO}{\overset{M^+}{\rightleftharpoons}} Ph\bar{C}H{-}CH{=}CH_2 \rightleftharpoons PhCH{=}CH{-}CH_3$$

$$PhCH{=}CH{-}\bar{C}H_2$$

(106)

Metal cation	k_{rel}
Li^+	1
Na^+	9
K^+	1000
Rb^+	2500
Cs^+	3900

Ion pairing by the smaller metal ions evidently reduces the effectiveness of the base.

    If the reaction is carried out in deuteroethanol, the product and also unreacted starting material are found to contain deuterium which is evidence that the carbanion **106** is an intermediate and that the rearrangement is not a concerted one. A simple case of prototropy already discussed is keto–enol tautomerism (Section 4.7).

    Many simple olefins will undergo rearrangement in basic conditions to an equilibrium mixture of double-bond isomers, the amounts of which reflect their respective stabilities. A series of allylic carbanions are intermediates in the following equilibration of 2-methylpentene-1 (107):

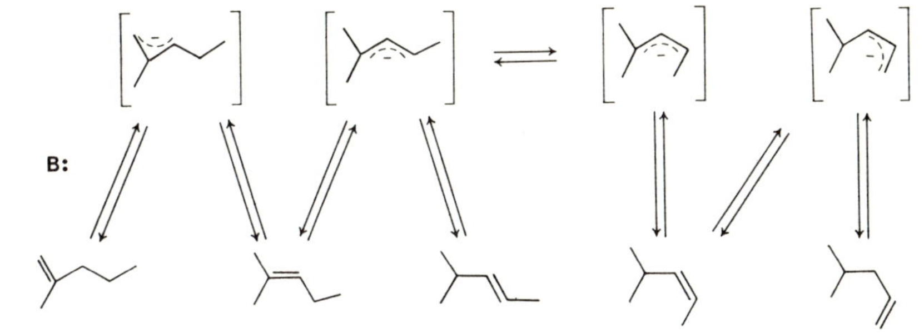

| (107) 11.3% | 80% | 7.2% | 1.2% | 0.3% |
| 2-methylpentene-1 | 2-methylpentene-2 | *trans*-4-methylpentene-2 | *cis*-4-methylpentene-2 | 4-methylpentene-1 |

The most highly substituted ethylene, 2-methylpentene-2, is the most stable isomer and predominates at equilibrium. Note the absence of skeletal rearrangements as would be observed in the analogous carbonium ions.

Even if rearrangement is not evident, prototropy must be occurring in the complete exchange of all protons in cyclohexene and in the partial exchange in an olefin such as 108:

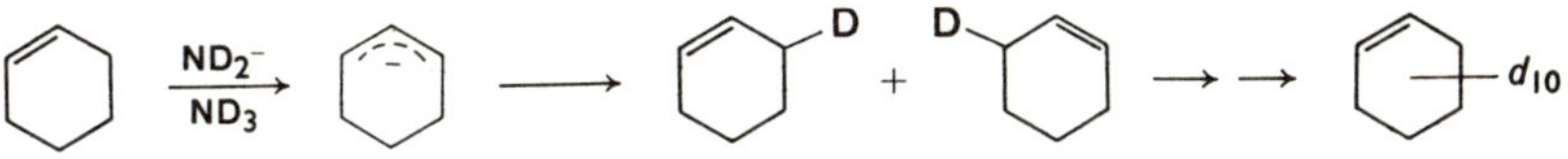

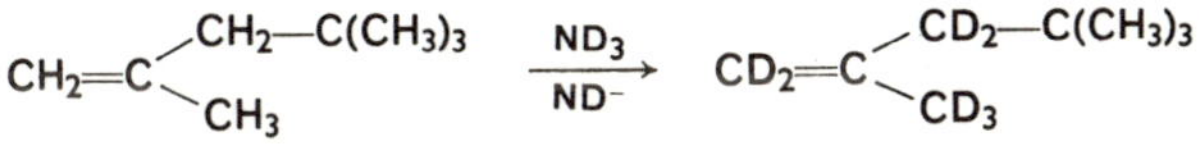

There is no mechanism for the terminal methyl protons of 108 to become allylic hence they do not exchange.

Electrocyclic Rearrangements

The family of open-chain to cyclic rearrangements includes examples from conjugated carbanions. Pentadienide anions (109) will cyclize to cyclopentenyl anions (compare the analogous rearrangements of carbonium ions, p. 179). This reaction and other related examples are controlled by orbital-symmetry restrictions (Section 1.3). The highest filled molecular orbital of the conjugated

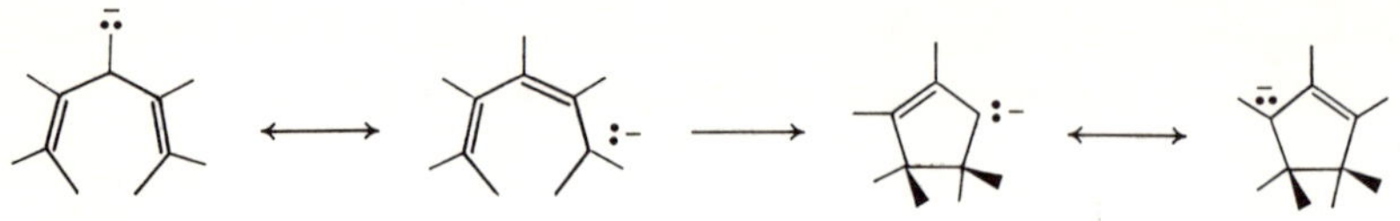

(109)

system interacts across its termini **(110)**, which necessitates that each rotates in the *opposite sense* (a disrotatory ring closure) in order that overlap between orbital lobes with the same sign of the wave function may occur. The result

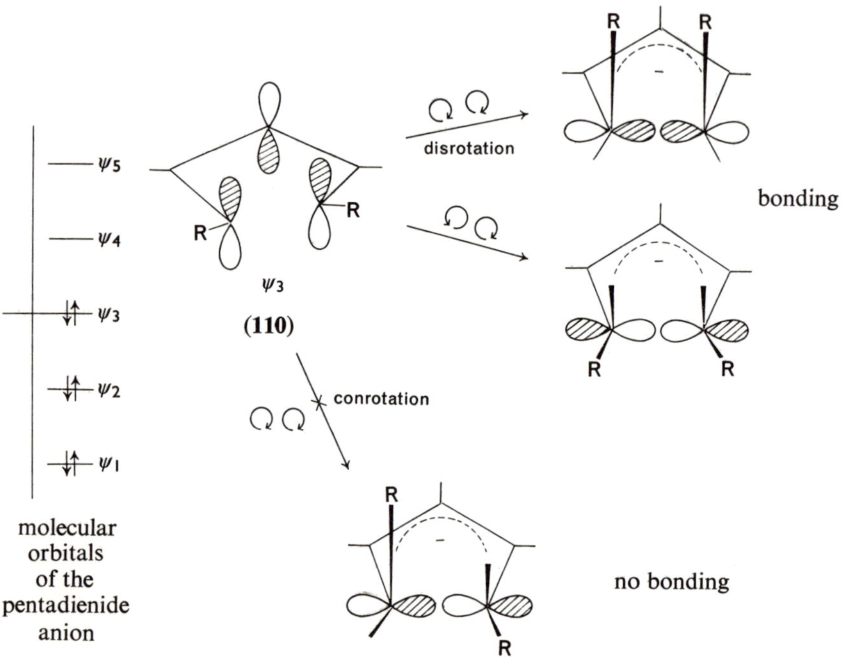

is a completely stereospecific conversion of a *trans,trans* diene **(111)**, to *cis*-substituted cyclopentene **(112)** and *cis,trans* diene, **(113)** to *trans*-cyclopentene **(114)**. An example which requires a further double-bond shift, is the conversion

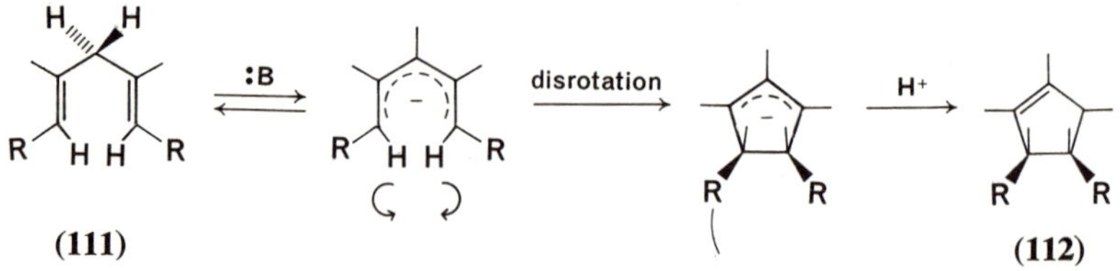

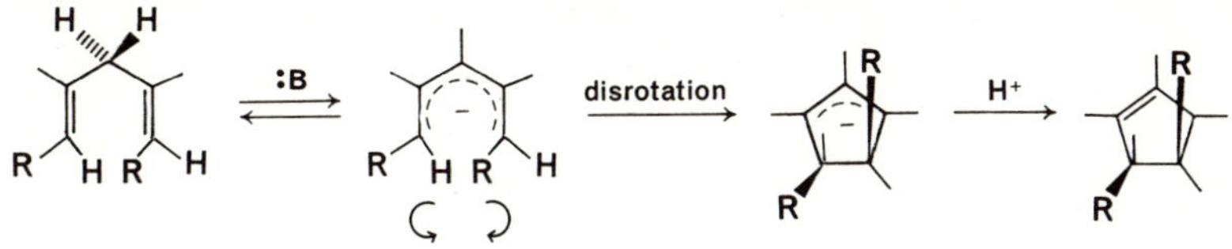

(113)   **(114)**

of cycloocta[1,5]diene (115) to bicyclo[3,3,0]octene (116). A homologous

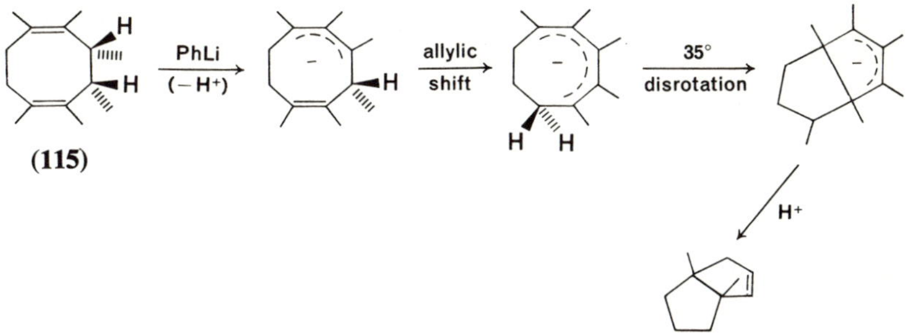

(115)

(116)

rearrangement of this type, the cyclization of a heptatrienide ion (117) requires
a *conrotatory* geometry to achieve bonding across the termini of the conjugated
system. The product is a cycloheptadiene.

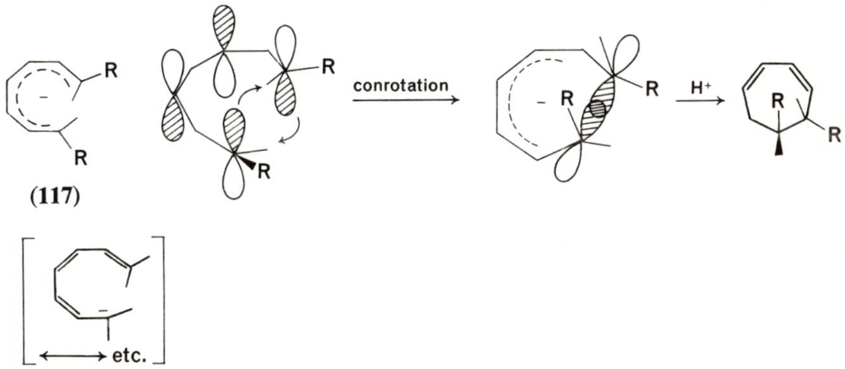

(117)

Another example of 'ring–chain' tautomerism which has been known for
many years is the equilibrium conversion of a 'homoallylic'* anion (118)

* $C=C-C-C- =$ homoallylic, i.e. one more carbon than the allyl system, $C=C-C$
(cf. p. 212).

to the isomeric cyclopropylmethylide (**119**):

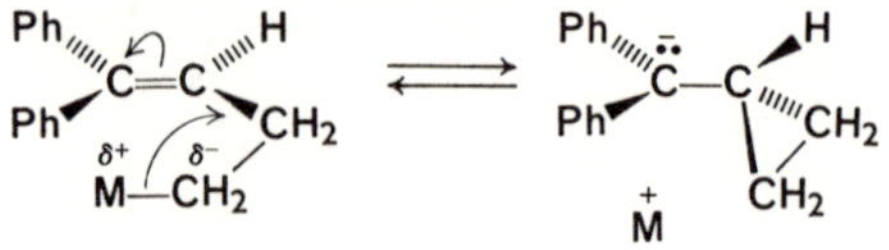

(118)   (119)

The equilibrium is shifted to the right by highly electropositive metals (M = Na, K) and by polar solvents, and to the left for M = Mg and less-polar solvents since the methylide is a more-ionic compound than the homoallylic system.

1,2-Anionic Shifts

Unlike carbonium ions, 1, 2-alkyl shifts in carbanions are unusual. There is no facile pathway for such a rearrangement since the movement of four electrons is involved and hence cannot fit into a 'quasi-aromatic' transition state ($4n$-electron systems are 'anti-aromatic' and destabilized) in the same way that the 2-electron transition state for the carbonium-ion rearrangement does. If we

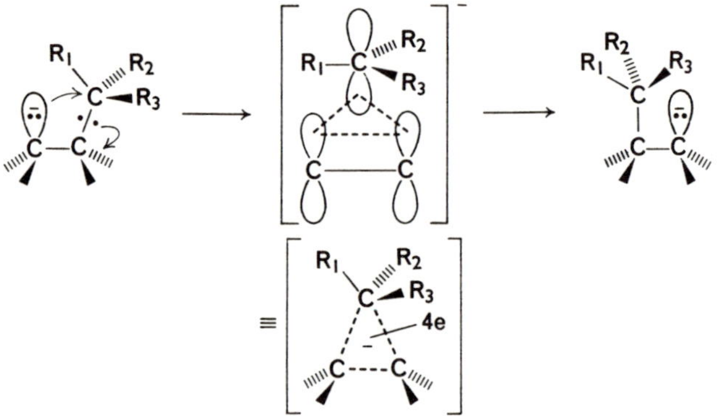

transition state

were to regard the process as a nucleophilic attack of the carbanion on the migrating alkyl group then the attack would be at the front face and hence unfavourable. From another viewpoint we cannot form a 'five-coordinate' carbanionic intermediate analogous to the cationic species (Section 2.29) since there would be four electrons to accommodate in the three-centre orbital.

Examples of 1,2-shifts are known but usually need forcing conditions and may have charged centres which, through becoming neutralized, provide a driving force.

The *Wittig Rearrangement* of benzyl ethers (**120**) requires a strong base:

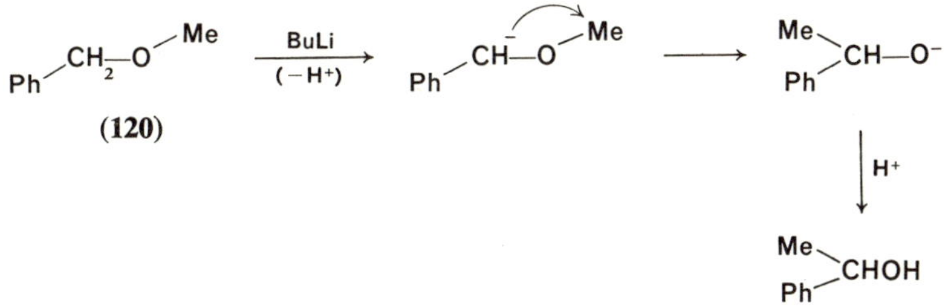

(**120**)

The *Stevens Rearrangement* occurs with retention of configuration if the migrating alkyl group is asymmetric,

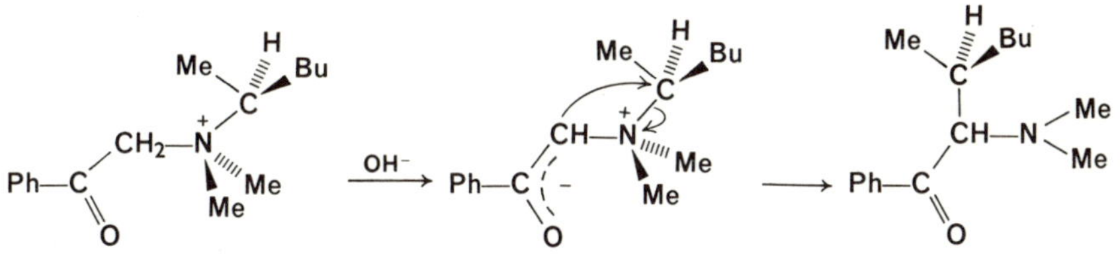

However, the recent demonstration of nuclear polarization (CIDNP—Section 5.9) in this reaction indicates radical character in the intermediates, the exact nature of which is at present somewhat in doubt.

4.22   **ELECTROPHILIC SUBSTITUTION AT SATURATED CARBON**

The rate-determining departure of an electrophile from a saturated carbon atom and subsequent replacement by a second electrophile is the carbanion counterpart to the more familiar nucleophilic substitution reaction (S_N) and may be designated an S_E process. The same dichotomy of mechanism exists in that the displacement may take place in a single concerted step (S_E2 process), or by ionization followed by electrophilic attack on a carbanion intermediate (S_E1 process). Some examples of S_E1 reactions have already been encountered under another heading, for example proton exchange in carbon acids and the Retro-aldol reaction (Section 4.17). In the field of organometallic

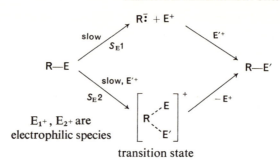

chemistry, metal-exchange reactions usually proceed by S_E2 process; however, the following reaction is well authenticated as an example of the unimolecular mechanism.

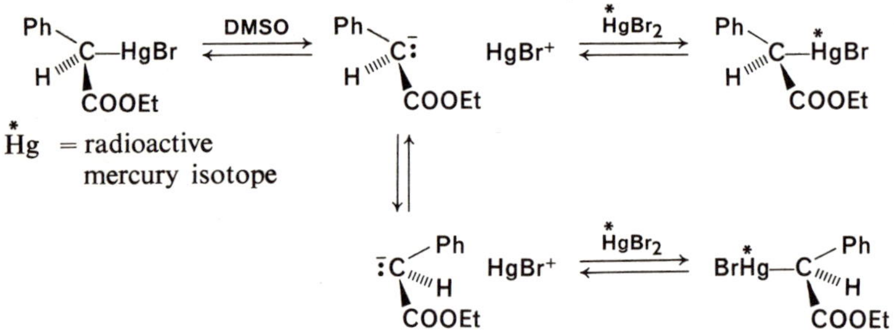

Radioactive-mercury exchange converts an optically active substrate to a racemic product, the rates of racemization and of radioactive-mercury exchange being the same. The kinetic form is unimolecular—rate $= k_1[\mathrm{RHgBr}]$—and is unaffected by changes in the concentration of the reagent (HgBr). These findings are consistent with the production of an essentially free carbanion (121) as intermediate in a rate-determining process. The reaction reverts to S_E2 characteristics—retention of configuration, bimolecular kinetics—in solvents less conducive to carbanion formation than dimethylsulphoxide.

The stereochemistry of the bimolecular reaction is very distinct from that of the nucleophilic counterpart; very commonly it is found that the S_E2 reaction leads to retention of configuration whereas the S_N2 reaction always proceeds with inversion. It has been suggested that an electrophile must interact with a *filled* orbital and that the most accessible one is the σ-bond which is to

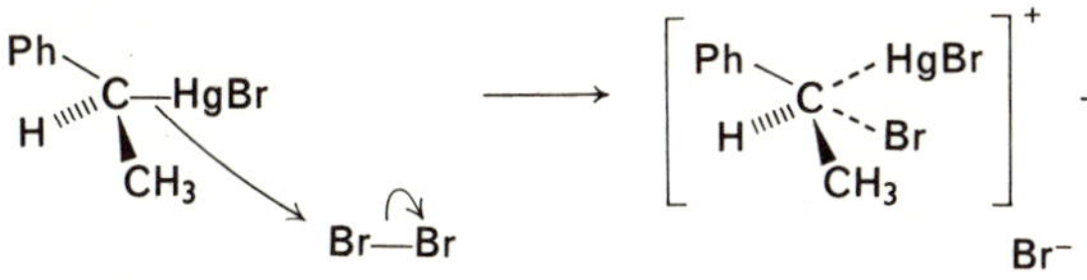

Electrophilic attack   transition state   100%
of Br$_2$ on carbon   retention
    $-S_E2$

be broken; on the other hand, a nucleophile must interact in the first instance with a *vacant* orbital and the most available one is the antibonding σ-orbital which lies mainly at the rear side of the carbon centre.

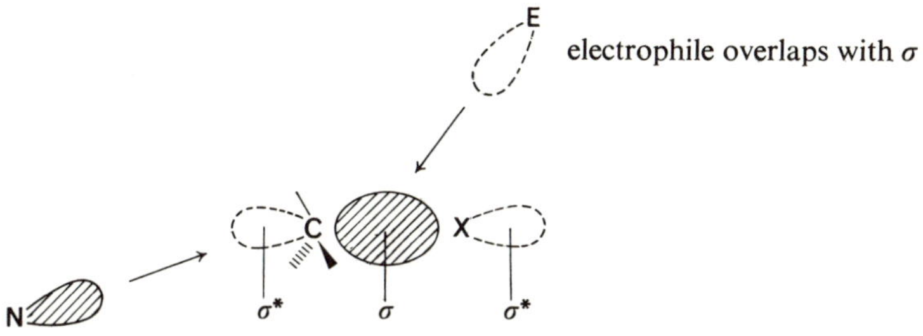

electrophile overlaps with σ

nucleophile
overlaps with σ*

4.23   NUCLEOPHILIC SUBSTITUTION AT AN AROMATIC RING

Whereas simple aryl halides such as chlorobenzene are extremely resistant to hydrolysis and other nucleophilic displacements, these reactions are greatly facilitated by strong electron-withdrawing groups (—M groups) in the *ortho* or *para* positions. The chlorine in 1-chloro-2,4-dinitrobenzene (**122**) is sufficiently labile to be displaced by a secondary amine such as piperidine. The mechanism of this reaction is believed to be of the addition–elimination type, parallel to electrophilic substitutions at the aromatic ring (Section 2.14). The evidence for a two-step mechanism is compelling: first, the activating effect of the nitro groups is easily rationalized as a conjugative stabilization of an intermediate pentadienide anion (**123**). Stabilization would be most effective from the *ortho* or *para* positions, the charge being largely resident on oxygen (**123b**). Then, the ease of displacement of halide ion falls in the order F > Cl > Br > I—i.e. the halogen which forms the *strongest* C—hal bond is the

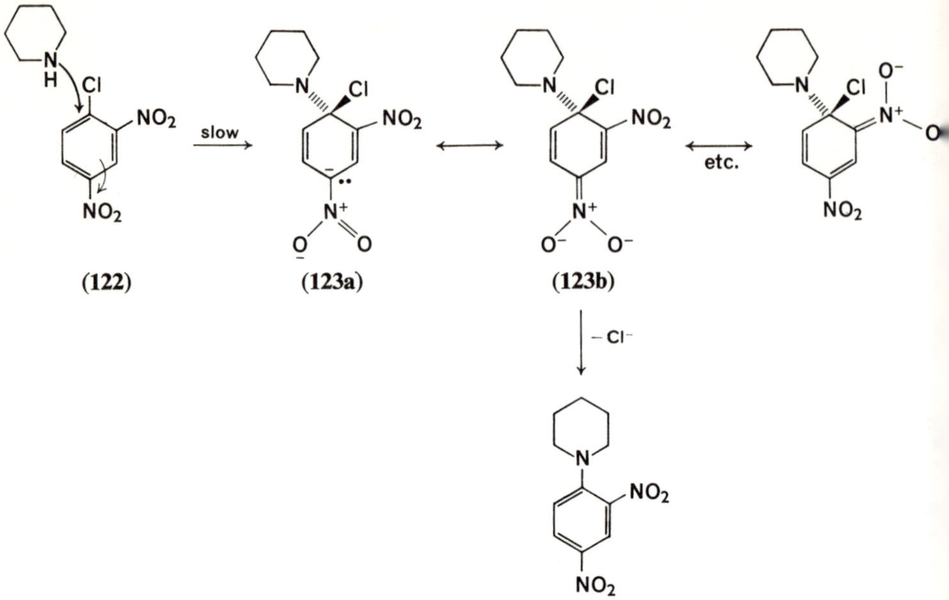

(122)   (123a)   (123b)

most easily displaced. It is unlikely that the C—hal bond, therefore, is being broken in the slow step but instead occurs in a subsequent rapid step (implying necessarily more than one step). The increased reactivity of fluorine may be explained in terms of its high electronegativity (inductive effect) which makes the ring more susceptible to nucleophilic attack. Finally, compounds with structures analogous to the postulated intermediate **123** have been known for 70 years under the name of 'Meisenheimer complexes'. X-ray diffraction has shown the structure of **124** to be a conjugated, planar, pentadienide anion.

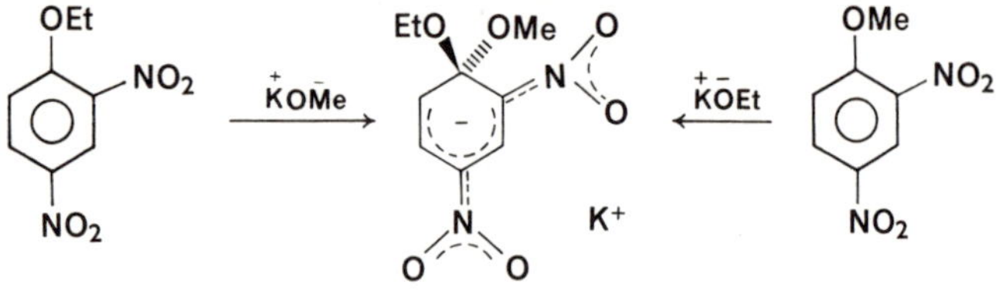

(124)

4.24   **AUTOXIDATION OF CARBANIONS**

Carbanions are often sensitive to oxygen with which peroxy anions are formed:

$$R^- + O_2 \longrightarrow R\text{—}O\text{—}O^-$$

However, the reaction is probably not as simple as this might suggest. Since oxygen in the ground state is a triplet species (i.e. containing two unpaired electrons) while the other species are singlets, a simple addition would require a change of multiplicity and hence would be 'forbidden'—of low probability. It is more likely that the carbanion is converted to a radical which combines with oxygen (Section 5.9) and is subsequently reduced back to a peroxy anion. Likely vehicles for this catalysis are the nitro compounds which greatly accelerate the reaction:

$$Ph_3C\overset{..}{:} + PhNO_2 \longrightarrow Ph_3C\cdot + PhN\overset{.}{O_2}{}^-$$

$$Ph_3C\cdot + \overset{.}{O}—\overset{.}{O} \longrightarrow Ph_3C—O—\overset{.}{O}$$

$$Ph_3C—O—\overset{.}{O} + PhN\overset{.}{O_2}{}^- \longrightarrow Ph_3C—O—\overset{..}{O}{:}^- + PhNO_2$$

4.25   YLIDES

The term 'ylide' denotes a compound which contains a negative carbon adjacent to a heteroatom which bears a formal positive charge. In the best investigated examples, the heteroatom is phosphorus or sulphur, although nitrogen, arsenic and antimony ylides are known. Except in the case of nitrogen, stabilization of the carbanion and partial neutralization of the charge can occur by *d*-orbital overlap (Section 4.9). A typical example is **125** which can be represented as triphenylphosphonium meth*ylide* (**125a**) or methylene-triphenylphosphorane (**125b**), a derivative of five-valent phosphorus. All ylides of second-row elements may be given structures expressed by two such canonical forms. A great many ylides have been characterized, especially phosphorus

$$\underset{(125\,a)}{\overset{\displaystyle Ph}{\underset{\displaystyle Ph}{\overset{\displaystyle Ph\text{\tiny ''''}}{\diagup}}}P^+—\bar{C}H_2} \longleftrightarrow \underset{(125\,b)}{\overset{\displaystyle Ph}{\underset{\displaystyle Ph}{\overset{\displaystyle Ph\text{\tiny ''''}}{\diagup}}}P{=}CH_2}$$

ylides, bearing a wide variety of substituent groups attached both to the heteroatom and to the carbanionic centre. Frequently these compounds are capable of isolation as high melting, usually yellow, solids although in carrying out the useful reactions which they exhibit, the ylides occur as reactive intermediates.

Formation of Ylides

Although a variety of methods may be used for the preparation of ylides, three of the most useful are mentioned here.

(a) Proton removal from the corresponding 'onium' salt: as mentioned previously, a positive centre acidifies an α-proton although a very strong base may be needed to cause ionization. Although not necessary for their formation, it is convenient to have only one α-hydrogen-bearing substituent at the heteroatom, otherwise a mixture of ylides may result. The ease of ylide formation is

$$Ph_3\overset{+}{P}-CH_3 \xrightarrow[\text{ether}]{\overset{-}{Bu}-\overset{+}{Li}} Ph_3\overset{+}{P}-\overset{-}{C}H_2 + BuH$$

$$Ph_2\overset{+}{S}-CH_2Ph \xrightarrow[\text{ether}]{\overset{-}{Bu}-\overset{+}{Li}} Ph_2\overset{+}{S}-\overset{-}{C}HPh + BuH$$

in accordance with carbanion stabilities as discussed above. Thus, the phosphonium ester **126** forms its ylide on treatment with sodium carbonate in

$$Ph_3\overset{+}{P}-CH_2COOEt \xrightarrow{OH^-} Ph_3\overset{+}{P}-\overset{-}{C}HCOOEt$$

$$\textbf{(126)}$$

water and the fluorenyl ylide **127** forms by the action of ammonia on its conjugate acid.

$$\textbf{(127)}$$

(b) The addition of a carbene to a phosphine: phosphorus ylides may be prepared by the direct addition of a carbene or carbenoid species (Chapter 6) to a phosphine. This is especially useful for the production of halo- and alkoxy-substituted ylides (**128, 129**).

$$Ph_3P: + :CCl_2 \longrightarrow Ph_3\overset{+}{P}-\overset{-}{C}Cl_2$$

$$\textbf{(128)}$$

$$\uparrow$$

$$CHCl_3 + tBuO^-$$

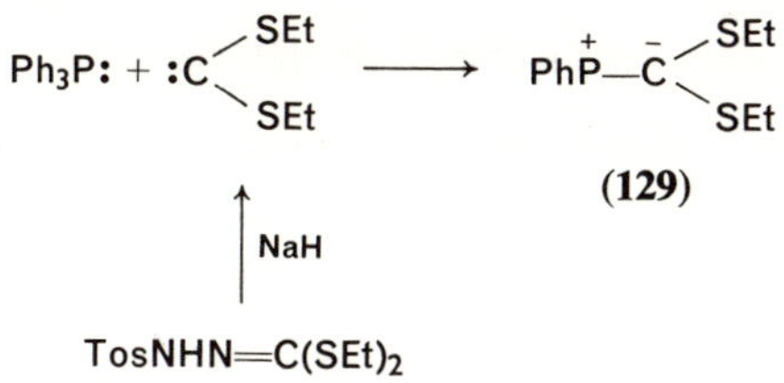

$$Ph_3P: + :C \overset{SEt}{\underset{SEt}{\diagdown}} \longrightarrow PhP\overset{+}{-}\overset{-}{C} \overset{SEt}{\underset{SEt}{\diagdown}}$$

(129)

↑ NaH

$$TosNHN=C(SEt)_2$$

(c) The addition of a carbanion to a vinylphosphonium ion results in the formation of an ylide (130) by a process similar to a Michael addition, (Section 4.15).

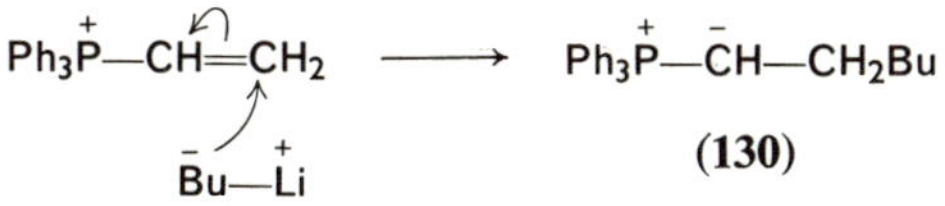

$$\overset{+}{Ph_3P}-CH{=}CH_2 \longrightarrow \overset{+}{Ph_3P}-\overset{-}{CH}-CH_2Bu$$

$$\overset{-}{Bu}-\overset{+}{Li}$$

(130)

The Structure of Ylides

X-ray crystallographic structures for phosphorus ylides, e.g. **131**, show that the carbon is planar and the C—P distance (1.7 Å) rather shorter than a C—P single bond (1.87 Å), which is indicative of considerable dp-π-overlap between phosphorus and carbon. This is also inferred from the observation that ylides

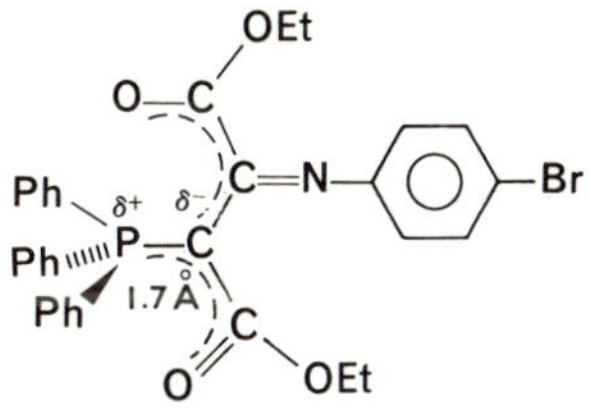

(131)

are stabilized far more by second-row elements than by first (i.e. phosphorus and sulphur ylides are far more stable than nitrogen ylides).

Reactions of Ylides

Protonation

The reverse of the dissociation leading to the ylide occurs in sufficiently acidic solution.

Reactions With Aldehydes and Ketones; *Wittig Reaction*

Attack on carbonyl carbon by the carbanion occurs readily but the reaction continues by the coordination of phosphorus to oxygen in the intermediate betaine with the resulting formation of a phosphine oxide and olefin, the driving force being the formation of the extremely strong P—O bond (ca 560 kJ mol^{-1}). Kinetic analysis shows that the first step is reversible and, in

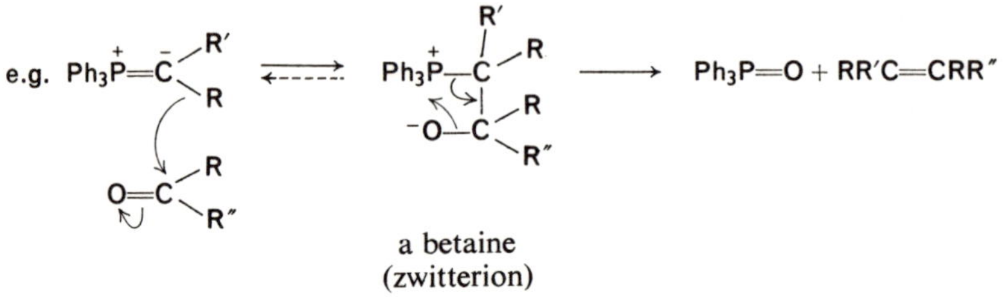

a betaine
(zwitterion)

some instances, the intermediate betaine is isolable. The reaction is of wide applicability and fails only when the ylide is excessively stabilized and lacks sufficient nucleophilic character such as the fluorenylide **127**. By the use of an asymmetric ylide it may be shown that the oxygen coordinates to phosphorus on the same side from which carbon leaves. The reaction may be extended to nitrogen analogous of the carbonyl compound as shown in the following examples:

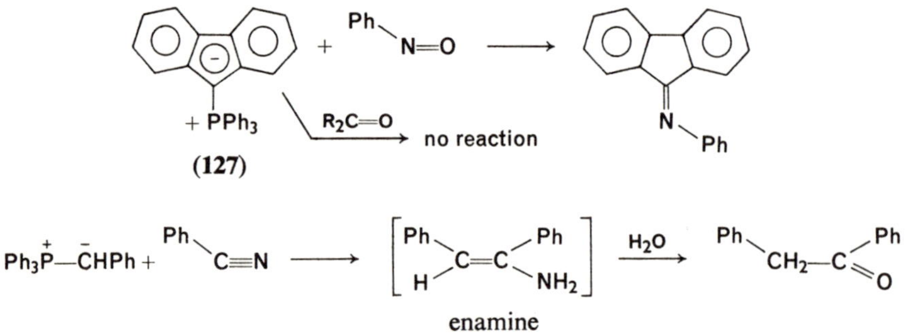

enamine

Sulphur ylides behave differently towards carbonyl compounds. The initial betaine formation takes place, but sulphur has less affinity for oxygen than phosphorus so that an internal S_N2 reaction occurs with the displacement of sulphide by the oxide ion and the formation of an epoxide (**132**). In a similar

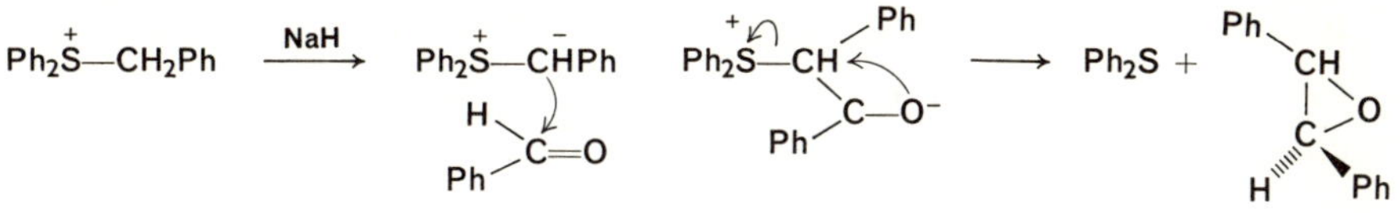

(132)

manner, a thioketone **(133)** forms an episulphide **(134)**. Sulphur ylides may add

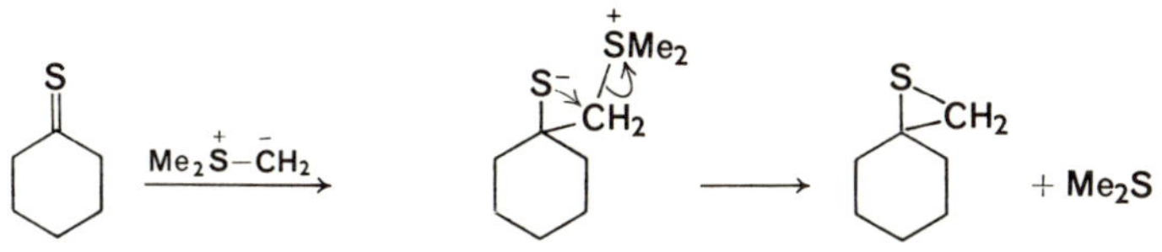

(133)   **(134)**

as nucleophiles to olefinic double bonds in a process essentially similar to the
Michael reaction (Section 4.15), but leading eventually to cyclic products.

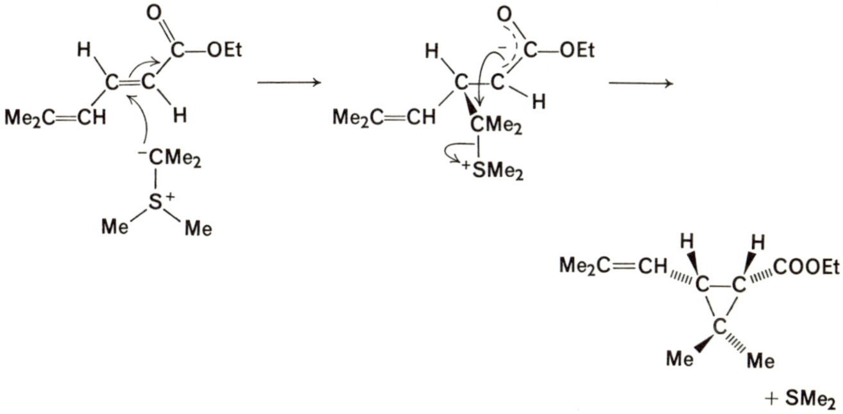

chrysanthemic acid
(ethyl ester)

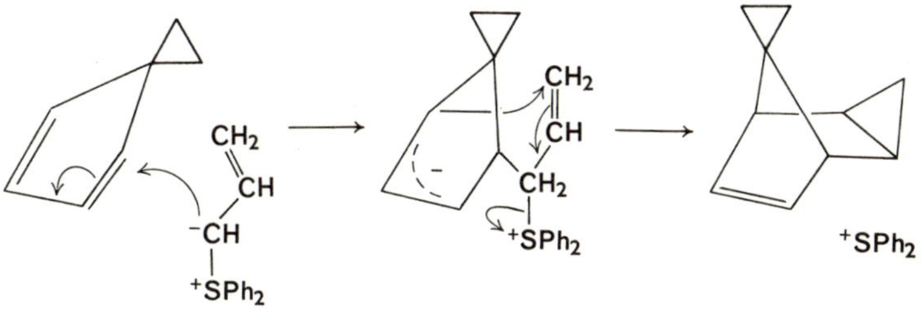

Reduction

Lithium aluminium hydride will reduce an ylide with the explusion of one of the groups attached to the heteroatom. The reaction evidently does not involve the

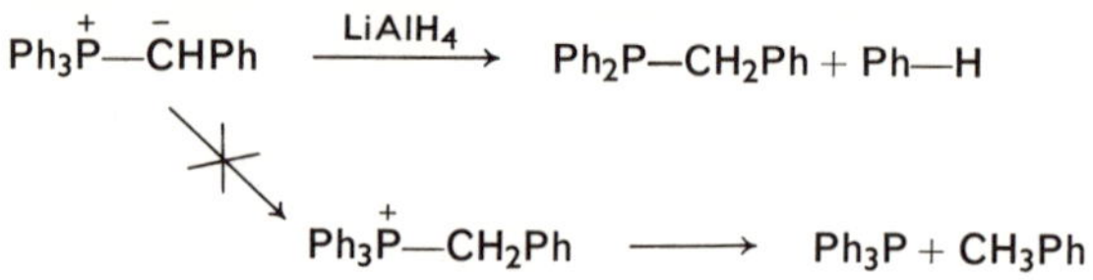
Ph3P—CHPh
LiAlH4
Ph2P—CH2Ph + Ph—H
Ph3P—CH2Ph
Ph3P + CH3Ph

benzyltriphenyl
phosphonium ion

formation of a phosphonium salt and expulsion of an anion since in that case the benzyl anion would be lost as it is more stable than a phenyl anion. The reaction is apparently more akin to the displacement of the phenyl cation by a hydride ion at saturated phosphorus (**135**).

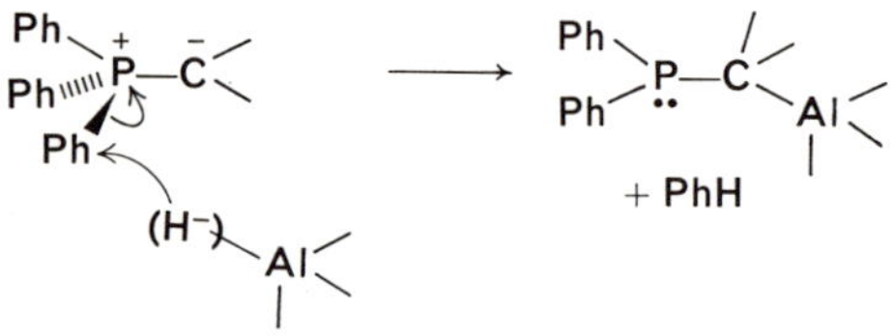
Ph
Ph
Ph
(H⁻)
Al
+ PhH

(135)

Oxidation

Oxygen (or a peracid) cleaves an ylide with the formation of phosphine oxide and a carbonyl compound which, if the ylide is in excess, reacts further in a Wittig reaction to give the symmetrical olefin **136**.

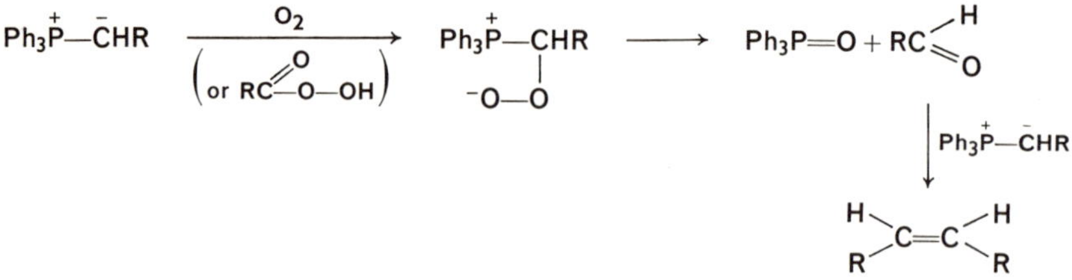
Ph3P—CHR
O2
or RC—O—OH
Ph3P—CHR
⁻O—O
Ph3P=O + RC
Ph3P—CHR

(136)

Hydrolysis

Ylides react with water with varying degrees of readiness to give initially a phosphonium hydroxide (137) which then apparently expels the most stable carbanion with the resultant formation of a hydrocarbon and phosphine oxide. The loss of groups from phosphorus follows the order of preference: fluorenyl

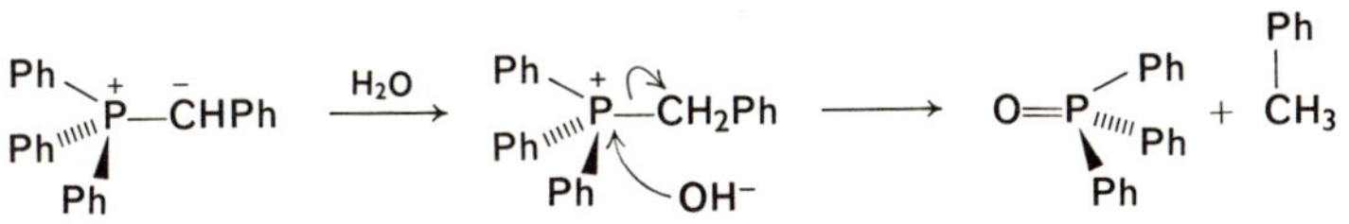

(137)

> benzyl > phenyl > alkyl. This process may be represented as a kind of S_N2 reaction at phosphorus, the displacement of a strong nucleophile (carbanion) by hydroxide ion receiving the driving force it needs from the stability of the phosphine oxide.

Rearrangement

Allyl ylides will undergo a [2,3]-sigmatropic rearrangement (p. 131) with inversion of the allylic group. This is formally analogous to the Cope rearrangement of hexa-(1,5)-dienes. It has been shown that two pathways

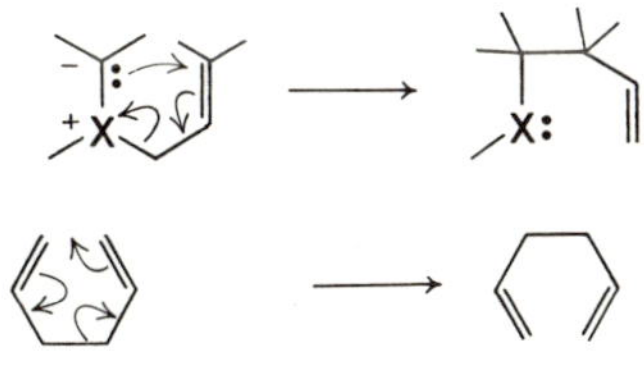

Cope rearrangement

operate in the following example. The orbital-symmetry controlled sigmatropic shift leads only to 138 but a small proportion of a radical process, giving initially the radical-pair 139 leads to both 138 and 140. The radical component in this reaction is apparent from the CIDNP emission signals in the product.

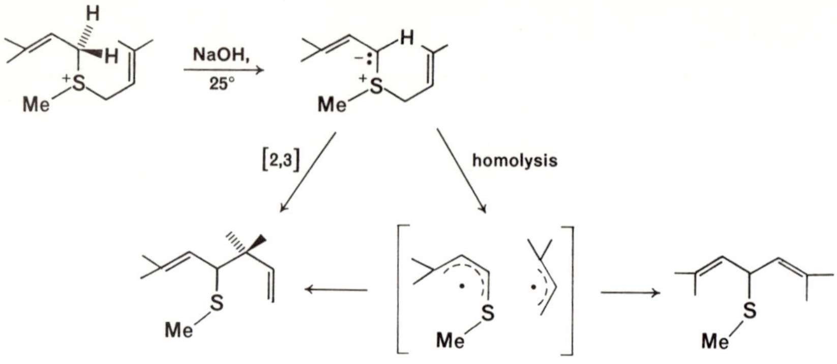

(138)   (139)   (140)

97%   radical pair   3%

Nitrogen Ylides

Quaternary ammonium ions are far less easily dissociated as acids than are phosphonium of sulphonium ions. Tetramethylammonium bromide (141) will react in ether with phenyl lithium (but not the weaker base benzyl sodium) to form trimethylammonium methylide (142) as a lithium bromide complex. Attempted removal of the lithium bromide by placing the complex in a

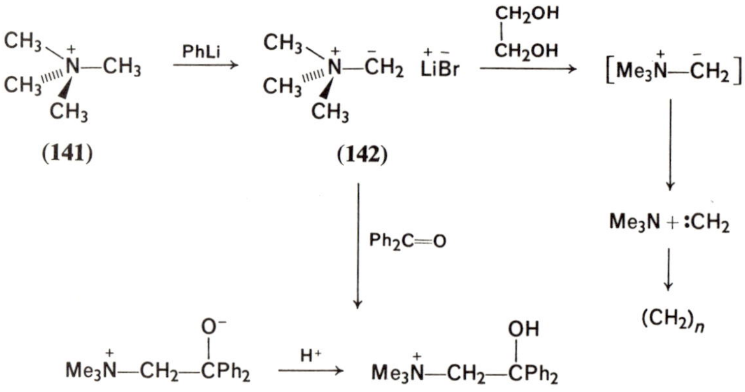

(141)   (142)

(143)

hydroxylic solvent leads to decomposition of the ylide, methylene being expelled and polymerizing. The complex reacts with a ketone but the product is a quaternary ammonium alcohol (143).

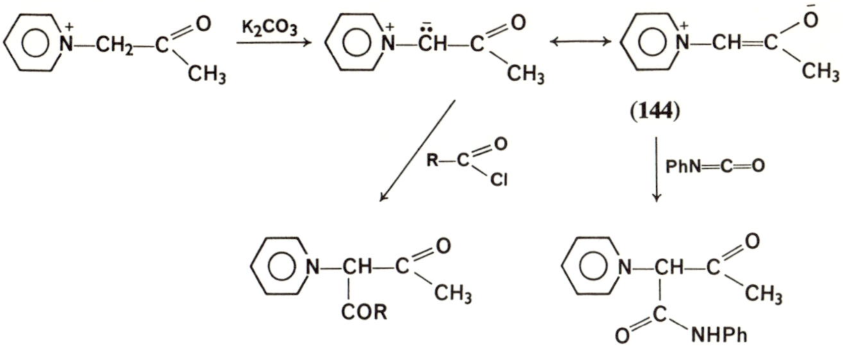

(144)

Pyridinium acylmethylides (144) are much more stable. They behave as strong nucleophiles in a variety of reactions. The *Stevens Rearrangement* (Section 4.21) takes place via a nitrogen ylide of this type.

SUGGESTIONS FOR FURTHER READING

General texts, especially J. Hine, E. Gould (see reading list for Chapter 1).

D. Cram, *Fundamentals of Carbanion Chemistry*, Academic Press, New York, 1965.

A. W. Johnson, *Ylid Chemistry*, Academic Press, New York, 1966.

J. R. Jones, 'Highly Basic Media', *Chem. Brit.*, **7**, 336, (1971).

J. R. Jones, 'Proton Transfer Reactions in Highly Basic Media', *Prog. Phys. Org. Chem.* **9** (1972).

A. I. Shatenshtein, 'Hydrogen Isotope Exchange Reactions of Organic Compounds in Liquid Ammonia', *Adv. Phys. Org. Chem.*, **1**, 156 (1963).

A. Streitweiser and J. H. Hammons, 'Acidity of Hydrocarbons', *Progr. Phys. Org. Chem.* **3**, 41 (1965).

Chapter 5

Radicals

INTRODUCTION

During the early growth of organic chemistry, 'binary' compounds such as methyl iodide, etc., were prepared and characterized and a major objective developed with the aim of isolating the parent organic 'radicals', e.g. 'methyl', in much the same way that new elements such as sodium and potassium had been isolated from their salts. This hope, of course, was not realized although the concept of a 'free radical' persisted and in 1900 the first of these compounds was shown to have a definite existence in solution—the yellow compound triphenylmethyl, $Ph_3C\cdot$. This discovery opened the way to the study of free radicals of many types which were gradually recognized as reactive intermediates in a wide variety of reactions.

Today, the term 'radical' applies to any species which possesses an unpaired electron which is more or less localized on a non-metallic element, especially carbon (this excludes metallic compounds with half-filled d-orbitals). Typically, organic radicals possess a carbon atom which is trivalent and has seven electrons in its valence shell (1); one electron must necessarily be

$$
\begin{array}{c}
R \\
\cdot | \cdot \\
C\cdot \\
\diagup \quad \diagdown \\
R' \quad R''
\end{array}
$$

(1)

unpaired. Simple examples of radicals are methyl (2), phenyl (3) and hydroxyl (4). The nomenclature is straightforward; the normal systematic or trivial

$$\dot{C}H_3$$

$$\cdot \ddot{O}H$$

(2)   **(3)**   **(4)**

name of the group is used and the word 'radical' appended. In formulae used in this chapter the unpaired electron is emphasized by a dot and the movements of single electrons indicated by singly barbed arrows.

Types of Radicals

A great variety of structural types of radicals are now known, though many are very transient species. There are simple alkyl radicals, $R_3C\cdot$, and aryls, e.g. **2**, **3**, and their derivatives such as alkoxy, $RO\cdot$, hydroxyalkyl, $R\dot{C}HOH$, or haloalkyl, $Cl_3C\cdot$. Radicals may in addition bear charges as in the cation radical and anion radical derived from naphthalene by removal and donation of one electron (**5** and **6** respectively); the unpaired electrons in these species occupy π-orbitals of the aromatic framework. Many atoms are radicals since

9 π-electrons,	10 π-electrons,	11 π-electrons,
cation radical	naphthalene	anion radical
(5)		**(6)**

they bear an unpaired electron; these include the alkali metals, $Na\cdot$, and the halogens, $F\cdot$, $Cl\cdot$, $Br\cdot$ and $I\cdot$. Nitric oxide, NO, and nitrogen dioxide, NO_2, are examples of odd-electron molecules; molecular oxygen in its ground state is a diradical (triplet state) with two unpaired electrons. This is due to the highest filled molecular orbitals being degenerate; according to Hund's rules, such orbitals are singly occupied.

5.2   METHODS OF PRODUCTION OF RADICALS

Thermolysis

The homolytic fission of a covalent bond leads to a pair of radicals being formed. In contrast to the other mode of bond breaking, heterolysis, which usually requires a polar solvent to separate the ions, homolysis is the preferred reaction in the gas phase or in solvents of low polarity and tends to be unaffected by the nature of the solvent, indicating that there is little energy of interaction between a radical and the medium. The energy profile for homolytic bond fission is

$$R\!-\!R \xrightarrow{\;\Delta(heat)\;} R\cdot + R\cdot$$

shown schematically in Figure 1. Since the reverse reaction, radical recombination, occurs with no activation-energy requirement, the energy of homolysis

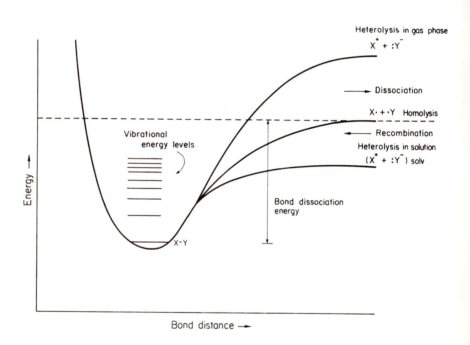

Figure 1. Schematic energy diagram for bond fission.

is synonymous with bond strength. In Table 1 are set out some representative bond-dissociation energies. Homonuclear bonds, e.g. –O—O–, N—N, hal—hal, are not only somewhat weak but are particularly prone to dissociate homolytically since the same element is not usually willing to take both positive and negative charge, as would be required for heterolysis. Peroxides, for example, are notoriously ready to undergo decomposition and provide a convenient source of radicals.

Table 1. Representative homolytic bond dissociation energies (kJ mol^{-1})

H—H	430	C—Cl	325
F—F	155	C—Br	272
Cl—Cl	240	C—I	240
Br—Br	190	C—H	410
I—I	150	H—I	295
-O—O-	140	H—Br	365
N—N	150	H—Cl	430
C—C	325–335	H—F	565
H—S	335	H—O-	460

Azo compounds such as azoisobutyronitrile, AIBN (7), also undergo homolysis at a low temperature since the bond-breaking energy is partly

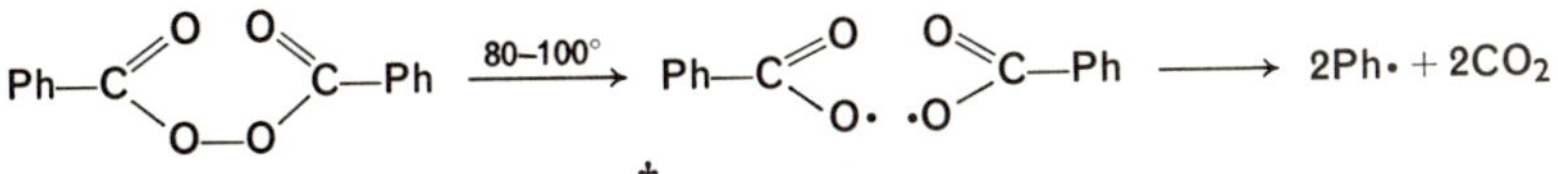

Ph—C(O)(O—O)C—Ph $\xrightarrow{80-100°}$ Ph—C(O)(O·) ·(O)C—Ph $\longrightarrow$ 2Ph· + 2CO$_2$

dibenzoyl peroxide

$E_A = 125$ kJ mol^{-1}

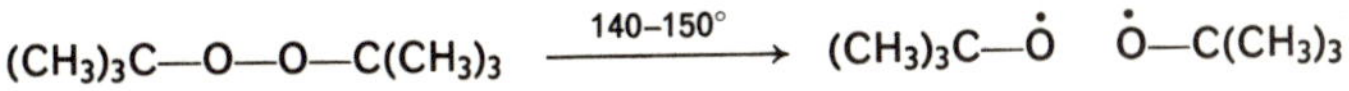

$(CH_3)_3C—O—O—C(CH_3)_3$ $\xrightarrow{140-150°}$ $(CH_3)_3C—\dot{O}$ $\dot{O}—C(CH_3)_3$

t-butyl peroxide

$E_A = 155$ kJ mol^{-1}

provided by the formation of a strong N≡N bond. Rates of decomposition of both the peroxides and AIBN are almost unchanged in solvents ranging in

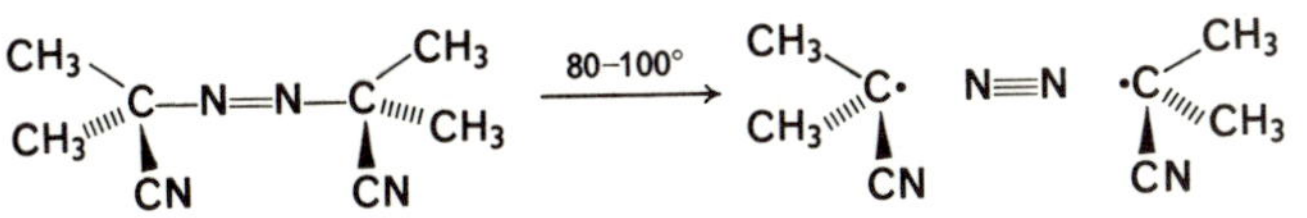

(CH$_3$)(CH$_3$)C(CN)—N=N—C(CH$_3$)(CH$_3$)(CN) $\xrightarrow{80-100°}$ (CH$_3$)(CH$_3$)Ċ(CN) N≡N Ċ(CH$_3$)(CH$_3$)(CN) $E_A = 132$ kJ mol^{-1}

(7)

polarity from acetic acid and alcohols to paraffin hydrocarbons, a characteristic of non-ionic processes.

In the gas phase at least, and given sufficient energy, any covalent bond will dissociate to radicals; some will, of course, require rather high temperatures

† E_A = Arrhenius activation energy (Section 1.5) for the reaction.

for this to occur thermally (e.g. C—C bonds in alkanes at 600–700°, $E_A = 300$–350 kJ mol^{-1}).

Photolysis and Radiolysis

Energy sufficient for bond dissociation may be placed in a molecule by the absorption of a quantum of radiation. The energy, E, of electromagnetic radiation is related to wavelength, λ, by the equation,

$$E = \frac{\cdot hc}{\lambda} = \frac{12 \times 10^4}{\lambda \text{ (nm)}} \quad \text{where} \quad h = \text{Planck's constant}$$

(kJ mol^{-1}) $c = \text{velocity of light}$

whence

$$400 \text{ nm} \equiv \text{ca } 300 \text{ kJ mol}^{-1}$$
$$300 \text{ nm} \equiv \text{ca } 400 \text{ kJ mol}^{-1}$$
$$200 \text{ nm} \equiv \text{ca } 600 \text{ kJ mol}^{-1}$$

A mercury-discharge lamp emitting much of its light at 253.7 nm is capable of causing homolysis of most of the bonds listed in Table 1, if absorbed, and is frequently used for this purpose. Acetone vapour (absorbing at about 310 nm) is first promoted to an excited state and then dissociated into methyl and acetyl radicals. The bombardment of a compound by an electron beam (radiolysis)

* = in an excited electronic state

provides a further means of delivering energy to the molecule. The primary act in many cases is the removal of a valence electron; the resulting cation radical may further fragment. This is also the mechanism of molecular fragmentation in the mass spectrometer where the interest is centred upon the ions produced.

$$\text{n-C}_4\text{H}_{10} + e^- \xrightarrow[\text{energy}]{\text{kinetic}} \text{n-C}_4\text{H}_{10}^{+\bullet} \longrightarrow \text{C}_2\text{H}_5\bullet + \text{C}_2\text{H}_5^+$$
$$+ 2e^-$$

In aqueous solution, hydroxyl radicals are formed which can generate secondary radicals from substrates in solution by hydrogen abstraction.

$$H_2O \xrightarrow{\ e^-\ } H_2O^{\cdot+} \longrightarrow \ \cdot OH + (H^+) \xrightarrow{\ RH\ } R\cdot + H_2O$$

Redox Reactions

The donation or removal of one electron from a spin-paired molecule (reduction and oxidation respectively) will in either case lead to the formation of a radical. Ti^{II} is a powerful one-electron reducing agent which reacts with hydrogen peroxide to form the hydroxyl radical,

$$Ti^{2+} + H_2O_2 \longrightarrow Ti^{3+} + OH\cdot + OH^-$$

and Co^{III} is a powerful one-electron oxidant capable of removing an hydrogen atom from an aromatic side-chain,

$$Co^{III} \diagup H - CH_2Ph \longrightarrow Co^{II} + H^+ + \cdot CH_2Ph$$

$$\text{benzyl radical}$$

Anodic oxidation can produce radicals as in the Kolbe electrosynthesis (Section 5.9) in which the primary reaction is the formation of carboxylate radicals from the corresponding anions.

$$RCOO^{\bar{:}} \xrightarrow[\text{anode}]{\ -e^-\ } (RCOO\cdot) \longrightarrow \text{products}$$

5.3   METHODS OF DETECTION OF RADICALS

Although more detailed examples will be found throughout this chapter, the general methods available for the detection of free radicals are discussed here. The production of radicals in a reaction may be inferred by their direct spectroscopic observation or by the characteristics of the reaction and the nature of the products formed.

General Characteristics of Reactions Involving Transient Radicals

Solvent-Polarity Effects

The reaction in general is indifferent to the polar nature of the solvent and often takes place readily in the gas phase, since no ionic intermediates are formed which can interact strongly with the solvent. This characteristic is shared by non-ionic reactions in general (including concerted processes such as the Diels–Alder reaction) and is therefore often indicative, but not diagnostic, of a radical reaction. Similarly, radical processes are not usually influenced by either acid or base catalysis.

Induced Decompositions

Radical decompositions may, however, be strongly influenced by solvents which are themselves capable of forming a rather stable radical by loss of hydrogen. Thus, ethers induce the decomposition of dibenzoyl peroxide (**6**) by a radical mechanism (see also p. 15).

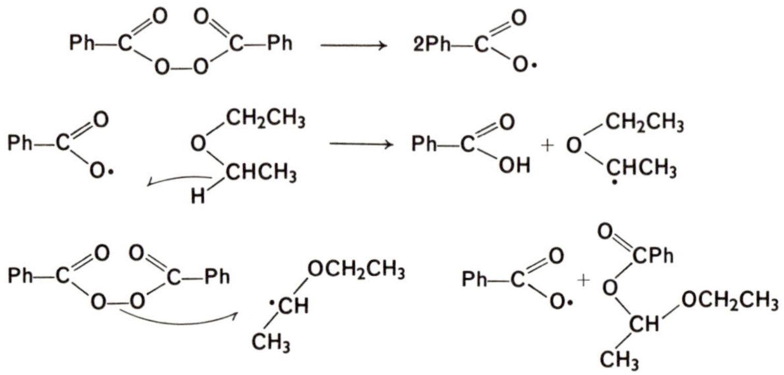

Initiation and Chain Reactions

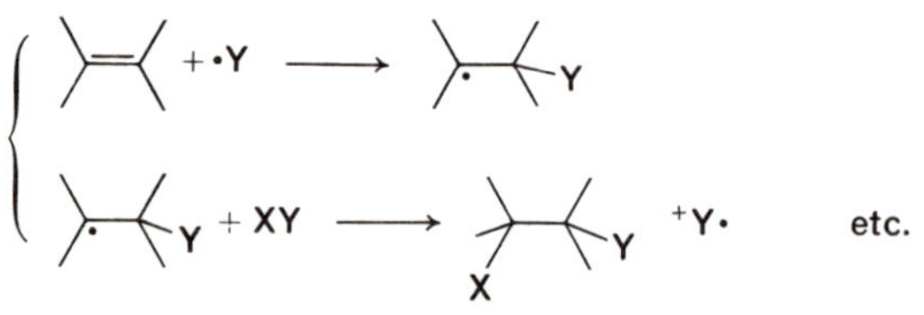

Radical reactions often partake of a chain mechanism, which needs to be started by an independent source of radicals (such as decomposing peroxide or light), and then continues, the formation of each product molecule generating a further radical to continue the chain. The addition of a species X—Y to a double bond may be taken as illustrative. The species Y· acts as the chain-carrying radical. In practice, X—Y may be H—Br, Cl_3C—Br, H—SR, etc. The

Initiation

$$X{-}Y \xrightarrow{\Delta,\ h\nu,\ \text{etc.}} X\cdot + \cdot Y$$

$$\text{or } X{-}Y \xrightarrow{R\cdot} X{-}R + \cdot Y$$

Propagation

chains may be very long (e.g. up to 10^6 additions per initiator radical) or quite short; they are terminated by the destruction of radicals by dimerization or disproportionation. Chain reactions are not unknown in ionic reactions but are far less common. Inevitably, the kinetics of these processes are complex. Thus the addition of chlorine to olefins by a radical chain (X—Y = Cl—Cl) has been found to obey a rate law of the approximate form,

$$\text{rate} = kI^{0.5}[\text{olefin}]^{0.5}[\text{Cl}_2]^{1.0-1.5}$$

where I refers to the intensity of the light used to initiate the reaction (creating chlorine atoms as primary radicals, $Y\cdot = Cl\cdot$). A kinetic analysis of a chain reaction is given in Section 1.5.

Substituent Effects

Organic cations are well known to be stabilized by electron-donating substituents and anions by electron-withdrawing groups. This leads to a universal order of substituent effects for ionic reactions (the Hammett relationship, Section 1.10). Radicals on the other hand are often found to be stabilized by both types (e.g. by both –OMe and –NO$_2$ groups) relative to hydrogen which leads to very poor linear free-energy relationships. Some radical reactions do have a somewhat polar nature in addition but ρ-values are usually close to zero. The decomposition of *meta*- and *para*-substituted dibenzoyl peroxides (8) gives a reasonably good Hammett plot with $\rho = -1.4$ to -0.5 (i.e. slight assistance from electron-donating polar groups).

(8)

Inhibition of Reaction

The rates of radical reactions are frequently very sensitive to certain impurities such as oxygen, amines, thiols, phenols. These are compounds which are capable of removing free radicals to form less-reactive radicals and hence can scavenge the initiator radicals of a chain reaction and prevent it occurring. The presence of *inhibitors* such as these causes an induction period or delay at the initial stage of the reaction during which the inhibitor is consumed by reaction; the desired reaction does not start until the inhibitor is eliminated. The introduction of inhibitors such as hydroquinine is often used to prevent undesirable radical chain reactions.

Typical Reactions of Radicals

Dimerization

At ordinary temperatures, and with few exceptions, radicals are unstable with respect to their dimers which, if different from starting materials, may be observed as characteristic products. For example, methyl radicals generated in the gas phase by the decomposition of lead tetramethyl (**9**) form ethane as

$$Pb(CH_3)_4 \longrightarrow Pb + 4CH_3\cdot$$
$$(9)$$

$$CH_3\cdot + CH_3\cdot \longrightarrow CH_3CH_3$$

the principal product. Dimerization leads to loss of unpaired electrons in the system.

Disproportionation and Exchange

A highly reactive radical may abstract an atom, frequently hydrogen, from a molecule capable of forming a more stable radical (i.e. if the process is exothermic). Toluene is often added to trap radicals in the gas phase which can react with it to produce benzyl radicals stabilized by conjugation. These in turn are inferred from the dibenzyl (1,2-diphenylethane) **10** isolated. Two radicals

$$CH_3\cdot + H\!-\!CH_2Ph \longrightarrow CH_4 + \cdot CH_2Ph$$

$$2\ \cdot CH_2Ph \longrightarrow PhCH_2CH_2Ph$$
$$(10)$$
$$\text{dibenzyl}$$

may interact and exchange a hydrogen atom producing one alkane and one alkene molecule (disproportionation). Like dimerization, this leads to the loss of radicals. These two processes commonly terminate radical chains.

$$
\begin{array}{c}
(CH_3)_2C\!-\!COOMe \\
| \\
N \\
\| \\
N \\
| \\
(CH_3)_2C\!-\!COOMe
\end{array}
\quad \xrightarrow{\ \Delta\ } \quad 2(CH_3)_2\dot{C}\!-\!COOMe
$$

methyl azoisobutyrate

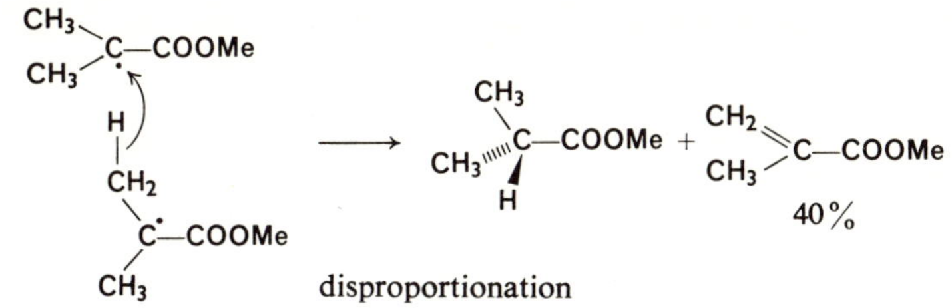

disproportionation

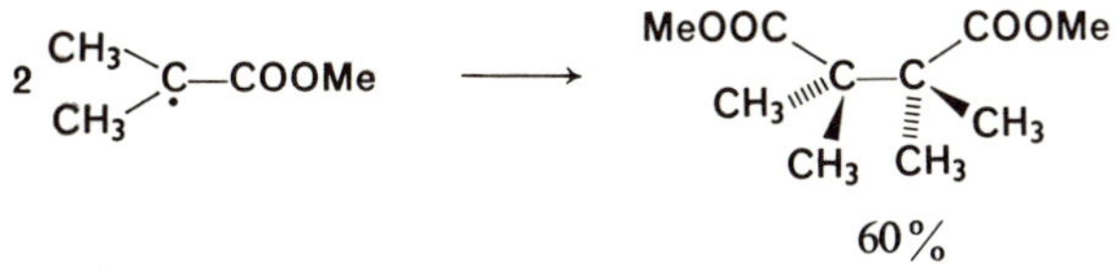

dimerization

Addition to an Unsaturated System

A radical may be accepted by an olefinic or aromatic system to form a more complex radical. Typically, the radicals derived from olefins either add a second fragment or polymerize leading to saturation of the double bonds while those from aromatic molecules lose a hydrogen atom leading to substitution and regeneration of the aromatic character.

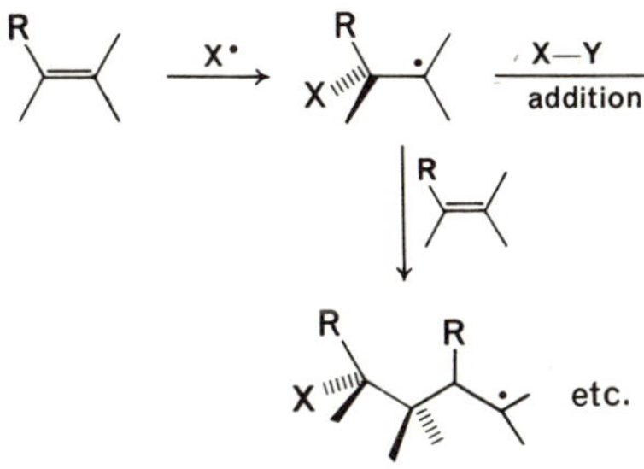

etc.   polymerization

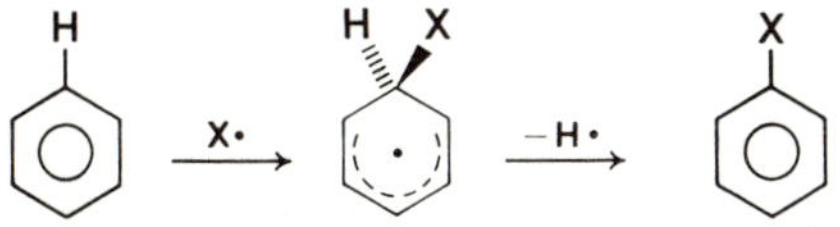

substitution

Oxidation and Reduction

The removal of an electron from, or donation to, a radical results in the formation of a carbonium ion or a carbanion, respectively, which then undergoes its characteristic reaction. Examples of these reactions will be discussed below. The isolation of characteristic products from a radical reaction may be regarded as an example of trapping of the radical.

Spin Trapping

See Section 5.5.

5.4　　　　PHYSICAL DETECTION OF RADICALS

Radicals which are stable or which have a moderately long lifetime ($>10^{-3}$ s) or, if transient, can be generated at a rate great enough to maintain a sufficient stationary concentration, may be observed directly and provide the most unambiguous evidence for the existence of free radicals. By far the most important physical method for the study of radicals is electron spin resonance.

Electron Spin Resonance, e.s.r., also known as Electron Paramagnetic Resonance, e.p.r.

The unique feature of a radical is its unpaired electron which gives rise to a resultant magnetic moment or paramagnetism. When placed in a magnetic field, the radical may exist in one of two possible energy states due to the quantum-mechanical restriction on the orientation of the electronic magnetic moment—either with or against the applied field. The separation of the energy states, ΔE, increases with the applied field, H, according to the relation, $\Delta E = g\beta H$, where g, the gyromagnetic ratio for the electron (magnetic moment/angular momentum), equals 2.0, and β is the Bohr magneton, a constant. Resonance absorption of radiation, promoting a molecule from the lower to the upper spin state, occurs when the condition,

$$h\nu = g\beta H$$

is satisfied. At an applied field of 3000 gauss, this corresponds to a frequency of 9000 MHz or a wavelength of 3 cm. E.s.r. spectroscopy therefore is carried out in the microwave (radar) region of the spectrum. An e.s.r. signal (absorption of radiation) is obtained by irradiating a radical sample with 3 cm radiation within a microwave 'cavity' while maintaining it in a magnetic field of 3000 G which is varied over fairly narrow limits until resonance occurs. The signal confirms the presence of unpaired electrons (Figure 2). However, far more information than this may be obtainable with the high-resolution equipment now available. If the unpaired electron spends any of its time in the vicinity of a magnetic nucleus (e.g. ^1H, ^{13}C, ^{14}N) the applied field will be modified by that of the nucleus in its immediate vicinity and the resonance condition slightly

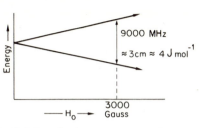

9000 MHz

$\approx 3cm \approx 4 J mol^{-1}$

3000
Gauss

$— H_o \longrightarrow$

Energy $\longrightarrow$

For technical reasons e.s.r
spectra are presented as
first-derivative curves
rather than a plot of absorption
against magnetic field

(a) Splitting of spin energy states of
an election in a magnetic field

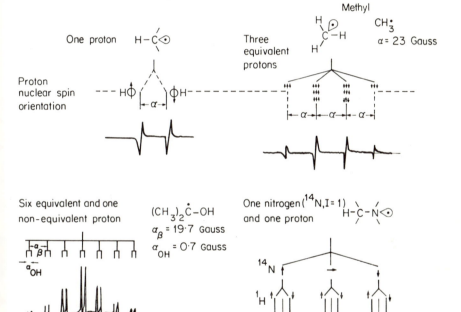

(b) Nuclear coupling of electron spins

Figure 2. Principles of electron spin resonance.

altered. Since each magnetic nucleus can have two or more permitted orientations in the magnetic field, the e.s.r. signal will be split into this number of
components—in general $(2I + 1)$ components where I is the spin number of
the nucleus (for ^1H, ^{13}C, ^{19}F, $I = \frac{1}{2}$; for ^2H, ^{14}N, $I = 1$). If the resonance is
split by n equivalent nuclei the multiplicity is $(2nI + 1)$ and if split by n nuclei

of one type and m nuclei of a second chemical type, the multiplicity of the signal is

$$(2nI_n + 1)(2mI_m + 1)\ldots*$$

An analysis of the splitting pattern of the spectrum can give a great deal of information on the structure of the radical. Some simple examples are set out in Figure 2.

Splitting Factors

Although the number of lines of a given spectrum can be simply calculated, another factor determines the appearance of the spectrum, namely the separation of lines caused by interaction with each group of magnetic nuclei, known as the splitting factor, α. The value of α (in units of magnetic-field strength) is a measure of the strength of the interaction between the unpaired electron and the magnetic nucleus, or a measure of the time which the electron spends in its vicinity. A nucleus too distant for interaction may be considered to have $\alpha = 0$. In simple saturated alkyl radicals it is usually found that protons which are one or two bonds distant from the radical centre couple with α about 22–25 G. Protons further removed do not couple unless the structure is conjugated. Thus the methyl radical appears as a triplet ($\alpha = 23$ G) and the t-butyl radical a decuplet (10 lines) ($\alpha = 25$ G). More complex radicals may have spectra so complicated that computer analysis is necessary.

g-Factors

The gyromagnetic ratio, g, for the free electron has the more exact value 2.0023. This varies to a small extent among free radicals depending on their chemical nature. Since the value of g is measured with high accuracy by the position of the centre of the spectrum along the base line (measured relative to a standard radical) this parameter has some value for characterizing the type of radical being observed.

---

* The intensity ratios of the lines of a multiplet spectrum are given, as for n.m.r. spectra, by the appropriate numbers in Pascal's triangle,

$$1$$

For multiplicity of 2, ratio of lines =	1 1
For multiplicity of 3, ratio of lines =	1 2 1
For multiplicity of 4, ratio of lines =	1 3 3 1
For multiplicity of 5, ratio of lines =	1 4 6 4 1
For multiplicity of 6, ratio of lines =	1 5 10 10 5 1
	etc.

Pascal's Triangle

Table 2. g-Factors of free radicals

Radical type	g
Free electron	2.0023
Alkyl radicals	2.00255
Nitrogen radicals $R_2N\cdot$	2.0032
Oxygen radicals $RO\cdot$	2.015
Nitroxides $R_2NO\cdot$	2.006

^{13}C-splitting

^{13}C with a spin number $I = \frac{1}{2}$ has a natural abundance of only 1 % in carbon. This means that further splitting of the e.s.r. signal by each carbon atom with whose magnetic field the electron interacts will occur, but this splitting is of low intensity. When observable, ^{13}C-splitting gives a measure of the unpaired-electron density on the various carbon atoms of the molecule which may, in conjugated radicals, be compared with theoretical predictions.

5.5   SOME EXAMPLES OF DIRECTLY OBSERVABLE RADICALS

Many radicals, including a wide variety of structural types, have been directly observed by e.s.r. spectroscopy. This section serves to exemplify the characteristics of a selection of radicals and the techniques used to produce and observe them. While many species are known which are stable in solution or even in the crystalline state, the majority have short lifetimes and must be continuously generated, maintained at low temperature or in an inert matrix, for spectroscopic observation.

Methyl and Arylmethyl Radicals

The series $CH_3\cdot$, $PhCH_2\cdot$, $Ph_2CH\cdot$ and $Ph_3C\cdot$, provides a gradation of properties from highly reactive to highly stable radicals.

The methyl radical, $CH_3\cdot$, may be generated by the following reactions:

(a) The thermolysis of metal methyls such as tetramethyl lead (**11**). This

$$PbMe_4 \xrightarrow{450°} Pb + 4Me\cdot$$

(11)

method is of historical interest in that it was used by Paneth in the first un-equivocal demonstration of radicals in the gas phase. A stream of tetramethyl lead vapour was pyrolysed in a glass tube and the reaction products swept out of the heated zone and past a deposit of lead on the wall of the tube some distance from the oven. It was observed that the lead deposit was removed by an agency with which it formed lead tetramethyl (by a reversal of the radical-forming reaction). The interpretation that this agency was methyl radicals could not be avoided.

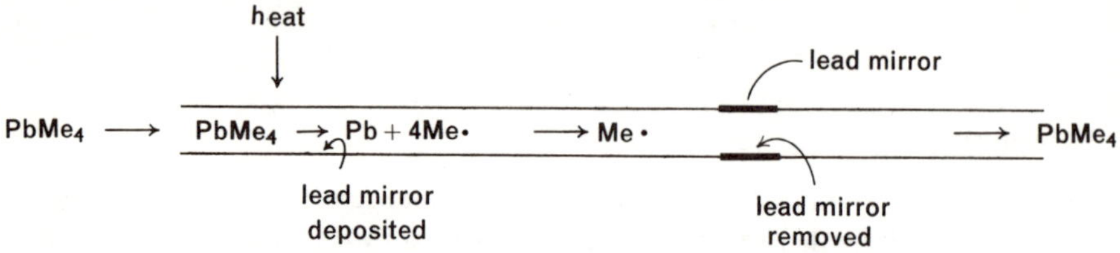

(b) The photolysis of acetone by ultraviolet light ($\lambda_{max} = 310$ nm) yields two methyl radicals and carbon monoxide in the principal decomposition route,

$$CH_3COCH_3 \xrightarrow{h\nu} CH_3\cdot + \cdot COCH_3 \longrightarrow CO + 2\,CH_3\cdot$$

(c) The thermal decomposition of certain peroxides, whose primary oxygen radicals can expel a methyl radical and form a stable carbonyl compound. Diacetyl peroxide (12), di-*t*-butyl peroxide (13) and cumene hydroperoxide (14), are convenient sources of methyl radicals. Methyl radicals

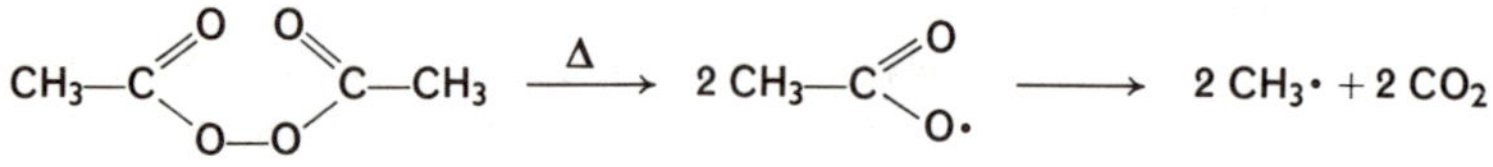

(12)

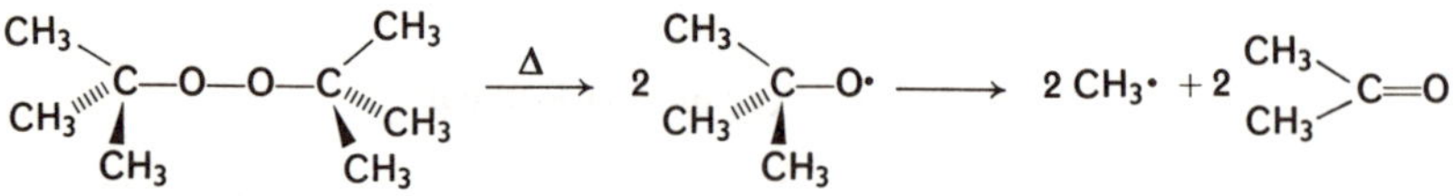

(13)

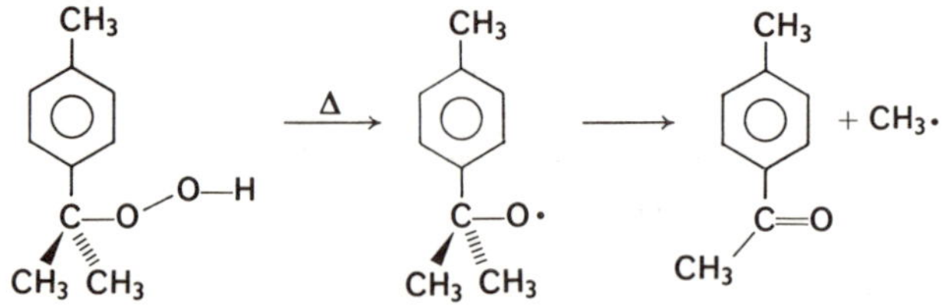

(14)

generated by any of these methods are highly reactive and rapidly dimerize to ethane or abstract hydrogen from available donors, e.g. toluene, which may be present. The e.s.r. spectrum of the methyl radical may be observed by rapidly

$$CH_3\!-\!CH_3 \xrightarrow{\ 2\times\ } CH_3\cdot \xrightarrow{\ PhCH_3\ } CH_4 + PhCH_2\cdot$$

sweeping it through the spectrometer cavity immediately after its formation in a continuous-flow system. The principal feature of the spectrum is a quartet of lines in the intensity ratio, $1:3:3:1$, $\alpha = 23$ G, indicating the odd electron coupled to three equivalent hydrogen atoms (Figure 2).

The benzyl radical, $PhCH_2\cdot$, is much more stable than methyl as shown by its formation ($\Delta H = 80$ kJ mol^{-1}) from methyl radicals and toluene. Other methods which have been used include the photolysis of benzyl chloride,

$$PhCH_2Cl \xrightarrow{\ h\nu\ } PhCH_2\cdot + Cl\cdot$$

reaction of benzyl chloride with sodium in the vapour phase,

$$PhCH_2Cl + Na\cdot \longrightarrow PhCH_2\cdot + NaCl$$

and, in aqueous solution, the reaction of hydroxyl radicals with toluene or phenylacetic acid.

$$PhCH_3 + OH\cdot \longrightarrow PhCH_2\cdot + H_2O$$

$$PhCH_2COOH + OH\cdot \longrightarrow PhCH_2COO\cdot + H_2O$$

$$\downarrow$$

$$PhCH_2\cdot + CO_2$$

The benzyl radical is also far too reactive for prolonged existence and a flow system must be used to record its e.s.r. spectrum. This shows 54 lines indicating coupling of the unpaired electron to all protons and to *ortho* and *para* more than to *meta*. Evidently canonical structures of the type **15a–d** are important

a   b   c   d

(15)

although the major contributing structures are **a**, **b** and **c**.

$$\alpha_H = \qquad 16.4 \qquad\qquad 5.2 \qquad\qquad 1.8 \qquad\qquad 6.2 \ G$$

$$-CH_2 \qquad\quad 2\text{-}o\text{-}H \qquad\quad 2\text{-}m\text{-}H \qquad\quad 1\text{-}p\text{-}H$$

Multiplicity: $\Pi (2nI + 1) = 3 \times 3 \times 3 \times 2 = 54$

The splitting factors show that the unpaired electron spends most of its time (70%) on the methylene carbon and the remaining 30% in the π-system of the ring.

The next member of the series, benzhydryl, $Ph_2\overset{.}{C}H$ (**16**) shows a further increase in stability despite the fact that the radical cannot be completely planar owing to steric interactions between *ortho* hydrogens.

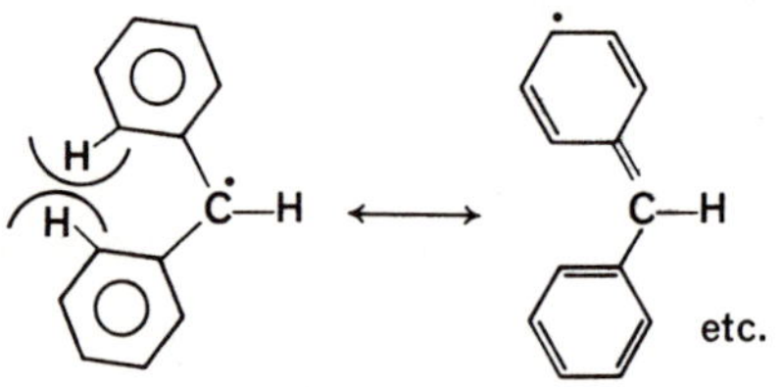

	$(2nI + 1)$
methine	2
4 *ortho*	5
4 *meta*	5
2 *para*	3

66%   (**16**)   34%—all forms $2 \times 5 \times 5 \times 3 = 150$ lines

The theoretical 150-line spectrum has been observed in solution and indicates that the unpaired electron is still mainly located on the methine carbon (66%).

The importance of steric interactions in stabilizing a radical is apparent from the observation that o,o'-disubstituted benzhydryl radicals (**17**) are stable with respect to their dimers (**18**). Evidently non-bonding repulsions between aryl groups on neighbouring carbon atoms can so weaken the central bond that it undergoes homolysis at ordinary temperatures.

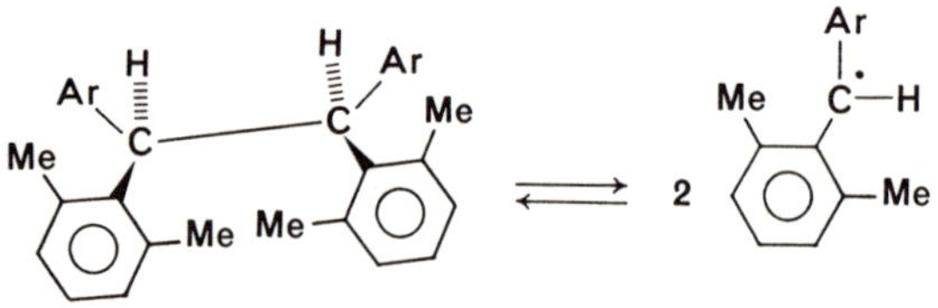

(**18**)   (**17**)

The final member of the series, triphenylmethyl (or, colloquially, 'trityl'), is stable in solution and is one of the best known radicals. This was, in fact, the first radical to be characterized and was produced, in equilibrium with its dimer, by Gomberg in 1901. One dimer, hexaphenylethane (**19**), is sterically crowded while the radical is especially stable due to the extended delocalized structure which is possible, and these two factors combine so that the radical is in equilibrium with the dimer to the extent of some 2–10% in solution. More

Ph_3C—CPh_3 ⇌ 2 $Ph_3C\cdot$ ⇌

(**19**)   triphenylmethyl   (**20**)
colourless   (trityl)   colourless
yellow
$\lambda_{max} = 400$ nm

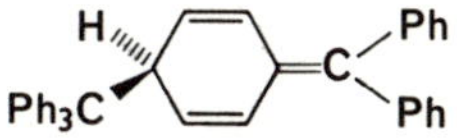

recently, the more stable dimer has been shown to be **20** rather than hexa-phenylethane. Following their discovery, a large number of substituted triphenylmethyl radicals were prepared and characterized. The following reactions are usually applicable. Treatment of the triarylmethyl halide with silver (the Gomberg method), zinc or a Grignard reagent:

$$Ph_3C—Cl + Ag \longrightarrow Ph_3C\cdot + AgCl$$

$$Ph_3C—Cl + PhMgBr \longrightarrow Ph_3C\cdot + Ph\cdot + MgBrCl$$

$$\downarrow$$

$$Ph—Ph$$

The triphenylmethyl cation (**21**) may be reduced to the radical by powerful reducing agents, e.g. vanadium(II):

$$Ph_3COH \xrightarrow{H^+} Ph_3C^+ \xrightarrow{V^{2+}} Ph_3C\cdot + V^{3+}$$
$$+ H_2O$$
(**21**)

and the triphenylmethyl anion (**22**) may be oxidized to the radical by, for example, silver ion:

$$Ph_3C\bar{:} + Ag^+ \longrightarrow Ph_3C\cdot + Ag^\circ$$
(**22**)

The thermal decomposition of various trityl species such as the ester (23) and the azo compound (24) are of use in the preparation of these radicals.

$$Ph_3C-C\underset{OCPh_3}{\overset{O}{\diagdown}} \xrightarrow{\Delta} Ph_3C\cdot + CO_2 + \cdot CPh_3$$

(23)

$$Ph_3C\diagdown \underset{}{N{=}N}\diagup ^{CPh_3} \xrightarrow{\Delta} Ph_3C\cdot + N_2 + \cdot CPh_3$$

(24)

The e.s.r. spectra of triarylmethyl radicals are readily observed in solution. Air must, however, be excluded or the peroxy radical (25) will form and confuse the spectrum. The parent radical, triphenylmethyl (trityl) shows the

$$Ph_3C\cdot + O_2 \longrightarrow Ph_3C-O-O\cdot$$

(25)

theoretical 196 lines, hence the unpaired electron is completely delocalized though spending most of its time on the central carbon.

Ion Radicals of the Aromatic Hydrocarbons

Aromatic hydrocarbons including benzene, naphthalene and polycyclic homologues, are both capable of donating and of accepting one electron, the products being cation radicals or anion radicals, respectively. The cation radicals may be formed by dissolving the aromatic hydrocarbon in concentrated sulphuric acid, which acts both as an oxidizing agent and a suitable medium for supporting a highly electrophilic cation. Iodine and lead(IV)

$$2\,ArH + 5\,H_2SO_4 \longrightarrow 2\,ArH^{+\cdot}\ HSO_4^- + 2\,HSO_4^- + 2\,H_3O^+ + SO_2$$

acetate can also be used as oxidants. The anion radicals are formed by treatment of the hydrocarbon with sodium in a solvent such as tetrahydrofuran. The

$$ArH + Na \longrightarrow ArH^{-\cdot} + Na^+$$

structures and π-orbital occupancy of the ion radicals derived from naphthalene are shown in Figure 3. The e.s.r. spectra of both species are very similar and

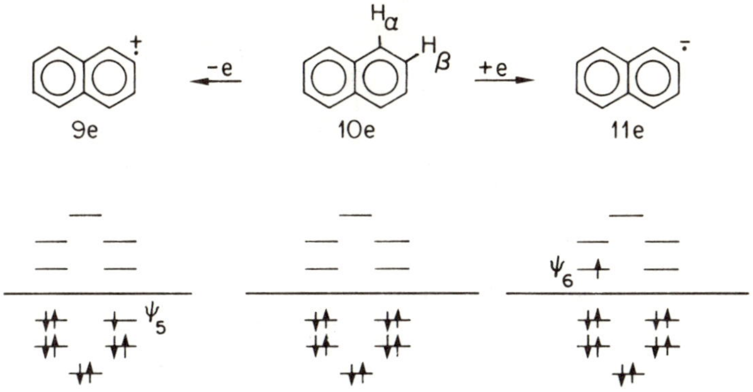

Figure 3. Structures and π-orbital occupancy in naphthalene ion radicals.

consist each of 25 lines when examined under high resolution (Figure 4). This is confirmatory evidence for the symmetry of each species; in each case the unpaired electron is coupled to four equivalent α-protons and four β-protons and values of the two splitting factors are the same for cation and anion radicals, showing that the orbital occupied by the odd electron in either ion has the same distribution over the aromatic framework. This is expected theoretically even though one is a bonding orbital and the other a corresponding antibonding orbital, and constitutes important evidence for the validity of simple Hückel molecular-orbital theory in these systems. The splitting factors for these two groups of protons is also in the predicted ratio—that of the magnitude of the square of the wave function of the occupied orbital at each type of carbon (Figure 4). The ^{13}C-splitting factors would provide an even better check on theory.

In a similar way, the anthracene ion radicals (26) show coupling to the four α-, four β- and two *meso* protons. The resulting 75-line spectrum requires high resolution to discern all the lines since many coincide owing to the splitting factors being nearly in the ratio $2:1:4$, as predicted. Alkylation of the anion

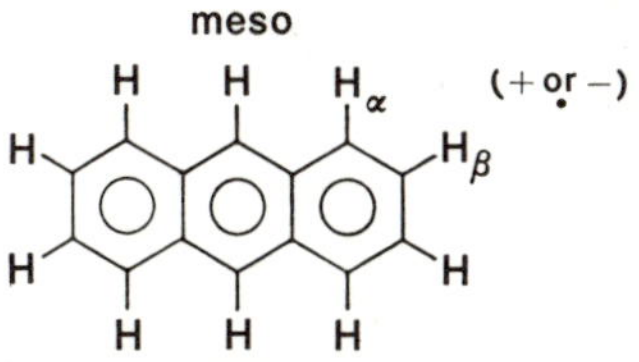

(26)

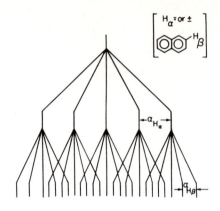

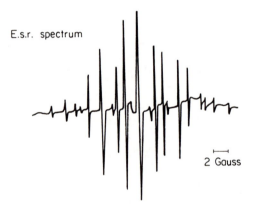

Figure 4. E.s.r. spectra of the naphthalene ion
radicals.

radicals, often specifically, can be accomplished readily on account of their
highly nucleophilic character, e.g.

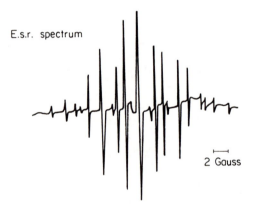

Anion radicals of aryl-substituted ethylenes are also known (27); they are
formed by the action of alkali metal on the olefin and are in equilibrium with

their dimers, the dianions (28). The position of equilibrium is very much dependent upon the metal ion.

$$Ph_2C{=}CPh_2 + M \xrightleftharpoons[\text{ether}]{} Ph_2\bar{C}{-}\dot{C}Ph_2.M^+ \underset{\underline{}}{\xrightleftharpoons{K}} Ph_2\bar{C}{-}CPh_2{-}CPh_2{-}\bar{C}Ph_2$$

$$\qquad\qquad\qquad\qquad\qquad\quad (27) \qquad\qquad\qquad\qquad\qquad (28)$$

M	K
Li	1
Na	0.1
K	45
Cs	2700

The corresponding species, 29, may also be formed from acetylenes. These

$$PhC{\equiv}CPh \xrightarrow{\overset{\displaystyle [\text{naphthalene}]^{\bar{\cdot}}}{}} Ph\bar{C}{=}\dot{C}Ph \longrightarrow$$

$$\qquad\qquad\qquad\qquad\qquad (29)$$

$$Ph\bar{C}{=}\underset{\underset{Ph}{|}}{\overset{\overset{Ph}{|}}{C}}{-}\bar{C}Ph \xrightarrow{2H^+} PhCH{=}\underset{\underset{Ph}{|}}{\overset{\overset{Ph}{|}}{C}}{-}C{=}CHPh$$

anion radicals are capable of reducing alkyl halides; the following two routes have been identified in the production of butane from butyl chloride in dimethyl ether as solvent,

$$BuCl + [\text{naphthalene}]^{\bar{\cdot}} \longrightarrow \dot{Bu} + Cl^- + [\text{naphthalene}]$$

(a) $\dot{Bu} + CH_3OCH_3 \longrightarrow BuH + \dot{C}H_2OCH_3$

(b) $\dot{Bu} + [\text{naphthalene}]^{\bar{\cdot}} \longrightarrow {}^-Bu^- + [\text{naphthalene}]$

$\qquad Bu^- + CH_3OCH_3 \longrightarrow BuH + \bar{C}H_2OCH_3$

Semiquinones

The facile reversible oxidation of hydroquinone to benzoquinone (30) is a two-electron process. Electron transfer takes place in two one-electron steps and the intermediate radical, a semiquinone (31), is a stable deep-red com-

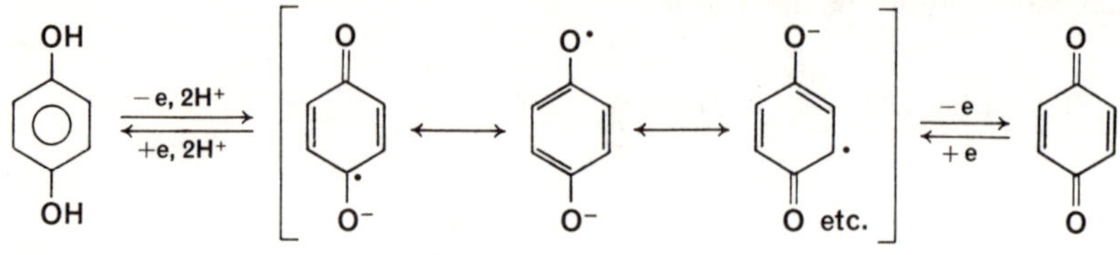

(31)   (30)

hydroquinone   semiquinone   benzoquinone

pound. Similar radicals may be obtained from a variety of 1,2- and 1,4-dihydroxybenzenes. A familiar example is pyrogallol (1,2,3-trihydroxybenzene) which absorbs oxygen in alkaline solution to give, initially, the radical 32.

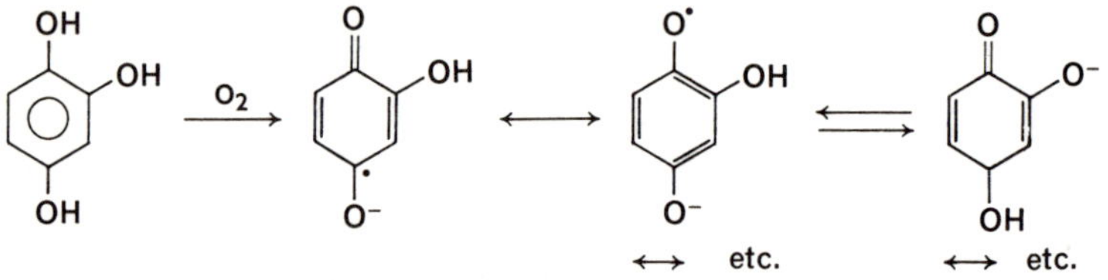

(32)

Ketyls

Aromatic ketones develop a dark-blue colour on exposure to sodium in the absence of air. These blue compounds are the one-electron reduction products of the ketones, radical anions known as ketyls (33). The ketyls are stable in

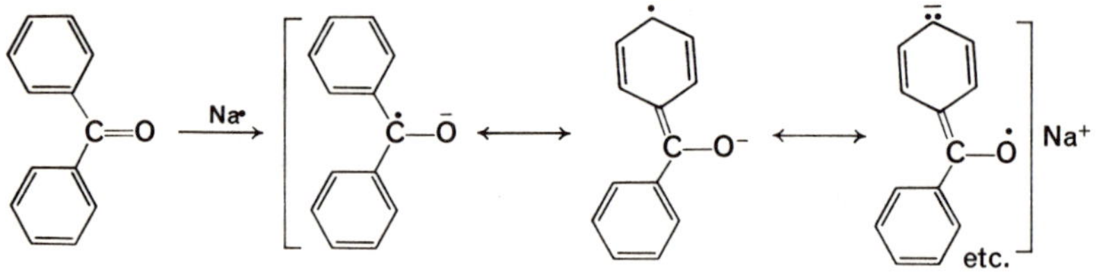

(33)

benzophenone   benzophenone ketyl $\lambda_{max} = 645$ nm
                                (sodium salt)

basic solution but dimerize to pinacols (34) on neutralization. The stability of ketyls results in part from the extensive delocalization of both the unpaired spin and the charge, and in part to crowding and electrostatic repulsion in the dimer, the pinacol dianion (**35**). The e.s.r. spectra confirm the delocalization of

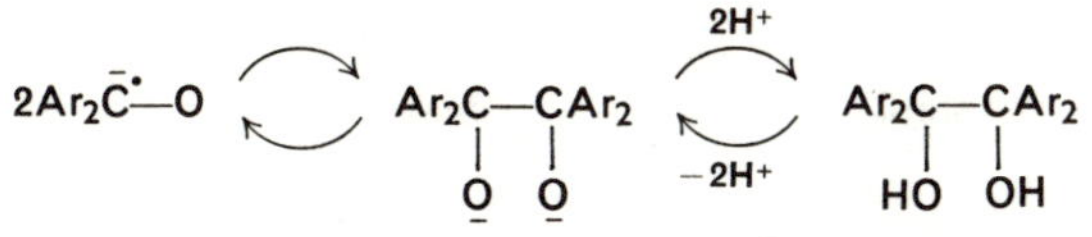

(**35**)   (**34**)

the unpaired electron and reveal that it exists mainly on the carbonyl carbon, and to a lesser extent in the aromatic rings, there principally at *ortho* and *para* positions; (α (*ortho*) = 4.6, α (*meta*) = 1.3 and α (*para*) = 6.4 G). Solutions containing ketyl radicals are also capable of removing traces of oxygen from a gas stream with high efficiency:

$$Ph_2\overset{\cdot}{C}-\overset{-}{O} + O_2 \longrightarrow Ph_2C{=}O + O_2^{\overset{-}{\cdot}}$$

Alkylation in the aromatic ring readily occurs.

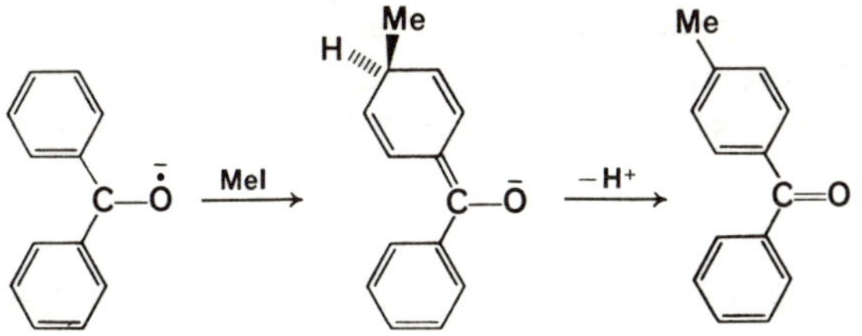

Aryloxy Radicals

Phenols are notably susceptible to oxidation and it is likely that abstraction of the phenolic hydrogen atom is the first step. The resultant aryloxy radicals are frequently stable in solution, particularly if substituted at the *ortho* positions. This reduces their tendency to dimerize by creating steric interactions. 2,4,6-Tri-*t*-butylphenol (**36**), for example, may be oxidized to the stable radical **37**, a deep-blue species. Another well-known and highly stable radical

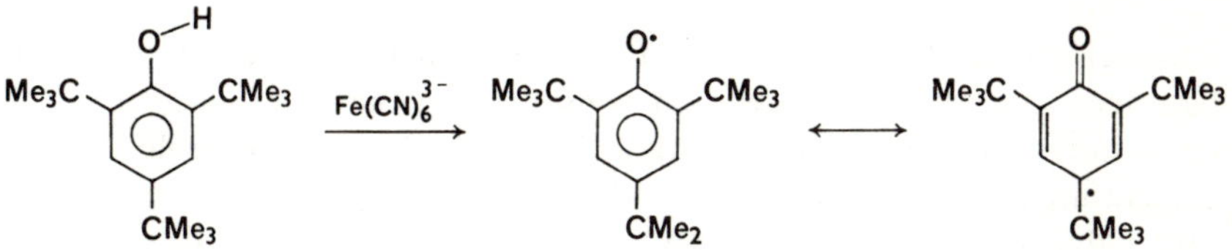

(**36**)   (**37**)

dark blue

λ_{max} = 400 and 615 nm

is galvinoxyl (**38**). The e.s.r. spectra of these radicals are extremely complex but confirm the delocalization of spin over the whole structure.

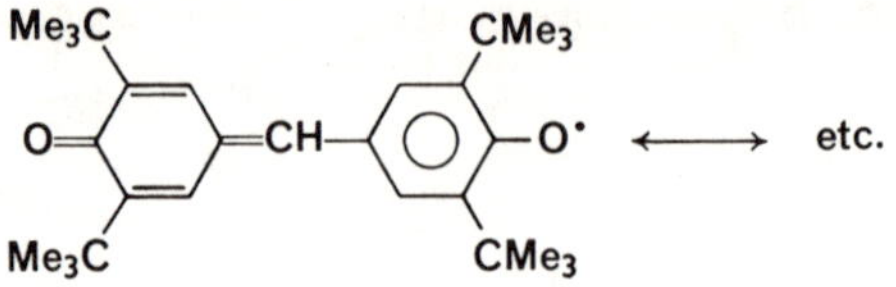

(**38**)

λ_{max} = 650 and 850 nm

Nitrogen-containing Radicals

Triphenylhydrazine (**39**) is oxidized by lead dioxide at low temperature to a blue paramagnetic species, isolable as a green solid, which is stable at −80°. This is evidently the triphenylhydrazyl radical (**40**); on warming, it dimerizes to the red tetrazene (**41**) and will couple with the trityl radical to give **42**. The

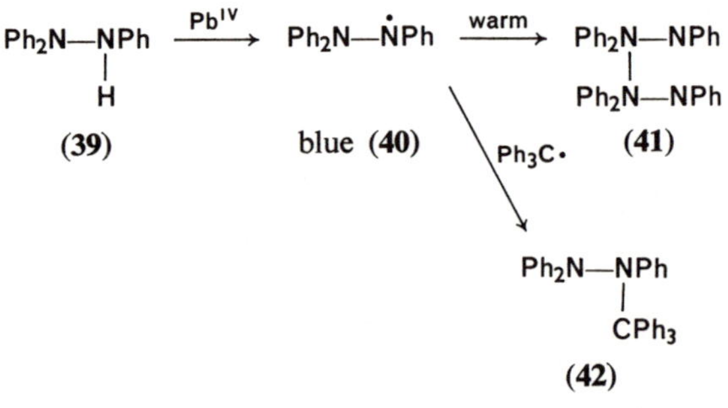

radical character of the hydrazyl, first prepared about 1920, was inferred from these reactions and has since been confirmed by the e.s.r. spectrum. The trinitro analogue, diphenylpicrylhydrazyl, DPPH, (**43**), is quite stable at room temperature as a violet crystalline solid and has no tendency to dimerize. The extra stability may be attributed both to the capability of the nitro groups to accept the unpaired electron and to the increased steric hindrance to dimerization.

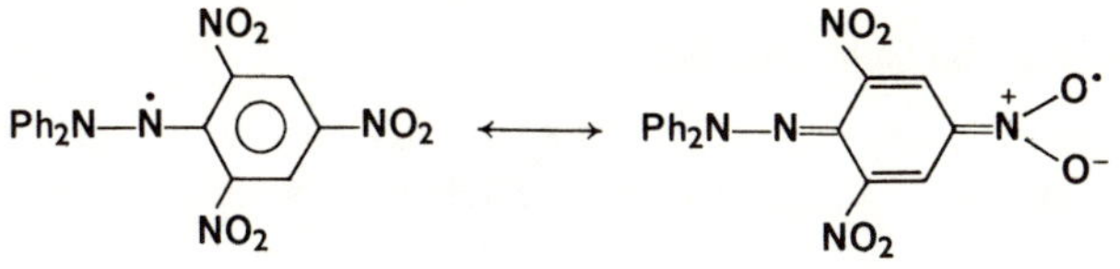

(43)

diphenylpicrylhydrazyl (DPPH)

$\lambda_{max} = 520$ nm

Nitric oxide possesses an odd number of electrons and will scavenge radicals in the gas phase:

$$CH_3^{\bullet} + NO^{\bullet} \longrightarrow CH_3NO$$

Nitroxides, R_2NO, which may be regarded as derivatives also have radical character. Fremy's salt, $K_4(SO_3)_2NO_2$, is an inorganic example known since 1845, while organic derivatives were prepared in the 1920s. The usual route is the oxidation of an hydroxylamine:

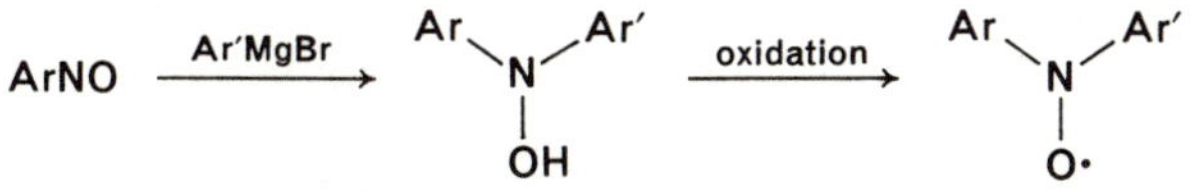

diarylhydroxylamine   nitroxide

The diarylnitroxides are stable in the solid state. Alkyl nitroxides are less stable but may still be isolated if stabilizing groups are present, as in 44 and 45. The

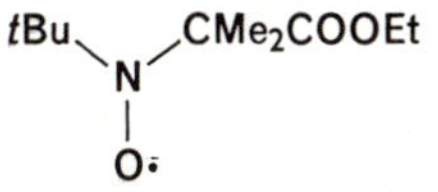

(44) 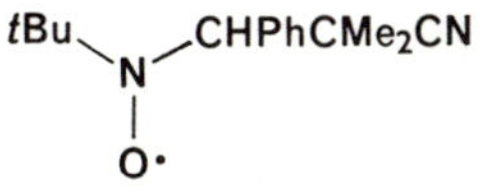   (45)

e.s.r. spectra of many nitroxides have been recorded. The main feature is invariably a large triplet splitting due to interaction of the unpaired electron with nitrogen $(I = 1)$. This is the sole feature in the spectrum of t-butylnitroxide. Arylnitroxides show splitting due to the protons and hence have a more delocalized structure.

Spin Trapping and Spin Labelling

Nitroso compounds may trap transient radicals with the production of long-lived nitroxides, which may then be detected. Nitrosoisobutane, Me_3CNO,

$$R\cdot + R'NO \longrightarrow \underset{\underset{O\cdot}{\overset{|}{N}}}{\overset{R\quad R'}{\diagdown\diagup}}$$

has been used extensively for this purpose. The direct photolysis of aryl and alkyl lead, mercury and tin compounds in the presence of this reagent, leads to long-lived and recognizable adducts of aryl and alkyl radicals with the nitroso compound,

$$HgPh_2 \xrightarrow{h\nu} Hg + 2Ph\cdot$$

$$Me_3C\!-\!N\!=\!O + \cdot Ph \longrightarrow Me_3C\!-\!N\diagup^{O\cdot}_{\diagdown Ph}$$

Similarly, the hydroxymethyl radical has been recognized as the nitroxide (46), when hydrogen peroxide in methanol is photolysed. Nitroxide radicals

$$H_2O_2 \longrightarrow 2\cdot OH$$

$$CH_3OH + \cdot OH \longrightarrow \cdot CH_2OH + H_2O$$

$$Me_3C\!-\!N\!=\!O + \cdot CH_2OH \longrightarrow Me_3CN\diagup^{O\cdot}_{\diagdown CH_2OH}$$

(46)

are also generated by the addition of radicals to a nitrone. Phenyl *t*-butylnitrone (47) has been used as a spin-trapping reagent, and the adduct with the acetyl radical (48), formed by photolysis of biacetyl, recognized.

$$CH_3COCOCH_3 \xrightarrow{h\nu} 2\,CH_3CO\cdot$$

$$PhCH\!=\!\overset{+}{N}\diagup^{O^-}_{\diagdown tBu} + CH_3CO\cdot \longrightarrow \underset{\underset{COCH_3}{\overset{|}{}}}{PhCH\!-\!N}\diagup^{O\cdot}_{\diagdown tBu}$$

(47)   (48)

Reagents have been developed which will coordinate to a given compound, a ketone for instance, and give a product easily converted to a nitroxide radical. Since the sensitivity of e.s.r. spectroscopy is very high, this 'spin-labelling' technique can be a sensitive analytical tool. Aminoisopropanol (49) will give an oxazolidine (50) with a ketone, readily oxidized to a nitroxide radical (51).

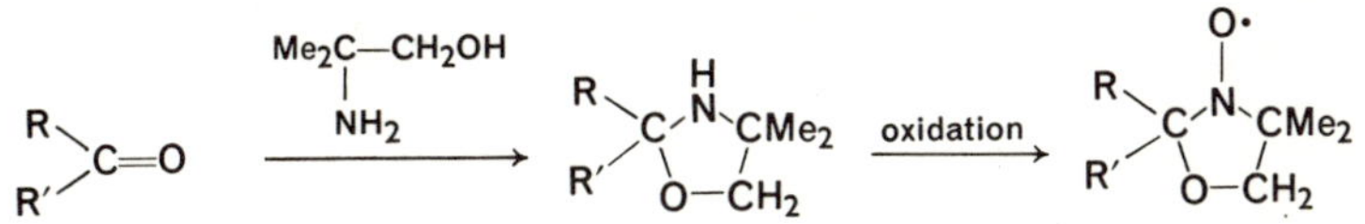

ketone to be detected

(49) (50) (51)

The oxidation of aromatic amines is rather similar to that of phenols and often leads to stable radicals. Triphenylamine may be oxidized by halogens, or electrochemically, to 52 which may be observed but rapidly dimerizes to 53. This in turn may be oxidized to the much more stable radical cation 54, known as a Würster salt. The prototype of these radicals is Würster's blue (55) made

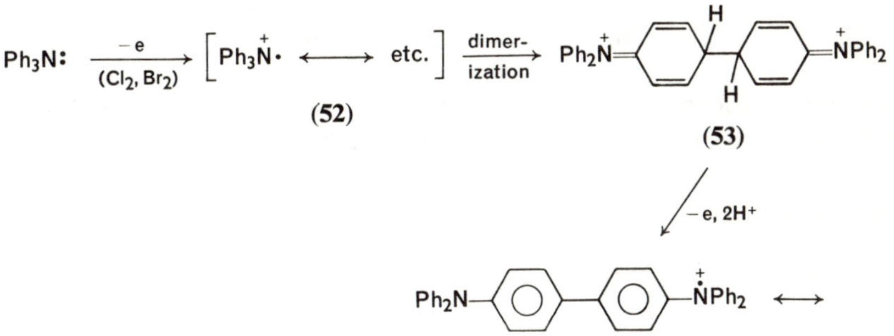

(52)

(53)

$-e, 2H^+$

(54)

by oxidizing tetramethyl-p-phenylenediamine (56); it is the nitrogen analogue of a semiquinone. The ready formation of stable radicals makes amines, like polyphenols, attractive radical chain inhibitors with commercial applications such as antioxidants.

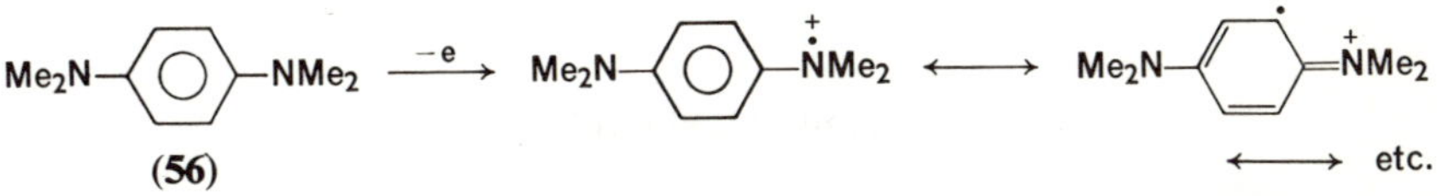

(56)

$\longleftrightarrow$ etc.

(55)

Würster's blue

Nitro compounds are susceptible to one-electron reduction by alkali metals or electrolysis forming anion radicals such as **57**, whose e.s.r. spectrum shows

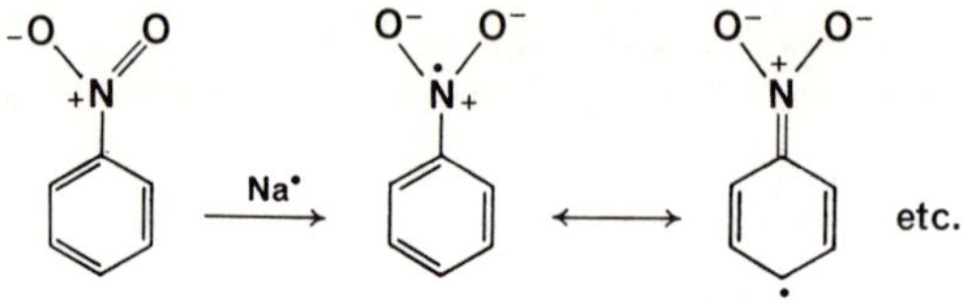

(57)

65 % of the unpaired electron to be on the nitro group. Tetracyanoethylene (**58**) a powerful electron acceptor, readily forms an anion radical (**59**) with even mild oxidizing agents. The e.s.r. spectrum of this radical, stable in solution, shows nine lines due to coupling of the odd electron with four equivalent nitrogens, $(2nI + 1) = (2 \times 4 \times 1) + 1 = 9$. Pyridinium salts undergo one-electron re-

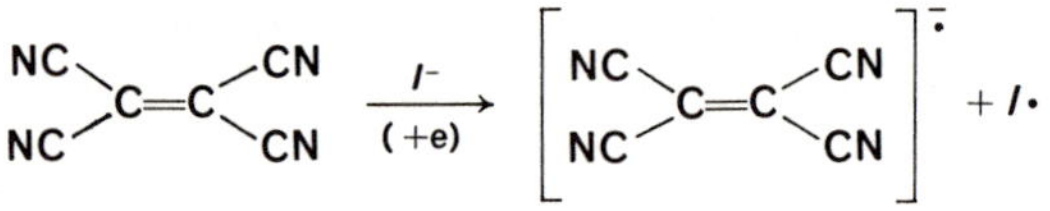

(58)   (59)

duction to coloured, stable radicals, e.g. **60**. The useful herbicides *Paraquat* (**61**) and related dipyridinium salts also exhibit this behaviour. In the plant, the radical **62** is able to interfere with the respiratory enzyme system and produce hydrogen peroxide, which is the actual lethal compound.

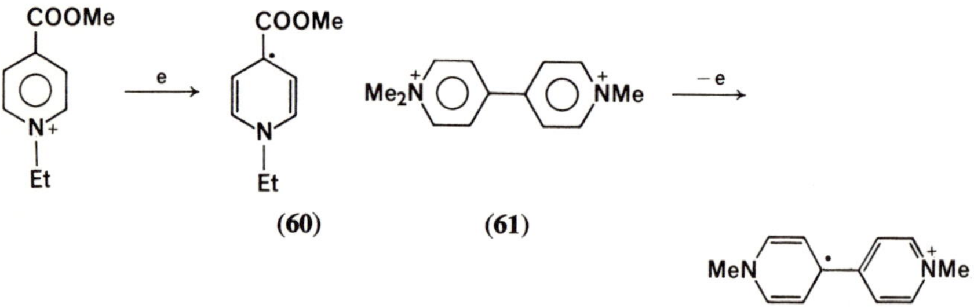

(60)   (61)

(62)

Simple Alkyl Radicals

Despite their high reactivity and normally short lifetime, several techniques have been successful for obtaining the e.s.r. spectra of simple alkyl radicals.

Flow methods permit the rapid mixing of reagents which produce the desired radical in solution, the product passing immediately into the microwave cavity to yield a stationary concentration of radicals within the cavity. This is maximized by adjusting the flow rates of the reagents. Redox methods are particularly useful for generating the radicals, and a considerable variety have been observed in this way by Waters and by Norman. Hydroxyl radicals are

$$Ti^{3+} + CH_3C\!\!\underset{O-O^-}{\overset{O}{<}} \longrightarrow Ti^{4+}(OH) + CH_3C\!\!\underset{O\cdot}{\overset{O}{<}}$$

$$\text{peracetate ion} \qquad\qquad \text{acetoxy radical}$$

$$CH_3C\!\!\underset{O\cdot}{\overset{O}{<}} \longrightarrow CH_3\cdot + CO_2$$

$$Ti^{3+} + (CH_3)_3C-OOH \longrightarrow Ti^{4+}(OH) + (CH_3)_3C-O\cdot$$

$$(CH_3)_3C-O\cdot \longrightarrow CH_3\cdot + (CH_3)_2C=O$$

highly reactive towards abstraction and easily generated in solution by $Ti^{III}-H_2O_2$ or $Fe^{II}-H_2O_2$ (Fenton's reagent); alcohols in such systems lose preferentially their α-*hydrogens* to form hydroxyalkyl radicals (**63, 64**), which may be observed in a flow system. The e.s.r. spectrum of $\cdot CH_2OH$ shows six

$$Ti^{3+} + H_2O_2 \longrightarrow Ti^{4+}(OH) + OH\cdot$$

$$CH_3OH + OH\cdot \longrightarrow \cdot CH_2OH + H_2O$$

$$(63)$$

$$CH_3CH_2OH + OH\cdot \longrightarrow CH_3\dot{C}HOH + H_2O$$

$$(64)$$

lines, a triplet of doublets quite distinct from the *quartet* which would have been expected for the isomeric methoxyl radical, $CH_3O\cdot$. Many simple alkyl radicals have been generated in detectable concentration by Fessenden and Schuler who irradiated suitable compounds with a 2.8 MeV electron beam directly in the spectrometer cavity. Liquid ethane at $-180°$ yielded the ethyl radical and isobutane the *t*-butyl, for example.

$$R-H \xrightarrow[\text{energy}]{e^-} R\cdot + H\cdot$$

Yet another successful method, due to Bennett and Thomas, allowed beams of sodium vapour and alkyl vapour to impinge on the opposite sides of a rapidly rotating drum ('rotating-cryostat' technique) cooled to –180°. The reaction,

$$\text{R—Cl} + \text{Na}\cdot \longrightarrow \text{R}\cdot + \text{NaCl}$$

occurred but the radicals trapped in a solid matrix of the reagents were unable to dimerize and could be observed. The e.s.r. spectra were substantially the same as those produced by other methods.

The bombardment of olefins by hydrogen atoms at low temperature has enabled alkyl radicals to be produced and detected by e.s.r.

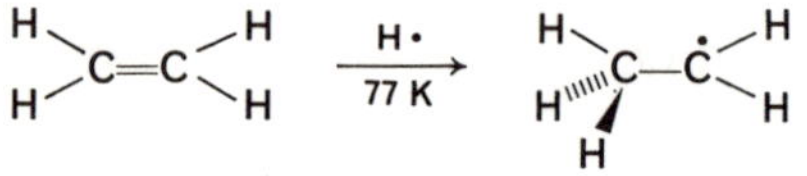

Infrared Spectra of Radicals

Some examples of infrared bands associated with radicals have been obtained from the sodium-chloride matrices containing radicals formed in a rotating cryostat. The $\cdot CF_3$ radical shows absorption at 1240 and 710 cm^{-1}, and $\cdot CCl_3$ absorbs at 897 cm^{-1}. Similar spectra have been recorded from the gas phase.

5.6 THE GEOMETRY OF RADICALS

Chemical evidence as to whether a planar (sp^2-hybridized) or pyramidal (sp^3) structure is preferred by simple alkyl radicals is somewhat ambiguous. Radicals produced from optically active precursers (e.g. **65**) are found to lead to racemic products (**66**) if the radical has a reasonable lifetime. This is explicable by either a planar (**67**) or rapidly inverting pyramidal (**68**) structure.

(**65**) optically active

(**68**) or

(**66**) racemic

(**67**)

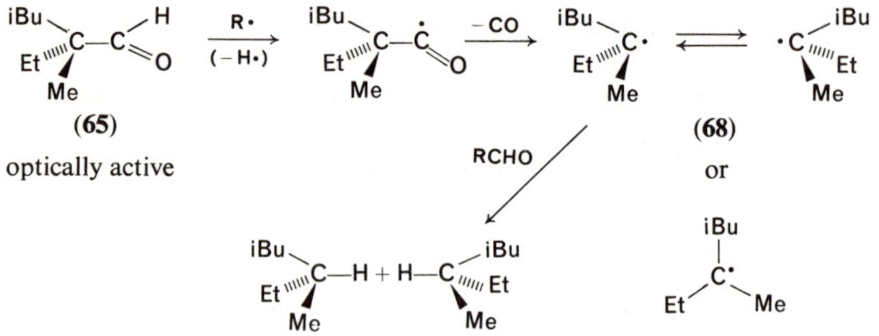

An argument in favour of the pyramidal configuration can be made from the observation that the stabilities of radicals constrained in a pyramidal framework ('bridgehead' positions in bicyclic systems) do not differ appreciably from those which are free to adopt either geometry, in contrast to the carbonium ions (Section 2.11). One of the most sensitive tests of this kind is to examine the relative rates of decomposition and chlorine-atom abstraction by acyl radicals (69). Any instability associated with the radical R· should show up in a low

$$R-C\overset{O}{\underset{H}{\diagdown}} \xrightarrow{tBuO} R-\overset{\cdot}{C}=O \xrightarrow[-CO]{k_1} R\cdot \xrightarrow{CCl_4} RCl$$

(69)

$$+ tBuOH \searrow^{k_2}$$

R—COCl

CCl_4

(Cl· abstraction)

value of k_1/k_2 (i.e. RCl:RCOCl). As is seen by the following values, the formation of bridgehead radicals does not appear to be accompanied by any adverse energy considerations.

R	k_1/k_2	Relative stability of R·
Me$_3$C–	12.3	1.0
	30.5	2.5
	15.2	1.2

The most unequivocal evidence for the sterochemistry of simple radicals comes from e.s.r. studies. The methyl radical for example is found to have the unpaired electron coupled to the nuclear spins of the protons and ^{13}C with splitting factors of 23 and 38.5 G respectively. Now there are two mechanisms whereby electron–nuclear spin coupling can occur. The first depends upon the electron residing in an orbital which has finite density (probability) at the nucleus with which it couples. Pure p-orbitals have zero density at a carbon nucleus but s- and sp-hybrid types will have a finite density depending upon the amount of s-character. It has been calculated that the ^{13}C splitting constant expected if the electron were occupying a pure carbon s-orbital would be 1200 G and an sp^3 hybrid should be a quarter of this, i.e. 300 G. The observed value is only a tenth of the latter value and is consistent with the unpaired electron occupying a p-orbital i.e. a *planar radical*. The small amount of coupling which is observed is a contribution from the second mechanism, the spin polarization of the intervening bonds. The electronic configuration in which bonding electrons in the vicinity of carbon have the same spin as the unpaired electron $\overset{\uparrow}{C}\overset{\uparrow\downarrow}{::}H$ is slightly preferred to that in which they have the opposite, $\overset{\uparrow}{C}\overset{\downarrow\uparrow}{::}H$. Since the bonding electrons occupy hybrid orbitals with s-character, a small amount of resultant spin is induced at the nucleus. The same explanation holds for the small observed coupling to the protons.

Conjugated radicals have an even greater reason to be planar since only in that configuration can effective conjugation with adjacent π-orbitals occur. This is only applicable where not precluded on steric grounds, as is certainly the case for the trityl radical, which although planar at the central carbon, has the aromatic rings twisted in the plane. It appears then, that alkyl radicals are planar by preference but the energy barrier to becoming pyramidal is not very high.

5.7   RADICAL DYNAMICS

Some radicals undergo reversible isomerism to species differing in bonding or in geometry, but the consequences of this equilibrium on the appearance of the e.s.r. spectrum will depend on the rates of the forward and reverse reactions. If both are slow compared to the time resolution of the method (ca 10^{-3} s), two superimposed spectra may be seen. If the equilibration is rapid, an average spectrum will be observed. By determining the temperature at which an average spectrum just separates into two discrete ones and by analysis of the line shapes of the spectra, the energy of the transition can be calculated.

The semiquinone **70** shows line broadening due to restricted rotation about

$$E_A = 42 \text{ kJ mol}^{-1}$$

(70)

the C—O bonds, which on account of overlap by the unshared electrons have some π-character. Restricted rotation in the tetrahydroxynaphthalene cation

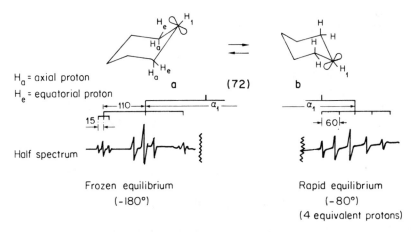

$E_A = 17 \text{ kJ mol}^{-1}$

(71)

radical (71) is evidently due to interchange of hydrogen-bonded partners. The cyclohexyl radical **72**, at $-196°$, shows the unpaired electron coupled appreciably ($\alpha = 110$ G) to only one pair of adjacent hydrogens and very slightly coupled ($\alpha = 15$ G) to another pair. At $-80°$ it is coupled to four equivalent hydrogens with an intermediate splitting factor (60 G). At the low temperature the interchange of chair forms **72 a** and **b** is slow and the odd electron couples to the axial and equatorial pairs of α-protons to very different extents. At $-80°$, the conformational interchange is more rapid and all four protons are equivalent (see Figure 5). Line broadening of the resonance of the hydroxymethyl radical, $\cdot CH_2OH$, is dependent on pH showing that hydrogen exchange is occurring at the hydroxyl group by a rapid acid-catalysed process whose rate

H_a = axial proton
H_e = equatorial proton

Half spectrum

(72)

Frozen equilibrium
$(-180°)$

Rapid equilibrium
$(-80°)$
(4 equivalent protons)

Figure 5. E.s.r. spectrum of the cyclohexyl radical.

has been estimated at $10^8 \, 1 \, mol^{-1} \, s^{-1}$. The e.s.r. spectrum of sodium benzo-phenone ketyl (**33**) shows additional splitting (by sodium, $I = 3/2$) when in a

$$CH_3OH \xrightarrow{\text{OH} \cdot} \cdot CH_2OH \underset{\longleftarrow}{\overset{H_3O^+}{\rightleftharpoons}} \cdot CH_2\overset{+}{O}H_4$$

non-polar solvent than when it is in a polar solvent. The two ions are separate in the polar solvent but in close proximity (intimate ion pair) in the non-polar medium. This constitutes direct evidence for the existence of ion pairs (Section 1.20).

Geometrical isomerism in allyl radicals may be studied by e.s.r. The allyl radical itself (**73**), generated by hydrogen abstraction from liquid propene directly in the spectrometer at $-130°$, possesses two distinct types of terminal hydrogen, with splitting factors of 13.90 and 14.81 G. The central hydrogen is also coupled to a smaller extent ($\alpha = 4.06$ G) giving an 18-line spectrum. This must be explained in terms of a structure (**74**) in which rotation about the C—C bonds (which are partial double bonds) is slow. At higher temperatures, rapid rotation evidently occurs since all four terminal hydrogens become equivalent, as shown by the 10-line spectrum which is then observed. The results confirm

(73)   (74)

the non-linear geometry of the allyl radical and show that the unpaired electron resides mainly on the termini, as predicted from simple theory. Similarly, two geometrical isomers of the 1-methylallyl radical (**75 a** and **b**) can be separately observed at $-130°$ but coalesce at $-40°$.

(75a)   (75b)

5.8   **STABILITIES OF RADICALS**

Stability in the sense of unreactivity or of a negative heat of formation (thermo-dynamic stability) are concepts which should be distinguished but frequently are parallel. Some generalizations on radical stabilities may be made and present a fairly coherent picture from a variety of evidence.

Tertiary radicals are more stable than secondary which in turn are more stable than primary and methyl. This follows from the appropriate bond strengths of a series of alkanes which are typically as follows:

	CH_3—H	RCH_2—H	R_2CH—H	R_3C—H
$\Delta H(R = Me)$	435	406	395	380 kJ mol^{-1}
CH				

The expulsion of the more stable radical from a tertiary alkoxide radical shows that ethyl is more stable than methyl:

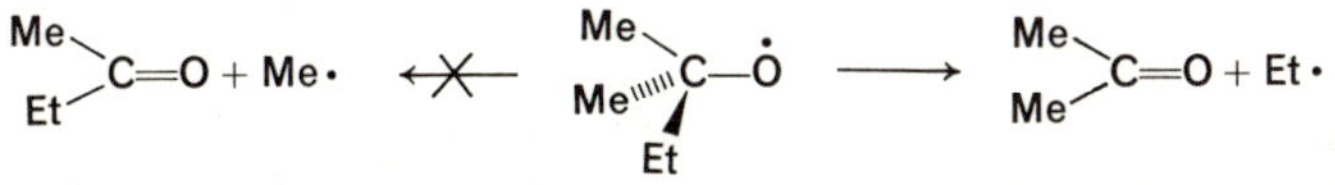

For the same reason, the rates of hydrogen abstraction by alkyl radicals from toluene decrease in the order primary > secondary > tertiary and at the same time selectivity increases with the less-reactive radicals. Conjugation with

$$PhCH_3 + R \cdot \longrightarrow PhCH_2 \cdot + RH$$

π-systems enhances radical stability. The unpaired electron in (76) occupies a molecular orbital embracing the whole of the conjugated structure. The more

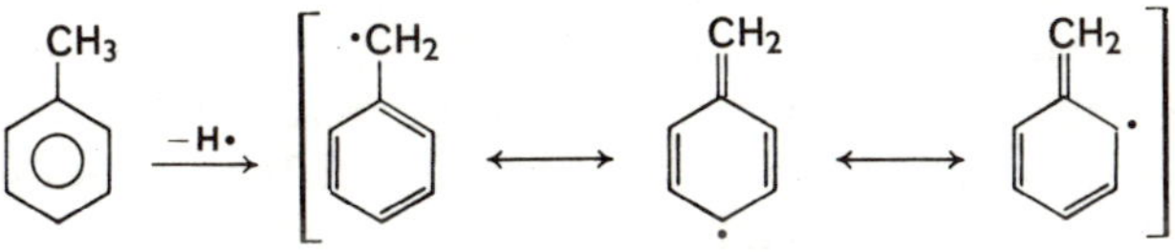

$$\Delta H_{CH} = 355 \text{ kJ mol}^{-1}$$

(76)

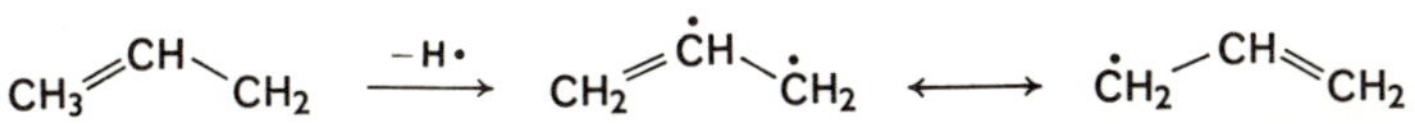

$$\Delta H_{CH} = 355 \text{ kJ mol}^{-1}$$

aryl groups conjugated, the more stable the radical despite the fact that steric interactions prevent the benzhydryl or trityl system from being entirely planar,

$$Ph_3C—H \longrightarrow Ph_3C \cdot + H \cdot \quad \Delta H_{CH} = 314 \text{ k J mol}^{-1}$$

The great stability of triarylmethyl radicals stems at least equally from steric hindrance to dimerization or further reaction.

Substituent groups with heteroatoms are in some cases able to stabilize radicals irrespective of their polar nature. The rates of phenyl-radical attack on substituted benzenes (77) (which depend upon the stability of the intermediate cyclohexadienyl radical 78) fall in the following order:

X	OMe	Me	H	Ph	Br	CN	NO$_2$
k_{rel}	1.2	1.9	1.00	4.0	1.75	3.7	4.0

It should be noted that the range of rates nonetheless is very small, much less than would be experienced in an ionic reaction (a factor of about 10^8 would not be uncommon). In some circumstances, radical stabilities do reflect polar influences of substituents, e.g. the decompositions of dibenzoyl peroxides (8). A moderately good linear free-energy relationship is found but the Hammett ρ-value is very low (−0.38) showing that polar stabilization of the radicals is only of minor importance. Halogenation is especially sensitive to polar factors, e.g. the chlorination of substituted toluenes, $\rho = -0.87$ and bromination, $\rho = -1.05$.

Adjacent oxygen stabilizes an alkyl radical ($\cdot CH_2OH > \cdot CH_3$) as do conjugating nitro and cyano groups. Aryl (79), and vinyl (80) radicals (like the corresponding cations and anions), on the other hand, are extremely reactive since the unpaired electron cannot be extensively delocalized. These radicals are less easily formed even than methyl as reflected in the following heats of homolysis:

Reaction		ΔH kJ mol^{-1}
$CH_3{-}CH_3 \longrightarrow 2 \cdot CH_3$		368
$Ph{-}Ph \longrightarrow 2\,Ph\cdot$	(79)	420
$HC{\equiv}C{-}C{\equiv}CH \longrightarrow 2\cdot C{\equiv}CH$	(81)	627
$CH_2{=}CH{-}CH{=}CH_2 \longrightarrow 2\cdot CH{=}CH_2$	(80)	420

Ethynyl radicals (**81**) are even more reactive. The vinyl radicals **83 a** and **b**, produced by thermal decomposition of *cis* and *trans* methoxycrotyl peresters **82 a** and **b**, do not undergo interconversion during their lifetime. Thus, the *cis*

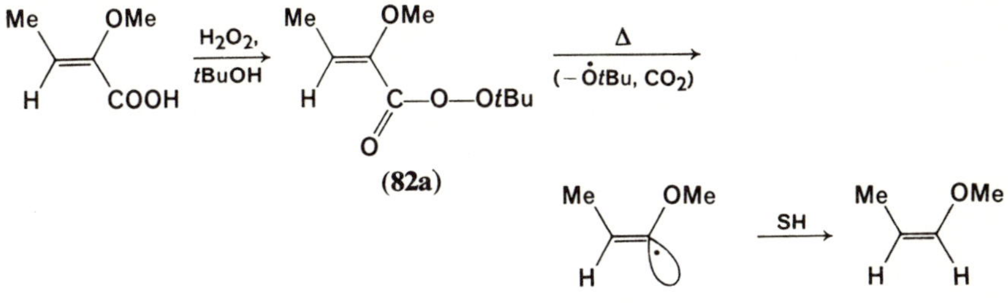

(82a)

(83a)

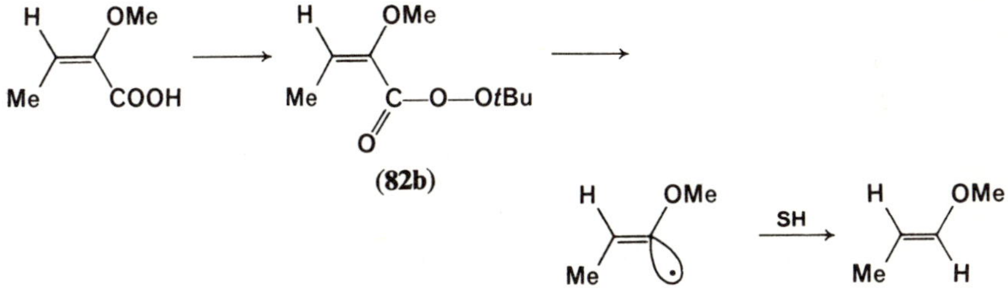

(82b)

(83b)

ester gives *cis* olefin, and *trans* ester gives *trans* olefin only. This indicates that the double bond remains intact in a vinyl radical.

5.9   REACTIONS WITH RADICAL INTERMEDIATES

Radical Halogenation

Saturated hydrocarbons and aromatic side-chains undergo substitution by chlorine in a typical radical chain process. The reaction is catalysed by light, peroxides and other radical sources and is inhibited by oxygen, thiols, hydroquinone etc. The following mechanism accounts for these facts:

	Reaction		ΔH kJ mol^{-1}
Initiation	(a) Cl—Cl	$\xrightarrow{h\nu}$ 2 Cl•	+243
Chain propagation	(b) R—H + •Cl $\longrightarrow$ R• + H—Cl		−5
	(c) R• + Cl—Cl $\longrightarrow$ R—Cl + •Cl		−96
Overall reaction	C—H + Cl$_2$ $\longrightarrow$ C—Cl + H—Cl		

Chlorine atoms, produced by photodissociation or interaction of chlorine with a radical (a) abstract hydrogen from the hydrocarbon (b) and the resultant radical in turn abstracts chlorine (c) to generate another chlorine atom and continue the chain. Only the chain propagation steps are exothermic, by ca 100 kJ mol^{-1}, but this will be the overall exothermicity of the reaction only if the chains are very long—i.e. one initiative step produces a large number of product molecules. Chlorine atoms are highly reactive and rather unselective reagents. One can usually expect every possible position in an alkane to be attacked with more or less statistical probability. The mixtures of products which usually result make this a reaction of limited application as far as the production of pure compounds is concerned. The distribution of products from the radical chlorination of some typical substrates is shown in Table 3.

Table 3. Product ratios from free-radical halogenations (gas phase). Relative rates of hydrogen abstraction by halogen atoms

Chlorination	Temperature	Average k_{rel}		
		$-CH_2$	$>CH_2$	$>CH$
$CH_2{-}CH_2{-}CH_2{-}CH_2$				
3.6 1.0				
	100°	1	4.3	7.0
$Ph{-}CH_2{-}CH_2{-}CH_2{-}CH_2$	200°	1	3.7	5.4
20 10 13 1				
$F{-}CH_2{-}CH_2{-}CH_2{-}CH_3$				
0.9 1.7 3.7 1.0				
Bromination				
$F{-}CH_2{-}CH_2{-}CH_2{-}CH_3$				
10 9 88 1				

In these examples there is a preferred order of attack at tertiary > secondary > primary and benzylic > non-benzylic carbons, which is consistent with the formation of the most stable radical. Chlorination chains are typically very long ($\sim 10^6$). At higher temperatures, the selectivity of the reagent is reduced and product ratios tend more towards the statistical as expected for concurrent processes with different activation energies (which in this reaction are quite low, of the order of 20 kJ mol^{-1}). The same substrates yield similar products with other chlorinating agents such as t-butyl hypochlorite (**84**), which is presumed to act in a similar way:

$$tBuO—Cl \xrightarrow{hv} tBuO\cdot + \cdot Cl$$

$$(84)$$

$$RH + \cdot Cl \longrightarrow R\cdot + HCl$$

$$R\cdot + Cl—OtBu \longrightarrow RCl + tBuO\cdot \quad etc.$$

In all these reactions, moderate to large primary isotope effects are observed confirming C—H bond breaking in the slow step: e.g. toluene-α-d with Cl·, $k_H/k_D = 2.2$; with acetate radicals, $CH_3COO\cdot$, $k_H/k_D = 9.9$. Relative rates of chlorination are measured by allowing a mixture of two substrates to compete for a small amount of chlorinating agent. Some typical results with the lower paraffins are as follows:

$$
\begin{array}{cccccc}
& \underset{H}{\overset{CH_3}{|}} & \underset{H}{\overset{CH_2—CH_3}{|}} & \underset{H}{\overset{CH_2—CH—CH_3}{|}} \ \underset{H}{\overset{}{|}} & \underset{H}{\overset{CH_2—C(CH_3)_2}{|}} \ \underset{H}{\overset{}{|}} \\
k_{rel} & 1.00 & 111 & 108 \quad 462 & 85 \quad 590
\end{array}
$$

Average reactivities for primary : secondary : tertiary

(relative to $CH_4 = 1.0$)   100   400   600

There is a slight acceleration by electron-donating substituents but Hammett σ-values are low as expected for a radical process (e.g. for the chlorination of m- and p-substituted toluenes in the side-chain, $\rho = -0.87$). Also, the hydrogen-isotope effect is low; k_H/k_D for chlorination of toluenes ≈ 2. Evidently in the transition state, the hydrogen is unequally bonded to C and Cl. Bromination

$$\overset{\diagdown}{\underset{\diagup}{C}} \cdots H \cdots\cdots\cdots\cdots Cl$$

by an analogous radical chain mechanism also occurs but there are important differences between this and chlorination. Bromination chains are much shorter since hydrogen abstraction by bromine atoms (b) is a reversible process on account of the relative weakness of the H—Br bond. Bromine atoms

$$\Delta H \text{ kJ mol}^{-1}$$

$$\text{(a) } Br—Br \xrightarrow{hv} 2Br \qquad\qquad 190$$

Propagation $\begin{cases} \text{(b) } R—H + \cdot Br \rightleftharpoons R\cdot + H—Br \qquad\qquad 46 \\ \text{(c) } R\cdot + Br—Br \longrightarrow R—Br + Br\cdot \qquad\qquad -82 \end{cases}$

are consequently much less reactive than chlorine atoms and are more selective in their attack, typical relative rates of attack at primary, secondary and tertiary positions, relative to CH_4 being $10^3:2.2 \times 10^5:2 \times 10^7$. Where the desired products are those favoured, bromination can be a useful preparative method. For example benzylic positions on an aromatic side-chain can be readily brominated to the exclusion of more remote positions.

Chlorosulphonation

If free-radical chlorination is carried out in the presence of sulphur dioxide, a sulphonyl chloride is produced by the propagation steps,

$$RH + \cdot Cl \longrightarrow R \cdot + HCl$$
$$R \cdot + SO_2 \longrightarrow R\text{—}SO_2 \cdot$$
$$RSO_2 \cdot + Cl\text{—}Cl \longrightarrow RSO_2Cl + \cdot Cl$$

The process was once used in the production of long-chain alkyl sulphonic acids for detergents.

$$RSO_2Cl + H_2O \longrightarrow RSO_2OH + HCl$$

Autoxidation

The direct reaction of oxygen with organic compounds is termed 'autoxidation', the primary products being peroxy compounds (85), which are often isolable.

$$RH + O_2 \longrightarrow R\overset{O\text{—}O}{\diagup}\diagdown H$$
(85)

This reaction is of great practical importance; useful autoxidations include the 'drying' of oils in paints and varnishes, while on the debit side the deterioration of rubber and spoilage of food in air occurs by this type of reaction. The autoxidation of benzaldehyde has been studied in detail and exemplifies the characteristics of many autoxidations. A sample of benzaldehyde when exposed to air is slowly converted to crystals of benzoic acid. The reaction is inhibited by common radical traps—hydroquinone or α-naphthol—and is accelerated by light with a quantum yield of 10^4 (i.e. each quantum of light absorbed produces 10^4 product molecules). These are characteristics of a radical chain reaction, originally formulated by Bäckstrom, thus:

$$Ph-C\overset{O}{\underset{H}{\lessgtr}} + O_2 \longrightarrow Ph-\overset{\cdot}{C}=O + \cdot O-OH$$

or $h\nu$ **(86)**

$$Ph-\overset{\cdot}{C}=O + O_2 \longrightarrow Ph-C\overset{O}{\underset{O-O\cdot}{\lessgtr}}$$

$$Ph-C\overset{O}{\underset{O-O\cdot}{\lessgtr}} + \overset{O}{\underset{H}{\gtrless}}C-Ph \longrightarrow Ph-C\overset{O}{\underset{O-OH}{\lessgtr}} + Ph-\overset{\cdot}{C}=O$$

(87)

perbenzoic acid

The chain is propagated by the benzoyl radical **(86)**, and the initial product is perbenzoic acid **(87)**. The oxidizing properties of the latter were detected in solutions of oxygenated benzaldehyde as early as 1900. The subsequent decomposition to benzoic acid occurs by an acid-catalysed polar process, probably thus:

$$Ph-C\overset{O}{\underset{O-OH}{\lessgtr}} \underset{}{\overset{(H^+)}{\rightleftharpoons}} Ph-C\overset{O}{\underset{\underset{H}{O-OH}}{\lessgtr}} \overset{H_2O}{\longrightarrow}$$

$$Ph-\overset{O^-}{\underset{\underset{H}{\overset{|}{O}}}{\overset{|}{C}}}\overset{}{\underset{\overset{+}{H}}{O}}-OH \longrightarrow Ph-C\overset{O}{\underset{OH}{\lessgtr}} + HOOH$$

$$+ (H^+)$$

Autoxidations are fairly selective and are especially prone to occur when a tertiary, allylic or benzylic hydrogen is available for abstraction. Decalin **(88)** and cumene **(89)** readily form hydroperoxides in almost quantitative yield. The

(88)

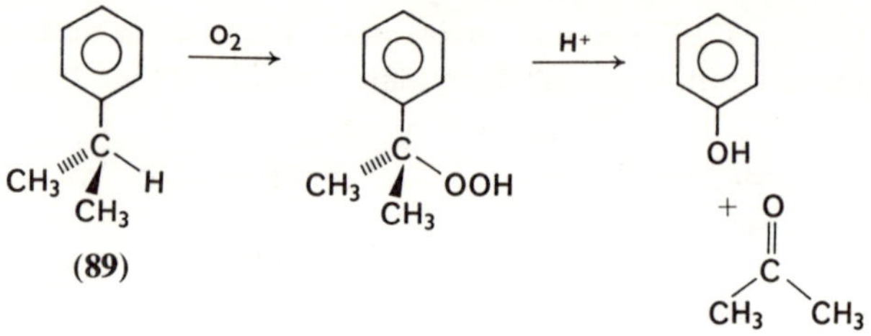

(89)

latter is prepared on a large scale and decomposed to acetone and phenol, both desirable products. Olefins are notoriously easily oxidized on exposure to air. Commercial samples of cyclohexene, for instance, frequently react vigorously with sodium on account of the cyclohexenol (90) present. Ethers autoxidize to

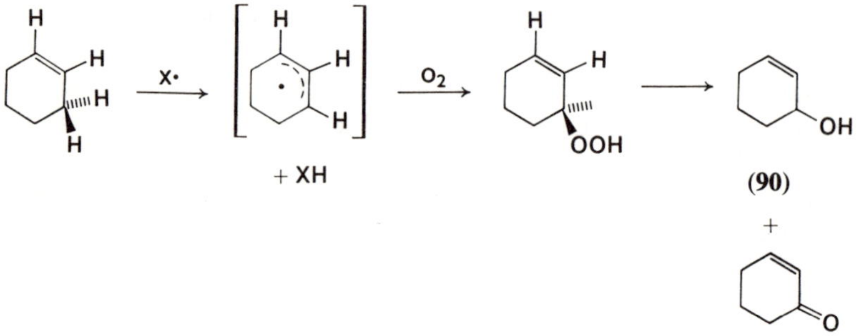

(90)

hydroperoxides, especially if possessing a tertiary hydrogen, e.g. diisopropyl ether (91 and 92). The peroxides accumulate on evaporation of the ether and can result in violent explosions during distillation. Heavy metals catalyse

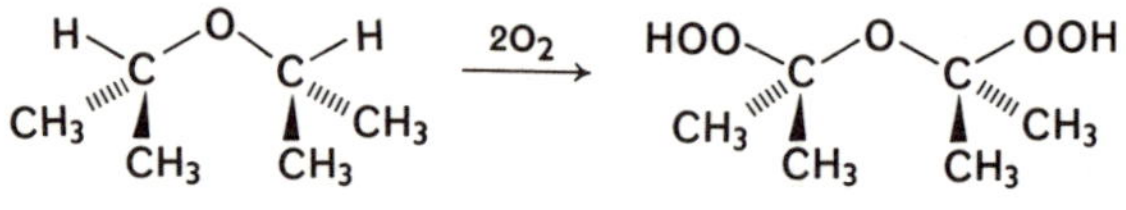

(91)   (92)

autoxidations; it is believed that they cause the decomposition of the hydroperoxides to further radicals which initiate fresh oxidation chains.

$$Fe^{2+} + ROOH \longrightarrow Fe^{3+} + RO\cdot + OH^-$$

$$Fe^{3+} + ROOH \longrightarrow Fe^{2+} + R\dot{O}_2 + H^+$$

Net reaction: $2\ ROOH \longrightarrow RO\cdot + RO_2\cdot + H_2O$

Radical Addition to Double Bonds

Hydrogen Bromide

The polar addition of hydrogen chloride to olefins has already been discussed (Section 2.13) and the orientation of addition rationalized in terms of carbonium-ion stabilities. The addition of hydrogen bromide was long ago found to show anomalies. Allyl bromide, for example, was found to form rather variable mixtures of 1,2-dibromopropane (the expected product) and 1,3-dibromopropane. It was finally demonstrated by Kharasch, a pioneer of radical chemistry, that highly purified allyl bromide (**93**) and HBr, in the absence of air and light, gives only 1,2-dibromopropane while in the presence of radical sources including light, increasingly large amounts of the 1,3-isomer may be obtained. Indeed, when dibenzoyl peroxide is added to the mixture,

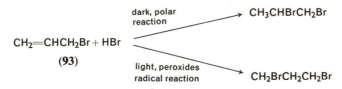

this is almost the exclusive product. The free-radical addition of HBr evidently proceeds by the following chain reaction:

$$
\begin{array}{l}
\text{H—Br} \xrightarrow{\; h\nu \text{ or } \text{R}\cdot \;} \text{Br}\cdot \\[4pt]
\text{CH}_2\!\!=\!\!\text{CHR} + \text{Br}\cdot \longrightarrow \text{CH}\overset{\displaystyle\cdot}{-}\text{CHR} \\[2pt]
\hspace{6.5cm} \overset{|}{\text{Br}} \\[6pt]
\text{CH}\overset{\displaystyle\cdot}{-}\text{CHR} + \text{H—Br} \longrightarrow \text{CH—CHR} + \text{Br}\cdot \\[2pt]
\overset{|}{\text{Br}} \hspace{4.2cm} \overset{|}{\text{Br}}\ \ \overset{|}{\text{H}}
\end{array}
$$

Bromine atoms, the chain carriers, produced by photodissociation or hydrogen abstraction from HBr, attack the olefin at the terminal position to produce the more stable radical. This in turn abstracts hydrogen from HBr generating a bromine atom to continue the chain. The difference in orientation between radical and polar reactions reflects the *order* in which the components add; Br before H in the radical process and the reverse in the polar reaction. In each case, the more stable trivalent-carbon derivative, radical or carbonium ion, forms—in the order: tertiary > secondary > primary. The ability to reverse the orientation of addition of HBr is a useful preparative feature. One may ask why HBr, alone among the hydrogen halides, exhibits this behaviour. The answer may be sought in the enthalpies of each step in the chain-propagation

Table 4. Energetics of radical addition of hydrogen halides to an olefin

Reaction		ΔH (kJ mol^{-1})			
	hal =	F	Cl	Br	I
(a) hal· + olefin ⟶ product		−167	−67	−12	+21
(b) radical + H—hal ⟶ product + hal·		+155	+21	−46	−113

process which may be estimated by summing the changes in bond energies (Table 4). HF and HCl do not readily undergo homolysis (in step b, Table 4) since their covalent bonds, especially of the former, are very strong. HI, at the other extreme, dissociates very readily but the iodine atom is not very reactive and addition to the double bond is endothermic. HBr alone of the four halogen acids is exothermic for both the steps. Considerable thermal activation is required to force HCl to add by a radical process. This may occur with ethylene at 150° in the presence of *t*-butyl peroxide, although the chains are short.

Carbon Tetrachloride

Radical chain addition of carbon tetrachloride to an olefin, adding the elements of CCl$_3$ and Cl to the double bond, occurs in the presence of radical sources. The C—Cl bond in carbon tetrachloride is rather weak and the trichloromethyl radical a moderately stabilized one. The reaction proceeds by the initial addition of this radical to a terminal position:

$$Cl_3C—Cl \xrightarrow{h\nu,\ R\cdot} Cl_3C\cdot + \cdot Cl\ \text{or}\ R—Cl$$

$$C_6H_{13}CH=CH_2 + \cdot CCl_3 \cdot \longrightarrow C_6H_{13}\overset{\cdot}{C}H—CH_2$$
$$|$$
$$CCl_3$$

oct-1-ene

$$C_6H_{13}\overset{\cdot}{C}H—CH_2 + Cl—CCl_3 \longrightarrow C_6H_{13}CH—CH_2CCl_3 + \cdot CCl_3$$
$$|\qquad\qquad\qquad\qquad\qquad\qquad\qquad\qquad\quad |$$
$$CCl_3\qquad\qquad\qquad\qquad\qquad\qquad\qquad\quad Cl$$

90%

Good yields of 1,1,1,3-tetrachloroalkanes may be obtained in this way. Carbon tetrabromide and several mixed halides have been found to behave similarly:

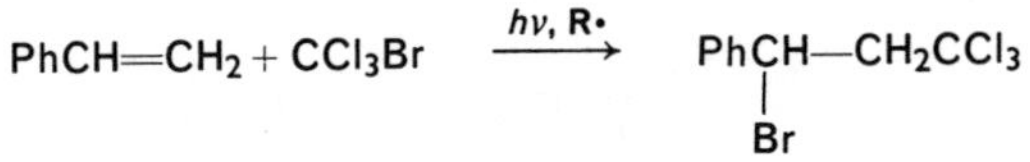

$$PhCH\!=\!CH_2 + CCl_3Br \xrightarrow{h\nu,\ R\cdot} PhCH\!-\!CH_2CCl_3$$
$$\underset{Br}{|}$$

(C—Br bond is weaker than C—Cl.)

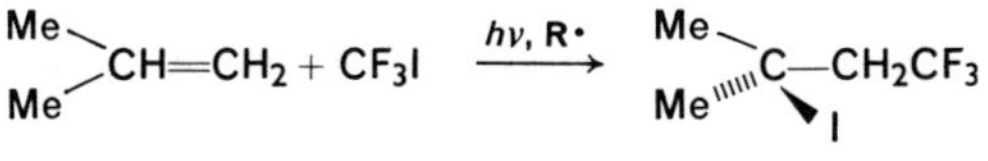

$$\underset{Me}{\overset{Me}{\diagdown}}CH\!=\!CH_2 + CF_3I \xrightarrow{h\nu,\ R\cdot} \underset{Me^{\backslash\backslash\backslash}}{\overset{Me}{\diagdown}}C\!-\!CH_2CF_3$$

By-products in the reaction include low polymers

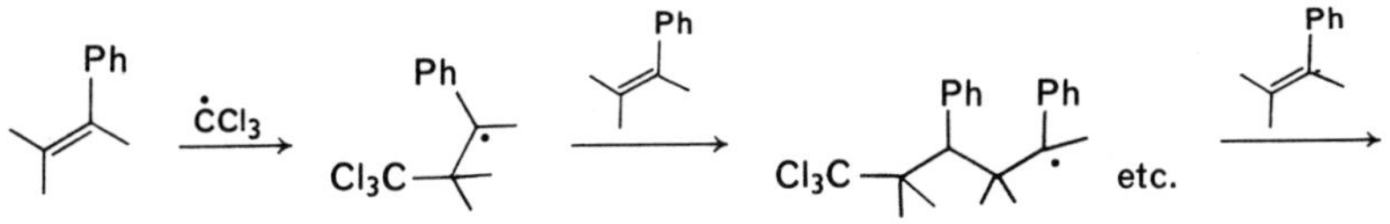

and allylic chlorides formed from abstraction reactions which compete with addition.

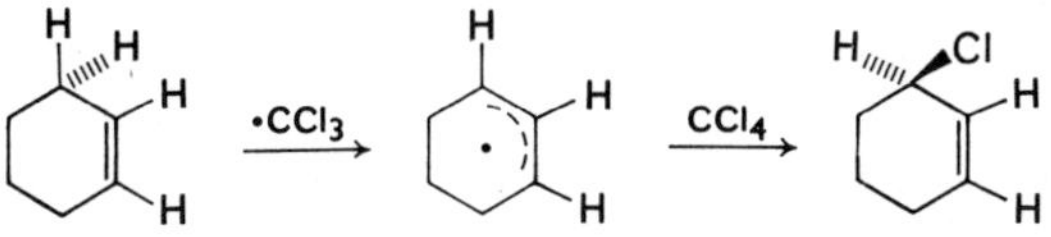

Thiols

Alkyl thiols, RSH, and related compounds such as thiol acids, RCOSH, sulphenyl chlorides, RSCl, and sulphonyl chlorides, RSO₂Cl may all be added to olefins in the presence of peroxides by radical chain reactions similar to those discussed above. Both steps in the chain are exothermic and long radical

$$2\,PhSH + Bz_2O_2 \xrightarrow{\Delta} 2\,PhS\cdot + 2\,BzOH$$

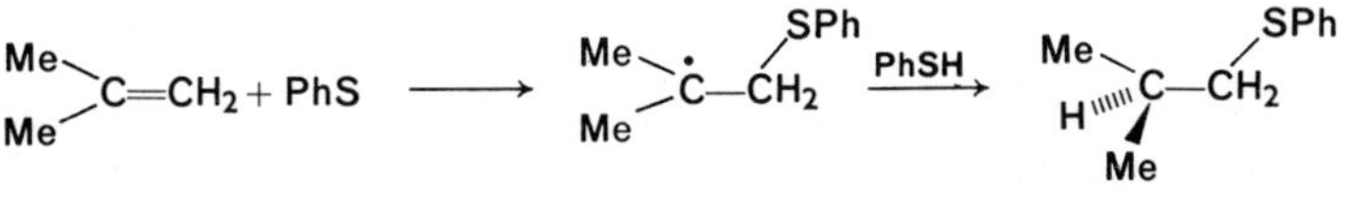

chains result. As with HBr, the addition of thiols is 'anti-Markownikov' and competes with an ionic addition which occurs with the opposite orientation, but is usually a good deal slower.

Alcohols

In contrast to thiols, the weakest bonds in alcohols are the α-CH bonds. In the presence of peroxide, alcohols will add to olefins in the following manner:

$$Me_2CHOH \xrightarrow[tBu_2O_2, 140°]{Bz_2O_2 \text{ or}} Me_2\overset{\bullet}{C}—OH$$

$$C_6H_{13}CH{=}CH_2 + Me_2\overset{\bullet}{C}—OH \longrightarrow C_6H_{13}\overset{\bullet}{C}H—CH_2\overset{OH}{\overset{|}{C}Me_2}$$

$$C_6H_{13}\overset{\bullet}{C}H—CH_2CMe_2OH + Me_2CHOH \longrightarrow C_6H_{13}CH_2—CH_2CMe_2OH + Me_2\overset{\bullet}{C}OH$$

$$50\%$$

The chains tend to be rather short and the yield low in reactions of this type. Carboxylic acids, ketones, amines, esters, aldehydes and even paraffins, may all be added to olefins by radical processes and with varying yields. Some typical examples are:

$$H—X + C_6H_{13}CH{=}CH_2 \xrightarrow{\text{peroxide}} C_6H_{13}CH_2—CH_2X$$
$$\text{1-octene}$$

HX	X	Yield (%)
Acetic acid	—CH_2COOH	70
Acetone	—CH_2COCH_3	30
Ethyl chloroacetate	—$CHClCOOEt$	70
Cyclohexane	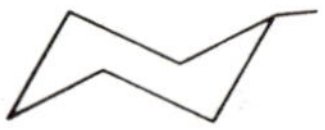	40
n-Butylamine	—CHC_3H_7 $|$ NH_2	36
Acetaldehyde	—$COCH_3$	36

Rates of free-radical additions are favoured by radical-stabilizing features in the olefin component as shown by the following reactivity order, although steric factors also play a part in determining rates:

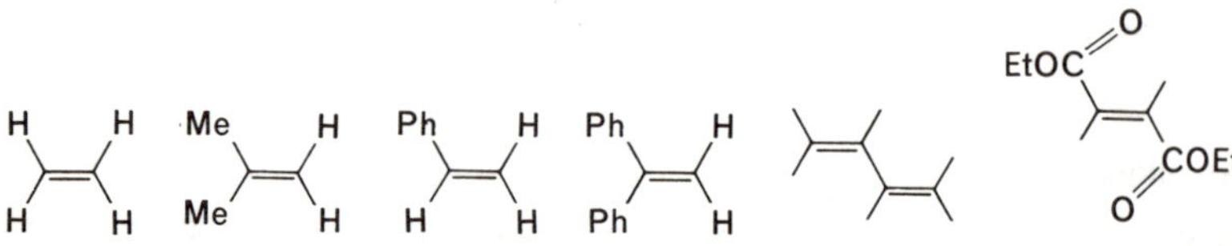

$$k_{rel} = 21 \qquad 36 \qquad 795 \qquad 1500 \qquad 2015 \qquad 2000$$

Reactive radicals trapped in matrices formed in the rotating cryostat may be directly observed to add to olefins. For example, on allowing 1,1-dideutero-ethylene to come into contact with trapped phenyl radicals, a mixture of **94** and **95** was observed by e.s.r. spectrometry. The separate spectra showed that the addition had occurred preferentially at the deuterated end by a factor of 3.7—a large secondary isotope effect.

$$Ph\cdot + CH_2{=}CD_2 \longrightarrow PhCH_2{-}\overset{\cdot}{C}D_2 + PhCD_2{-}\overset{\cdot}{C}H_2$$

$$\qquad\qquad\qquad\qquad\qquad\quad \textbf{(94)} \qquad\qquad \textbf{(95)}$$

$$\qquad\qquad\qquad\qquad\qquad\quad 3.7 \quad : \quad 1$$

Radical Polymerization

The formation of linear polymers from certain ethylenic compounds (monomers) by a radical chain mechanism, is one of the most typical of such reactions and, on account of the practical importance of the products as plastics materials, the most intensively studied. The steps involved in the conversion of monomer, the reaction being initiated by the decompositions of dibenzoyl peroxide, are given below.

(a) Formation of
    initiating radical

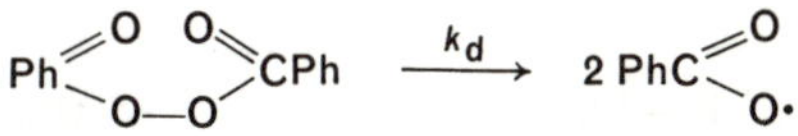

(b) Initiation

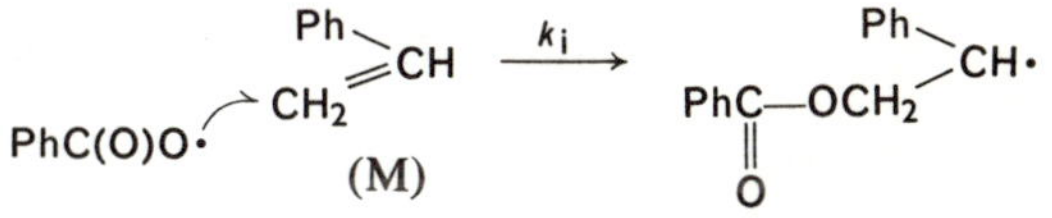

(c) Chain propagation

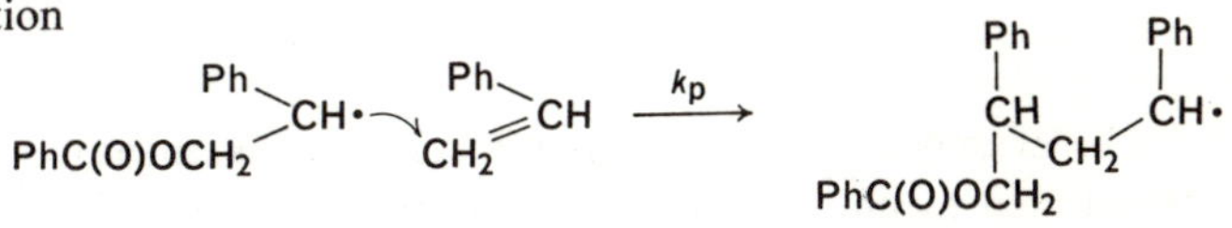

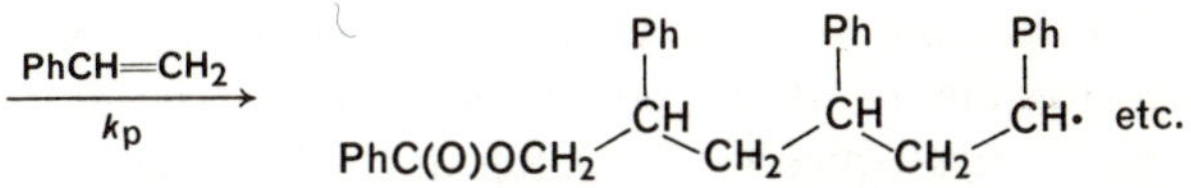

(d) Chain termination: Dimerization

Disproportionation

Polymerization, step (c), occurs with great rapidity and builds up very long chains (typically some 10^6 units long). Termination of this process is brought about by the interaction of the radical ends of growing polymer chains. This can result in dimerization or disproportionation (exchange of a hydrogen atom) but in either case two free radicals are destroyed. This scheme is followed for polymerization of all the monomers set out in Table 5.

Table 5. Some typical monomers and rate parameters for polymerization

Monomer				Rate constant for polymerization, k_p (mol^{-1} s^{-1}, 30°)
$RCH=CH_2$	R = Ph		Styrene	50
	Cl		Vinyl chloride	6800
	OAc		Vinyl acetate	990
	$CH=CH_2$		Butadiene	25
	$CMe=CH_2$		Isoprene	11
$R_1 \! \diagdown \atop R_2 \! \diagup \! C=CH_2$	R_1	R_2		
	Me	COOMe	Methyl methacrylate	350
	H	COOMe	Methyl acrylate	720
	Me	CN	Methacrylonitrile	29

The evidence for radical chain mechanism is unequivocal. Polymerization is initiated by any of a number of known sources of free radicals: peroxides, azo compounds, electrolysis, light, high-energy radiation. Indeed, the polymerization of styrene by a reacting system can be taken as diagnostic of free radicals in the system. The usual radical scavengers such as oxygen, hydroquinone, thiols, etc., inhibit polymerization and are usually added to monomers to preserve them. Residual radicals may be trapped in the entangling mass of polymer chains, unable to dimerize or disproportionate, and may then be detected by e.s.r. Careful analysis (using radioactive initiators as tracers) shows that each polymer chain contains 1–2 initiator residues as expected (there would be one at each end of the polymer chain if termination occurred by dimerization, and only one if by disproportionation). The mode of addition is extremely specific and all links between monomer units are 'head-to-tail' there being no more 'head-to-head' links than can be accounted for by dimerization. There is, however, no stereospecificity with respect to the configuration at each phenyl-bearing carbon. The phenyl groups are distributed randomly on either side of the carbon 'backbone' of the polymer (**96**). Such a polymer is said to be 'atactic' (cf. Section 10.6). Both the initiator–monomer

H Ph H Ph Ph H H PhPhH

(**96**)

radical (**97**) and the polystyryl radical (**98**) have been detected using spin trapping with tBuNO. The kinetics of polymerization are complex but add confirmation to the mechanism. Using the abbreviations for the reactive species which appear in brackets in the scheme, and rate constants for the individual steps, an overall rate equation may be deduced as follows.

$$\text{Disappearance of monomer} = \frac{-\text{d}}{\text{d}t}[M] = \underbrace{k_i[\text{In}\cdot][M]}_{(A)} + \underbrace{k_p[M\cdot][M]}_{(B)}$$

The term A may be neglected if the chains are very long implying that step (c) effectively accounts for the disappearance of all the monomer.

We assume a constant concentration of radicals present, then

$$\frac{\text{d}}{\text{d}t}[M\cdot] = k_i[\text{In}\cdot][M] - 2k_t[M\cdot]^2 = 0 \tag{5.1}$$

where k_t, the total rate of termination, equals $(k_{t'} + k_{t''})$.

We further assume a constant concentration of initiator radicals, whence,

$$\frac{\text{d}}{\text{d}t}[\text{In}\cdot] = 2fk_d[\text{In}] - k_i[\text{In}\cdot][M]$$

Here, f is a factor which expresses the fraction of initiator radicals which are successful in initiating a polymer chain; the figure 2 appears since each initiator molecule produces two radicals on decomposition.

From the above steady-state expressions, the concentrations of the radical species may be obtained;

$$[M\cdot] = \left(\frac{k_i[In\cdot][M]}{2k_t}\right)^{0.5} \quad \text{and} \quad [In\cdot] = \frac{2fk_d[In]}{k_i[M]}$$

substituting in equation 5.1 and putting the factor $f = 1$ (which has been found to be approximately true),

$$\frac{d}{dt}[M] = R_p = k_p[M]\left(\frac{k_d[In]}{k_t}\right)^{0.5}$$

or

$$R_p = \frac{k_p[M]}{2k_t^{0.5}} R_i^{0.5}$$

where R_p = rate of polymerization and R_i = rate of initiation.

This expression predicts that the rate of polymerization is proportional to monomer concentration to the first power and to the rate of initiation to the power 0.5. It has been found to hold with a wide range of monomers and initiators, including light (for which $R_i \propto I^{0.5}$; I = intensity), at least in the early stages of polymerization when the radicals are still more or less mobile. Values of the individual rate constants may only be evaluated by sophisticated techniques which are beyond the scope of this book. Some typical values are given in Table 5, from which it will be seen that the actual specific rates of these radical processes are very large indeed. Energies of activation are typically quite low (25–40 kJ mol^{-1}).

Copolymerization

A mixture of two monomers will usually polymerize together, the individual polymer molecules containing both types of unit. This process is called 'copolymerization'. If the two monomers are of very similar chemical properties, e.g. styrene and p-methylstyrene, the arrangement of monomer units in the chain is more or less random. Two monomers of very different chemical nature, e.g. a nucleophilic olefin such as styrene and electrophilic olefin such as methyl acrylate (99), show a strong tendency to form an alternating polymer chain. This effect is evidently in response to polar factors. Olefins which will not polymerize alone may, nonetheless, copolymerize. For

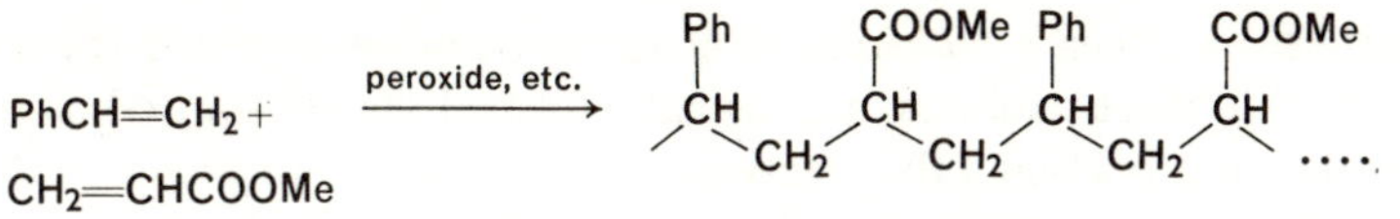

PhCH=CH₂ +

CH₂=CHCOOMe

(99)

$\xrightarrow{\text{peroxide, etc.}}$

styrene — methyl acrylate copolymer

example, maleic anhydride (100) will copolymerize with styrene, though it will not undergo homopolymerization, that is polymerization with itself. Since the

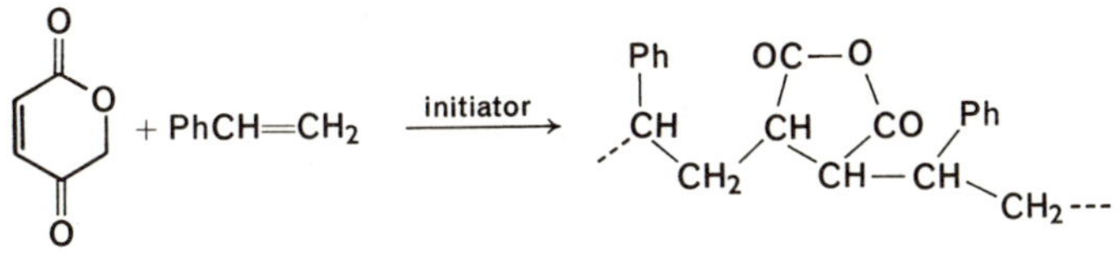

(100)

composition of the mixture of monomers may be varied continuously, copolymerization provides a means of preparing polymers with graded properties; many copolymers are important commercial materials such as synthetic rubber (basically 25% styrene, 75% butadiene), oil-resistant rubber (75% butadiene, 25% acrylonitrile) and *Saran* wrapping material (10% vinyl chloride, 90% vinylidene chloride). More complex copolymers are also used such as the terpolymer 'ABS'—acrylonitrile–butadiene–styrene—used for a wide variety of rigid mouldings.

Homolytic Aromatic Substitution

Arylation

When dibenzoyl peroxide is decomposed at 70–80° in the presence of an aromatic substrate, substitution of the ring occurs by a process which shows the expected characteristics of attack by a phenyl radical. There is little

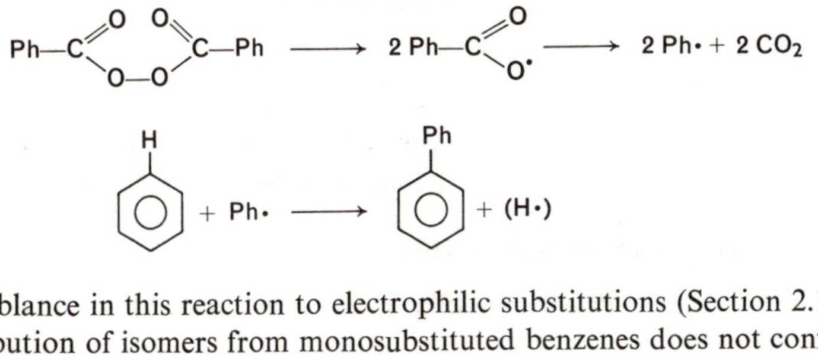

Ph—C(=O)O—OC(=O)—Ph $\longrightarrow$ 2 Ph—C(=O)O· $\longrightarrow$ 2 Ph· + 2 CO₂

resemblance in this reaction to electrophilic substitutions (Section 2.14); the distribution of isomers from monosubstituted benzenes does not conform to the pattern of either *ortho–para* or *meta* substitution expected for a polar reaction. All three isomers are usually formed in substantial amounts, frequently approaching the statistical distribution $(o:m:p::40:40:20)$, although a preference towards *ortho* attack is often noted (Table 6).

Table 6. Isomer distributions in the phenylation of substituted benzenes by dibenzoyl peroxide at 80°

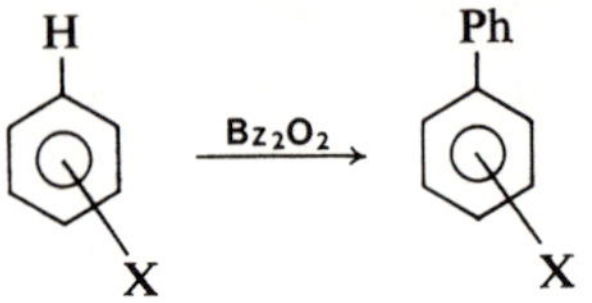

X	*o*	*m*	*p*
Me	66.5	19.3	14.2
Cl	50	32	18
Br	49	33	17
NO$_2$	62	10	28
COOMe	58	17	25
CF$_3$	30	41	29

Phenylation may be accomplished by a number of different reagents, but from a given substrate the *proportions of isomeric products are the same*, which argues for a common reactive intermediate. Phenylating agents are all compounds which most plausibly decompose to phenyl radicals. The following are typical.

Dibenzoyl peroxide $PhCOO\!-\!OCOPh \longrightarrow 2PhCOO\cdot \longrightarrow 2Ph\cdot + CO_2$

Diazonium ion decomposition

$$PhNH_2 \xrightarrow{HNO_2} Ph\!-\!\overset{+}{N}\!\!\equiv\!\!N \xrightarrow{Cu} Ph\cdot + N_2$$
diazonium salt

$\downarrow OH^-$

$$Ph\!-\!N\!\!=\!\!N\!-\!OH \longrightarrow Ph\cdot + N_2 + \cdot OH$$
diazohydroxide

From benzeneazotriphenylmethane (**101**)

$$Ph\!-\!N\!\!=\!\!N\!-\!CPh_3 \xrightarrow{\Delta} Ph\cdot + N_2 + \cdot CPh_3$$
(**101**)

From triphenyl bismuth (**102**)

$$Ph_3Bi \xrightarrow{\Delta} 3Ph\cdot + Bi$$

(**102**)

From lead (IV) benzoate (**103**)

$$Pb(OCOPh)_4 \xrightarrow{\Delta} Pb(OCOPh)_2 + 2PhCOO\cdot$$

(**103**) $2Ph\cdot + 2CO_2$

The nature of the solvent used in these reactions has very little effect upon either rates or product ratios. The rates of reaction do not vary systematically with the polar nature of substituents in the substrate; both electron-releasing groups (OMe) and electron-withdrawing groups (NO_2, CN, COOMe) appear to facilitate the rate of phenylation relative to H, but the rates span only a factor of ca 10 for the whole range of substituents, compared with a factor of say 10^8 for a typical electrophilic ionic substitution. Polar factors do have a slight effect, particularly when operating on the substituting radical rather than the substrate. Chlorophenyl and nitrophenyl radicals are more electrophilic than phenyl and tend to attack a substrate to a greater extent in the positions preferred for electrophilic attack than does phenyl. This is illustrated by the *changes* in isomer distribution brought about by attack of such an electrophilic radical on two substrates, chlorobenzene and nitrobenzene, relative to the distribution found for phenyl attack. Phenylation is frequently accom-

Change in per cent of each isomer formed by attack of as compared

with attack by

panied by the formation of tarry by-products from which terphenyls (**104**) have been isolated, presumably by further attack of phenyl radicals on the first-formed products. Side-chain hydrogen abstraction may also occur with the formation of diarylethane derivatives (**105**), and some benzoic-ester formation (**106**), is caused by the attack of benzoate radicals before decomposition to phenyl radicals. They can also abstract hydrogen from available sources to give benzoic acid (**107**). In addition to these characteristics, arylation occurs with a negligible isotope effect on the substituted hydrogen atom leaving little doubt that the reaction occurs in two stages, analogously to electrophilic substitution as follows:

$$(PhCOO)_2 \xrightarrow{\Delta} 2\ PhCOO\cdot \longrightarrow 2\ Ph\cdot + 2CO_2$$

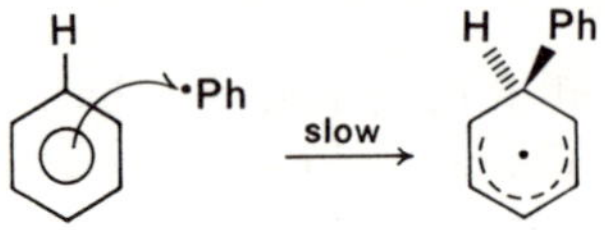

(**108**)

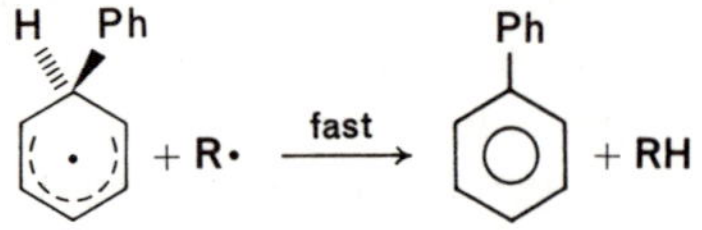

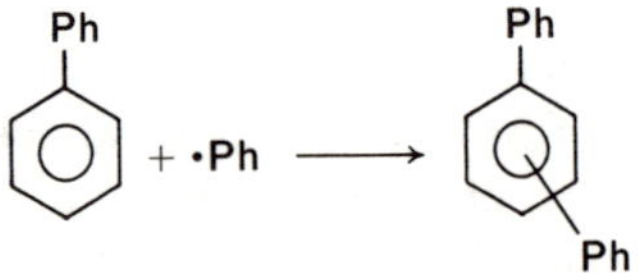

(**104**)

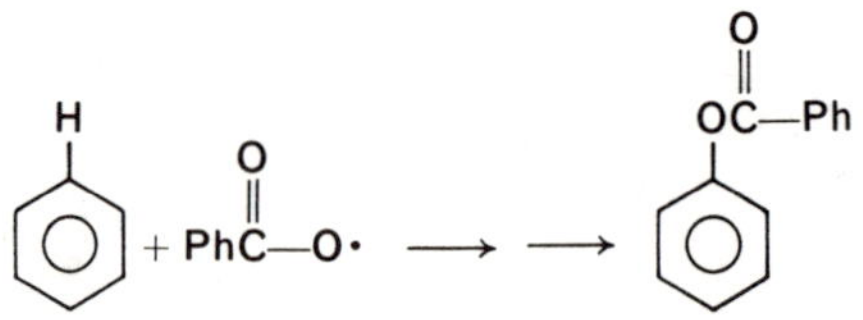

(**106**)

$$\underset{\text{(107)}}{Ph\overset{O}{\overset{\|}{C}}{-}O\cdot + R{-}H \longrightarrow Ph\overset{O}{\overset{\|}{C}}{-}OH + R}$$

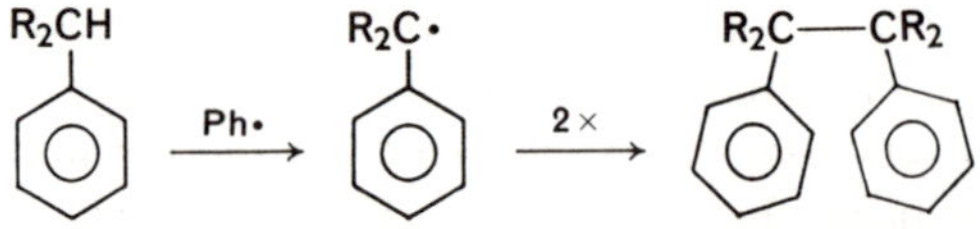

(**105**)

Further evidence for the intermediate formation of cyclohexadienyl radicals (**108**) is provided by the observation of **109**, by e.s.r.—a species formed by the coordination of a hydroxyl radical to benzene. It is presumed that the rate of

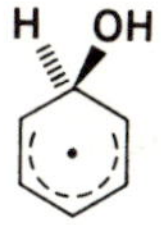

(**109**)

attack on a substrate with more than one reactive position is governed by the relative stabilities of the intermediate radicals. While the theory of radical stability is not as well understood as is that for benzenonium ions (Chapter 2), this appears to be the case for a number of polycyclic aromatic hydrocarbons which have been examined, and for which stabilities may be estimated for the intermediate formed by radical attack at various positions.

The Pschorr phenanthrene synthesis is an internal homolytic substitution in a diazotized *o*-amino-α-phenylcinnamic acid (**110**).

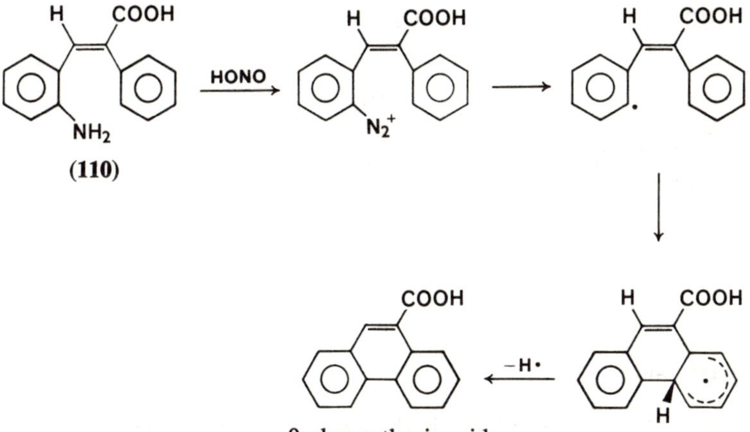

(**110**)

9-phenanthroic acid

Homolytic Alkylation

Alkyl radicals, generated by the thermal decomposition of peroxides (**111**) or lead (IV) carboxylates (**112**), are highly reactive and will substitute aromatic systems unspecifically. With the right choice of substrate, useful yields of products can sometimes be obtained, although in general this is not an attractive method.

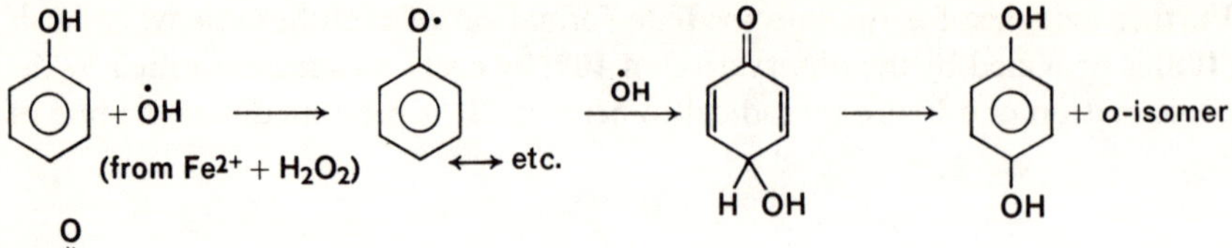

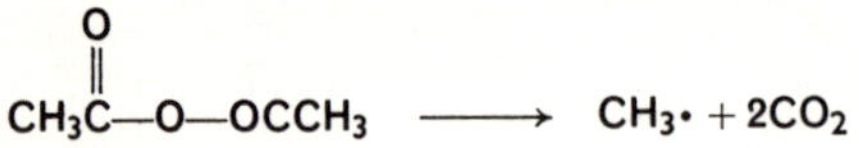

$$CH_3\overset{O}{\overset{\|}{C}}-O-OCCH_3 \longrightarrow CH_3\cdot + 2CO_2$$

(111)

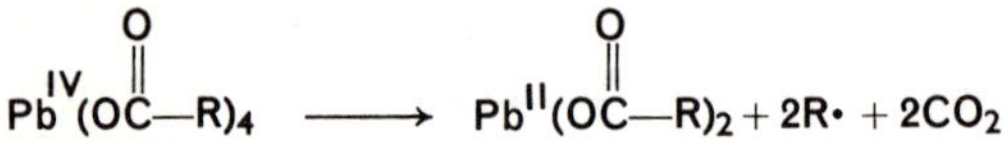

$$Pb^{IV}(O\overset{O}{\overset{\|}{C}}-R)_4 \longrightarrow Pb^{II}(O\overset{O}{\overset{\|}{C}}-R)_2 + 2R\cdot + 2CO_2$$

(112)

80%

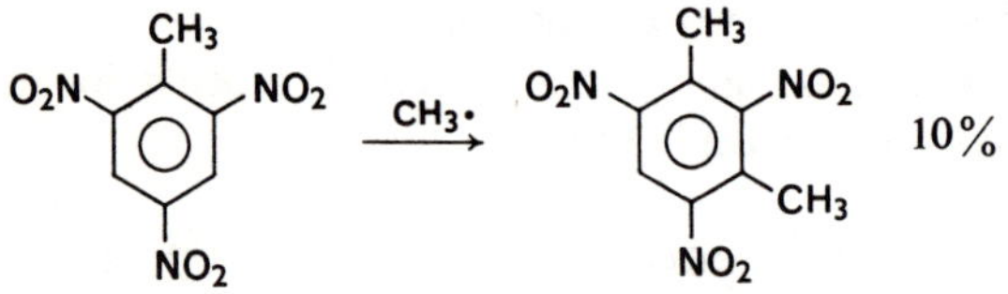

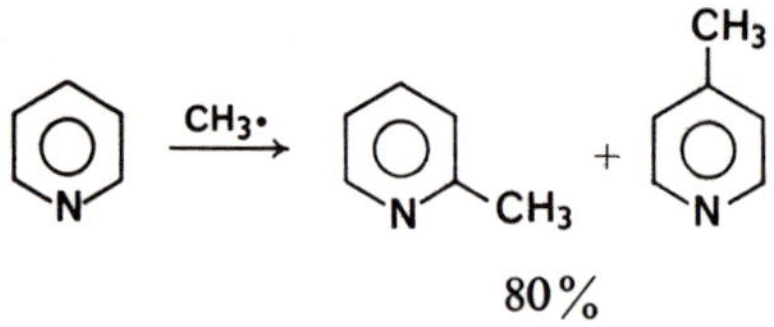

Hydroxylation

The hydroxyl radical (produced in the $Fe^{III}-H_2O_2$ or $Ti^{IV}-H_2O_2$ systems) will substitute aromatic substrates leading to phenols:

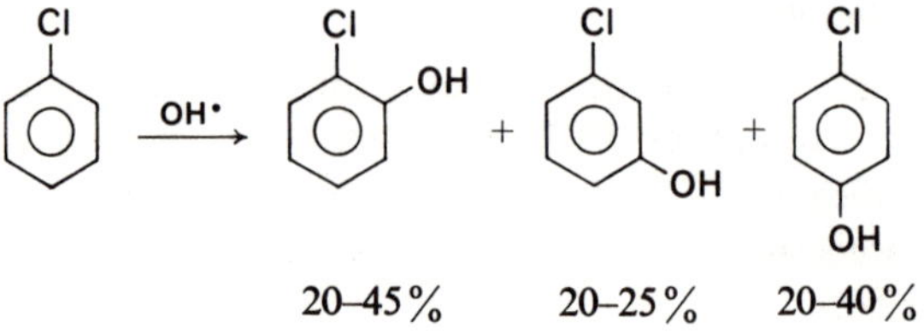

20–45%   20–25%   20–40%

Alkyl side-chains, if present, may also be attacked. On the whole, these reagents are too unselective to be of great use.

Phenols are hydroxylated by persulphate ion, $S_2O_8^{2-}$, (Elb's reaction); the reaction involves the homolysis of the peroxy bond in the persulphate ion by a one-electron transfer from the phenol. The reaction is found to be accelerated by electron-donating substituents:

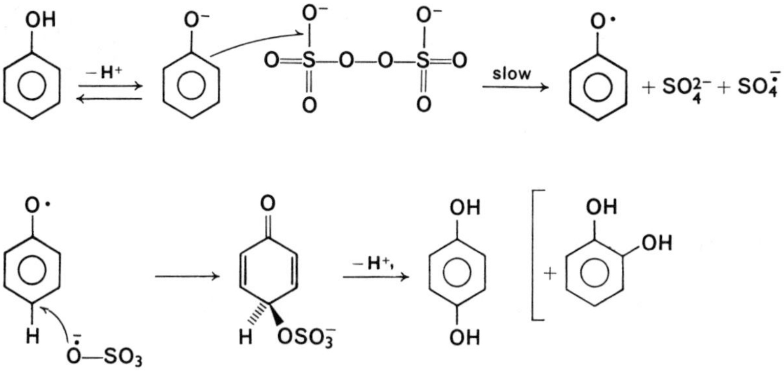

Radical mechanisms also operate in displacement reactions of *p*-nitrobenzyl halides and analogous compounds. Benzyl chloride undergoes a normal S_N2 displacement with the dinitroethide ion (113) leading to *o*-alkylation and this is also true for the *p*-cyano, *p*-trifluoromethyl and *m*-nitro compounds. The reaction product after hydrolysis is an aldehyde. However, the *p*-nitro isomer

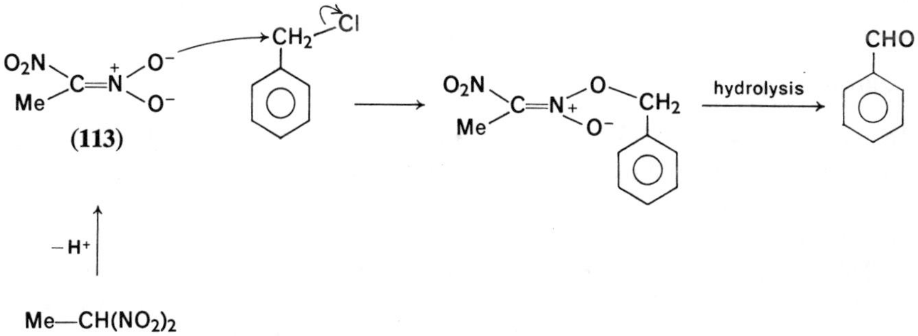

undergoes C-alkylation to 114 by a radical pathway which is suppressed by radical inhibitors. The change of mechanism is no doubt brought about by stabilization of the intermediate benzyl radical by the *p*-nitro group.

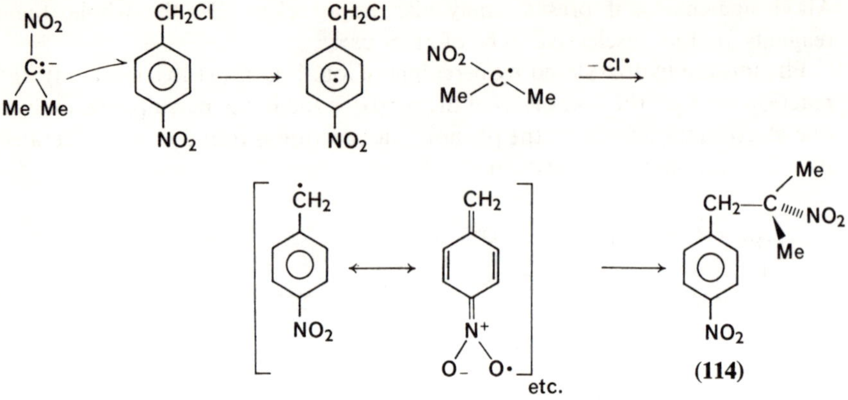

(114)

Amidation

Amidation by formamide in the presence of initiators has not been extensively examined but can give good yields, for example,

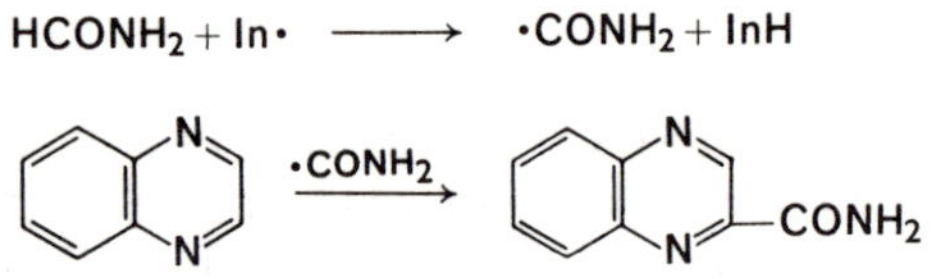

$$HCONH_2 + In \cdot \longrightarrow \cdot CONH_2 + InH$$

97%

Note that nitrogen heterocyclics, normally resistant to electrophilic substitution, are readily attacked by the radical.

Halogenation

In contrast to electrophilic halogenation by chlorine or bromine promoted by Lewis acids, radical halogenation leads to *addition* with the ultimate formation of a substituted product. The reaction is included with aromatic homolytic substitutions since the initial steps are essentially similar.

Benzene reacts with chlorine when irradiated with ultraviolet or sunlight to produce a mixture of isomeric hexachlorocyclohexanes (one of which has been used as the insecticide *Gammexane*). The intermediate dichlorodienes (115) and tetrachlorocyclohexenes are not isolable since they react with halogen atoms at a rate faster than does benzene. Although in principle a polar addition of chlorine to these intermediates is possible, it is generally believed that this is in fact a homolytic process also:

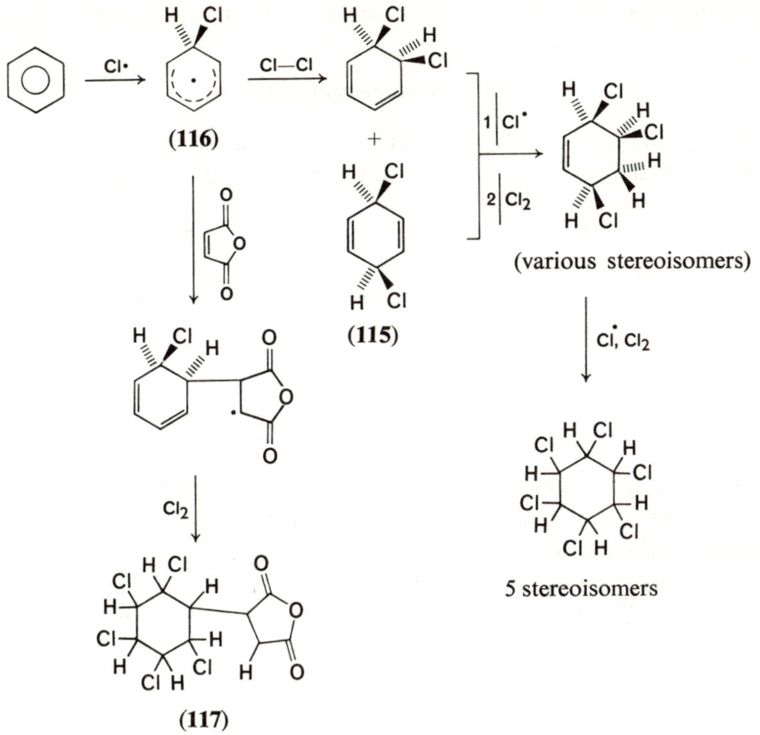

(116)

+

(115)

(various stereoisomers)

5 stereoisomers

(117)

The addition of maleic anhydride to the reaction mixture results in the trapping of the initial chlorocyclohexadiene radical (116) and the isolation of the compound 117. The addition of bromine follows an analogous path. Light-catalysed displacement of bromine by chlorine probably occurs by a similar mechanism,

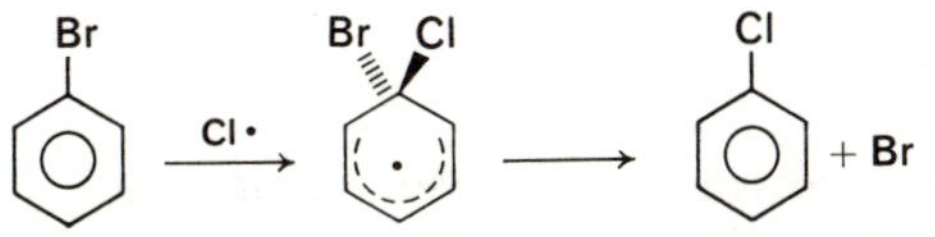

A radical pathway for the addition of chlorine to olefins is known. The reaction occurs in the gas phase or in non-polar solvents, is initiated by light and strongly inhibited by oxygen. The chain-carrying steps are,

$$RCH{=}CH_2 + Cl\cdot \longrightarrow R\dot{C}H{-}CH_2Cl$$

$$R\dot{C}H{-}CH_2Cl + Cl_2 \longrightarrow RCHCl{-}CH_2Cl + Cl\cdot$$

In the presence of moisture or polar surfaces the ionic addition is competitive.

Allylic Halogen Substitution

N-Bromosuccinimide (NBS) has been found to be a very useful reagent for the introduction of bromine into the allylic position, i.e. adjacent to a double bond.

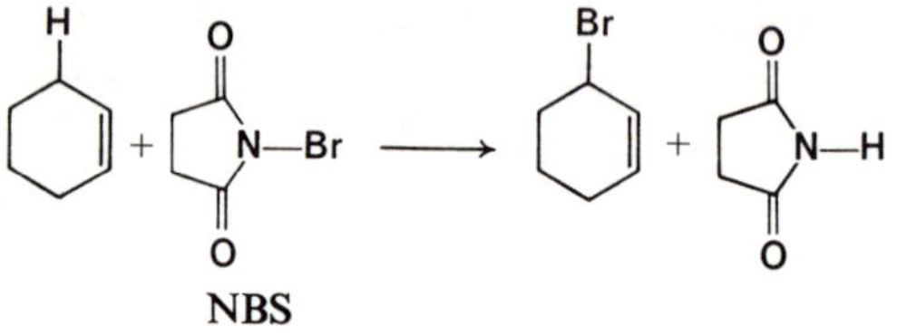

NBS

The reaction is accelerated by added peroxides and by light and is retarded by the common radical inhibitors. There seems little doubt that the mechanism is a radical chain reaction with the following propagation steps; allylic positions are especially susceptible to attack on account of the stabilized allyl radicals which result. There are, however, some unusual features to this reaction. The

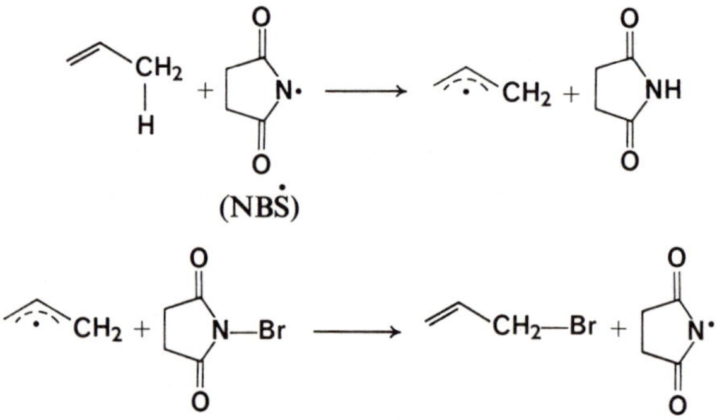

(NBS)

bromination is best carried out in carbon tetrachloride, a solvent in which N-bromosuccinimide is almost insoluble. The use of solvents in which the reagent is soluble leads to greatly reduced yields. It is believed that the radical process occurs at the surface of the crystal rather than in solution. This is substantiated by the observation that N-bromosuccinimide deposited on silica, and thus exposing a large surface area is far more reactive than the bulk solid material. The reagent will also brominate aromatic methyl groups since the intermediate benzyl radicals are stabilized. Electron-donating groups in the ring facilitate reaction, $\rho = -1.55$, indicating that the succinimide radical is rather electrophilic. *t*-Butyl hypochlorite, *t*BuO—Cl, is a somewhat similar

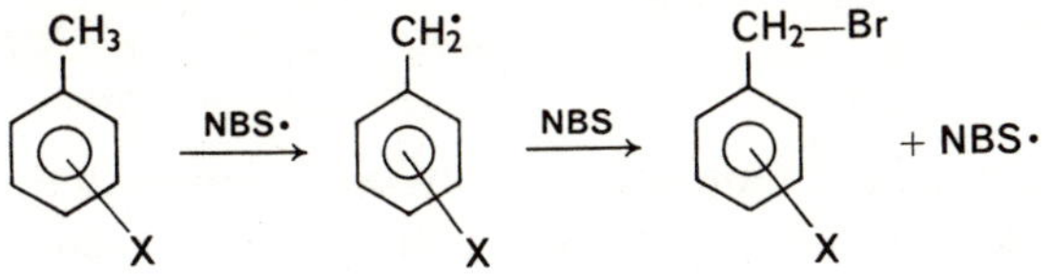

reagent which will chlorinate allylic or benzylic positions by an analogous radical chain reaction

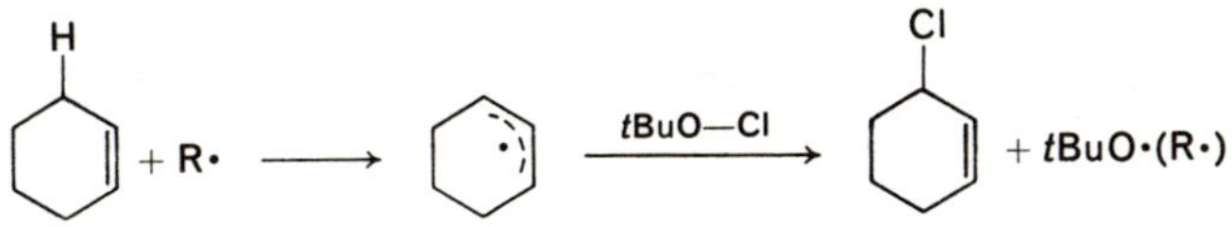

Phenol Coupling

The relatively easy one-electron oxidation of phenols (phenoxide ions) to phenoxy radicals has already been mentioned. A phenoxy radical may react further with another molecule of phenol resulting in coupling between the rings. Useful oxidizing agents for this purpose are ferricyanide or ferric ion, manganese dioxide, lead (IV) or cerium (III) compounds. The mechanism of coupling is analogous to the other radical substitutions discussed above.

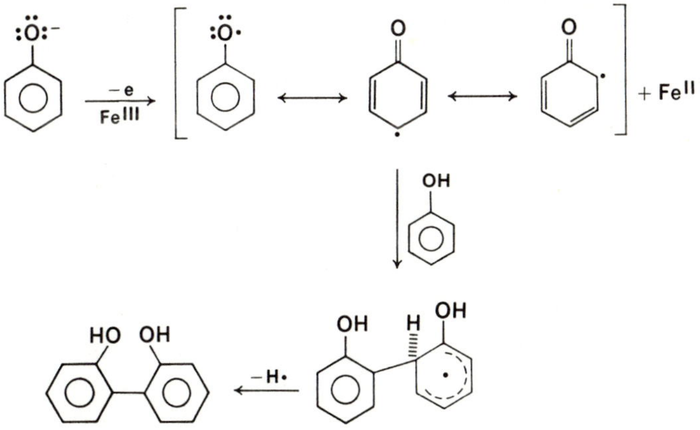

Phenol coupling is a very important process in the biogenesis of many natural products and can be induced *in vitro* by the enzyme peroxidase in the presence of hydrogen peroxide, for example,

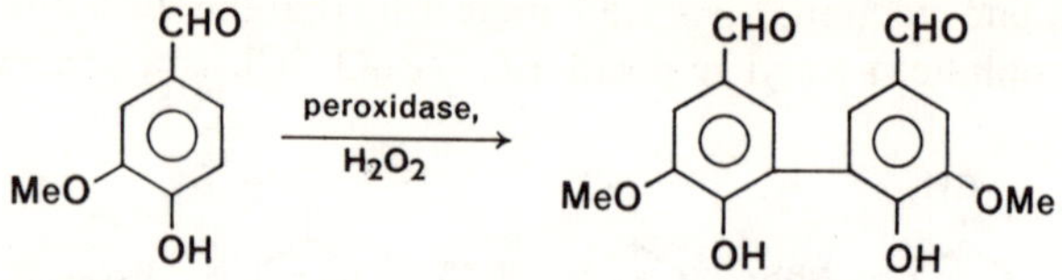

The intermediacy of free radicals, probably semiquinones in this reaction, has been demonstrated by e.s.r. spectroscopy using a flow system. As an example of phenol coupling in the formation of natural products we may consider the conversion of a typical benzylisoquinoline alkaloid, laudanosoline (118), a very common structural type to morphine (119) and to an aporphine alkaloid, glaucine (120). The following schemes are plausible though in the plant they are mediated by enzymes, and there is no direct evidence for radical intermediates.

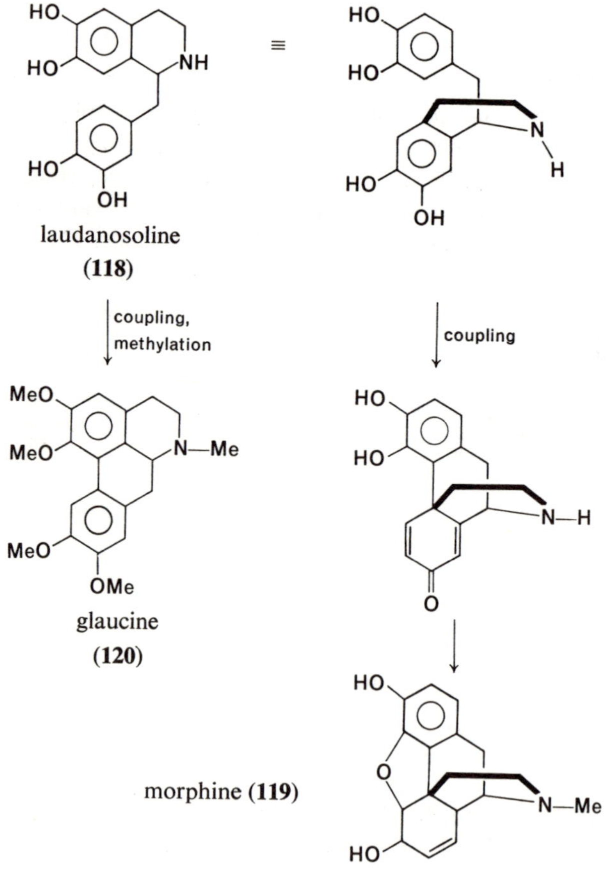

laudanosoline
(118)

coupling, methylation

coupling

glaucine
(120)

morphine (119)

The Hunsdiecker Reaction

Treatment of the silver salt of a carboxylic acid with bromine results in the formation of alkyl bromide and loss of CO_2 according to the scheme:

$$\underset{\text{O}}{\overset{\text{O}}{\underset{\|}{RC}}}\!-\!O^-\,Ag^+ + Br_2 \longrightarrow RBr + AgBr + CO_2$$

The reaction has the characteristics of a radical process. The initial step is the formation of an acyl hypobromite (**121**), which may be detected in solution and

$$R\!-\!C\!\!\underset{O^-\,Ag^+}{\overset{O}{\diagdown}} + Br\!-\!Br \longrightarrow R\!-\!C\!\!\underset{O-Br}{\overset{O}{\diagdown}} + AgBr$$

acyl hypobromite

(**121**)

Propagation
$$\left\{
\begin{array}{l}
R\!-\!C\!\!\underset{O-Br}{\overset{O}{\diagdown}} + R\cdot \longrightarrow R\!-\!C\!\!\underset{O\cdot}{\overset{O}{\diagdown}} + RBr \\[3ex]
R\!-\!C\!\!\underset{O\cdot}{\overset{O}{\diagdown}} \longrightarrow R\cdot + CO_2
\end{array}
\right.$$

which decomposes by a chain mechanism. The reaction is often accompanied by at least partial racemization of an asymmetric group R.

Pyrolytic Reactions in the Gas Phase

Hydrocarbons

Since C—C and C—H bonds are rather strong, saturated hydrocarbons normally require high temperatures to effect pyrolysis. Many paraffins have been studied in detail. The following primary processes are typical, the activation energies being essentially a measure of the strength of the bond which undergoes homolysis. The primary radicals will undergo dimerization, hydrogen abstraction and decomposition, leading ultimately to stable products. Toluene may be added as a radical scavenger so that the identities

Reaction		E_A–(kJ mol^{-1})
CH_4	$\longrightarrow$ $CH_3\cdot + H\cdot$	360
CH_3CH_3	$\longrightarrow$ $2CH_3\cdot$	342
$\begin{matrix} CH_3 \\ {}^{\diagdown} \\ CH_3{}^{\diagup} \end{matrix}CHCH_3$	$\longrightarrow$ $\begin{matrix} CH_3 \\ {}^{\diagdown} \\ CH_3{}^{\diagup} \end{matrix}CH\cdot + CH_3$	342
$\begin{matrix} CH_3 \\ \diagdown \\ CH_{3\text{\tiny IIII}}C-CH_3 \\ \diagup \\ CH_3 \end{matrix}$	$\longrightarrow$ $\begin{matrix} CH_3 \\ \mid \\ {}_{\diagup}C\cdot_{\diagdown} \\ CH_3 \quad CH_3 \end{matrix} + CH_3\cdot$	347
iPr—iPr	$\longrightarrow$ 2iPr$\cdot$	316
tBu—tBu	$\longrightarrow$ $2t$Bu$\cdot$	284

and amounts of various alkyl radicals may be determined from the products of the hydrogen exchange and the diphenylethane produced:

$$R\cdot + CH_3Ph \longrightarrow RH + \cdot CH_2Ph$$

$$2\ PhCH_2^\cdot \longrightarrow PhCH_2CH_2Ph$$

Alkyl radicals may stabilize by expelling a smaller alkyl radical or hydrogen atom which is adjacent to the unpaired electron,

$$CH_3\overset{\displaystyle CH_2}{\diagup}\overset{\displaystyle \cdot}{\underset{\diagdown CH_2}{}} \longrightarrow CH_3\cdot + CH_2{=}CH_2$$

$$CH_3\overset{\displaystyle \overset{\cdot}{CH}}{\diagup}\underset{\diagdown CH_3}{} \longrightarrow H\cdot + CH_2{=}CHCH_3$$

Isomerization by an internal hydrogen shift may occur to form a more stable radical, especially if a five- or six-membered transition state is possible.

$$CH_3{-}CH\overset{\diagup H}{\underset{\diagdown CH_2{-}CH_2}{}}\overset{\cdot}{CH_2} \longrightarrow CH_3{-}\overset{\cdot}{CH}\underset{\diagdown CH_2{-}CH_2}{}\overset{H\diagdown}{\diagup}CH_2$$

Homoallyl (but-3-enyl) radicals become symmetrical by a rapid rearrangement through the cyclopropylmethyl radical (122). Cyclic systems may open to give

$$CH_2{=}CH\overset{\diagup CH_2}{\underset{\diagdown \cdot CH_2}{}} \rightleftharpoons \overset{\cdot}{CH_2}{-}CH\overset{\diagup CH_2}{\underset{\diagdown CH_2}{\mid}} \longrightarrow CH_2{=}CH\underset{\diagdown CH_2}{}\overset{\cdot CH_2}{\diagup}$$

$$(122)$$

a diradical, a species with two non-interacting unpaired electrons. The following rearrangements exemplify this.

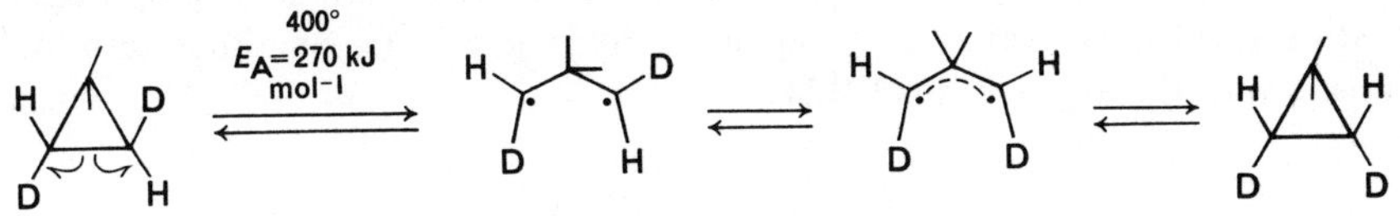

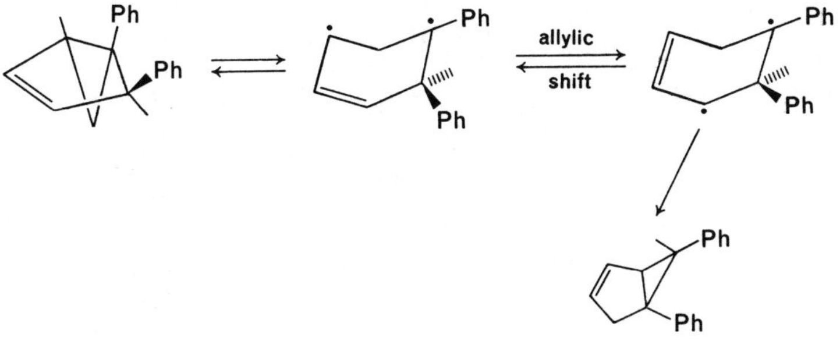

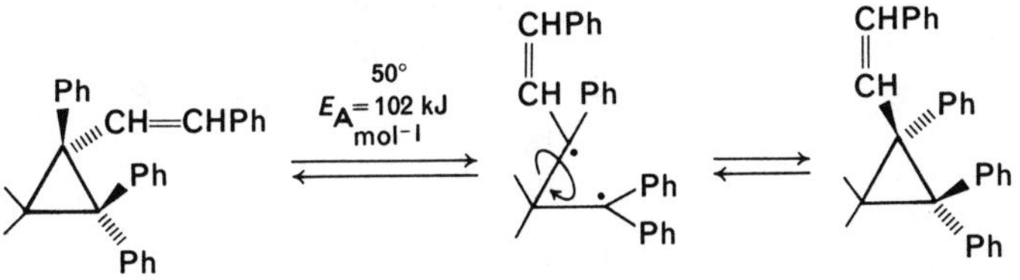

The strained molecule 2,2-*p*-cyclophane opens reversibly at 200° to give a diradical (**123**). This can be detected by using an optically active material

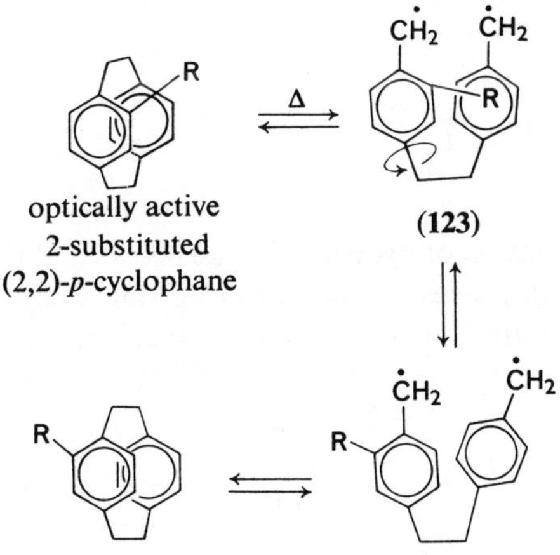

optically active
2-substituted
(2,2)-*p*-cyclophane

(**123**)

(**124**)

(e.g. R=COOEt) which is racemized. Isomerization of olefins occurs in the presence of iodine at ca 300°. At this temperature iodine atoms are formed which abstract hydrogen from the weakest (allylic) positions. The hydrogen iodide formed is available to replace a hydrogen atom at either terminal position of the allylic radical (124).

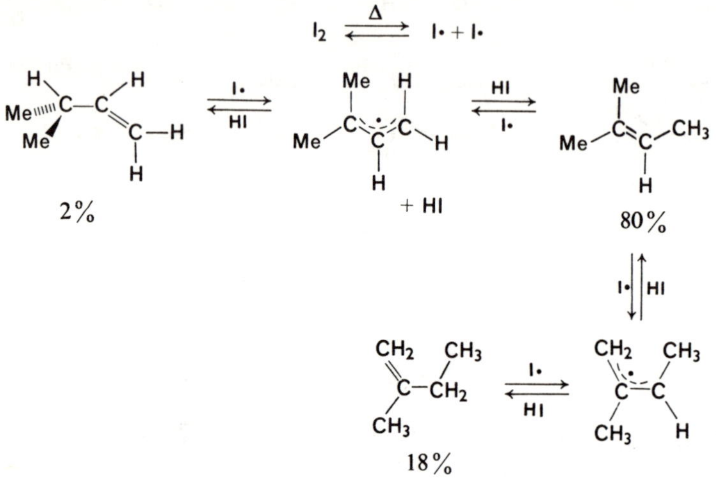

Alkyl Halides

Carbon–halogen bonds of alkyl halides which possess no β-hydrogen undergo homolysis at moderate temperatures (300–500°). If a β-hydrogen is available,

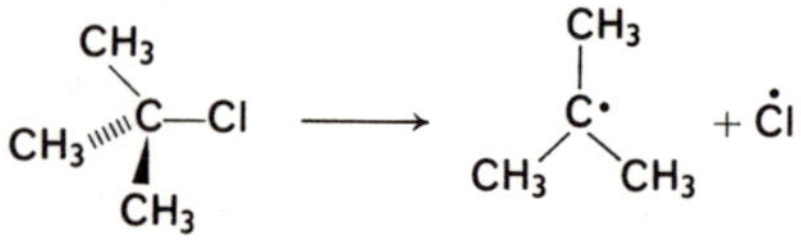

the molecular elimination of hydrogen halide occurs and the olefin so formed inhibits radical chain processes; the transition state (125) is probably highly polar. This does not apply to 1,2-dichloroethane, which decomposes after an

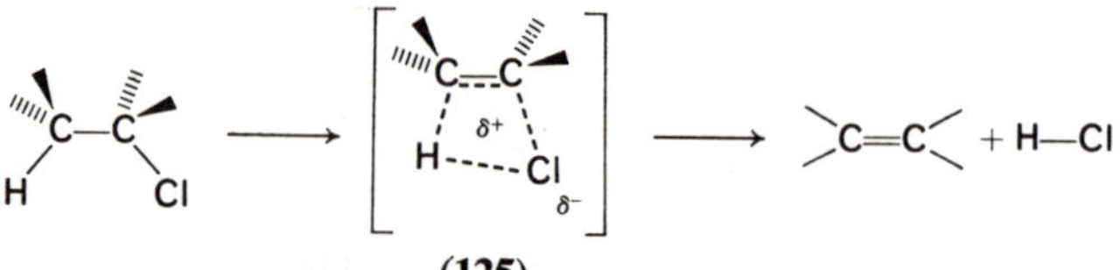

(125)

induction period and shows great sensitivity towards oxygen and other radical inhibitors.

$$\text{Cl}^\cdot + \begin{array}{cc} \text{CH}_2\text{—CH}_2 \\ | \quad\quad | \\ \text{Cl} \quad \text{Cl} \end{array} \longrightarrow \begin{array}{c} ^\cdot\text{CH—CH}_2\text{Cl} \\ | \\ \text{Cl} \end{array} + \text{HCl}$$

$$\begin{array}{c} ^\cdot\text{CH—CH}_2\text{Cl} \\ | \\ \text{Cl} \end{array} \longrightarrow \begin{array}{c} \text{CH}{=}\text{CH}_2 \\ | \\ \text{Cl} \end{array} + \text{Cl}^\cdot$$

Chemically-Induced Dynamic Nuclear Polarization (CIDNP)

An interesting extension of n.m.r. spectroscopy may often be used to provide unambiguous evidence for the intermediate presence of radicals in a reaction. The technique, CIDNP, depends upon the fact that products from a radical reaction may contain protons or other magnetic nuclei which are in a spin-state distribution significantly different from the equilibrium or Boltzmann distribution normally experienced. This then leads to anomalous intensities in the n.m.r. spectrum of the product. Indeed, some transitions may be negative, indicating emission rather than absorption of the radiofrequency radiation (Figure 6d).

The origin of this effect is as follows. Consider a radical dimerization reaction: as the unpaired spins begin to interact they will either be of like spin or unlike—either a triplet species or a singlet. It is necessary that a singlet pairing be formed in order for the reaction to proceed to product with the formation of a covalent bond. This will readily occur with the singlet radical pair but the triplet radical pair must first undergo a spin inversion, a relatively unlikely event but one which is affected by neighbouring magnetic nuclei. An adjacent proton which is coupled magnetically to the unpaired electron will facilitate or 'catalyse' the spin inversion, if in one *nuclear* spin state while retarding inversion if it is in the other (Figure 6b). The latter system may, in fact, not form a product at all before the radicals drift apart ('escape'). It follows that the product which is formed from the triplet radical pairs will have the coupled proton preferentially in a certain spin state, i.e. the spin state of the proton is other than the statistical which is given by the Boltzmann distribution law,

$$\frac{n_\alpha}{n_\beta} = e^{-(E^{\hat{\alpha}} - E^{\hat{\beta}})/kT} \approx e^{-\Delta E/600} \qquad \text{(at room temperature)}$$

where n_α and n_β are the mole fractions in spin states α and β with energies E_α, E_β. This means that normally there is only an excess of the lower β state by a factor

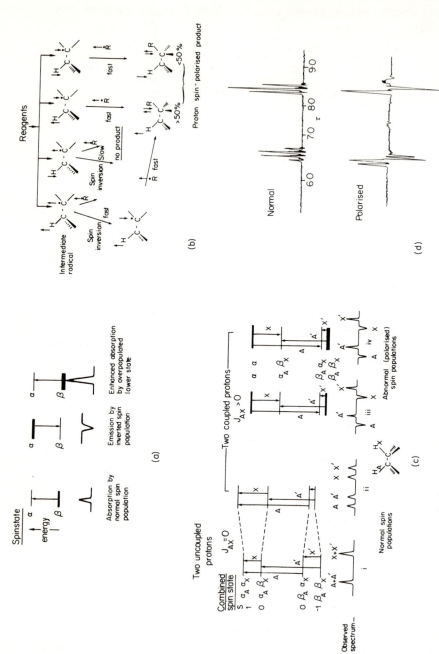

Figure 6. (a) N.m.r. signals from normal and abnormal (polarized) spin populations; (b) production of a spin-polarized product via a radical intermediate; (c) normal and spin polarized n.m.r. spectra of a two proton system; (d) normal and spin-polarized n.m.r. spectra of ethyl chloride.

of 1 in 10^6. This ratio is sufficient to observe net absorption of energy ΔE, but if the lower state is more populated than this ratio, the absorption will be more intense, while if the upper state is more populated than the lower, the net effect of the radiofrequency radiation will be to stimulate the emission of further radiation, reducing the population to its natural level and hence resulting in *emission* lines in the n.m.r. spectrum. This is indeed what is observed if the n.m.r. spectrum of a species resulting from a radical is obtained *during the course of the reaction*. Before quoting examples of this technique, the following analysis of the dynamic polarization of a two-spin system illustrates how both enhanced absorption and emission can occur in the same species.

Suppose the radical contains two protons, H_A and H_X, which are present in the product. The two-proton spin system may be described in terms of the energy levels in Figure 6c, and when magnetic coupling between protons occurs $(J > 0)$, the familiar double doublet n.m.r. spectrum is produced, i.e. there are four transitions each of different magnitude. Each level may be described by a total spin quantum number, S, which is the sum of the individual values, s $(= \pm\frac{1}{2})$ for each proton. It is possible that reaction of the radical proceeds to product when $S = \pm 1$, faster than when $S = 0$, hence the top and bottom spin states of the product will be overpopulated. This will lead to emission from the former and enhanced absorption from the latter and a polarized spectrum of the product as in Figure 6ciii, or 6civ if the populations in $S = -1$ and $S = 1$ levels are not equal. It is possible to predict the CIDNP spectrum from products arising from given radicals and hence to obtain considerably detailed information on the reaction mechanism. The detailed theory is beyond the scope of this account and the reader is referred to reviews of the topic.

The thermal decomposition of propionyl peroxide (**126**) in a halogen donor such as hexachloroacetone leads to ethyl chloride as a principal product by the following scheme.

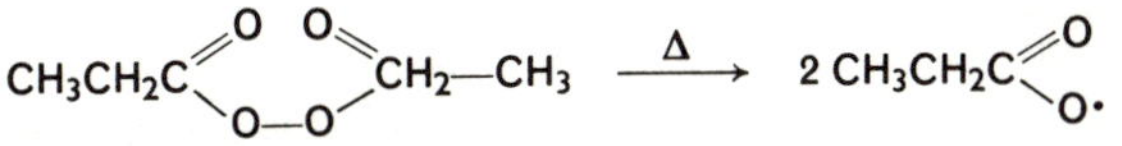

$$CH_3CH_2C\overset{O\quad O}{\underset{O-O}{\diagup\diagdown}}CH_2-CH_3 \;\xrightarrow{\;\Delta\;}\; 2\;CH_3CH_2C\overset{O}{\underset{O\cdot}{\diagup}}$$

(**126**)

$$CH_3CH_2C\overset{O}{\underset{O\cdot}{\diagup}} \;\longrightarrow\; CH_3CH_2\cdot + CO_2$$

$$CH_3CH_2\cdot + Cl_3CCOCCl_3 \;\longrightarrow\; CH_3CH_2Cl + Cl_2\dot{C}COCCl_3$$

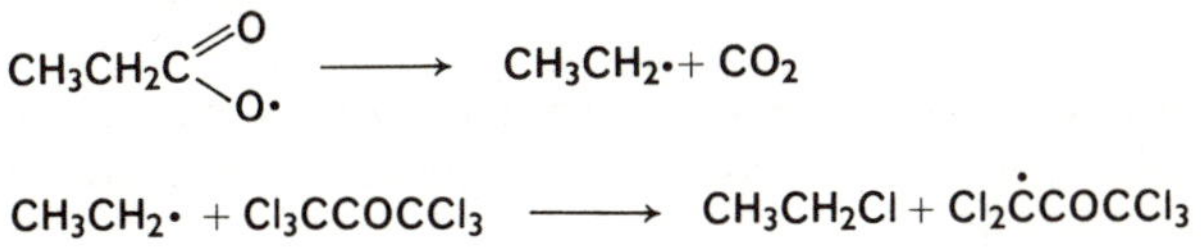

The polarized n.m.r. spectrum of ethyl chloride, compared with its normal spectrum, is shown in Figure 6d, confirming the radical mechanism. Many other peroxide decompositions have been examined and found to exhibit polarized n.m.r. spectra.

The photoreduction of ketones is a radical process; if benzophenone is irradiated in the presence of a hydrogen-atom donor such as toluene the product is diphenylbenzyl carbinol (127), believed to be formed by the following sequence:

$$Ph_2C{=}O \xrightarrow{\;h\nu\;} Ph_2\overset{\cdot}{C}{-}\overset{\cdot\cdot}{O}$$

(triplet,
excited)

$$\Big\downarrow PhCH_3$$

$$\left[\begin{array}{cc} Ph\overset{\cdot}{C}{\cdot} & \overset{\cdot}{C}H_2Ph \\ | & \\ OH & \end{array}\right] \longrightarrow \begin{array}{c} Ph_2C{-}CH_2Ph \\ | \\ OH \end{array}$$

radical pair   (127)

The system exhibits CIDNP confirming the intermediacy of a radical pair.

Ward, Lepley and others have also recorded nuclear polarization in the coupling reaction,

$$n\,Bu{-}Li + Bu{-}I \;\rightarrow\; LiBr + \text{butene, butane, octane.}$$

The signal due to the olefinic protons in the butene, which occurs as one of the products, is polarized and shows emission peaks. The mechanism is postulated to be:

Initiation	$Bu{-}I + Bu{-}Li \longrightarrow$	$2\,Bu\cdot + LiI$
Transfer	$Bu\cdot + BuI \longrightarrow$	$\overset{*}{B}uI + \overset{*}{B}u\cdot$
Termination	$2\,\overset{*}{B}u\cdot \longrightarrow$	$CH_3CH_2CH{=}\overset{*}{C}H_2 + Bu\overset{*}{H}$
		spin-polarized
$* =$ spin polarized		butene

Other polarized signals are assigned to starting materials and other products.

The Wurtz–Grignard reaction (an example with a rather low rate of reaction was chosen for study),

$$t\,BuBr + t\,BuMgBr \;\rightarrow\; \text{isobutene} + \text{isobutane} + Me_3C{-}CMe_3$$

was also observed to give rise to polarized products. Again, *t*-butyl radicals are proposed as intermediates. It seems likely that many organometallic reactions may have radical character. This was also proposed on chemical grounds by Bryce-Smith who obtained dicumyl in 10% yield from the coupling reaction carried out in the presence of cumene (isopropylbenzene), a well-known radical trap.

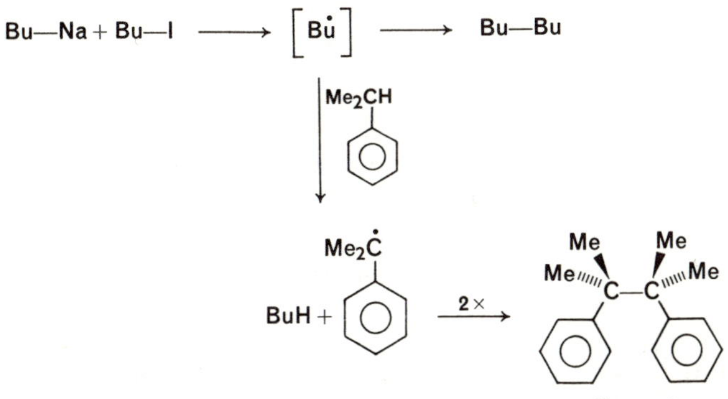

dicumyl

Radical Rearrangements

A 1,2-shift is potentially able to convert, for example, a primary radical (128) into the more stable secondary or tertiary isomer (130). Rearrangements of this

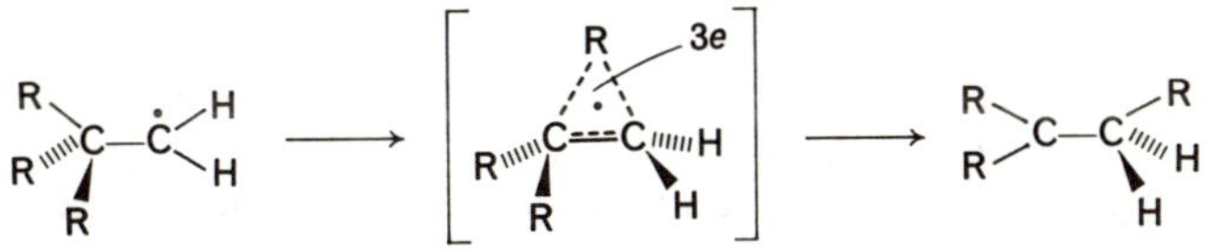

(128)   (129)   (130)

type are much less common in radical chemistry than with carbonium ions, since the transition state (129) requires accommodation for three delocalized electrons whereas the carbonium ion equivalent has only two and partakes of some aromatic character (Section 2.29).

   1,2-Migrations do occur in the radical field, the migrating group being almost exclusively an aromatic nucleus since it is able to form a more stable transition state by delocalization of the odd electron. Thus, the triphenyl-methoxy radical (131) rearranges in contrast to *t*-butoxy (132) which decom-

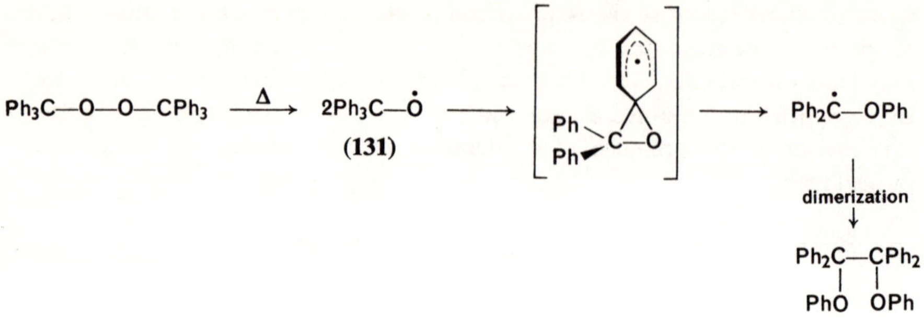

$$Ph_3C-O-O-CPh_3 \xrightarrow{\Delta} \underset{(131)}{2Ph_3C-\overset{\cdot}{O}} \longrightarrow \left[\underset{Ph}{\overset{Ph}{>}}C-O \right] \longrightarrow Ph_2\overset{\cdot}{C}-OPh$$

dimerization

$$Ph_2C-CPh_2$$
$$\underset{PhO}{\;} \quad \underset{OPh}{\;}$$

$$Me_3C-O\cdot \longrightarrow Me\cdot + Me_2C=O$$
$$(132)$$

poses. Further examples of phenyl migrations are,

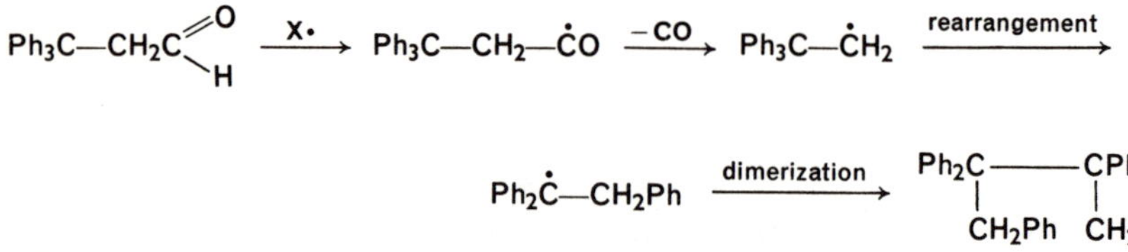

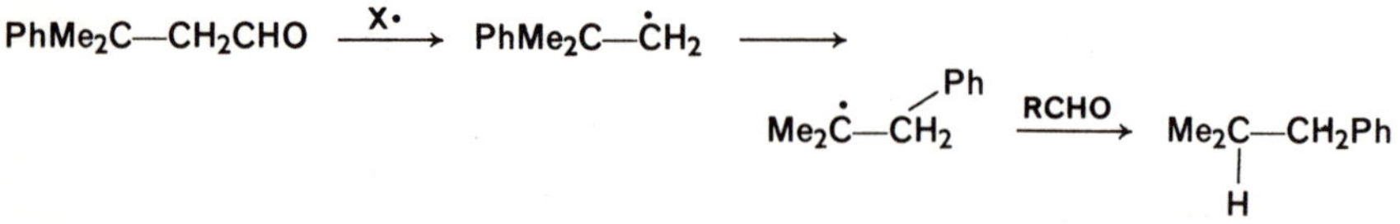

$$Ph_3C-CH_2C\overset{\nearrow O}{\underset{\searrow H}{}} \xrightarrow{X\cdot} Ph_3C-CH_2-\overset{\cdot}{C}O \xrightarrow{-CO} Ph_3C-\overset{\cdot}{C}H_2 \xrightarrow{rearrangement}$$

$$Ph_2\overset{\cdot}{C}-CH_2Ph \xrightarrow{dimerization} \underset{CH_2Ph \quad CH_2Ph}{Ph_2C-------CPh_2}$$

$$PhMe_2C-CH_2CHO \xrightarrow{X\cdot} PhMe_2C-\overset{\cdot}{C}H_2 \longrightarrow$$

$$\underset{Me_2\overset{\cdot}{C}-CH_2}{\overset{Ph}{\nearrow}} \xrightarrow{RCHO} \underset{H}{Me_2C-CH_2Ph}$$

Halogen migration also occurs quite readily. The reactions appear to be truly intramolecular since no reaction between added thiol or chlorine atoms can be detected.

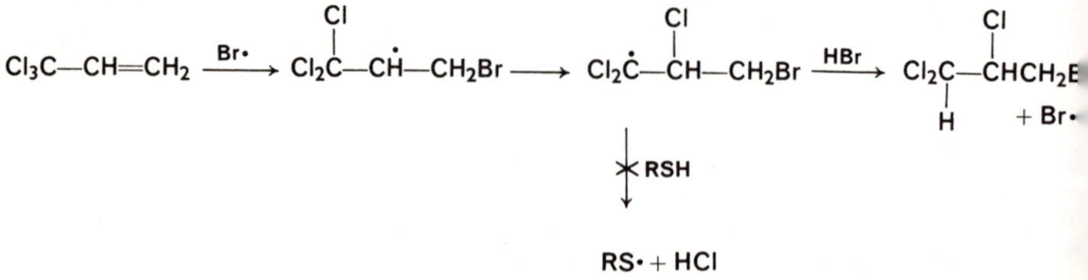

$$Cl_3C-CH=CH_2 \xrightarrow{Br\cdot} \underset{Cl_2C-\overset{\cdot}{C}H-CH_2Br}{\overset{Cl}{|}} \longrightarrow \underset{Cl_2\overset{\cdot}{C}-CH-CH_2Br}{\overset{Cl}{|}} \xrightarrow{HBr} \underset{Cl_2C-CHCH_2B}{\overset{Cl}{|}} + Br\cdot$$

$$\underset{RSH}{\times}\downarrow$$

$$RS\cdot + HCl$$

An acetoxy-radical shift is inferred in the following:

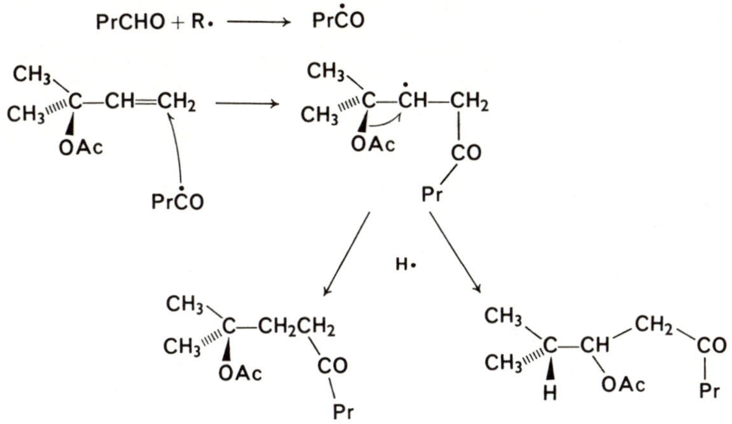

PrCHO + R· $\longrightarrow$ Pr$\overset{\cdot}{C}$O

80%   20%

The Barton Reaction

Photochemical decomposition of certain nitrite esters (133) was found by Barton to result in the interchange of a nitroso group with hydrogen from a spatially adjacent carbon. The reaction appears to be intramolecular and is believed to occur by an exchange of partners of a radical pair within a solvent 'cage'. This is a very useful reaction which provides a means of functionalizing

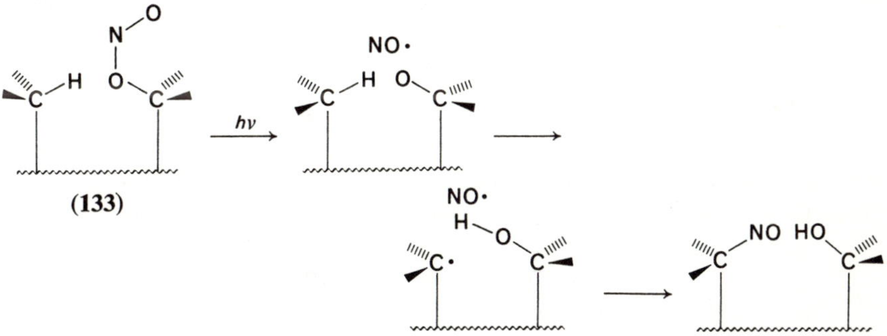

(133)

a methyl group when lying close to a hydroxyl function, even though only remotely connected by bonding. It has been particularly applied to the steroid field. Thus a nitrite group at any of the axial β positions, 2β, 4β or 6β, will bring about the nitrosation of the 19-methyl group. The hydrogen-atom migration occurs through a sterically favourable six-membered cyclic transition state.

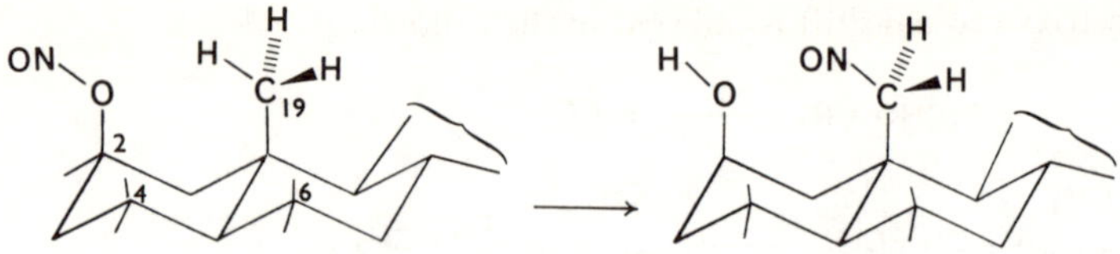

(134)

The nitroso compounds (134) produced in this way can be reduced to primary amines and converted to alcohols and other functions by standard means.

Somewhat similar is the 'hypoiodite' reaction in which the following sequence may convert an alcohol to a cyclic ether (135).

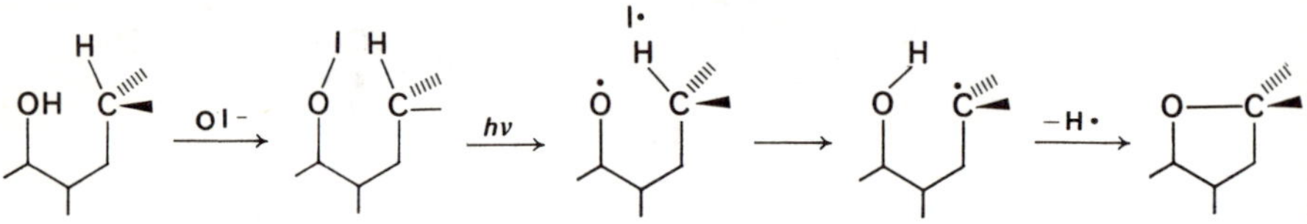

alkyl hypoiodite   (135)

The Kolbe Electrosynthesis

It has been known since 1849 that the passage of an electric current through a solution of a metal carboxylate results in the production of dimeric hydrocarbons at the anode, the carboxylate group being lost as CO_2. Cross-coupling can occur, leading to three products, when a mixture of two different salts is electrolysed. The mechanism is believed to commence with the anodic

$$2\ R.COO^- \xrightarrow{-2e} R\!-\!R + 2\ CO_2$$
$$(136)$$
$$R.COO^- + R'COO^- \xrightarrow{-2e} R\!-\!R + R\!-\!R' + R'\!-\!R' + 2\ CO_2$$

oxidation of carboxylate ions to the corresponding radicals (136), which then lose CO_2 and dimerize. In support of this radical mechanism, an electrolysing

$$CH_3COO^- \xrightarrow{-e} CH_3COO\cdot$$
$$CH_3COO\cdot \longrightarrow CH_3\cdot + CO_2$$
$$2\ CH_3\cdot \longrightarrow CH_3CH_3$$
$$90\%$$

solution of a metal carboxylate is found to initiate the polymerization of styrene, generally taken to be diagnostic of the presence of radicals. The following examples illustrate the scope of the Kolbe reaction. Straight-chain fatty acids and half-esters of dicarboxylic acids give good yields. Chain branching, particularly at the α-position, causes low yields but even so may be acceptable in the synthesis of natural products (137). Electronegative substituents also tend to reduce the yields.

$$CH_3(CH_2)COOH \longrightarrow n\text{-}C_{10}H_{22} \qquad\qquad 90\%$$

$$MeO_2C(CH_2)_4COOH \longrightarrow MeO_2C(CH_2)_8COOMe \qquad 90\%$$

$$Cl(CH_2)_4COOH \longrightarrow Cl(CH_2)_8Cl \qquad\qquad 50\%$$

$$\underset{+\ HOOC(CH_2)_8COOEt}{CH_3(CH_2)_7\overset{\overset{\displaystyle Me}{|}}{C}HCOOH} \longrightarrow \underset{(137)}{CH_3(CH_2)_7\overset{\overset{\displaystyle Me}{|}}{C}H(CH_2)_8COOH}$$

$$\pm \text{ tuberculostearic acid}$$

Reductions by Alkali Metals

Alkali metals are one-electron reducing agents and in many of their reactions with organic compounds, free radicals are probably transient intermediates.

The Acyloin Synthesis

The treatment of carboxylic esters with liquid sodium in an inert solvent, such as boiling xylene, results in the formation of a dimeric α-ketol (acyloin 138).

$$2RCOOEt \xrightarrow{\ Na\ } \underset{(138)}{RCOCHOHR}$$

The reaction is believed to commence with the one-electron reduction of the ester to an anion radical analogous to a ketyl (139). This then dimerizes to a

$$R-\overset{\overset{\displaystyle O}{\|}}{C}\diagdown_{OEt} + Na^{\cdot} \longrightarrow R-\overset{\overset{\displaystyle O^{\cdot}}{}}{C}\diagdown_{OEt} \longleftrightarrow R-\overset{\overset{\displaystyle O^{-}}{}}{\underset{\cdot}{C}}\diagdown_{OEt}$$

$$(139)$$

dianion which expels two ethoxide ions to form an α-diketone (140). This is in turn reduced by sodium to the ketol by way of its enolate (141). The reaction is useful for the synthesis of medium and large rings from dicarboxylic esters; the intramolecular reaction is promoted by using very high dilution. It is probable

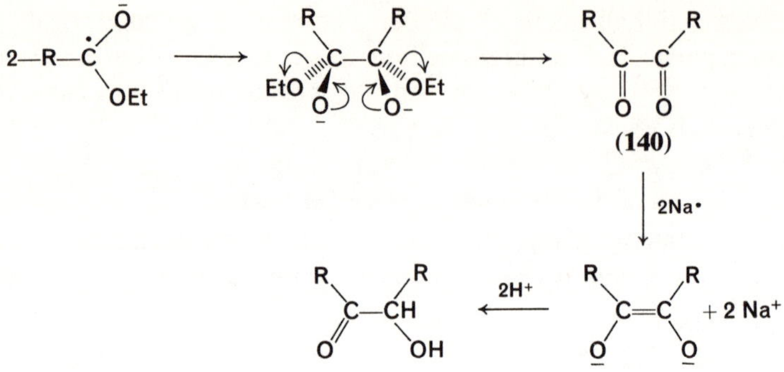

(140)

(141)

that the reduction occurs on the surface of the molten sodium. The Bouveault–Blanc reduction of esters to primary alcohols by sodium in boiling alcohol may

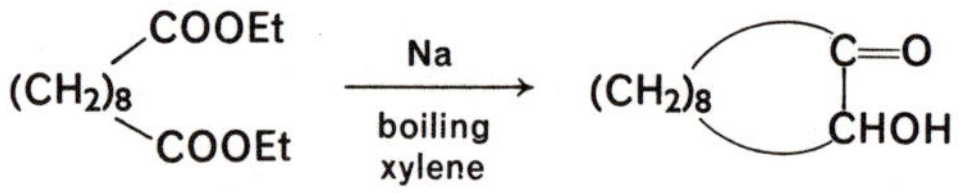

diethyl sebacate   sebacoin

also be initiated by formation of the anion radical **139**. In the presence of a protic solvent, however, this is protonated and reduced further to the acetal oxidation level (equivalent to the aldehyde **142**). The sequence is then repeated until the lowest oxidation level, that of alcohol, is reached (**143**).

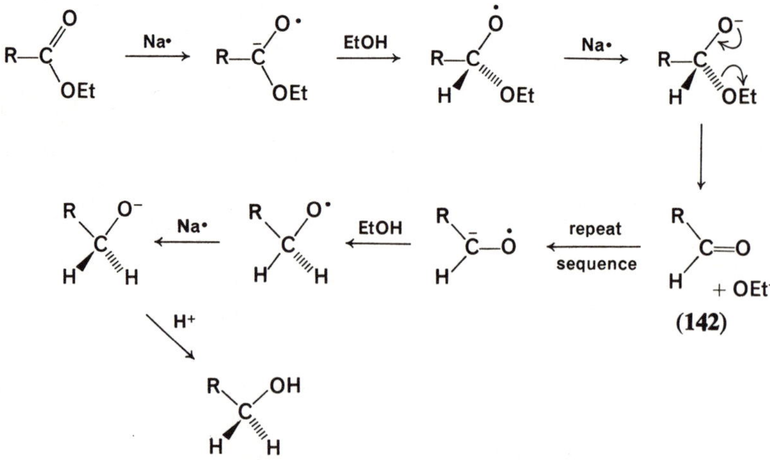

(142)

(143)

Birch Reduction

When sodium dissolves in liquid ammonia a blue solution results which is paramagnetic and contains solvated electrons,

$$Na^{\cdot} + NH_3 \longrightarrow Na^+ + (\bar{e})NH_3$$

The solvated electrons are capable of coordinating to an aromatic system to form the hexadiene radical anion (144), which in the presence of acids (e.g. methanol) accept a proton. The resulting radical is further reduced by a second electron leading ultimately to a 1,4-dihydrobenzene (145). This is the most convenient procedure for effecting partial reduction of a benzene ring.

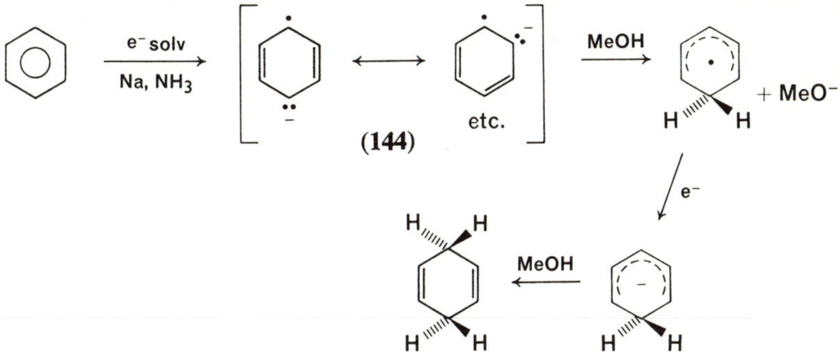

(144)

(145)

Further examples of this reaction, the Birch reduction, are as follows: lithium may replace sodium, and amines such as ethylamine replace ammonia for greater convenience and better yield and also the higher reduction potential available.

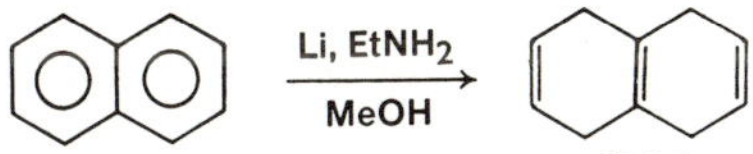

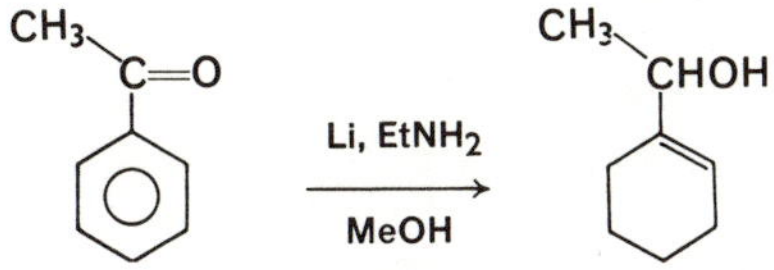

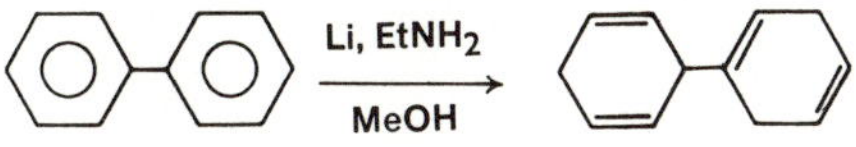

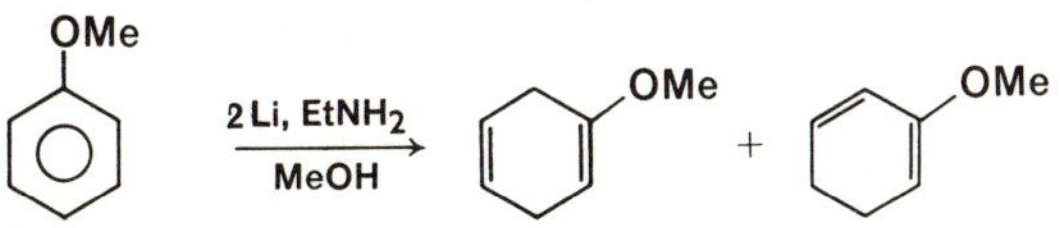

Copper-Catalysed Decompositions

The conversion of a diazonium ion to aryl chloride by the action of cuprous chloride (Sandmeyer reaction) has the characteristics of a radical reaction. The yields are generally poor with much tar formation and the reaction mixture will initiate polymerization. Radicals present may be trapped by the addition of nitrobenzene (Meerwein reaction). The reaction probably involves initially

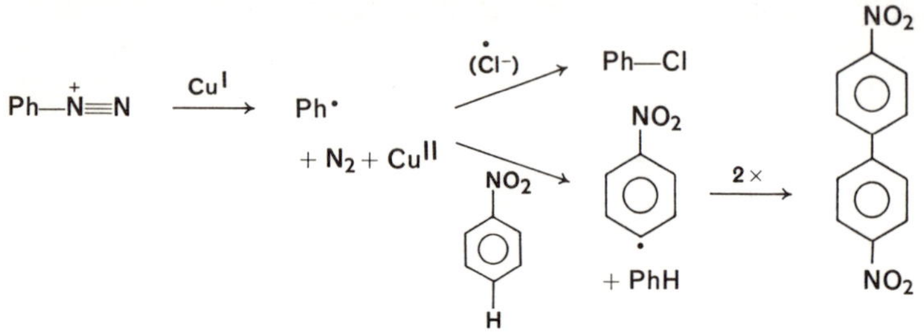

the one-electron reduction of the diazonium ion by univalent copper, in the form $CuCl_2^-$, to give aryl radicals which abstract chlorine atoms from copper (II) chloride. A high concentration of HCl inhibits the reaction, possibly due to the formation of an inactive chloro complex, e.g. $CuCl_3^{2-}$:

$$Ar\overset{+}{N}\!\!\equiv\!\!N + Cu^ICl_2^- \longrightarrow Ar\cdot + N_2 + Cu^{II}Cl_2$$

$$Ar\cdot + ClCuCl \longrightarrow ArCl + Cu^ICl$$

Aryl radicals are shown to be involved by diverting their attack to added acrylonitrile or other olefin when products from the addition of Ar and Cl across the double bond are obtained.

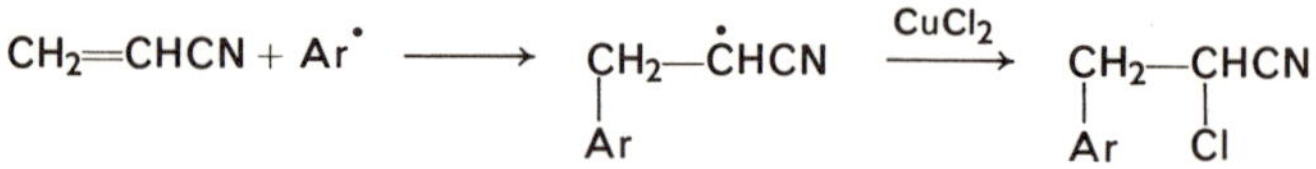

$$CH_2\!\!=\!\!CHCN + Ar\overset{\bullet}{} \longrightarrow \underset{Ar}{CH_2\!\!-\!\!\overset{\bullet}{C}HCN} \overset{CuCl_2}{\longrightarrow} \underset{Ar\quad Cl}{CH_2\!\!-\!\!CHCN}$$

Aryl radicals may be intermediates in the Ullmann reaction, the conversion of an aryl halide to the symmetrical biphenyl by heating with copper powder, e.g.

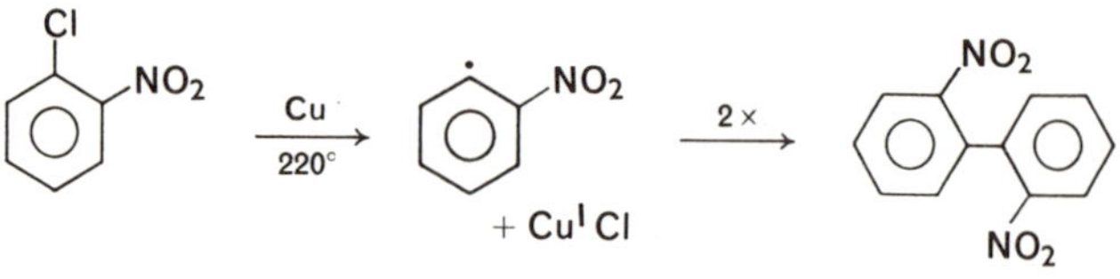

50%

Oxidative carboxylation of carboxylic acids by lead tetraacetate is greatly accelerated and the yield improved by the addition of copper salts (Kochi reaction). The reaction is believed to proceed via alkyl radicals from the thermal decomposition of a lead (IV) carboxylate and the further oxidation of the radicals to carbonium ions by intermediate lead (III) compounds. This step is more efficiently performed by copper (II) which can be in far greater concentration than the unstable lead (III). The carbonium ion **146** then loses a

$$CH_3(CH_2)_3COOH + Pb(OCOCH_3)_4 \rightleftharpoons CH_3(CH_2)_3COOPb^{IV}(OCOCH_3)_4$$
$$+ CH_3COOH$$

$$\longrightarrow CH_3(CH_2)_3COO\cdot + Pb^{III}(OCOCH_3)_3$$

$$CH_3(CH_2)_3COO\cdot \longrightarrow CH_3CH_2CH_2CH_2\cdot + CO_2$$

$$CH_3CH_2CH_2CH_2\cdot + Cu^{II} \longrightarrow CH_3CH_2CH_2CH_2^+ + Cu^I$$
$$(Pb^{III}) \qquad\qquad (146) \qquad\qquad (Pb^{II})$$

β-proton with the formation of olefin as the major product. In the presence of

$$CH_3CH_2CH_2CH_2^+ \xrightarrow{-H^+} CH_3CH_2CH=CH_2$$

$$\downarrow Li^+Cl^-$$

$$CH_3CH_2CH_2CH_2Cl$$

lithium chloride, the alkyl chloride is formed in good yield.

REFERENCES FOR FURTHER READING

A. M. Bass and H. P. Broida (Eds), *Formation and Trapping of Free Radicals*, Academic Press, New York, 1960.

J. E. Bennett, B. Mile, A. Thomas and B. Ward, 'The study of Free Radicals and Their Reactions at Low Temperature using a Rotating Cryostat', *Adv. Phys. Org. Chem.*, **8**, 1 (1970).

J. Betts, 'Hydrocarbon Autoxidation in the Liquid Phase', *Quart. Rev.*, **1971**, 265.

A. Carrington, 'E.s.r. Spectra of Aromatic Radicals and Radical-Ions', *Quart. Rev.*, **1963**, 67.

R. S. Davidson, 'Hydrogen Abstraction in the Liquid Phase by Free Radicals', *Quart. Rev.*, **1967**, 249.

A. R. Forrester, J. M. Hay and R. H. Thomson, *Organic Chemistry of Stable Free Radicals*, Academic Press, New York, 1968.

E. G. Janzen, 'Spin Trapping', *Accounts Chem. Res.*, **4**, 31 (1971).

R. G. Lawler, 'C.I.D.N.P. the Radical Pair Model', *Accounts Chem. Res.*, **25** (1972).

R. O. C. Norman and B. C. Gilbert, 'E.S.R. Studies of Short-lived Organic Radicals', *Adv. Phys. Org. Chem.*, **5** (1967).

S. H. Pine, 'C.I.D.N.P.', *J. Chem. Ed.*, **1972**, 664.

W. A. Pryor, *Free Radicals*, MacGraw-Hill, New York, 1966.

Reactions of Free Radicals in the Gas Phase, Chemical Society Special Publication No. 9 (1957).

W. V. Sherman, 'Free Radical Intermediates in the Radiation Chemistry of Organic Compounds', *Adv. Free Radical Chem.*, **3**, 1 (1969).

M. R. C. Symons, 'The Identification of Organic Free Radicals by E.s.r.', *Adv. Phys. Org. Chem.*, **1** (1962).

C. Walling, *Free Radicals in Solution*, John Wiley, New York, 1957.

H. R. Ward, 'C.I.D.N.P. Examples and Applications', *Accounts Chem. Res.*, **18** (1972).

W. A. Waters, *Modern Developments in Free Radical Chemistry*, Chemical Society Special Publication No. 19 (1965).

F. Williams, 'Radiation Chemistry of Hydrocarbons', *Quart. Rev.*, **1963**, 101.

G. H. Williams, *Homolytic Aromatic Substitution*, Pergamon Press, London, 1960.

Chapter 6

Carbenes (Methylenes)

6.1 **INTRODUCTION**

The nucleophilic-substitution reactions of the chloromethanes show some strange features (Table 1). For instance, the hydrolysis in water or reactions with weak nucleophiles in general occur at rates in the following order:

$$CH_3Cl \gg CH_2Cl_2 > CHCl_3 \gg CCl_4$$

while in reactivity towards strong bases such as OH^- and OR^- the sequence

$$CH_3Cl > CH_2Cl_2 \ll CHCl_3 \gg CCl_4$$

holds. It must be concluded that the reactions are occurring by two or perhaps more mechanisms. The first sequence can be attributed to normal S_N2 reactions (Section 2.3) in which successive chlorine atoms retard the reaction by making it more difficult for the chloride ion to leave. However, the second sequence

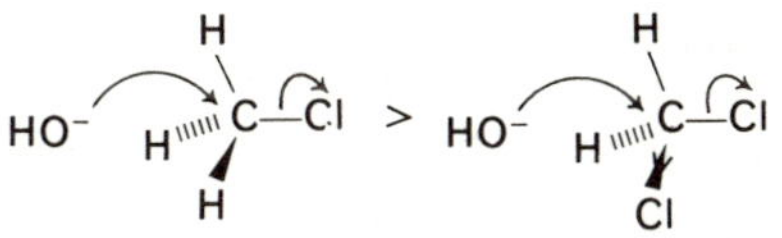

Table 1. Relative rates of reaction of chloro-methanes

	k_{rel}; NaOH in aqueous dioxan (strong base)	k_{rel}; Piperidine in ethanol (weak base)
CH_3Cl	0.0013	87
CH_2Cl_2	0.0002	4
$CHCl_3$	1.0	1.0
CCl_4	0.0007	0.01

indicates that chloroform is anomalously reactive and suggests that another mechanism is operating. It has been known since 1936 that in D_2O/OD^-, chloroform exchanges its proton at a rapid rate compared with its hydrolysis. The relatively high acidity of this proton is attributable to stabilization of the carbanion (1) by the vacant d-orbitals on chlorine (Section 4.10). The mech-

(1)

anism of hydrolysis was proposed by Hine (1950) to commence with this acidic dissociation. The trichloromethyl anion was then supposed to lose a chloride ion to form the species: CCl_2 (2), dichloromethylene, which rapidly reacted with water to form the observed products which included formate ion and carbon monoxide. In reality, the suggestion of $:CCl_2$ as a reactive intermediate

(1)   (2)

$$Cl_2C: + H_2O \xrightarrow{\text{fast}} Cl_2C-OH \longrightarrow HCO_2^-, CO$$

dates back to the nineteenth century when it was proposed by Geuther (1862) to account for the hydrolysis of chloroform. Later, in 1895, Nef proposed the parent compound, CH_2, as an intermediate in the decomposition of diazomethane, CH_2N_2. Both mechanisms are now believed to be substantially correct though they were rejected for more than half a century as chemists became preoccupied with the octet rule for carbon.

There are further observations which support the mechanism of hydrolysis of chloroform given above.

(a) Added chloride ion markedly reduces the rate of hydrolysis in aqueous-alcoholic medium. The effect is much greater than that produced by equivalent amounts of perchlorate or nitrate ions and therefore is not a Brønsted salt effect (Section 1.20) but is specific to Cl^-. It indicates that chloride ions is being produced reversibly in the slow step in accordance with the proposed dissociation of CCl_3^- to CCl_2 and Cl^-.

(b) The chloride-ion effect is an instance of the capture or 'trapping' of the reactive intermediate, in this instance to regenerate its precursor. Yet there are other reagents which will accomplish this to give products which are more easily recognized. Bromide and iodide ion will lead to the formation of bromo-dichloromethane and iododichloromethane, respectively, at the same time greatly reducing the rate of hydrolysis of the chloroform. It is unlikely that these products could arise from attack of bromide and iodide upon the chloroform.

$$:CCl_2 + Br^- \longrightarrow BrCCl_2^- \xrightarrow{H_2O} BrCHCl_2$$

$$:CCl_2 + I^- \longrightarrow ICCl_2^- \longrightarrow ICHCl_2$$

(c) More characteristically, the addition of an olefin such as cyclohexene gives rise to the formation of some dichloronorcarane (3), 7,7-dichloro-bicyclo[4,1,0]heptane. This is most plausibly seen as the capture of dichloro-methylene by the olefin; no other species likely to be present would react in this way. In non-aqueous systems (e.g. the addition of chloroform at low tempera-

(3)

tures to potassium t-butoxide in cyclohexene), this adduct may be obtained in excellent yield.

With the recognition of dichloromethylene as an intermediate in the reactions of chloroform with strong bases, many other examples of this interesting class of compounds have been sought and discovered.

Nomenclature

The terms 'methylene' and 'carbene' are applied indiscriminately to the species :CH$_2$, and its homologues are described as derivatives of it. Thus :CCl$_2$ may be named 'dichloromethylene' or 'dichlorocarbene'. CH$_3$—C̈—CH$_3$ may be named 'dimethylcarbene' or 'isopropylidene'.

Routes to Carbenes

There are two main types of reaction which have been found to yield carbenes:

(a) The thermal or photodecomposition of molecules R$_2$C=X where X is a small, stable molecule.

$$R_2C{=}X \xrightarrow[\substack{or \\ h\nu}]{\Delta} R_2C: + X$$

(b) The successful removal of a nucleophilic and an electrophilic group, in either order, from the carbon atom,

$$R_2C\begin{smallmatrix}N\\E\end{smallmatrix} \xrightarrow{\hspace{1cm}} \begin{array}{c} R_2\overset{+}{C}-E + :N^- \xrightarrow{-E^+} \\ \\ R_2C-N + E^+ \xrightarrow{-:N^-} \end{array} R_2C:$$

This type of reaction is designated an α-elimination.

Fragmentation Routes to Carbenes

Photolysis or pyrolysis of diazomethane (4), $CH_2=N_2$, in the gas phase readily causes the expulsion of nitrogen and the formation of the parent compound, $:CH_2$. This primary product is exceedingly reactive. It dimerizes to ethylene which reacts with further methylene and the principal products are ethylene (63%), propylene (7%), but-1-ene (9%) and small quantities of higher hydrocarbons. The decomposition of other diazo compounds in the gas phase or in

$$CH_2=N_2 \xrightarrow[\text{or}\,\Delta]{h\nu} :CH_2 + N_2$$

(4)

$$:CH_2 \xrightarrow[\text{or }CH_2N_2]{2\times} (CH_2=CH_2)^* \longrightarrow CH_2=CH_2$$

* = vibrationally excited

$$\underset{H}{\overset{H}{>}}C=C\underset{H}{\overset{H}{<}} + :CH_2 \xrightarrow{\text{insertion}} \underset{H}{\overset{H}{>}}C=C\underset{H}{\overset{CH_2}{<}}\overset{H}{} \quad \text{etc.}$$

solution provides a general synthesis of carbenes. Carboxyl-substituted diazo compounds are rather stable and may be isolated and separately decomposed:

$$O=C\underset{OEt}{\overset{CH=N_2}{<}} \xrightarrow[\Delta,\,150°]{h\nu} O=C\underset{OEt}{\overset{CH:}{<}} + N_2$$

The rather explosive diazoalkanes may be prepared and decomposed *in situ* by the oxidation of an aldehyde or ketone hydrazone. Alternatively, alkali-

$$\underset{R'}{\overset{R}{>}}C=O \xrightarrow{N_2H_4} \underset{R'}{\overset{R}{>}}C=N-NH_2 \xrightarrow{HgO} \underset{R'}{\overset{R}{>}}C-N=N \xrightarrow[\Delta]{h\nu,} \underset{R'}{\overset{R}{>}}C: + N_2$$

metal salts of the toluenesulphonylhydrazones of aldehydes or ketones (**5**) may be cleaved thermally or photochemically (the Bamford–Stevens reaction).

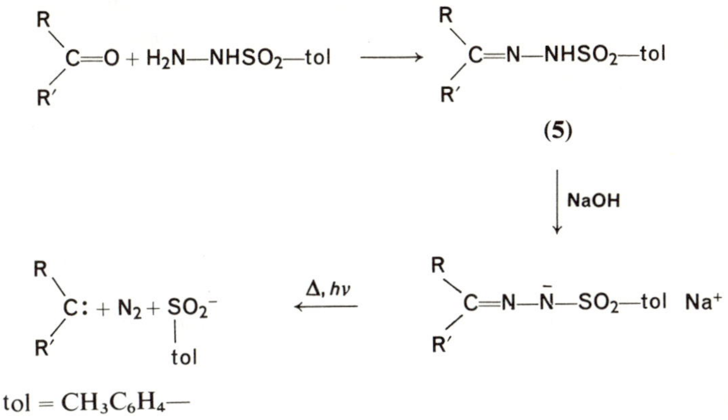

$$\underset{R'}{\overset{R}{>}}C{=}O + H_2N{-}NHSO_2{-}tol \longrightarrow \underset{R'}{\overset{R}{>}}C{=}N{-}NHSO_2{-}tol$$

(**5**)

$$\downarrow NaOH$$

$$\underset{R'}{\overset{R}{>}}C: + N_2 + \underset{tol}{\overset{|}{SO_2^-}} \xleftarrow{\ \Delta,\ h\nu\ } \underset{R'}{\overset{R}{>}}C{=}N{-}\bar{N}{-}SO_2{-}tol\ Na^+$$

$$tol = CH_3C_6H_4{-}$$

Diazoketones

Diazoketones (**6**) are readily available from the reaction of diazomethane with an acyl chloride; their decomposition to ketocarbenes (**7**), which rearrange to ketenes (**8**), is the basis of the Arndt–Eistert homologation procedure, by which a carboxylic acid is converted to its next higher homologue, e.g.

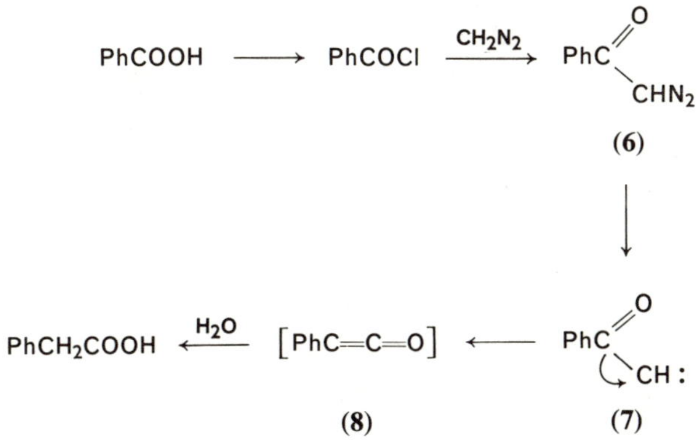

$$PhCOOH \longrightarrow PhCOCl \xrightarrow{CH_2N_2} PhC\overset{O}{\underset{CHN_2}{\diagup}}$$

(**6**)

$$\downarrow$$

$$PhCH_2COOH \xleftarrow{H_2O} [PhC{=}C{=}O] \longleftarrow PhC\overset{O}{\underset{CH:}{\diagup}}$$

(**8**) (**7**)

Diazirenes

Cyclic, three-membered azo compounds (**9**) also yield carbenes on decomposition. They are prepared from an aldehyde or ketone by reaction with ammonia and chloroamine followed by oxidation of the diaziridine (**10**), and are frequently explosive.

$$
\underset{R'}{\overset{R}{>}}C=O \xrightarrow[\text{NH}_2\text{Cl}]{\text{NH}_3} \underset{R'}{\overset{R}{>}}\underset{\text{NH}}{\overset{\text{NH}}{C}} \xrightarrow[hv]{\text{oxidation}} \underset{R'}{\overset{R}{>}}\underset{\text{N}}{\overset{\text{N}}{C}}
$$

$$
\textbf{(10)} \qquad\qquad \textbf{(9)}
$$

$$
\xrightarrow{hv}
$$

$$
\underset{R'}{\overset{R}{>}}C: + N_2
$$

Ketenes

Ketenes photolyse to carbenes; the parent compound, keten (**11**), yields

$$
CH_2\!=\!C\!=\!O \xrightarrow[\text{or}\,\Delta,\,700°]{hv} :CH_2 + CO
$$

$$
\textbf{(11)}
$$

methylene. Other fragmentation routes of minor importance are known, e.g. the photolysis of phenyl-substituted epoxides (**12**),

$$
\underset{NC}{\overset{Ph}{>}}\underset{\text{Ph}}{\overset{O}{C\!-\!C}}\underset{}{\overset{Ph}{<}} \xrightarrow{hv} \underset{NC}{\overset{Ph}{>}}C: + Ph_2CO
$$

$$
\textbf{(12)}
$$

and cyclopropanes (**13**),

$$
\underset{}{\overset{Ph}{>}}\underset{\text{CH}_2}{\overset{CH\!-\!CH_2}{\diagdown\diagup}} \xrightarrow{hv} PhCH\!=\!CH_2 + :CH_2
$$

$$
\textbf{(13)}
$$

α-Eliminations

The dissociation of an acidic proton from carbon followed by loss of a nucleophilic group from the same carbon leads to carbene formation. This type of reaction affords the most convenient route to the halocarbenes (see Section 6.1).

$$
\underset{Cl}{\overset{Cl}{>}}\underset{\text{Cl}}{\overset{C\!-\!H}{\diagup}} \xrightarrow{\bar{O}t\text{Bu}} \underset{Cl}{\overset{Cl}{>}}\overset{\ominus}{C}\!-\!Cl \longrightarrow \underset{Cl}{\overset{Cl}{>}}C: + Cl^-
$$

$$
+ t\text{BuOH}
$$

The electrophilic leaving group need not be a proton as in the following examples. It is not certain that carbanions are involved in all these reactions, some are probably concerted eliminations.

$$Cl_2C \overset{COO^-}{\underset{Cl}{\diagup}} \quad Na^+ \xrightarrow[150°]{\Delta} Cl_2\bar{C}\text{—}Cl \longrightarrow Cl_2C\colon + Cl^-$$

sodium trichloroacetate   (in diglyme) $+ CO_2$

$$Cl_2C \overset{CO.CCl_3}{\underset{Cl}{\diagup}} \xrightarrow[MeO^-]{MeOH} Cl_2C \overset{O^- \quad CCl_3}{\diagup}{\underset{Cl}{\diagdown}OMe} \longrightarrow Cl_2C\colon +$$

hexachloroacetone CCl_3COOMe

$+ Cl^-$

$$Cl_2C \overset{Cl}{\underset{Cl}{\diagup}} \xrightarrow{BuLi} Cl_2C\colon + BuCl + LiCl$$

$$Cl_2C \overset{HgPh}{\underset{Br}{\diagup}} \xrightarrow{> -15°} \colon CCl_2 + PhHgBr$$

Dibromocarbene and difluorocarbene can be made by essentially similar methods from the appropriate precursors, e.g. $CHBr_3$ or CF_2BrCOO^-. Mixed halogenocarbenes are also formed; the reactions are catalysed by silver ions. Methylcarbene and phenylcarbenes can be prepared from ethyl chloride

$$Br^{\text{'''''}}\overset{Cl}{\underset{Br}{\overset{|}{C}}}\text{—}H \xrightarrow[Ag^+]{\bar{O}tBu} \overset{Cl}{\underset{Br}{\diagdown}}C\colon + tBuOH + Ag^+ Br^-$$

and benzyl chloride, respectively. On account of the lower acidity of these halides, an alkyl lithium is required to act as base.

$$PhCH_2Cl + LiBu \longrightarrow Ph\overset{\cdot\cdot}{C}H + BuH + LiCl$$

$$CH_2CH_2Cl_2 + LiBu \longrightarrow CH_3\overset{\cdot\cdot}{C}H + BuH + LiCl$$

6.2   THE ELECTRONIC STRUCTURE OF CARBENES

Methylene is an electron-deficient species in that carbon is associated with only six electrons. Two structures may be considered according to whether the unshared electron pair occupies the same orbital or whether each electron occupies a separate orbital. In the first case the methylene is necessarily in the singlet spin state (all electrons paired) (**14**), while in the second it may be in a triplet state (two unpaired spins) (**15**). Singlet methylene may be said to use

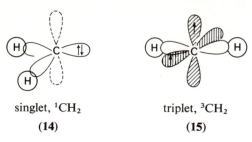

$$\text{singlet, } ^1\text{CH}_2 \qquad\qquad \text{triplet, } ^3\text{CH}_2$$
$$\textbf{(14)} \qquad\qquad\qquad \textbf{(15)}$$

sp^2 hybrid orbitals analogous to a carbonium ion, and so the bond angle should approximate to 120°. Triplet methylene, on the other hand, would use sp hybrid orbitals for bonding, the unpaired electrons each occupying a carbon-$2p$ orbital mutually at right angles. The geometry should be much nearer a linear one.

Both singlet and triplet carbenes are known. A mixture of the two states is produced when methylene is formed by photolysis of diazomethane. In the presence of an inert gas (M) the singlet species is converted to triplet by collisions. From this it appears that the triplet state is more stable than the singlet

$$^1\text{CH}_2 \xrightarrow{\ M\ } {}^3\text{CH}_2 \qquad \Delta H \approx -10 \text{ kJ mol}^{-1}$$

and is, indeed, the ground state of methylene and alkyl derivatives.

Photolysis of diazomethane in the presence of benzophenone yields only the triplet state since the decomposition occurs with conservation of spin by the interaction of triplet-excited benzophenone and diazomethane. On the other

$$^1\text{Ph}_2\text{C}{=}\text{O} \xrightarrow{\ h\nu\ } {}^3\text{Ph}_2\text{C}{=}\text{O} \xrightarrow{\ ^1\text{CH}_2\text{N}_2\ } {}^1\text{Ph}_2\text{C}{=}\text{O} + {}^3\text{CH}_2{:} + \text{N}_2$$

hand singlet methylene may be prepared by carrying out the photolysis in the presence of oxygen which scavenges the triplet state more rapidly than the singlet as it is itself a triplet molecule. The evidence for the spin state of methy-

$$^1{:}\text{CH}_2 + {}^3{:}\text{CH}_2 + {}^3\text{O}_2 \longrightarrow {}^1{:}\text{CH}_2 \ (+ \text{CH}_2\text{O etc.})$$

lene comes from observations of reactions such as this and also from direct observation of the e.s.r. spectra of methylenes (Chapter 5). Only the triplet species can show resonance absorption characteristic of two unpaired electrons ($g = 4$).

6.3   DIRECT OBSERVATION OF CARBENES

Carbenes generated in a solid inert matrix at low temperatures are long-lived and able to be examined by e.s.r. spectroscopy. The triplet nature is confirmed by the appearance of a signal in the $g = 4$ region (i.e. at half the magnetic field required for simple free radicals).* Aryl carbenes (16) give spectra which show

(16)   (17)

that one unpaired electron is delocalized over the ring while the other is localized. Quintet states (four unpaired spins) of some bis-carbenes have been observed, e.g. 17. Two rotational conformers of the naphthyl carbene (18) have been detected by this method. Visible spectra of methylene have been

(18)

measured. Since its lifetime is normally so short, the measurement is carried out by flash photolysis of diazomethane. Singlet methylene absorbs in the red–yellow region (around 590 nm) while the triplet state absorbs at shorter wavelength. Diphenylcarbene, Ph_2C:, absorbs throughout the visible region principally at 350–450, and weakly to 700 nm. The electronic spectra show rotational fine structure from which the geometries and force constants of the carbenes may be determined. These values may also be obtained from infrared spectra. For example, when carbon tetrabromide and lithium vapours are allowed to impinge on solid argon at 15 K, the tribromomethyl radical and

---

* The e.s.r. signal is the result of simultaneous inversion of *both* spins of the triplet. This requires twice the energy of a single inversion or, at a given energy of the radiation, occurs at half the field strength.

dibromocarbene are successively formed. The latter has symmetric and un-symmetric stretching vibrations at 595 and 640 cm^{-1}. From spectroscopic

$$CBr_4 + Li \xrightarrow{15K} \overset{\cdot}{C}Br_3 \xrightarrow{Li} :CBr_2 + LiBr$$
$$+ LiBr$$

$$\downarrow \text{warm}$$

$$Br_2C{=}CBr_2$$

measurements it has been shown that triplet methylene is a bent molecule (HCH angle 140–155°) rather than linear as predicted from simple theory, and the singlet species is much more bent (HCH 108–115°) with a bond length 1.03–1.12 Å. Halogeno carbenes such as :CCl$_2$ appear to be singlets in their ground states.

6.4   **REACTIONS OF CARBENES**

Proof of the existence of carbenes as transient intermediates depends usually on the isolation of stable characteristic products of their subsequent reactions; only in a few cases is direct spectrographic evidence available (Section 6.7).

Carbenes are intensely reactive and electrophilic entities, their reactivity being approximately in the order :CH$_2$ > :CR$_2$ > :CAr$_2$ > :Chal$_2$. Methylene has been described as the 'most indiscriminate reagent in organic chemistry'. Two main types of reaction are exhibited: insertion into single bonds,

$$X{-}Y + :C\overset{\diagup}{\diagdown} \longrightarrow X{-}\overset{|}{\underset{|}{C}}{-}Y$$

and addition to multiple bonds,

$$X{=}Y + :C\overset{\diagup}{\diagdown} \longrightarrow \begin{array}{c} X{-}Y \\ \diagdown \diagup \\ C \\ \diagup \end{array}$$

The products from reactions of carbenes may depend markedly on their spin state, i.e. whether singlet or triplet.

Insertion Reactions

Singlet methylene reacts with hydrocarbons indiscriminately inserting into all possible C—H bonds with virtually random distribution of products. (The

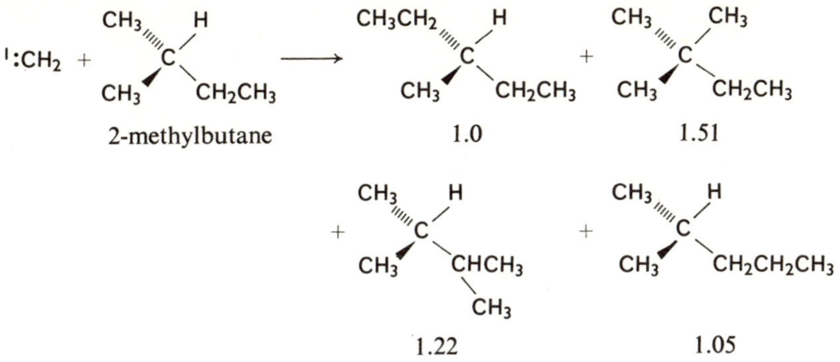

2-methylbutane   1.0   1.51

1.22   1.05

figures give the relative amounts of each product corrected statistically for the numbers of identical C—H bonds.)

There is evidently a slight preference for attack at a tertiary carbon followed by secondary, with the primary position least reactive. Two mechanisms for the insertion reaction have been considered, namely a one-step reaction via a three-membered transition state (cf. Section 2.30), or intermediate (a penta-coordinate carbon, Section 2.30),

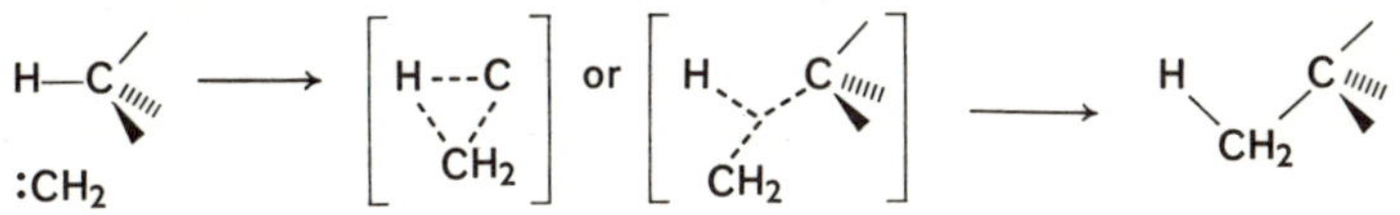

or a hydrogen-atom abstraction leading to a radical pair which subsequently combine together.

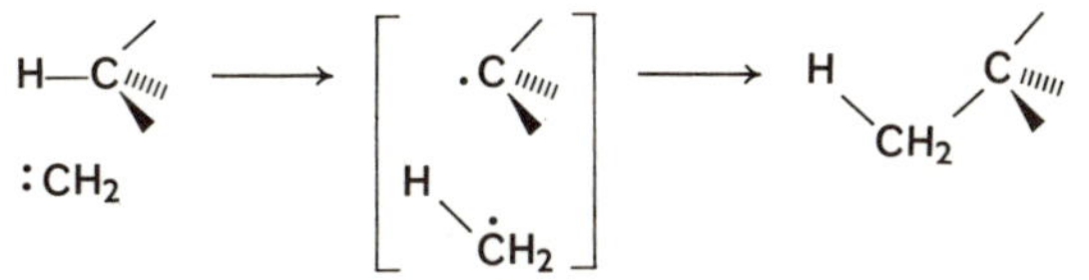

The former mechanism is supported by the observation that insertion into the allylic position of isobutene, labelled with ^{14}C at C-1, involves no isomerization of the label; a radical would be expected to undergo at least some allylic rearrangement before recombination since C-1 and C-3 would become equivalent. Also, insertion into a C—H bond at an asymmetric carbon occurs with retention of asymmetry.

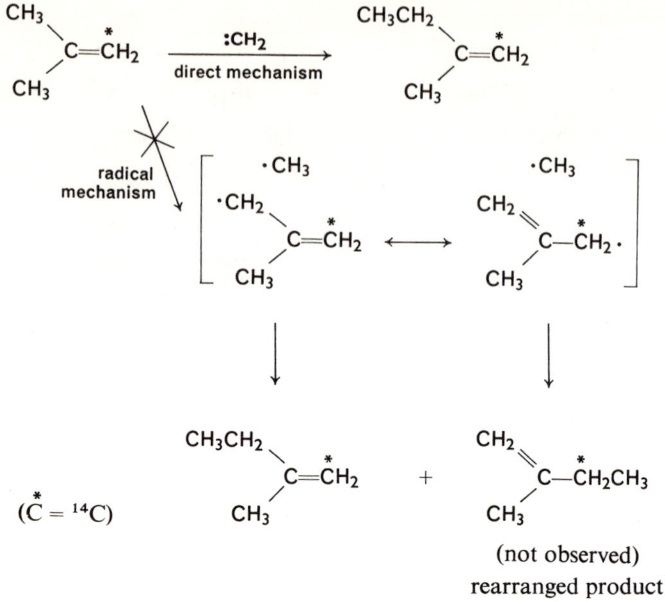

Triplet methylene is less reactive and more selective in its action than the singlet species. It attacks tertiary CH about four times faster than primary. Triplet methylene may insert by a diradical mechanism since, in the labelling experiment described above, some of the rearranged product is formed. These insertion reactions are very exothermic; ethene formed from methylene and methane, for example, is so vibrationally excited ('hot') that considerable dissociation to methyl radicals occurs before deactivation, as shown by the products from methane-d_4

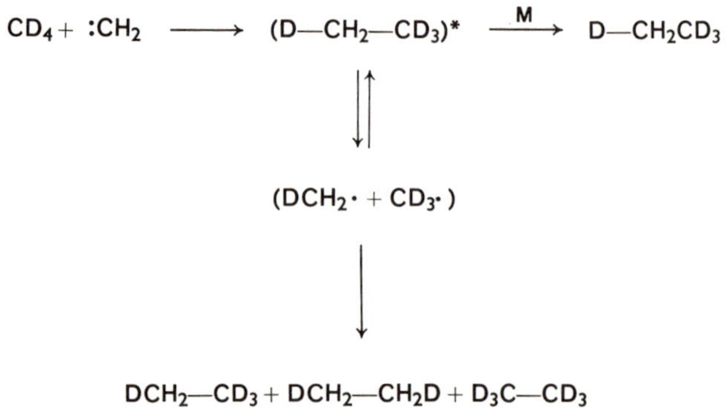

Alkylcarbenes preferentially undergo intramolecular insertion reactions, often at the β- or γ-carbons to form olefins and cyclopropanes respectively, e.g. isopropylmethylene (19) rearranges:

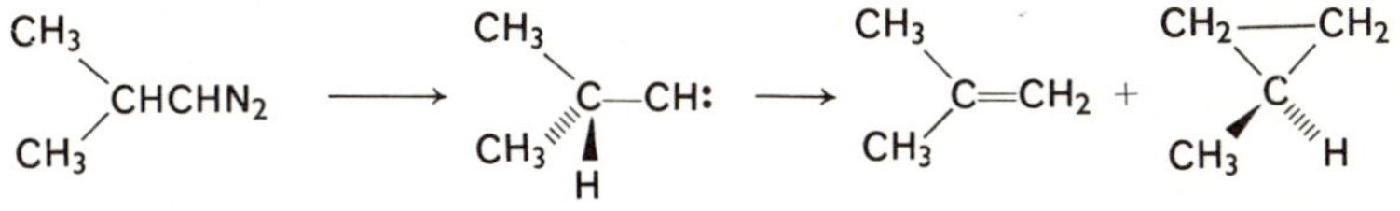

For this reason, the intermolecular reactions of alkylcarbenes are not easily studied. Highly strained molecules may be formed by these insertions. Carbonylcarbenes are much less reactive in insertion reactions and dihalocarbenes

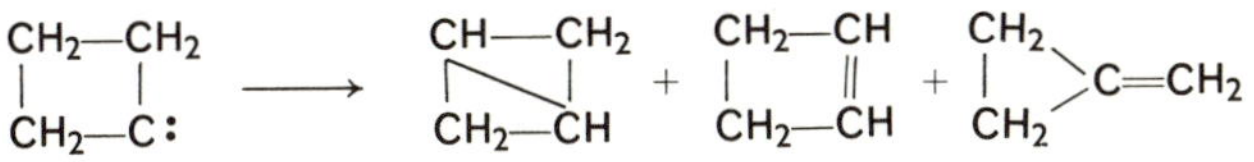

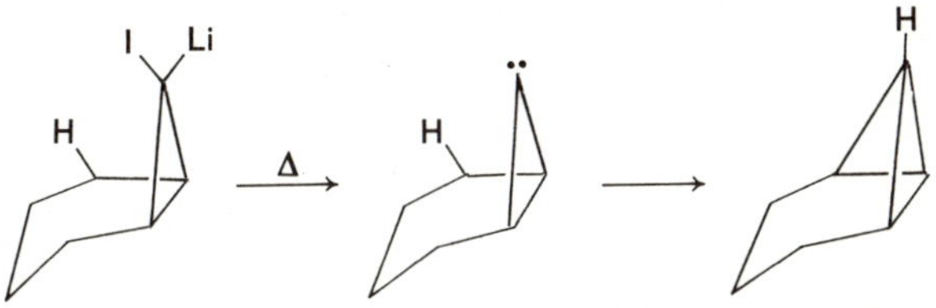

tricyclo[3,1,0,0,^{3,7}]heptane

less reactive still. Dicarboxycarbene, :C(COOEt)$_2$, for instance, exercises a preference for insertion into tertiary CH compared to primary of about 20:1. This greater selectivity may be due to a contribution to the transition state from dipolar structures, (20, 21) tertiary carbon being better able to carry positive charge than primary. Dichlorocarbene has only rarely been shown to undergo

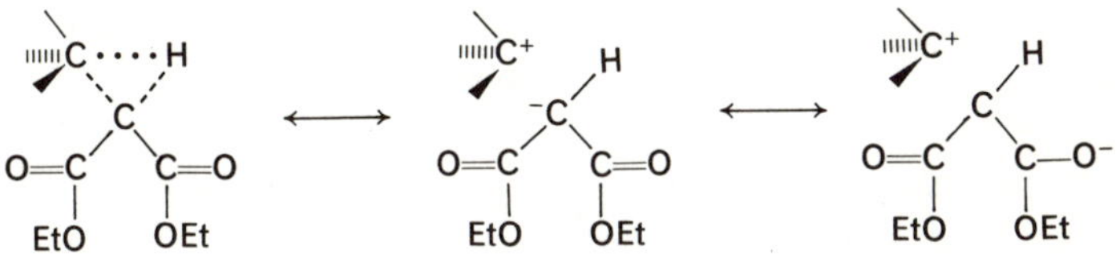

insertion but will attack benzylic CH. The reaction is accompanied by racemization at the benzylic carbon and therefore may occur by a dipolar mechanism.

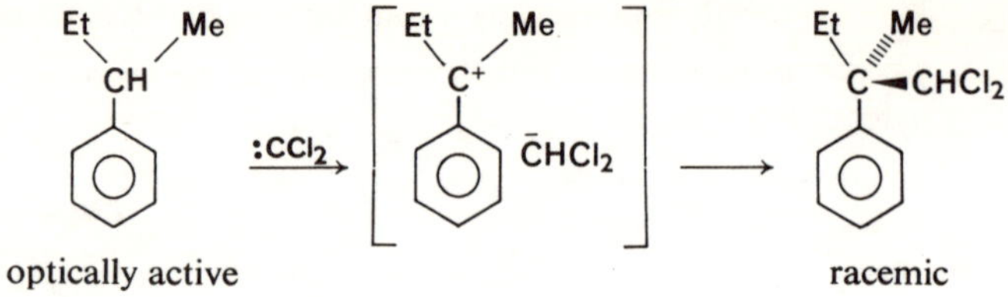

optically active   racemic

Insertions into carbon–chlorine, –bromine (but not –fluorine) and –oxygen bonds are also known to occur.

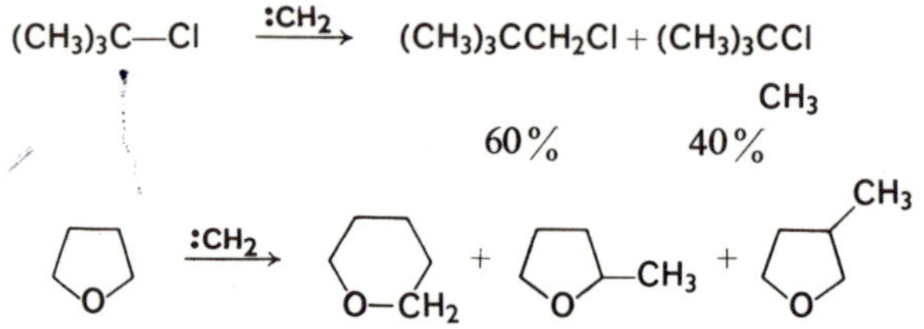

$$(CH_3)_3C—Cl \xrightarrow{:CH_2} (CH_3)_3CCH_2Cl + (CH_3)_3CCl$$

$$60\% \qquad 40\%$$

Additions to Multiple Bonds

A valuable diagnostic technique for testing the presence of carbenes in a reaction is trapping by means of an added olefin, a cyclopropane derivative being formed. The reaction with a singlet carbene is a concerted cycloaddition,

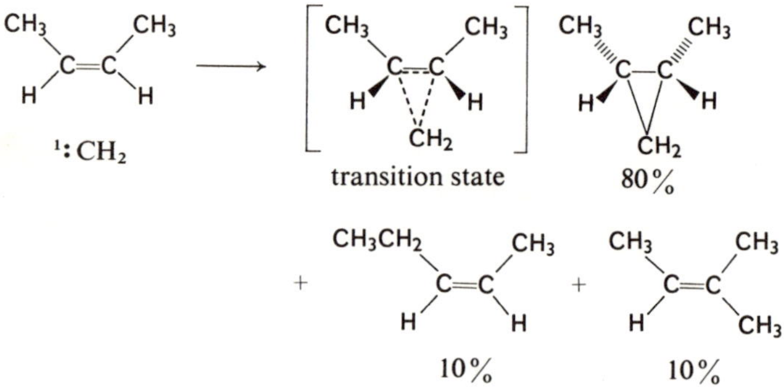

permitted by the Woodward–Hoffmann rules of orbital-symmetry conservation (Section 1.31). It is highly exothermic ($:CH_2 + CH_2{=}CH_2$, $\Delta H = -360$ kJ mol^{-1}) and stereospecific—groups *cis* to one another in the olefin appear *cis* also in the cyclopropane. In the example above, minor yields of both possible insertion products accompany the addition.

Triplet carbenes, on the other hand, add to olefins with considerable loss of stereochemical integrity. The reaction no doubt proceeds by way of a di-radical intermediate (Section 5.9) which may rotate before ring closure. Triplet methylene, for example, reacts with both *cis*- and *trans*-2-butenes to give *the same mixture* of *cis*- and *trans*-dimethylcyclopropanes, showing that rotation must be much faster than ring closure. Addition by triplet

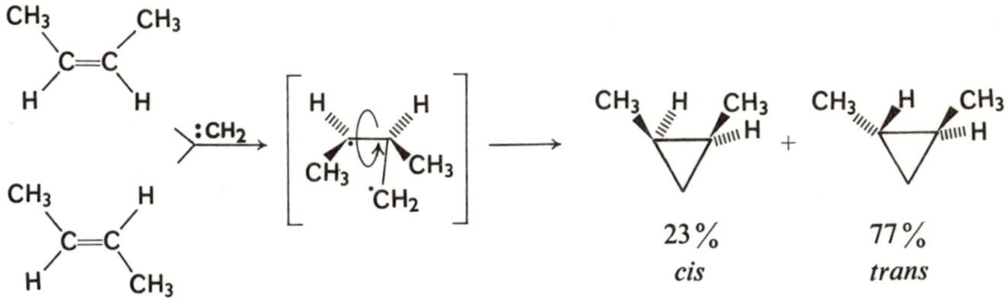

23%
cis

77%
trans

carbene takes place with virtually no insertion, as is the case with aryl and halo-

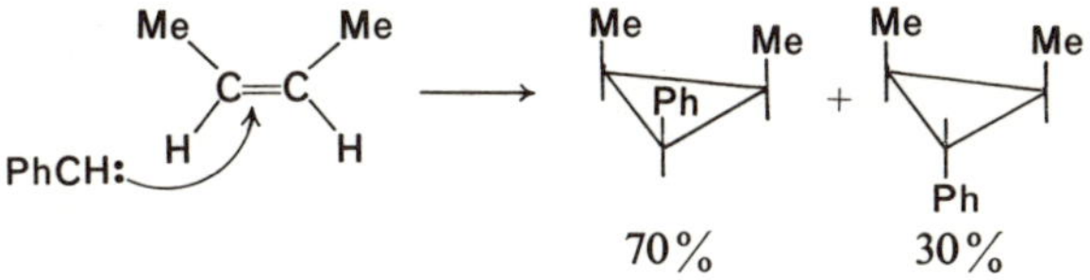

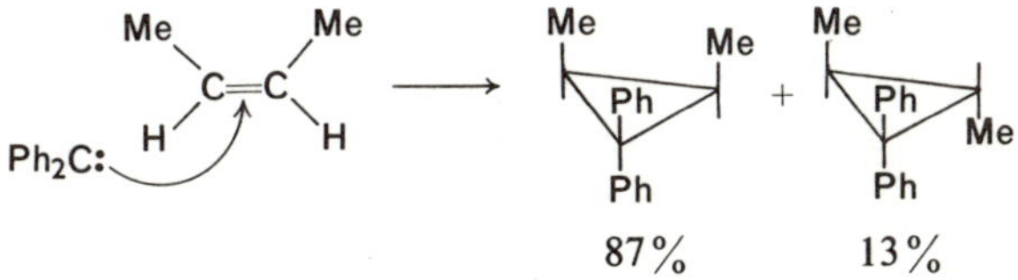

70%

30%

87%

13%

carbenes. The apparently *more* sterically hindered product forms preferenti-ally.

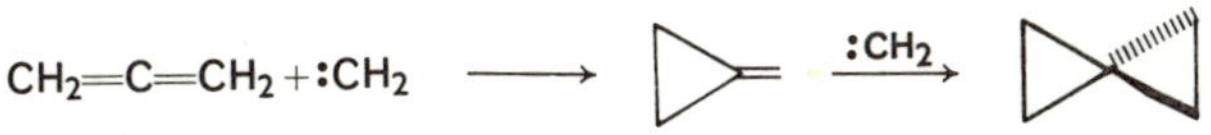

spiropentane

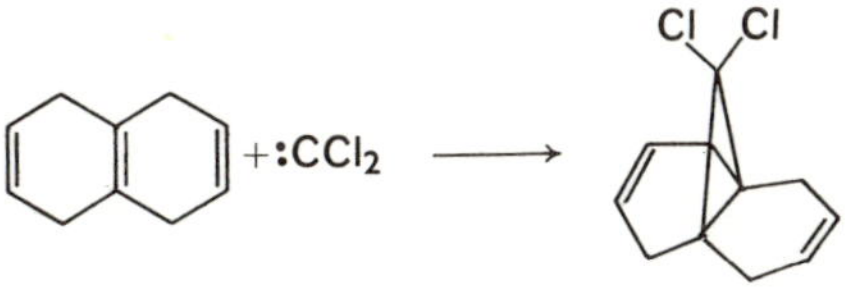

In the last example, the most highly substituted ethylenic bond is attacked. This is a general tendency since alkyl substitution tends to increase the nucleophilic character of the olefin (Table 2).

Table 2. Relative rates of addition to dichlorocarbene

Olefin	k_{rel}
Pent-1-ene	0.085
Pent-2-ene	1.0
Isobutene	5.8
2-Methyl-2-butene	14.2
2,3-Dimethyl-2-butene	52.0

The electrophilic nature of the carbene is shown by the rates of addition to substituted styrenes,

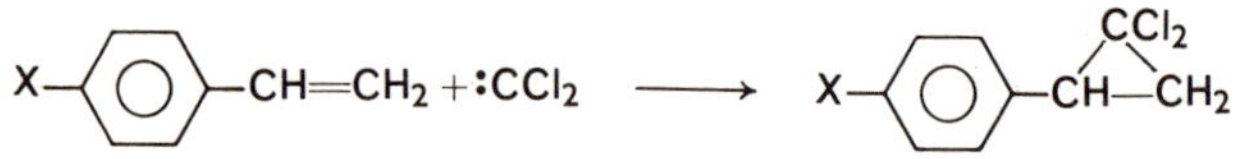

$$X-\!\!\bigcirc\!\!-CH\!\!=\!\!CH_2 + :CCl_2 \longrightarrow X-\!\!\bigcirc\!\!-\overset{\displaystyle CCl_2}{\underset{}{CH-CH_2}}$$

for which $\log k \propto \sigma^+$ (rather than the Hammet σ) and $\rho = -0.6$. Clearly, electron release facilitates the reaction and positive charge resides upon the benzylic carbon in the transition state.

The more reactive carbenes will even attack the aromatic ring, the primary bicyclic product undergoing valence tautomerism to a cycloheptatriene derivative, a reaction discovered by Bunsen in the Nineteenth century.

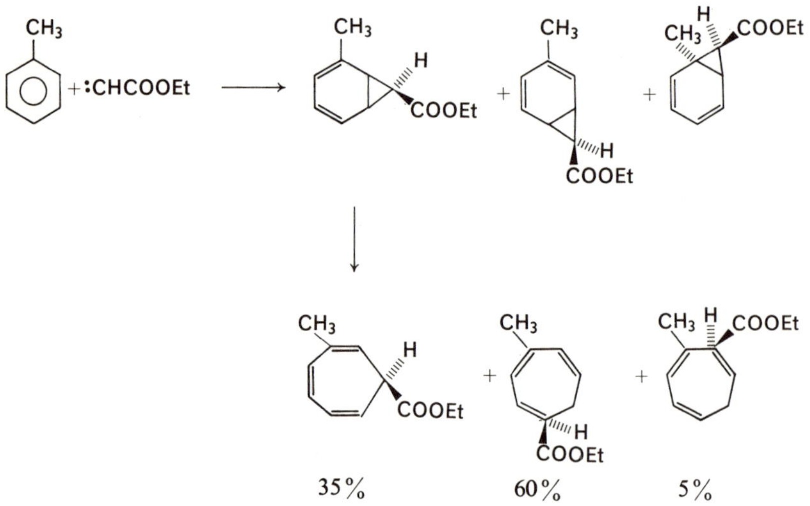

35%   60%   5%

Less reactive carbenes require the presence of nucleophilic substituents in the ring,

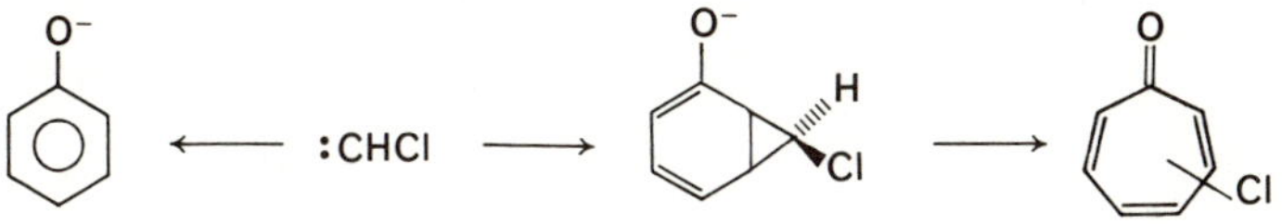

Acetylenes are attacked, but less readily than olefins; the reaction provides a route to cyclopropenones (22).

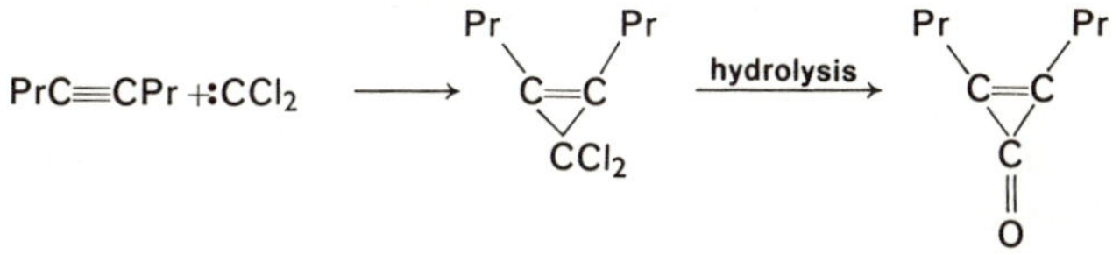

(22)

Dimerization

In principle, dimerization leads to symmetrical olefin formation.

$$2:CH_2 \longrightarrow CH_2=CH_2$$

In practice, the concentration of carbene is likely to be so low that a bimolecular reaction is improbable and 'dimeric' products probably arise from attack of the carbene on a precursor, thus:

$$:CH_2 + CH_2N_2 \longrightarrow CH_2=CH_2 + N_2$$

However, the formation of tetrabromoethylene from $:CBr_2$ on a solid argon matrix may be an example of a genuine dimerization.

Addition of CO and O_2

Carbon monoxide adds methylene to form a vibrationally excited ketene molecule which may re-dissociate or become collisionally deactivated, the reverse of the generation of methylene from ketene.

$$:CH_2 + CO \longrightarrow (CH_2=C=O)^* \xrightarrow{M} CH_2=C=O$$

Oxygen likewise yields a peroxide with triplet carbene.

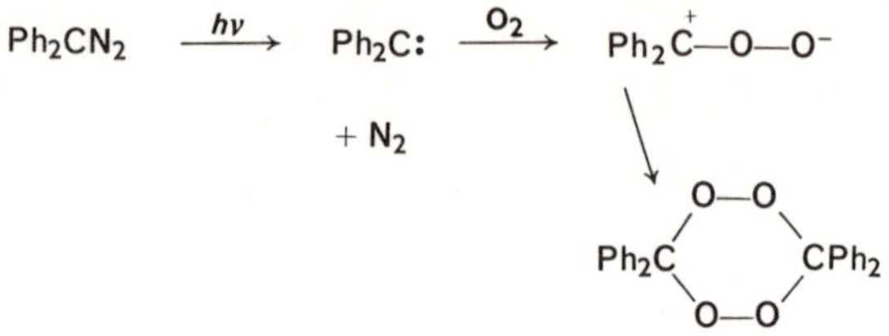

$$Ph_2CN_2 \xrightarrow{h\nu} Ph_2C: \xrightarrow{O_2} Ph_2\overset{+}{C}\text{—}O\text{—}O^-$$

$$+ N_2$$

Molecular Rearrangements

Rearrangements may occur by migration of an alkyl or aryl group or hydro-gen, a reaction which has already been mentioned as an intramolecular α-insertion. The order of migratory aptitude is H > aryl > alkyl, distinctly

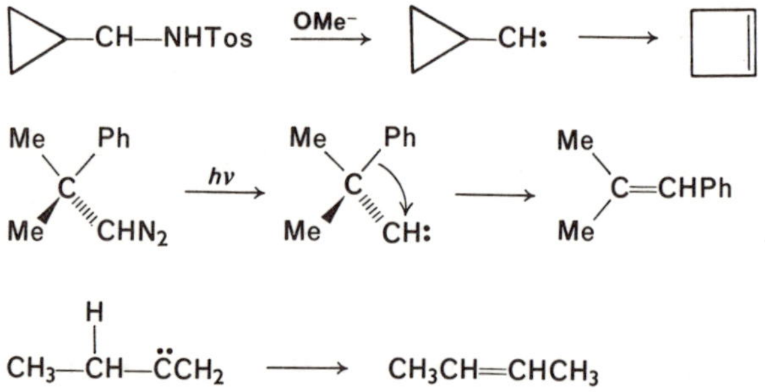

different to that observed in carbonium-ion rearrangements (Section 2.22 and 2.29) in which aryl > alkyl > H.

Ketocarbenes rearrange by a 1,2-alkyl (aryl) shift to form ketenes (Wolff rearrangement); the ketocarbenes may be prepared by the decomposition (thermal, photochemical or catalytic) of a diazoketone. The migrating group may be shown to retain its stereochemistry and the ketocarbene may sometimes be trapped, for example as an oxazole, in the presence of a cyanide (a 1,3-dipolar cycloaddition). Under normal reaction conditions, the ketene is decomposed by solvent to a carboxylic-acid derivative.

The yield of oxazole in the above example is low (0.4 %) but increases to 16 % in the presence of cuprous chloride. This suggests that the metal-catalysed decomposition goes much more via a free carbene than the thermal reaction in which the rearrangement may be at least partially concerted with nitrogen loss, as indicated by a carbon-isotope effect, $k_{12}/k_{14} = 1.015$. The photochemical

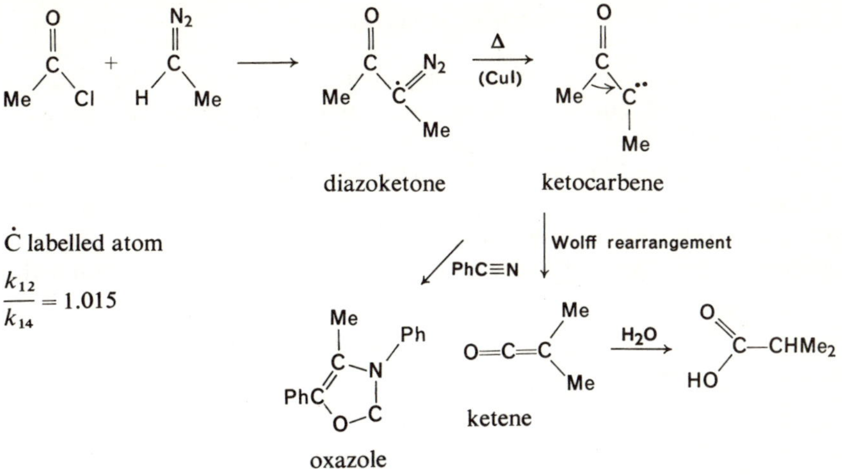

diazoketone   ketocarbene

$\dot{C}$ labelled atom

$\dfrac{k_{12}}{k_{14}} = 1.015$

PhC≡N   Wolff rearrangement

oxazole   ketene

Wolff rearrangement appears to proceed via an intermediate *oxirene* since in this reaction, carbon scrambling occurs,

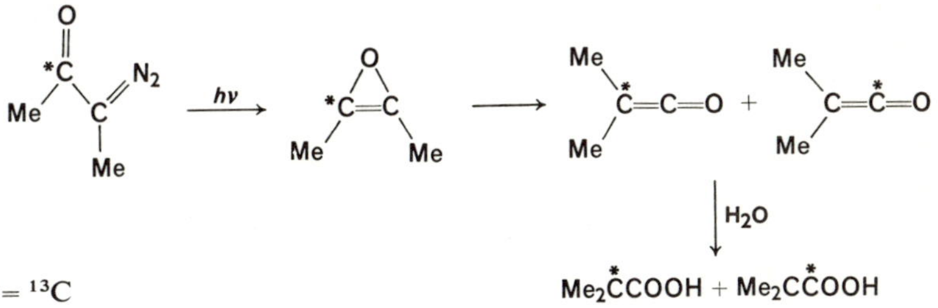

$^*C = {}^{13}C$

$Me_2\overset{*}{C}COOH + Me_2C\overset{*}{C}OOH$

This reaction is used in the Arndt–Eistert method for ascending a homologous series of carboxylic acids.

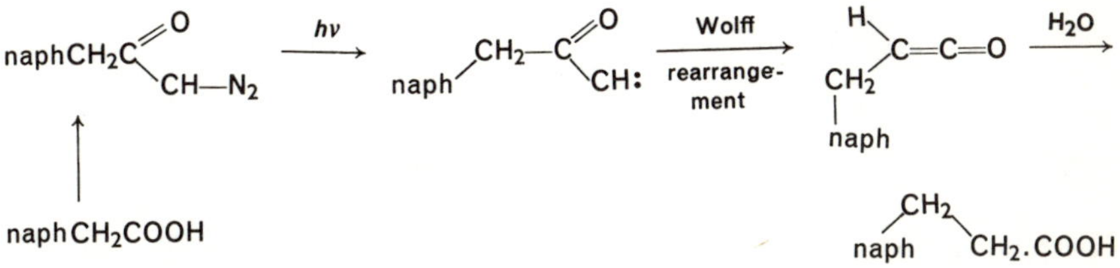

naph = 1-naphthyl

Addition of Nucleophiles

Carbenes, being electron deficient, add a variety of nucleophiles. The hydrolysis of chloroform (Section 6.1) provides two examples, the reversible addition of chloride ion to dichlorocarbene and the addition of water or hydroxide ion leading to hydrolysis products. For these reasons, adducts of carbenes with olefins are obtained in best yield in a non-nucleophilic solvent. Diarylcarbenes however can behave as nucleophiles and will coordinate a proton to form a carbonium ion. This benzhydryl cation was found to coordinate with added

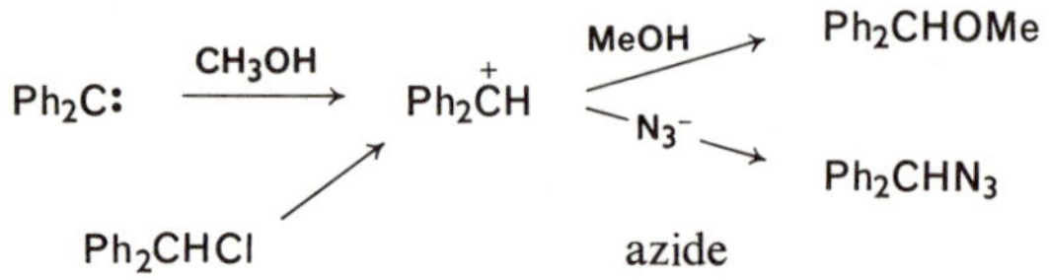

nucleophiles, e.g. azide ion, at the same relative rates as the species formed by the ionization of benzhydryl chloride, which was thereby inferred to be the same carbonium ion. The well-known Reimer–Tiemann reaction for the preparation of salicylaldehyde probably involves attack of dichlorocarbene on a phenolate ion.

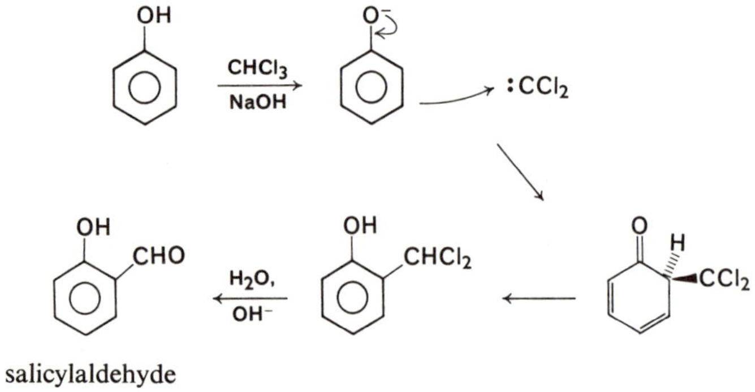

6.5   **ATOMIC CARBON**

Carbon vapour sublimed at a high temperature from a carbon filament contains the species C_1, C_2 and C_3, which behave as dicarbenes. These molecules may be trapped on a solid paraffin surface on which they are inert, and allowed to react with olefins to form bis-cyclopropanes.

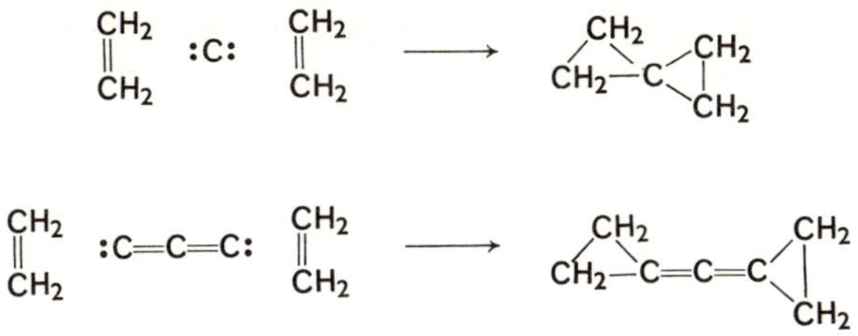

$$\underset{CH_2}{\overset{CH_2}{\|}} \quad :C: \quad \underset{CH_2}{\overset{CH_2}{\|}} \quad \longrightarrow \quad \text{cyclopropane structure}$$

$$\underset{CH_2}{\overset{CH_2}{\|}} \quad :C{=}C{=}C: \quad \underset{CH_2}{\overset{CH_2}{\|}} \quad \longrightarrow \quad \text{allene structure}$$

Oxygen abstraction also occurs, e.g. from epoxides.

$$CH_3\overset{\displaystyle}{CH}\!\!-\!\!\overset{\displaystyle}{CH_2} + :C: \quad \longrightarrow \quad CH_3CH{=}CH_2 + CO$$

Reactions of C_1 in the ^3P (ground) state appear to be stereospecific for the first cycloaddition (to give a triplet carbene) but non-stereospecific for the second.

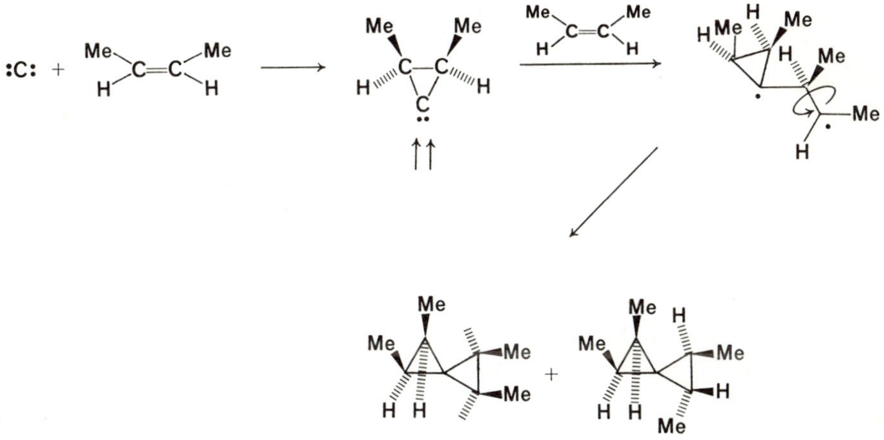

6.6   CARBENOIDS

The reactions discussed above are generally believed to proceed via free carbenes. There are, however, a number of other reactions with similar characteristics—many of them preparatively important—in which the actual reagent has a masked carbene function, often a metal derivative. These reagents are called carbenoids.

The Simmons–Smith reaction converts an olefin stereospecifically to a cyclopropane apparently by the addition of $:CH_2$ generated by the action of methylene diiodide and the zinc–copper couple. Unlike additions of methylene, CH bond insertion does not occur and the reagent is probably iodomethyl

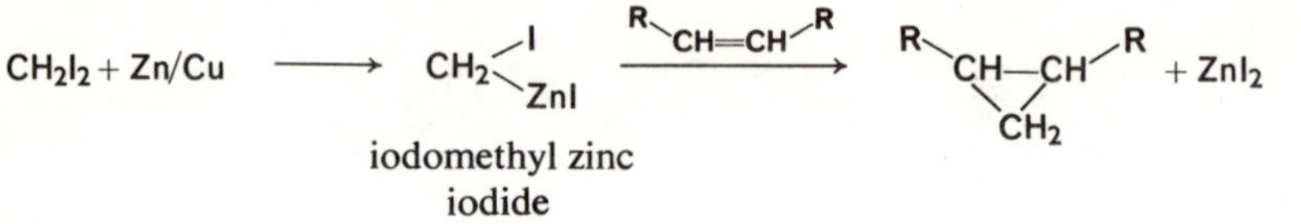

iodomethyl zinc
iodide

zinc iodide. The decomposition of diazo compounds is catalysed by copper metal or salts which are believed to give rise to carbene–copper complexes of lower reactivity than the free carbene and which will undergo addition but not

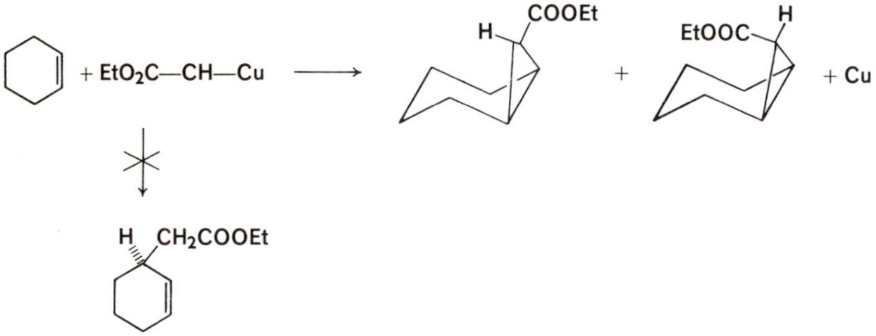

insertion reaction. In the presence of copper, ketocarbenes fail to rearrange to ketenes, a useful modification of this function.

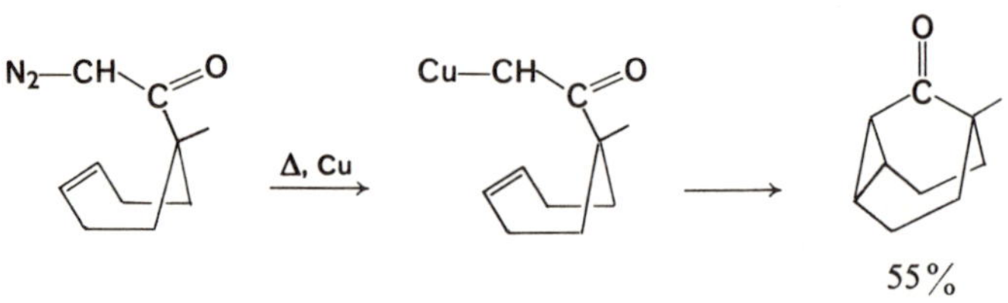

55%

The reactivity of phenylcarbene by photolysis of the appropriate azide and by α-elimination of benzyl bromide by lithium methyl, differs by a factor of two; in the latter case, the reagent is most likely to be a carbenoid.

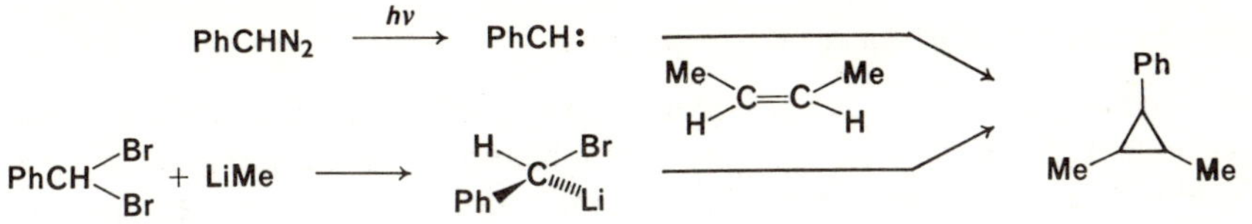

$$PhCHN_2 \xrightarrow{h\nu} PhCH:$$

$$PhCH\begin{smallmatrix}Br\\Br\end{smallmatrix} + LiMe \longrightarrow \underset{Ph}{\overset{H}{\diagdown}}C\begin{smallmatrix}Br\\\\Li\end{smallmatrix}$$

A number of organometallic complexes of carbenes have now been isolated, for example,

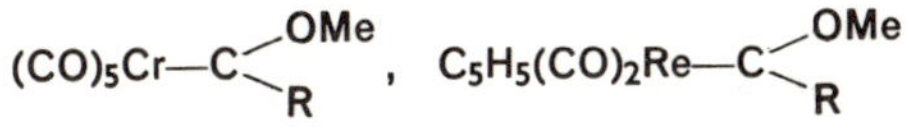

$$(CO)_5Cr\!-\!C\begin{smallmatrix}OMe\\\\R\end{smallmatrix} , \quad C_5H_5(CO)_2Re\!-\!C\begin{smallmatrix}OMe\\\\R\end{smallmatrix}$$

and

$$C_5H_5(CO)_2Fe\!-\!C\begin{smallmatrix}OH\\\\Me\end{smallmatrix}.$$

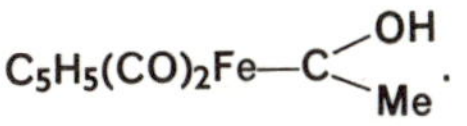

6.7   NITRENES

Organic azides are unstable substances, frequently explosive. The arylsulphonyl azides, e.g. 23, are among the more tractable examples but will decompose on heating with the loss of nitrogen. If the decomposition of 23 occurs in benzene, catalysed by iron carbonyl, a good yield of benzenesulphonanilide (24) is obtained. The following mechanism is proposed.

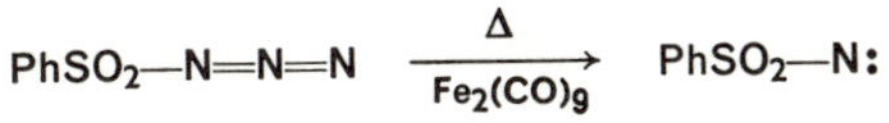

$$PhSO_2\!-\!N\!=\!N\!=\!N \xrightarrow[Fe_2(CO)_9]{\Delta} PhSO_2\!-\!N:$$

(23)   (24)

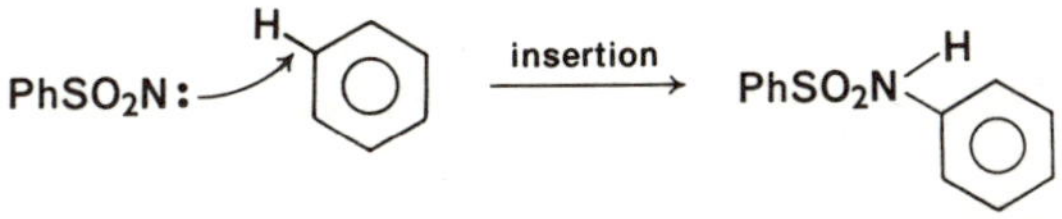

$$PhSO_2N: \xrightarrow{insertion} PhSO_2N\begin{smallmatrix}H\\\\\end{smallmatrix}$$

benzenesulphonanilide

The reactive intermediate is benzenesulphonylnitrene (24), a nitrogen analogue of a carbene, which undergoes an insertion reaction into the benzene C—H bond.

Nitrenes are derivatives of the molecule :NH and, like carbenes, are intensely reactive. Their methods of generation resemble those for carbene production and will be briefly summarized here.

Thermal and Photochemical Decomposition

A wide variety of azides have been decomposed in order to yield nitrenes. Usually a triplet sensitizer such as benzophenone is added, particularly if the azides do not absorb light in a convenient region (e.g. 254 nm mercury resonance light). The thermal decompositions of aryl azides show a large positive

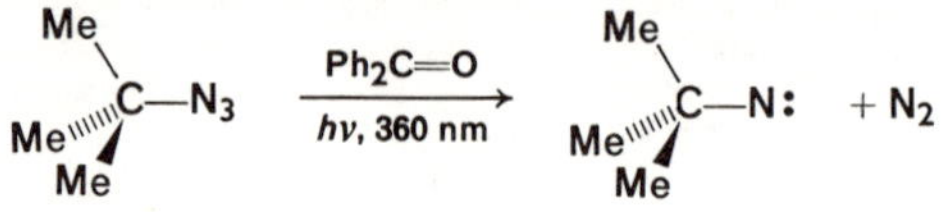

t-butylazide

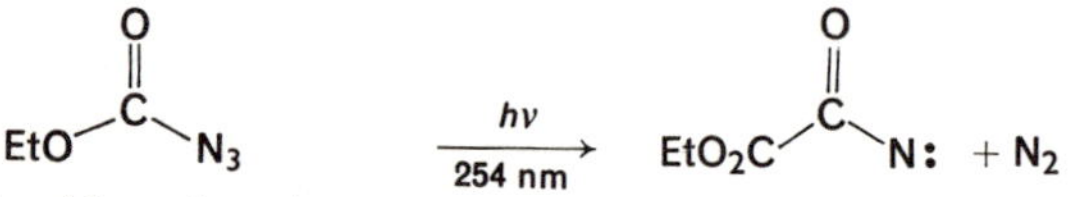

ethyl azidocarboxylate

entropy of activation (e.g. $PhN_3 \rightarrow PhN$; $\Delta G^{\neq} = 160$ kJ mol^{-1}, $\Delta S^{\neq} = +80$ kJ mol^{-1}) which is consistent with a unimolecular fragmentation in the slow step. Isocyanates (**25**) and oxaziridines (**26**) can also be photolysed to nitrenes, analogous to carbene production from ketenes and epoxides, respectively.

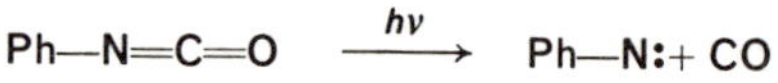

$$Ph-N=C=O \xrightarrow{h\nu} Ph-N: + CO$$
$$(25)$$

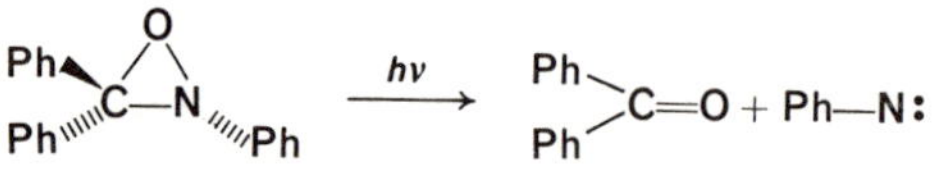

$$(26)$$

α-Elimination

These are relatively rare—one probable example is the following:

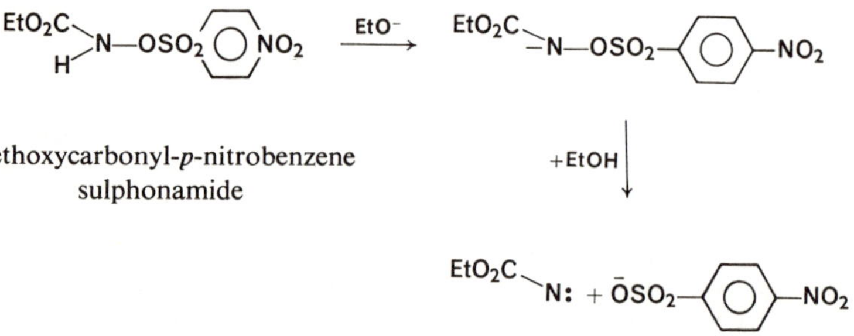

N-ethoxycarbonyl-*p*-nitrobenzene sulphonamide

+EtOH

Oxidation of Primary Amino Groups

The removal of two hydrogen atoms (or H$^-$ and H$^+$) from an –NH$_2$ group can potentially form a nitrene. It is likely that an aminonitrene (**27**) is formed in the lead (IV) acetate oxidation of the hydrazine derivative (**28**) since a species which adds to an olefin is formed.

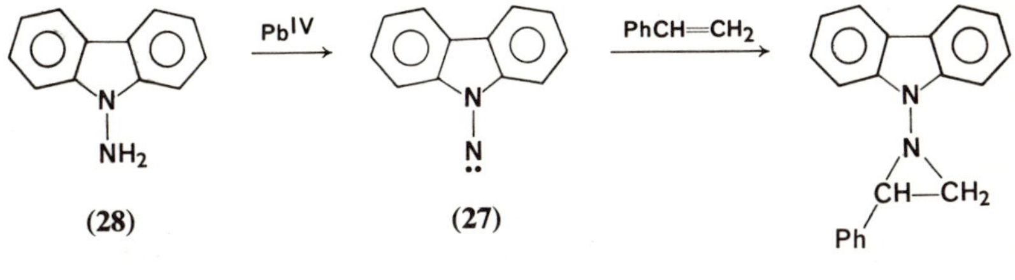

Deoxygenation of Nitrocompounds

Heating a nitrocompound with triethyl phosphite brings about a transfer of the oxygen to phosphorus (forming the phosphate) and may well involve a nitrene since the products of reaction are much the same as those from decomposition of the corresponding azide.

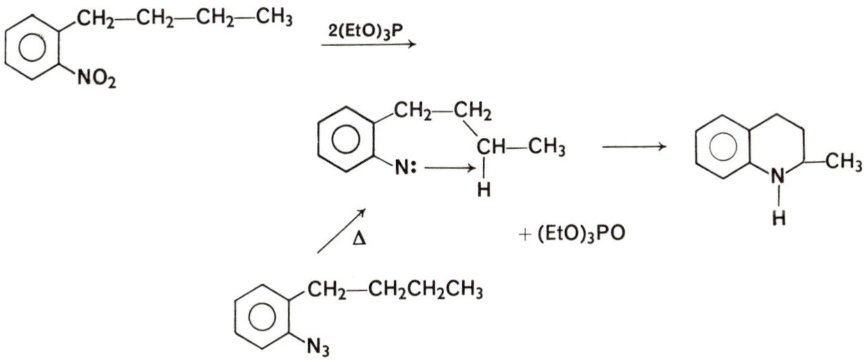

Reactions of Nitrenes

Additions to multiple bonds and insertion reactions typify the further reactions of nitrenes.

Additions to olefins occur with the formation of aziridines, e.g. **28**. When the nitrene is generated from an azide there may be an ambiguity in the mechanism since it is known that azides can add to olefins thermally to form triazolines

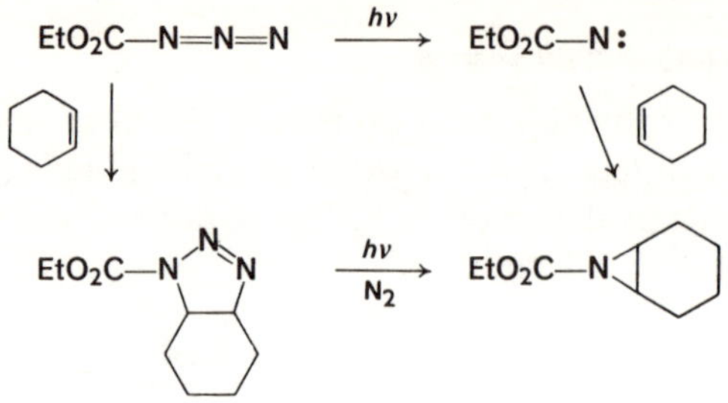

$$EtO_2C\!-\!N\!=\!N\!=\!N \xrightarrow{\ h\nu\ } EtO_2C\!-\!N\!:$$

$$EtO_2C\!-\!N \xrightarrow[N_2]{\ h\nu\ } EtO_2C\!-\!N$$

(29)

(29) and these decompose photochemically to aziridines, expelling nitrogen. However, addition reactions to olefins occur when the nitrene is formed from a precursor other than an azide. The addition appears to be stereospecific with singlet nitrene but non-specific via a diradical when a triplet nitrene adds. Since the latter is the ground state it is formed in increasing yield as the lifetime of the nitrene increases, hence the non-specific addition is more prominent in

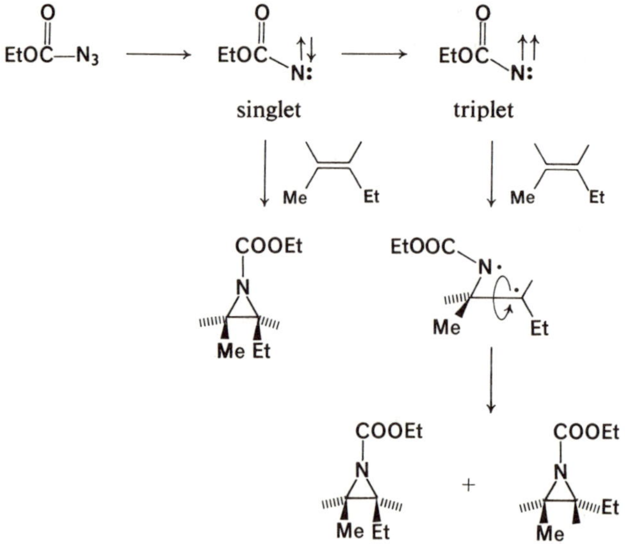

$$EtO\overset{O}{\overset{\|}{C}}\!-\!N_3 \longrightarrow EtO\overset{O}{\overset{\|}{C}}{\underset{N:}{\nwarrow}}\uparrow\!\downarrow \longrightarrow EtO\overset{O}{\overset{\|}{C}}{\underset{N:}{\nwarrow}}\uparrow\uparrow$$

singlet   triplet

dilute solution. Aromatic hydrocarbons add, the products being azepins (30).

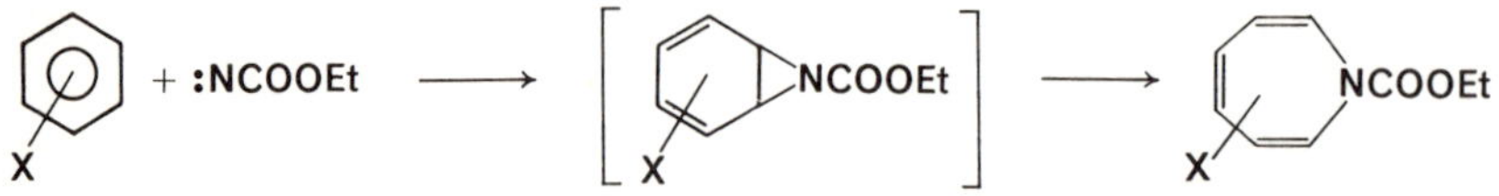

$$\underset{X}{\bigcirc} + :NCOOEt \longrightarrow \left[\underset{X}{\bigcirc}\!\!NCOOEt \right] \longrightarrow \underset{X}{\bigcirc}\!\!NCOOEt$$

(30)

The ρ-value for this reaction (-1.32) is greater than that for the corresponding carbene addition (-0.38), showing the nitrene to be more electrophilic than the carbene. Intramolecular C—H bond insertion will also occur; an α-insertion will lead to the formation of an imide which hydrolyses to give a carbonyl compound in addition to insertion at the δ-carbon (cyclization):

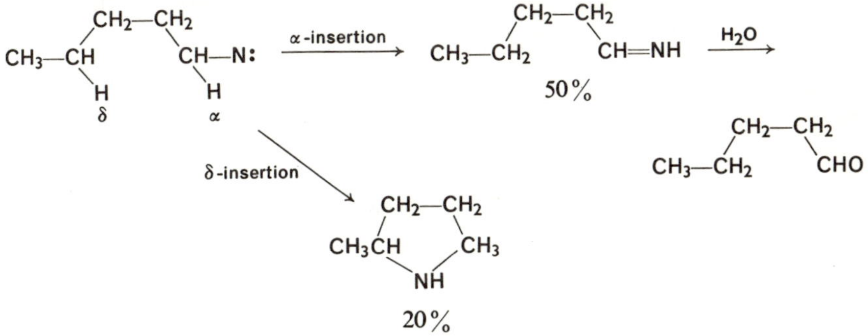

The ability to insert into paraffinic C—H bonds has a practical application in crosslinking polypropylene by the addition and decomposition of a bis-

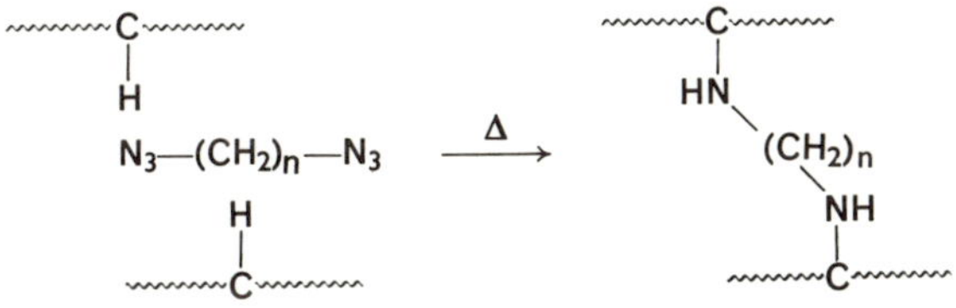

azide. α-Insertion into a C—Ph bond can occur—effectively the migration of phenyl from carbon to nitrogen.

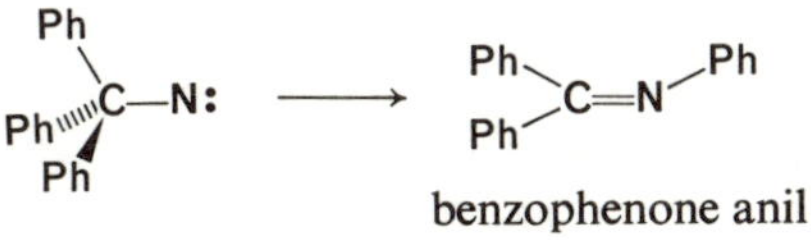

benzophenone anil

Nitrenes dimerize to azo compounds; arylnitrenes sometimes have a sufficient lifetime for the preparation of mixed azo compounds. There remains the

$$Ph—N_3 \atop An—N_3 \} \xrightarrow{\Delta} \quad {PhN: \atop + \atop AnN:} \quad \longrightarrow \quad \left\{ {Ph—N=NPh \atop {+ \atop {Ph—N=N—An \atop {+ \atop An—N=N—An}}}} \right.$$

$$An= \bigcirc\!\!\!-\!OMe \text{ or } PhN:+ PhN_3 \quad \longrightarrow \quad Ph—N=N—Ph + N_2$$

possibility that this reaction occurs by attack of the nitrene on the azo compound, which is difficult to disprove. Carbonyl and sulphonyl nitrenes behave

$$Ph—N{\overset{+}{\underset{\underset{Ph}{\overset{|}{N}}}{\overset{N}{=}}}}\ddot{N} \longrightarrow \left[Ph—N{\underset{\overset{|}{\underset{Ph}{+N}}}{\overset{N=\ddot{N}}{}}} \right] \longrightarrow Ph—N{\underset{\underset{Ph}{\overset{|}{N}}}{\overset{N\equiv N}{\overset{+}{}}}}$$

as 1,3-dipolar molecules (29, 30) and will add to, for example, cyanides at nitrogen and oxygen, to form interesting heterocyclic systems.

$$\underset{Ph}{\overset{N}{\underset{\|}{\overset{\|\|}{C}}}} + \left(\overset{:N}{\underset{O}{\diagdown}}\!\!\!{C—OEt} \longleftrightarrow \overset{+N}{\underset{-O}{\diagdown}}\!\!\!{C—OEt} \right) \longrightarrow \underset{Ph}{\overset{N—N}{\underset{O}{\diagup C \diagdown C}}}\!\!—OEt$$

3-ethoxy-5-phenyl
oxadiazole

The Direct Detection of Nitrenes

Many types of nitrene can be generated and rendered stable if a solution of the appropriate azide is irradiated at very low temperature in the clear glassy solution obtained by cooling, for example, methyltetrahydrofuran; owing to the rigidity of this matrix, the nitrene is unable to reach a reactive site and under these conditions may be detected spectroscopically. The observation of e.s.r. spectra confirms the triplet character of the nitrenes and allows structural information to be obtained, such as the geometry and the interaction between

$$R—\overset{+}{N}=N=N^- \xrightarrow[Ph_2C=O, \, h\nu]{20-100 \text{ K, } \diagup\!\!\!\!\bigcirc\!\!\!\diagdown O —Me} R—N: + N_2$$

the unpaired electrons. Alkylnitrenes have 92–95% of the unpaired electrons resident on nitrogen whereas arylnitrenes show that one electron is delocalized on the ring, similar to the situation with arylcarbenes.

These studies show that aryl and sulphonylnitrenes are more stable than alkyl and acyl analogues.

6.8 THE HOFMANN–CURTIUS–LOSSEN–SCHMIDT GROUP OF REARRANGEMENTS

This group of reactions is important preparatively for the conversion of a carbonyl function to primary amino with loss of the carbonyl carbon. They all

$$R—C \underset{NXY}{\overset{O}{<}} \longrightarrow R—NH_2$$

involve the loss of a group or groups from the nitrogen under circumstances which might yield a nitrene; this rearranges by migration of the alkyl group, R, onto nitrogen. The result is an isocyanate which hydrolyses in water to the amine and CO_2. This is the nitrogen analogue of the acylcarbene rearrangement.

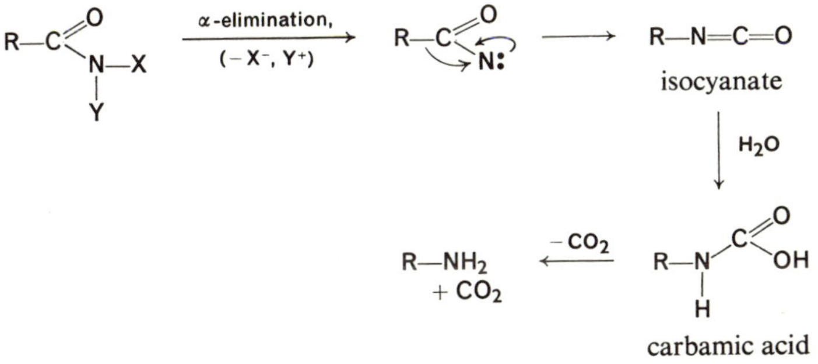

Hofmann Rearrangement

A bromoamide (31) made by treating an amide with bromine under mildly alkaline conditions, is treated with strong base to bring about an α-elimination:

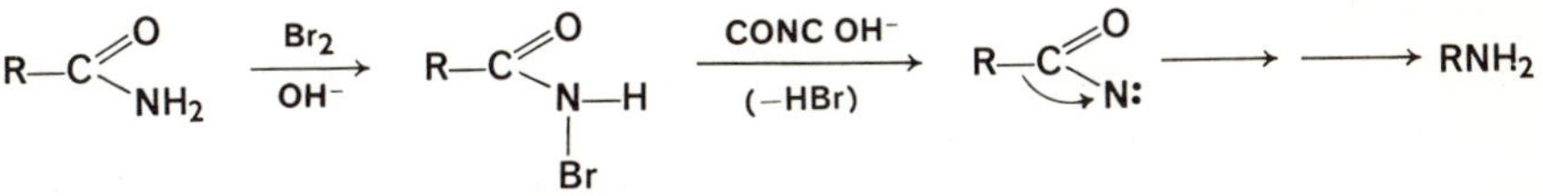

Lossen Rearrangement

An acylhydroxamic acid (**32**) is subjected to strong base with similar results,

$$R-C\overset{O}{\underset{\underset{OCOCH_3}{|}}{\overset{}{N-H}}}\quad\overset{OH^-}{\longrightarrow}\quad R-C\overset{O}{\underset{N:}{}}\quad\longrightarrow\quad\longrightarrow\quad R-NH_2$$

(32) $+\ \bar{O}COCH_3 + BH^+$

Schmidt Reaction

The carboxylic acid is treated with hydrazoic acid in strongly acidic solution at temperatures between 70 and 100°. An acyl azide (**33**) forms *in situ* and decomposes thermally.

$$R-C\overset{O}{\underset{OH}{}}\quad\overset{H^+}{\underset{\rightleftharpoons}{}}$$

$$R-C\overset{OH}{\underset{OH}{}}{+}\quad\overset{HN_3}{\longrightarrow}\quad R-\overset{OH}{\underset{OH}{\overset{|}{C}}}N=\overset{+}{N}-NH\quad\overset{-H_3O^+}{\longrightarrow}\quad R-\overset{O}{\overset{||}{C}}N=\overset{+}{N}=\bar{N}$$

(33)

Δ $\downarrow$

$$RNH_2\quad\longleftarrow\quad\longleftarrow\quad R-\overset{O}{\overset{||}{C}}N:\quad +\ N_2$$

Curtius Reaction

The acyl azide is prepared by reaction of sodium azide with an acyl halide, or diazotization of an acyl hydrazide, and is isolated and decomposed with results as for the Schmidt reaction. Both the Schmidt and Curtius reactions may be carried out in an aqueous medium in which case the amine is obtained, or in non-aqueous medium when the isocyanate may be isolated.

In all this group of reactions, whether or not there is a nitrene intermediate depends upon the timing of the alkyl migration. If this occurs concurrently with the loss of the groups attached to nitrogen there will be no discrete nitrene formed, whereas if migration occurs subsequent to the elimination there must be an intermediate nitrene.

Two-step mechanism—as in above examples:
Concerted mechanism,

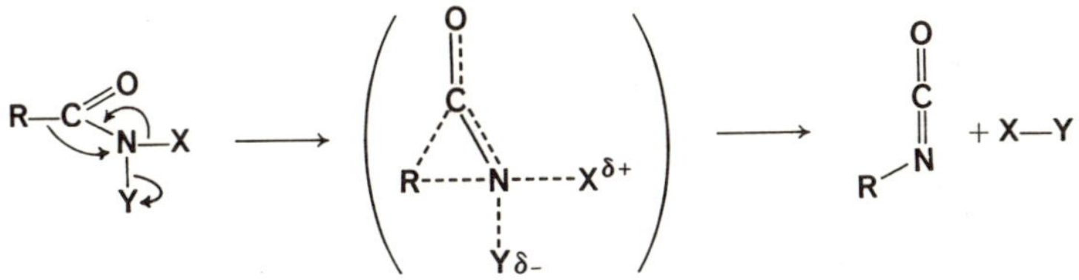

A distinction may be made by the carbon-isotope effect of the migrating alkyl group. The isotope effect for the Curtius, Lossen and Hofmann reactions of benzazide (34), X = $-N_2$, potassium benzoyl benzohydroxamate (35) X = $-$OCOPh and benzamide (36), X = H, with carbon labelled (^{14}C) at the point of attachment of the ring, have been measured,

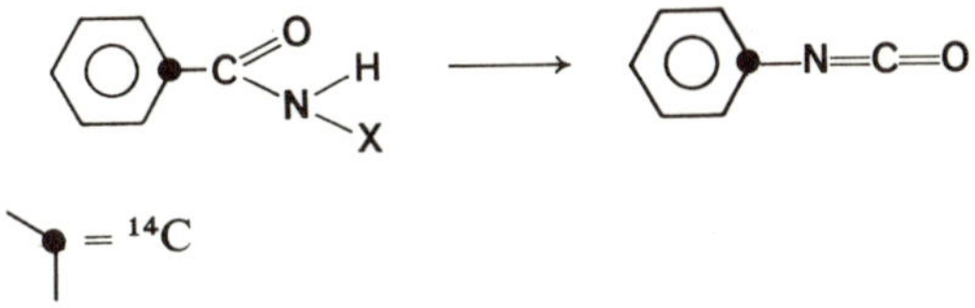

	X	k_{12}/k_{14}
(34)	$-N_2$	1.042
(35)	$-$OCOPh	1.044
(36)	$-$H	1.032

In all cases there is a substantial carbon primary-isotope effect, meaning that the C—C bond must be in the process of breaking in the slow step. The concerted mechanism must be correct and discrete nitrenes are not intermediates here.

6.9   THE CARBYLAMINE REACTION

Evil-smelling isocyanides (carbylamines) result from the reaction of primary amines with chloroform and alkali and have been used as a test for a primary amine. The reaction probably involves the carbene:

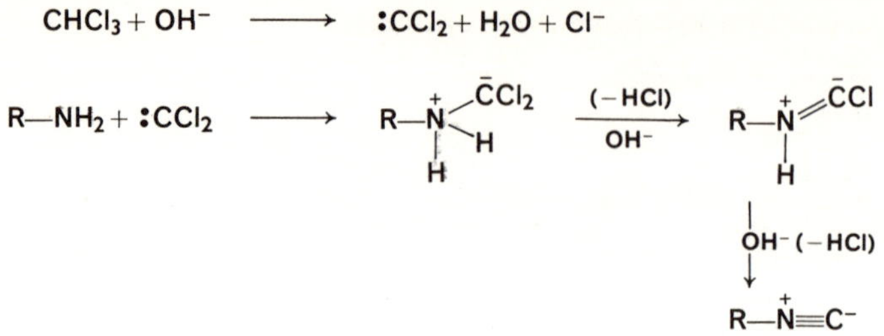

$$CHCl_3 + OH^- \longrightarrow\ :CCl_2 + H_2O + Cl^-$$

$$R—NH_2 + :CCl_2 \longrightarrow\ R—\overset{+}{N}\underset{\underset{H}{|}}{\overset{\overset{-}{C}Cl_2}{\diagdown}}{}_H \xrightarrow[OH^-]{(-HCl)} R—\overset{+}{\underset{H}{N}}{\diagup}^{\bar{C}Cl}$$

$$\downarrow OH^- (-HCl)$$

$$R—\overset{+}{N}\!\!\equiv\!\!C^-$$

6.10   CARBYNES

Photolysis of mercury bis-ethyl diazomalonate yields a species which exhibits an e.s.r. spectrum due to one unpaired electron and behaves chemically as carbethyoxycarbyne, the first member of a class of reactive intermediates of formula R—$\dot{C}$:

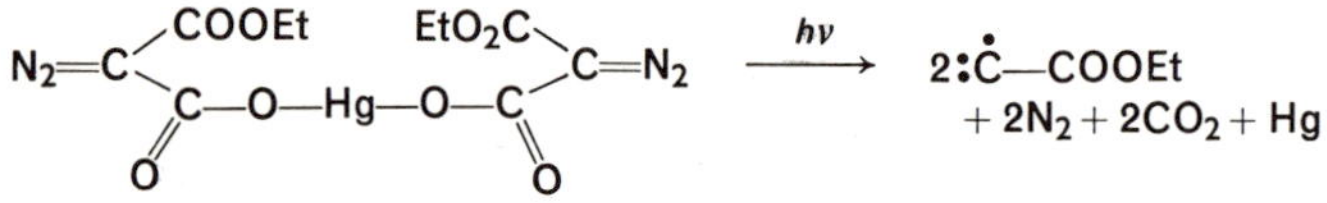

$$N_2\!\!=\!\!C\underset{\underset{O}{\overset{\|}{C}}—O—Hg—O—\underset{\underset{O}{\overset{\|}{C}}}{}}{\overset{COOEt}{\diagup}}C\!\!=\!\!N_2 \xrightarrow{h\nu}\ 2:\!\dot{C}—COOEt + 2N_2 + 2CO_2 + Hg$$

The transient compound both adds and inserts into cyclohexene, further abstracting a proton, since the primary product is a radical.

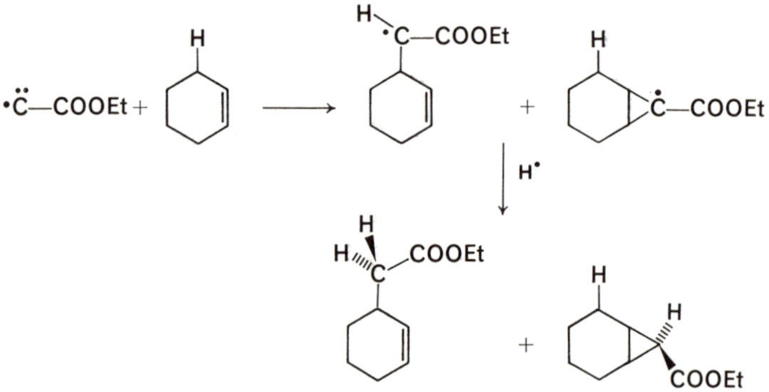

6.11   SULPHENES

These are reactive intermediates of general formula $R_2C\!\!=\!\!SO_2$. They are more analogous to ketenes than to carbenes but are conveniently considered here. Sulphenes are believed to be formed when a β-elimination of a sulphonyl halide is carried out in non-aqueous solution,

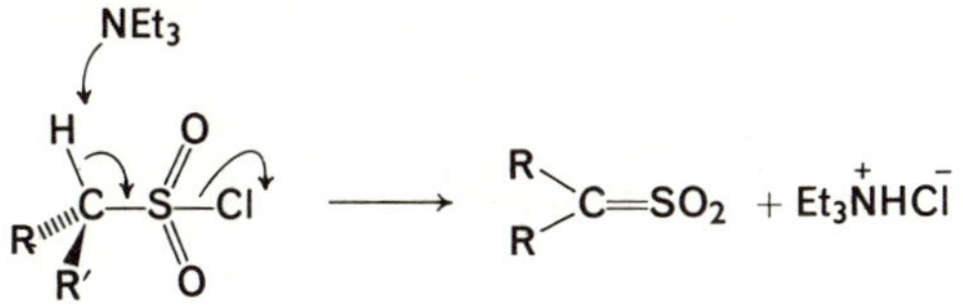

or by the rearrangement of sulphonylcarbenes (a sulphur analogue of the Wolff rearrangement, Section 6.4),

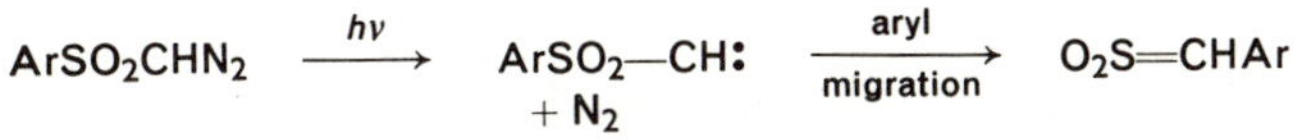

Sulphenes are highly reactive and will dimerize, polymerize, add nucleophiles such as water and alcohols, and will undergo cycloadditions related to the Diels–Alder reaction.

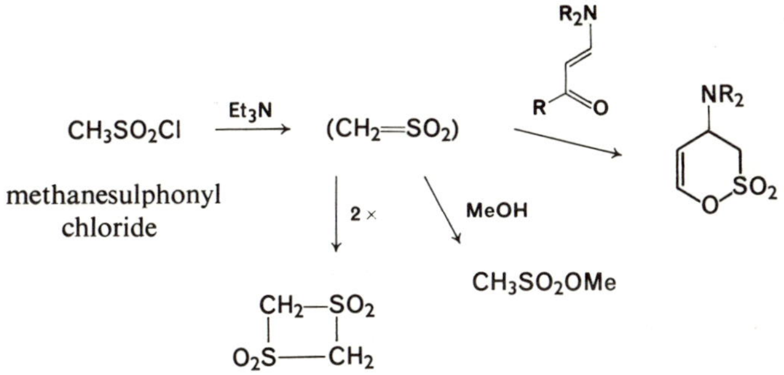

SUGGESTIONS FOR FURTHER READING

C. A. Buehler, 'Carbenes in Insertion and Addition Reactions'. *J. Chem. Ed.*, **49**, 239 (1972).

T. L. Gilchrist and C. W. Rees, *Carbenes, Nitrenes and Arynes*, Nelson, London, 1969.

J. Hine, *Divalent Carbon*, Ronald Press, New York, 1964.

M. Jones and R. A. Moss (Eds), *Carbenes*, Wiley–Interscience, New York, 1972.

W. Kirmse, *Carbene Chemistry*, Academic Press, New York, 1964.

E. Wasserman, 'Electron Spin Resonance of Nitrenes', *Progr. Phys. Org. Chem.*, **8**, 319 (1971).

Chapter 7

Dehydrobenzene (Benzyne) and Related Intermediates

7.1 **INTRODUCTION**

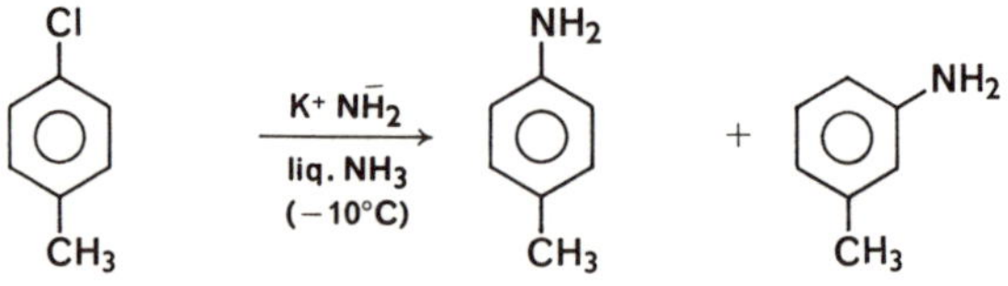

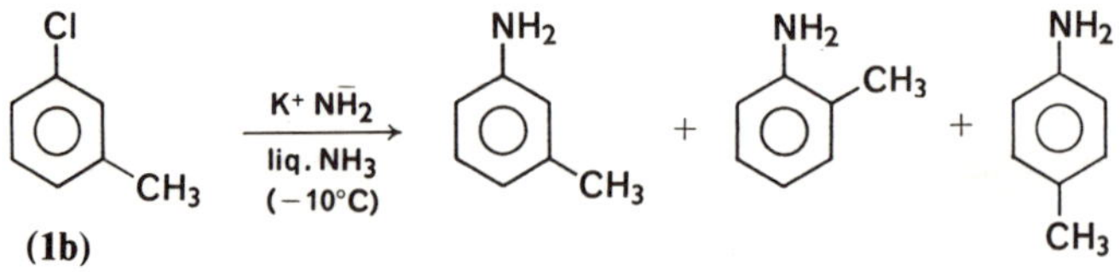

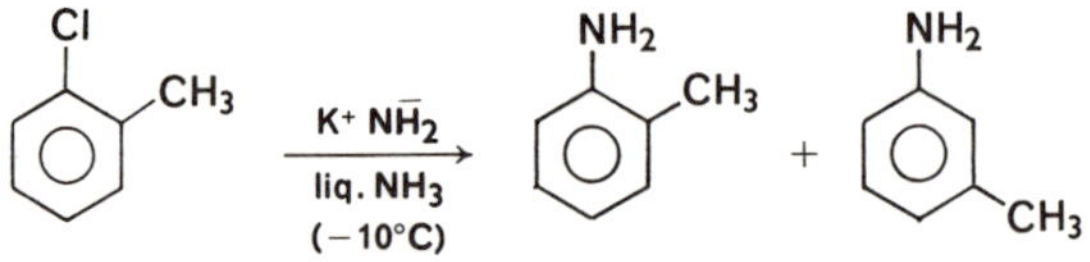

Nucleophilic substitutions at the aromatic nucleus have been shown (Section 4.23) to be somewhat difficult unless activated by, for example, nitro groups, and to occur under normal circumstances by the addition–elimination mechanism. It has long been known that certain substitutions carried out under forcing conditions or with very powerful bases lead to rearranged products. These were termed 'cine' substitutions by Bunnett. For example, the chlorotoluenes (**1a, b, c**) separately react with potassium amide in liquid ammonia to give a mixture of two toluidines; in each case the rearranged product has the entering group *ortho* to the departing chloride. Labelling the carbon to which the leaving group was attached confirmed the reaction of chlorobenzene as leading to 50% *ortho*-rearranged product, and this together with the observation that the rearranged product only loses the proton (deuterium) from the *ortho* position led to the conclusion that these reactions proceed by way of a symmetrical intermediate. Furthermore, the reaction shows a sizable

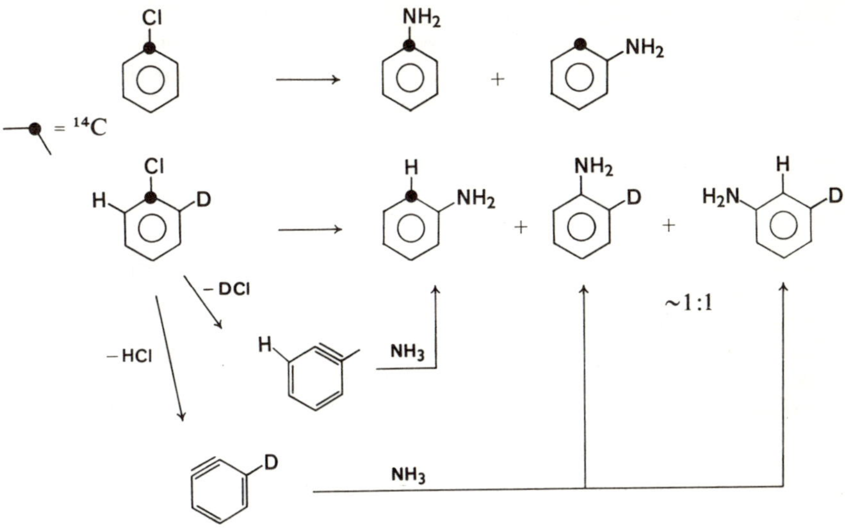

ortho-deuterium isotope effect which must be a primary effect ($k_H/k_D = 5$–6). This evidence is consistent with a rate-determining loss of *ortho* hydrogen to give a symmetric intermediate, either **2** or **3**. The former is ruled out by the observation that chlorine can be replaced by fluorine as a leaving group. A structure such as **2** is not possible for a first-row element as it would have ten valence electrons. Subsequent studies have tended to confirm the intermediate

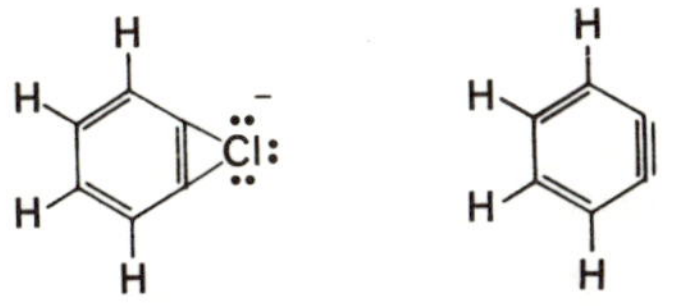

(2)   (3)

benzyne
(dehydrobenzene)

as the species C_6H_4 (**3**), or a derivative, which has been named 'dehydrobenzene' or 'benzyne'; generically, the term 'aryne' is often used. The parent species is available from the phenyl halides but many other routes are available.

Telling evidence for the existence of the intermediate arises from the observation that the products of reaction are independent of the method of generating the benzyne (Section 7.2).

7.2   GENERATION OF ARYNES

Since the recognition of benzyne as reactive intermediate, literally dozens of reactions have been shown to be capable of yielding arynes but in all cases the reaction can be summed up as the loss of electronegative and electropositive groups from adjacent positions,

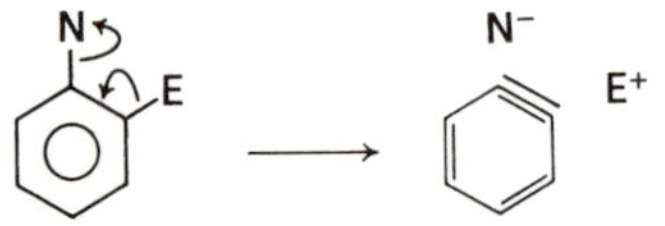

Dehydrohalogenation

As previously mentioned, the treatment of an aryl halide with a very strong base leads to aryne formation by a β-elimination which could be concerted or stepwise, initiated by proton loss. The reaction occurs with lithium alkyls

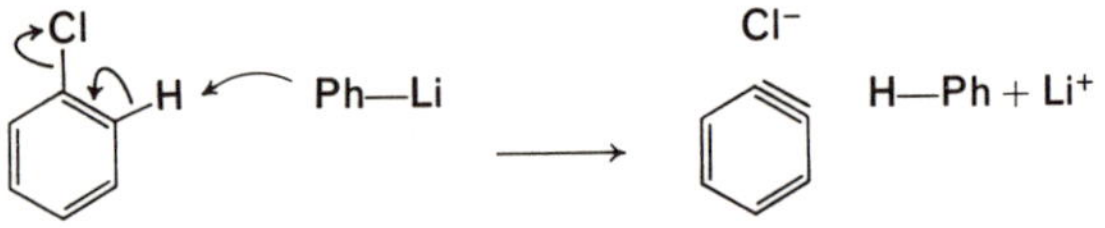

and aryls in ether at $-30°$, often in good yield. Fluoride is displaced more readily than chloride or bromide in many cases (though not with potassium amide as base), which indicates that the loss of halogen occurs in a fast step, the electronegative fluorine increasing the acidity of the *ortho* proton (cf. carbene formation, Chapter 6). A direct-displacement mechanism may also compete. Thus, the product ratio,

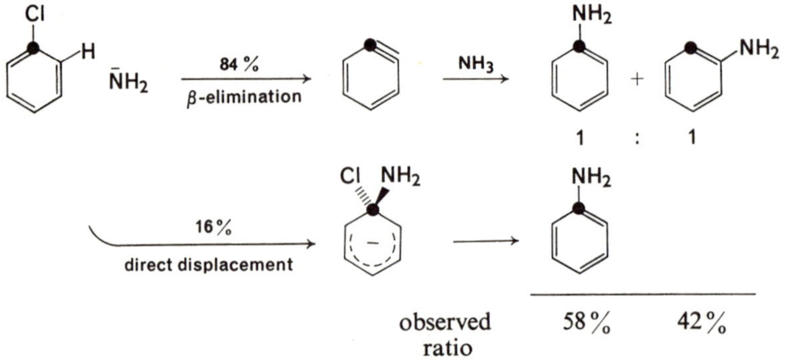

observed
ratio

indicates that this amounts to 16 % in the reaction of chlorobenzene with amide.

ortho-Halogeno Organometals

A very convenient preparation of benzynes is from dihalogeno aryls via the organolithium or magnesium compounds. The *ortho*-halogenoaryllithiums

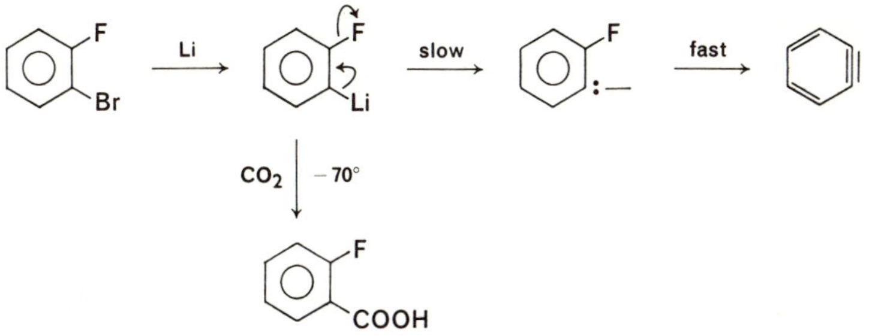

are fairly stable below −70°, but on warming above −60° decompose to the aryne. Fluorine compounds are more labile than chlorine or bromine hence loss of halide must be secondary and the carbon–metal bond weakened by the electronegativity of the halogen.

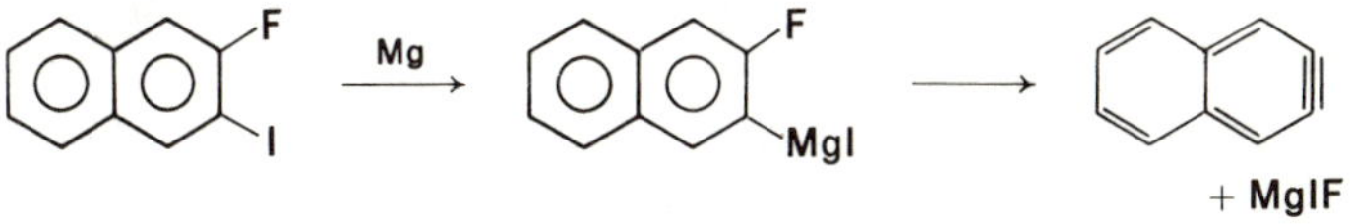

Loss of Neutral Leaving Groups

Formation of benzyne under very mild conditions occurs by the thermal decomposition of the diazonium carboxylate, or carboxylic acid (4). Both these compounds are isolable, the former being explosive at room temperature,

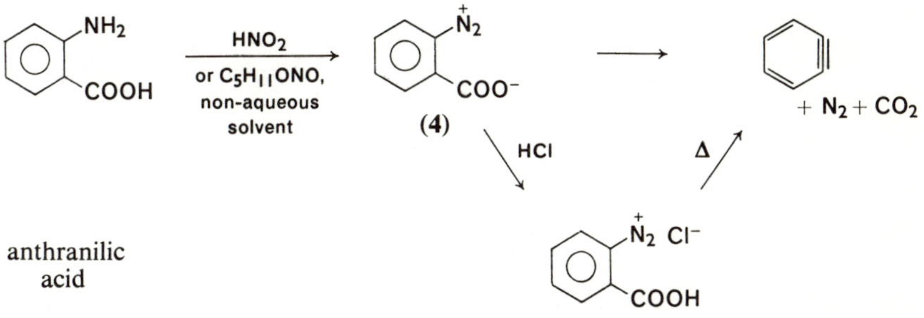

The following routes are analogous.

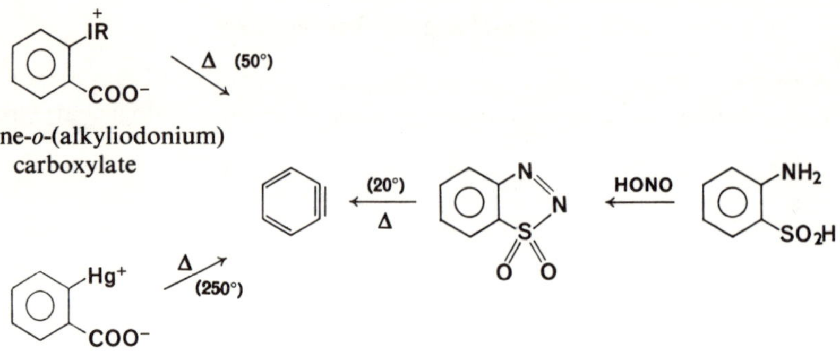

benzene-*o*-(alkyliodonium)
carboxylate

benzthiadiazole   *o*-aminobenzene-
dioxide   sulphinic acid

Decomposition of Cyclic Compounds

Benzotetrazoles (**5**) were found not to decompose to benzynes, but intermediates formed in the oxidation of the aminobenztriazoles (**6**) smoothly form the aryne.

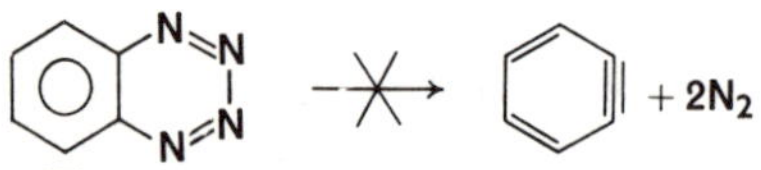

(**5**)

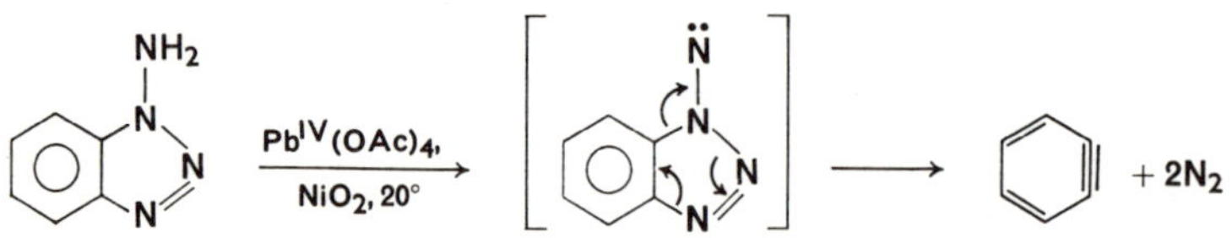

(**6**)

Photolytic Decompositions

Excitation by ultraviolet light causes the decomposition of many species to arynes; the following are typical.

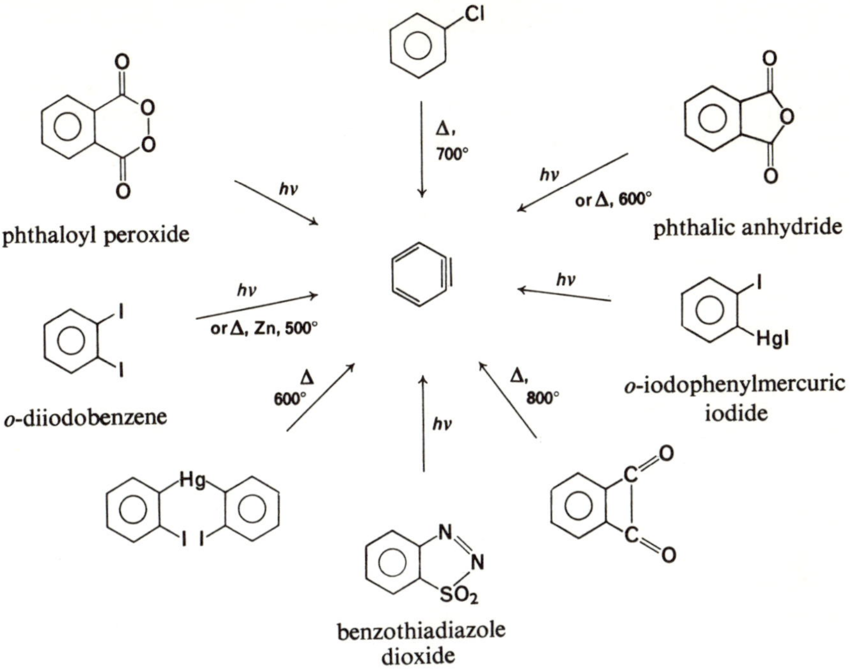
Cl
Δ,
700°
O
O
O
O
hν
phthaloyl peroxide
hν
or Δ, 600°
phthalic anhydride
hν
or Δ, Zn, 500°
hν
I
HgI
I
I
o-diiodobenzene
Δ
600°
hν
Δ,
800°
o-iodophenylmercuric
iodide
Hg
I I
N
N
SO2
C=O
C=O
benzothiadiazole
dioxide

Other Pyrolytic Methods

In addition to the unstable precursors mentioned, many stable aromatic compounds may be pyrolysed to benzyne at a sufficiently high temperature. Some methods used to generate benzyne in the gas phase at high temperatures are shown in the scheme above.

7.3 REACTIONS OF ARYNES

The formation of characteristic products from a transient aryne constitutes an example of 'trapping' the intermediate in the form of a stable recognizable entity and thus affords further proof of its existence. In carrying out the following reactions any appropriate method of generation of the aryne may be used; the products are the same irrespective of the source.

The reactions of arynes always involve additions to the 'triple' bond to restore the aromatic character in the product. The reactions may be polar in nature or concerted cycloadditions.

Nucleophilic Additions

The addition of ammonia has already been seen to be the fate of arynes generated in liquid ammonia as solvent. In substituted cases, a preference towards
one orientation of addition may be displayed.

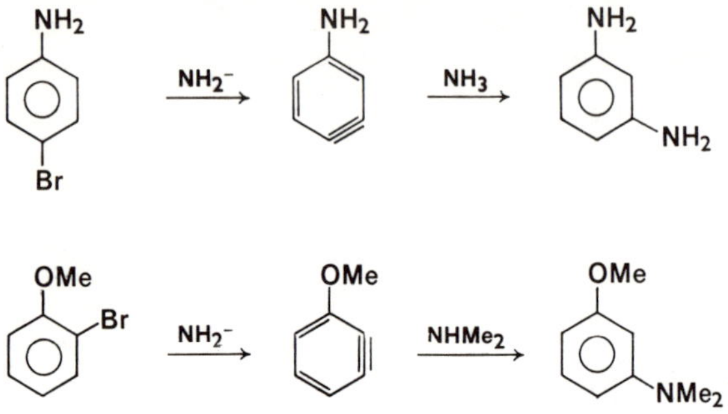

Alcohols and alkoxides, lithium alkyls and aryls and nucleophiles such as
silver carboxylates, halide ion, cyanide, etc., will readily add to the aryne.

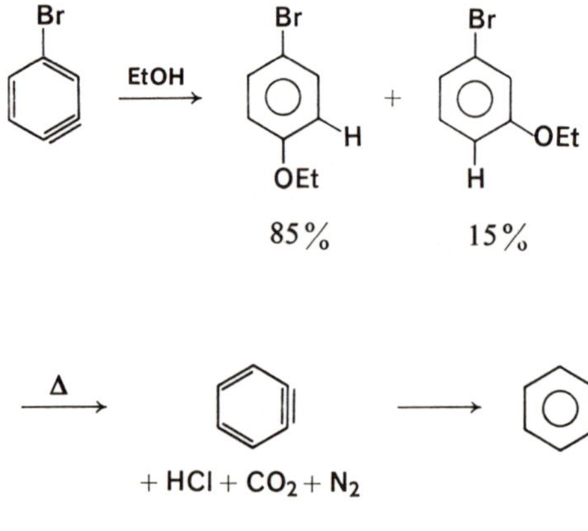

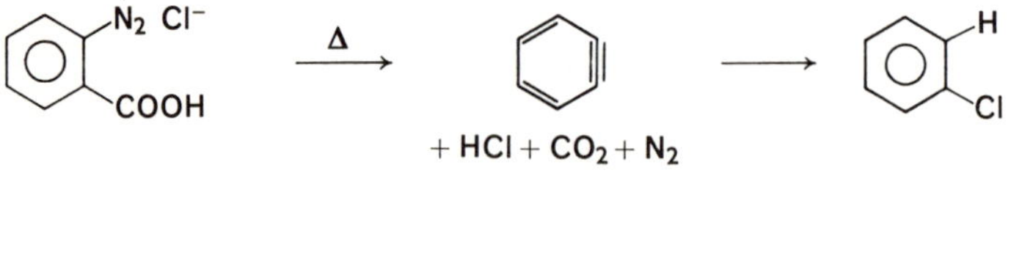

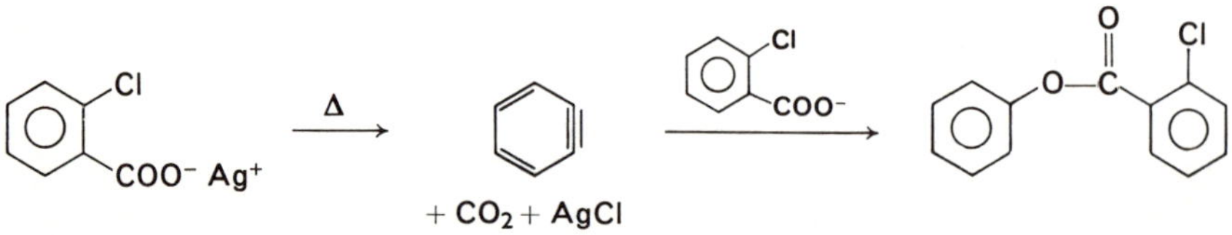

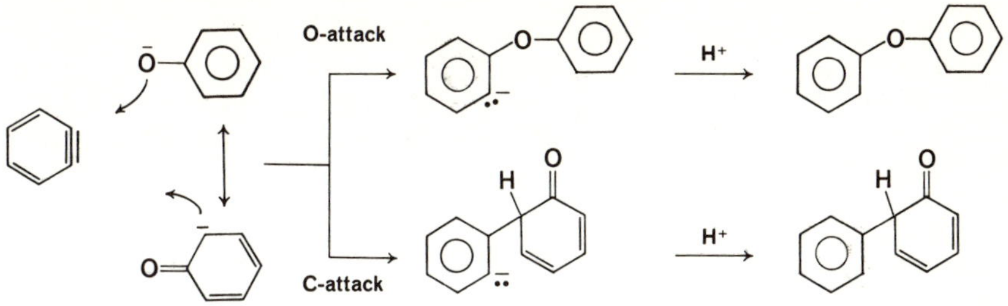

phenolate

Cyclization can occur to an enolizable ketone function:

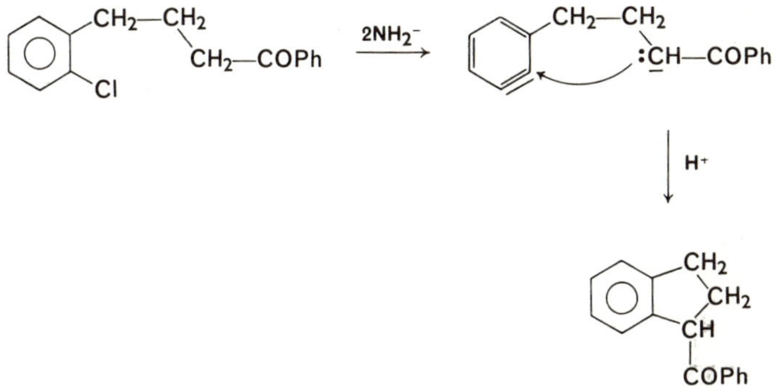

Alkyl sulphides will also add, but in strongly basic conditions elimination of one of the alkyl groups as an olefin may occur.

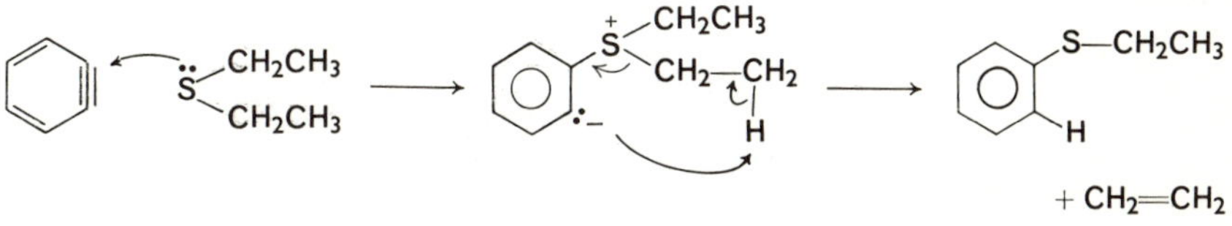

Electrophilic Additions

Arynes will add electrophilic species readily. Trialkylboranes are typical Lewis acids which add with a subsequent elimination of an alkene if a β-hydrogen is available.

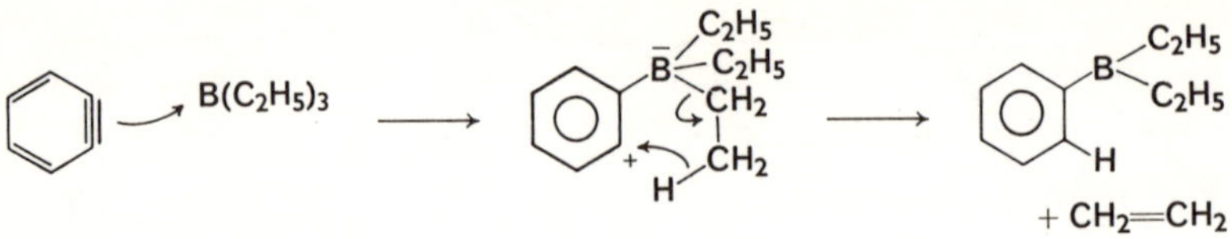

$$+ CH_2{=}CH_2$$

Halogens and halides of mercury, tin and silicon are also electrophilic agents which add to arynes,

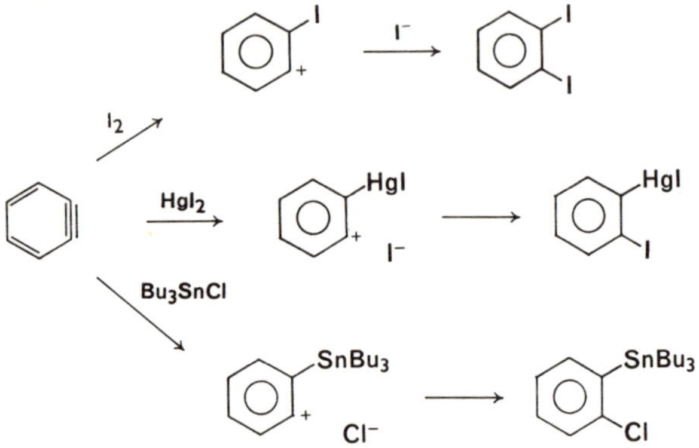

Cycloadditions

The reagents of choice for trapping arynes as recognizable, stable products, are those to which cycloadditions occur. Arynes dimerize to biphenylene derivatives with great facility. In the case of benzyne in the gas phase, the specific rate constant is very high and a negative entropy of activation is found as expected for a concerted cycloaddition.

$$k_2 \sim 10^9 \text{ mol}^{-1} \text{ s}^{-1}$$
$$\Delta S^{+} = -44 \text{ J K}^{-1}$$

Arynes act as highly reactive dienophiles in the Diels–Alder reaction and add to most 1,3-dienes such as cyclopentadiene, furan and tetraphenylcyclopentadienone (tetracyclone 7). The latter forms an initial adduct (8), which loses carbon monoxide, and tetraphenylnaphthalene (9) is isolated. Additions to a

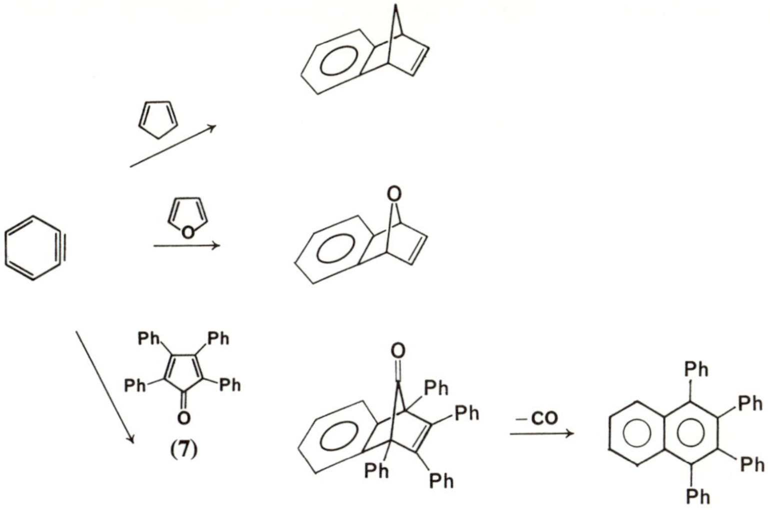

(7)   (8)   −CO   (9)

strained olefin such as norbornadiene (10) to form a four-membered ring also occur though yields are lower than with dienes, this being a 'forbidden' reaction which does not occur at all with simple olefins. A similar reaction

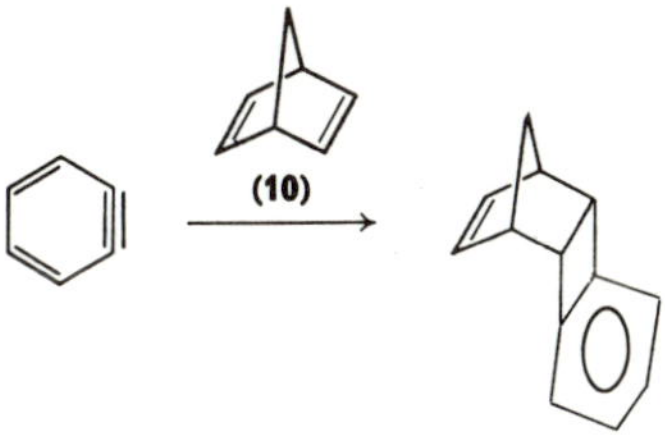

occurs with acetylene, though the initial adducts (11), being derivatives of cyclobutadiene, dimerize:

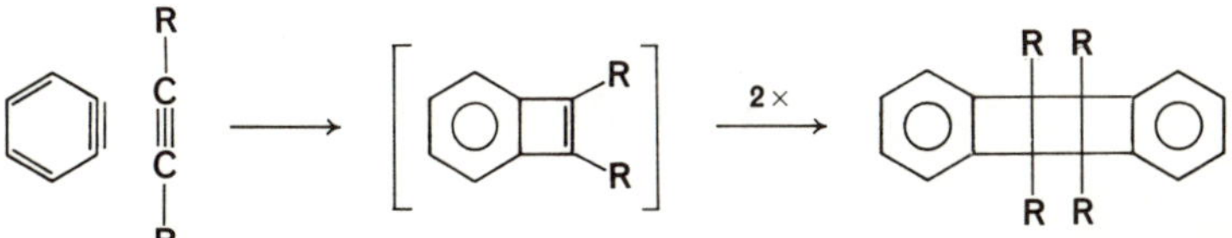

(11)

1,3-Dipolar species add; useful heterocycle syntheses result.

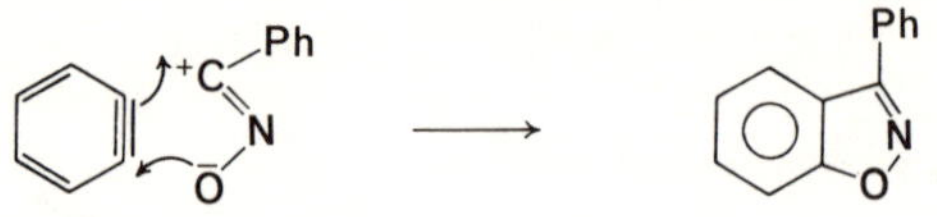

phenylnitrile oxide   5-phenylbenzoxazole

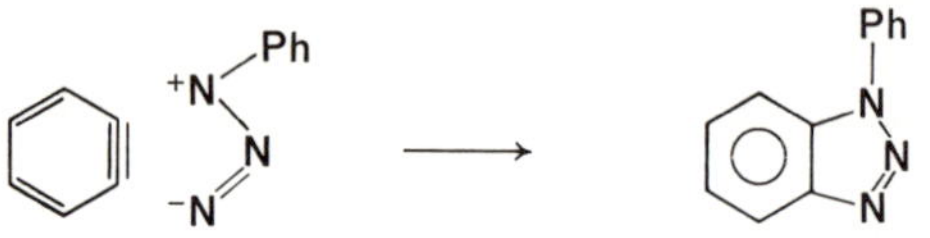

phenyl azide   1-phenylbenztriazole

Benzyne acts as the olefinic component in the 'ene' reaction with an allylic compound.

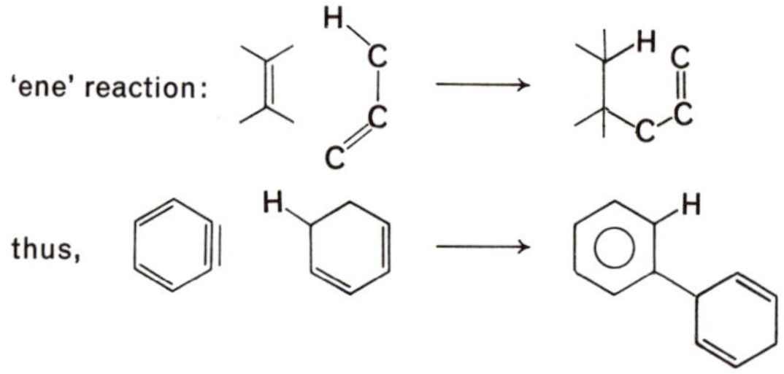

'ene' reaction:

thus,

Dehydrogenations may be included among this type of reaction; it is likely that the hydrogen is transferred in a single step.

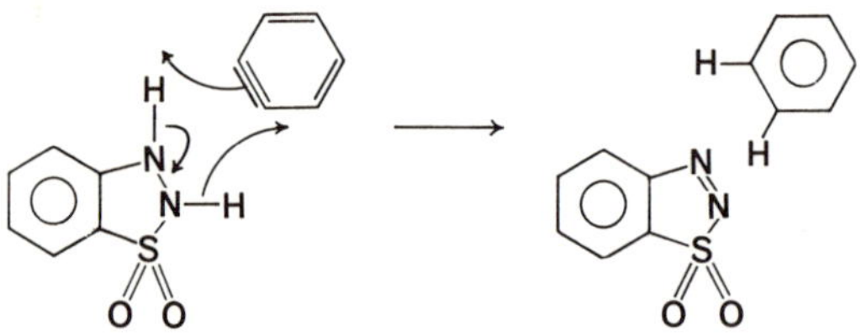

7.4   THE NATURE OF BENZYNE

While the structure of benzyne is conventionally written as possessing a triple bond, it is clear that this cannot be of the same type as found in an alkyne.

sp Hybridization at carbon requires a linear geometry which in the six-membered ring cannot be achieved. Other structures which have been suggested include **12** and **13**. These may indeed be contributing structures but neither is by itself satisfactory. Benzyne is not apparently a diradical or triplet species (as depicted in **13**) and attempts to detect radical character have not succeeded. Calculations have suggested the 'triple' bond length is of the order 1.29 Å, shorter than the normal aromatic bond, and it is reasonable to suppose that orbital overlap as in **14** is the cause.

Direct detection of benzyne is limited to its appearance in the mass spectrometer ($m/e = 76$). Using a time-of-flight mass spectrometer which enables very fast time-resolved mass spectra to be measured, it was found that the fragment resulting from the thermal decomposition of diiodobenzene and other suitable precursors disappeared with time and the peak due to the dimer (biphenylene, $m/e = 152$) appeared concurrently. From this, the rate constant for dimerization of benzyne in the gas phase was estimated at $4.6 \times 10^9 \text{ mol}^{-1} \text{ s}^{-1}$. This rate has been confirmed by generation of benzyne by flash photolysis of phthalic anhydride (**15**), and observation of the appearance of biphenylene as a function of time (in the μs range).

7.5   **OTHER RELATED INTERMEDIATES**

Halopyridines have been claimed to undergo dehydrohalogenation by strong base to the dehydropyridine. It seems clear, however, that these reactions are much less facile than in the benzene series and compete unfavourably with the addition–elimination mechanism for substitution. It appears that the reaction of potassium amide with 3-chloropyridine (**16**) proceeds to the extent of about 5% by the aryne pathway. The pyridine diazonium carboxylate (**17**) has been isolated and shown to decompose to the expected dimers of 3-pyridyne, **19** and **20**, although in much poorer yield than the analogous reaction to generate benzyne. Decomposition of the material in the time-of-flight mass spectrometer shows a product of $m/e = 77$, presumably the pyridyne (**18**).

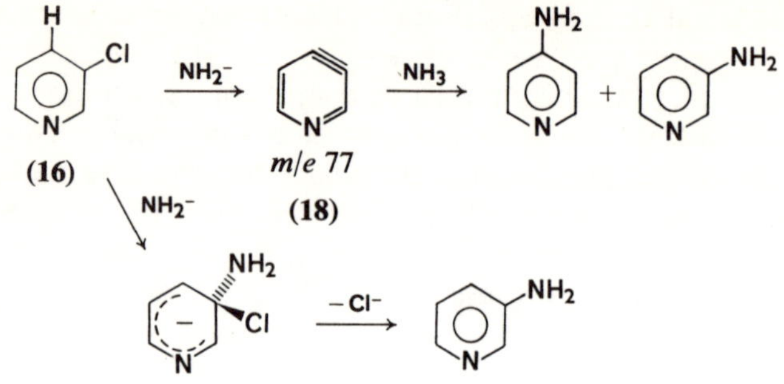

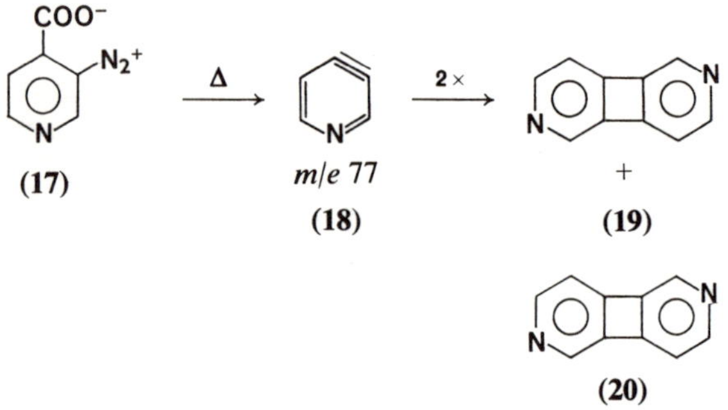

The quinoline analogue (**21**) is also implicated in the reaction,

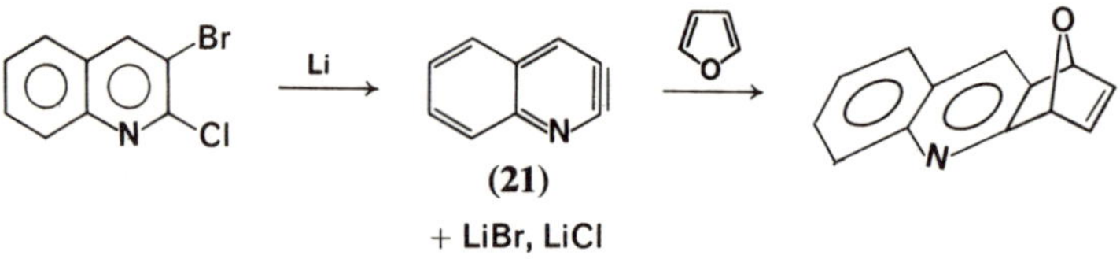

An aryne derivative of the cyclopentadienide ion (**22**) has been generated and characterized as its adduct with tetracyclone (**23**).

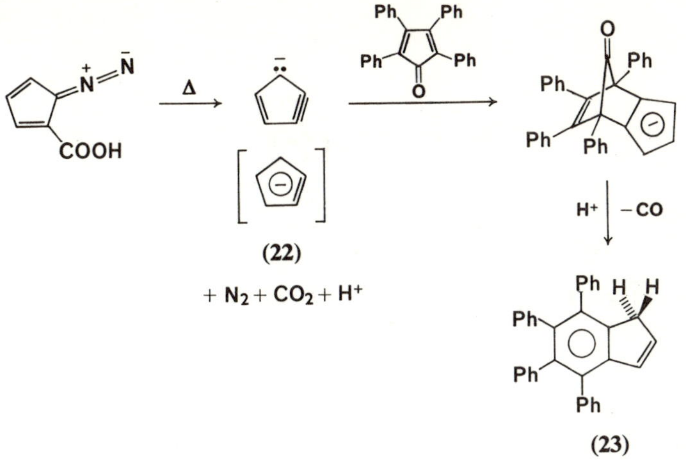

(22)

$+ N_2 + CO_2 + H^+$

$H^+ \mid -CO$

(23)

SUGGESTIONS FOR FURTHER READING

T. L. Gilchrist and C. W. Rees, *Carbenes, Nitrenes and Arynes*, Nelson, London, 1969.

H. Heaney, 'Benzyne and Related Intermediates', *Chem. Rev.*, **1962**, 81.

R. W. Hoffmann, *Dehydrobenzene and Cycloalkynes*, Academic Press, New York, 1967.

Chapter 8

Tetrahedral Intermediates in Reactions at the Carbonyl and Related Groups

8.1   **INTRODUCTION**

The carbonyl group, $C=O$, occurring in aldehydes, ketones, esters, carboxylic acids and similar compounds, is susceptible to attack both by nucleophiles, at carbon, and by electrophiles, at oxygen. This is due to the capacity of the oxygen atom to act as either a donor or an acceptor of an electron pair.

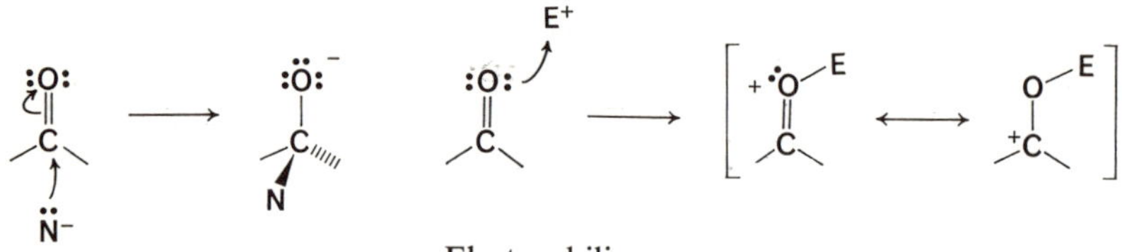

Nucleophilic attack   Electrophilic attack

Almost all the reactions at the carbonyl group commence by one or other of these bond-making processes. Many reactions require the attachment of both nucleophilic and electrophilic entities and then two mechanisms are available according to the order of attachment. If a nucleophile (e.g. OH^-) becomes attached first, the reaction may be base catalysed, while if the electrophile (especially H^+) becomes attached first, the reaction is acid catalysed. The final result of the reaction may be addition, substitution or condensation (substitution followed by elimination), which accounts for almost all of carbonyl chemistry. The following scheme illustrates the relationship between these types of reaction.

Substitution and condensation reactions typically proceed by way of a series of intermediates such as **1** in which the central carbon is in a tetrahedral (sp^3-hybridized) configuration. Some examples of these reactions will be discussed.

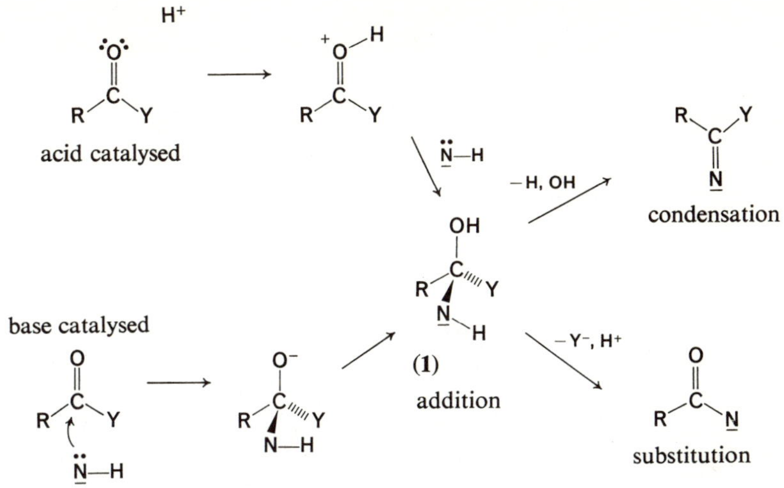

Note: $\underline{N}$—H = proton-bearing nucleophile, e.g. OH^-, NH_3, etc.

8.2   **BASE-CATALYSED ESTER HYDROLYSIS**

One of the common methods of hydrolysing a carboxylic ester is to treat it, usually at the boiling point, with aqueous alkali.

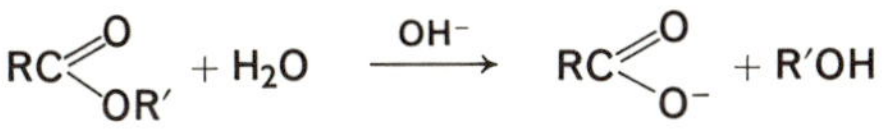

$$RC\overset{O}{\underset{OR'}{\big\langle}} + H_2O \xrightarrow{OH^-} RC\overset{O}{\underset{O^-}{\big\langle}} + R'OH$$

The kinetic law which is obeyed is of second order,

$$\text{rate} = k_2[\text{ester}][OH^-]$$

and it follows that the transition state contains one molecule of ester and one of OH^-. It is certain that the carboxyl group is attacked since several lines of evidence show that the alkoxy function remains intact, i.e. that acyl-oxygen fission occurs rather than alkyl-oxygen fission:

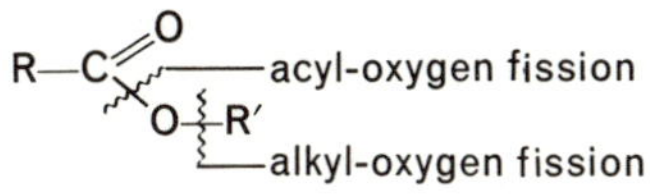

$$R-C\overset{O}{\underset{O\,-\,R'}{\big\langle}} \quad \begin{array}{l}\text{—acyl-oxygen fission}\\ \\ \text{—alkyl-oxygen fission}\end{array}$$

The evidence for this is as follows: if the water in which the hydrolysis is carried out is enriched in ^{18}O, none of this isotopic label appears in the alcohol produced. Hence, the alcohol must contain the intact alkoxy group present in the ester. This experiment was first carried out by Polanyi in 1934, and is

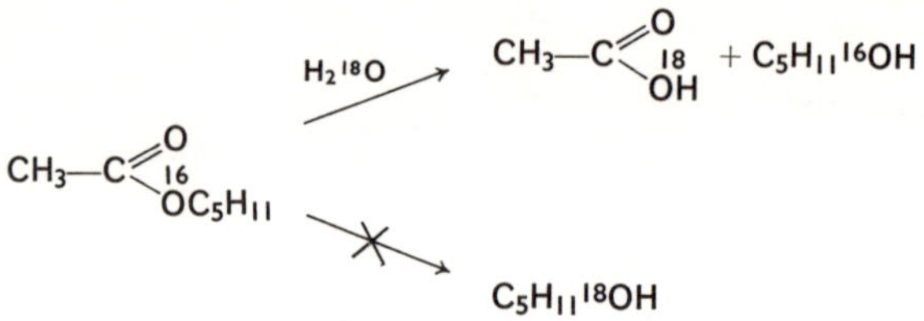

$$CH_3-C\overset{O}{\underset{OC_5H_{11}}{<}}{}_{16} \quad \xrightarrow{H_2^{18}O} \quad CH_3-C\overset{O}{\underset{OH}{<}}{}_{18} + C_5H_{11}{}^{16}OH$$

$$\not\longrightarrow \quad C_5H_{11}{}^{18}OH$$

confirmed by the observation that an ester formed from an optically active alcohol, e.g. menthol, gives the alcohol with complete retention of configuration upon hydrolysis. This means that the alkyl group in the alcohol moiety can never become free (as a carbonium ion, which would lead to racemization) or cannot suffer nucleophilic displacement (S_N2 reaction, leading to inversion of configuration), and implies that it remains bound to oxygen throughout the reaction. Furthermore, no rearrangements occur in the alcohol fragment even

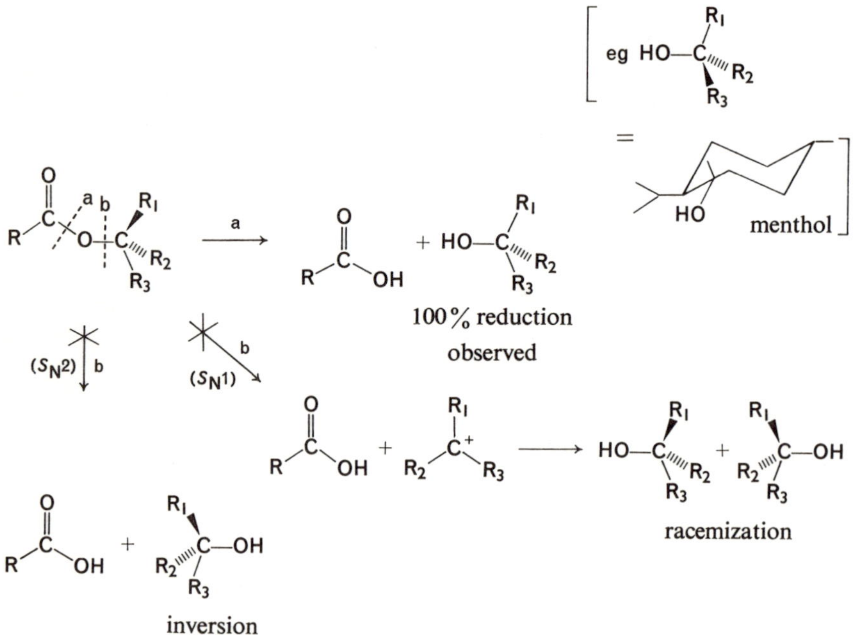

when, as in the case of neopentyl (2) and crotyl (3) esters, this would be expected if free carbonium ions formed. The inescapable conclusion is that

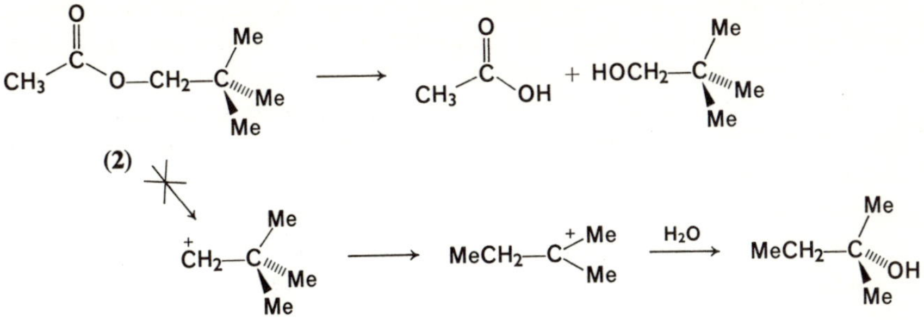

$$CH_3-C(=O)-O-CH_2-C(Me)(Me)_{Me} \longrightarrow CH_3-C(=O)-OH + HOCH_2-C(Me)(Me)_{Me}$$

(2)

$$^+CH_2-C(Me)(Me)_{Me} \longrightarrow MeCH_2-C^+(Me)_{Me} \xrightarrow{H_2O} MeCH_2-C(Me)(Me)_{OH}$$

reaction occurs without the intermediate formation of an alkyl cation. The alkaline hydrolysis of esters, then, is a nucleophilic substitution of the alkoxide

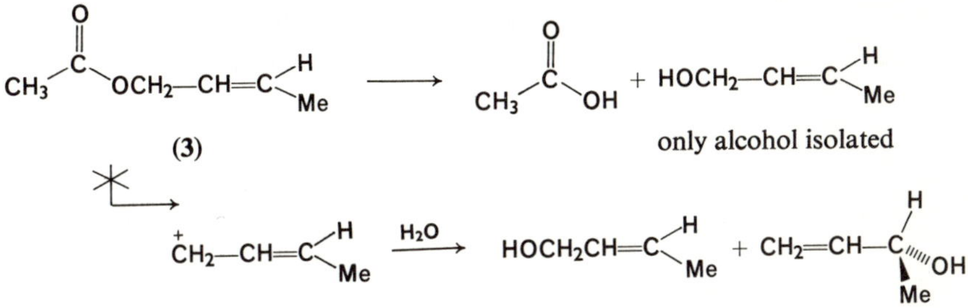

$$CH_3-C(=O)-OCH_2-CH=C(H)_{Me} \longrightarrow CH_3-C(=O)-OH + HOCH_2-CH=C(H)_{Me}$$

(3)   only alcohol isolated

$$^+CH_2-CH=C(H)_{Me} \xrightarrow{H_2O} HOCH_2CH=C(H)_{Me} + CH_2=CH-C(H)(Me)_{OH}$$

group by hydroxide ion. It remains to be established whether the displacement is synchronous or whether the addition of the hydroxide ion precedes the expulsion of the alkoxide ion, i.e. whether the tetrahedral species is a transition state (4), or a true intermediate (5) with a finite lifetime,

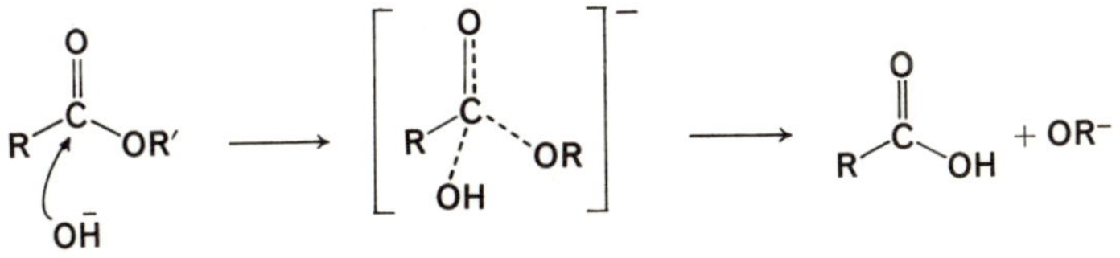

$$R-C(=O)-OR' \quad (OH^-) \longrightarrow \left[R-C(OR)(OH)-\overset{O}{\underset{}{}} \right]^- \longrightarrow R-C(=O)-OH + OR^-$$

concerted   transition state   (4)

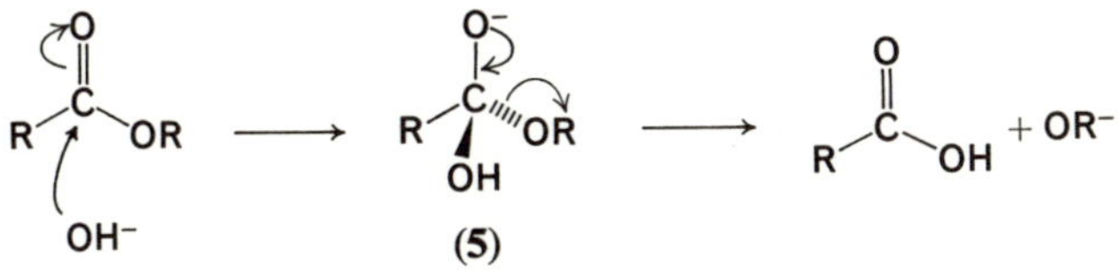

$$R-C(=O)-OR \quad (OH^-) \longrightarrow R-C(OR)(OH)-O^- \longrightarrow R-C(=O)-OH + OR^-$$

2-step   (5)

                         intermediate

This matter was settled by Bender in 1951 who showed that if an ester labelled with ^{18}O in the carbonyl position is allowed to undergo incomplete hydrolysis in water, the *unreacted* ester, isolated at various times during the reaction, gradually lost its ^{18}O label. Similarly, an ester hydrolysing in water which is enriched in ^{18}O gradually acquires the label in the unreacted ester. This ability of the ester to exchange its oxygen with water is readily explained by an addition–elimination reaction with a tetrahedral intermediate in which two oxygens become equivalent. In Ingold's terminology, this mechanism is called $B_{AC}2$ (Base-catalysed, acyl-oxygen fission, bimolecular).

$B_{AC}2$ Mechanism of Ester Hydrolysis

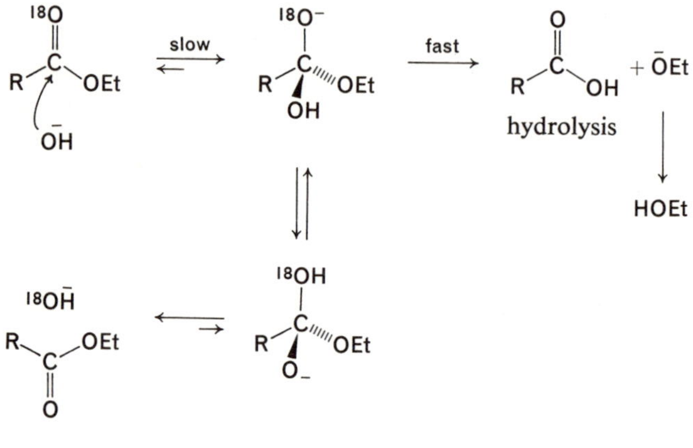

O-exchange

Since the slow step involves coordination of a nucleophile (acceptance of an electron pair), electron-withdrawing substituents on the ester facilitate hydrolysis and vice versa (Table 1). Likewise, the Hammett reaction constant, ρ

Table 1. Rates of hydrolysis (base-catalysed) of some substituted acetate esters

$$RCOOEt + OH^- \xrightarrow[25°]{H_2O} RCOO^- + EtOH$$

R	k_{rel}
Me	1
MeOCH$_2$	12
ClCH$_2$	760
Cl$_2$CH	16000
MeC(=O)—CH$_2$	10000
MeO$_2$CCH$_2$	170000

(Section 1.10), for the hydrolysis of substituted benzoate esters is positive (1.8 to 2.8), which has the same interpretation.

8.3   ACID-CATALYSED ESTER HYDROLYSIS

As the pH of the solution is lowered, the rate of ester hydrolysis decreases, in accordance with the rate law, to a minimum around neutrality but then increases rapidly as the solution becomes progressively more acidic. Evidently a second mechanism is coming into play and this takes the kinetic form,

$$\text{rate} = k_2[\text{ester}][H_3O^+]$$

The acid-catalysed mechanism occurs by acyl-oxygen fission as judged by the previous criteria—retention of stereochemistry in the alcohol fragment and of labelled oxygen, and failure to observe any carbonium-ion characteristics. Also carbonyl-oxygen exchange with the water occurs prior to reaction confirming the presence of a tetrahedral intermediate. The entropies of activation are large and negative suggesting an associated, highly solvated transition state; e.g. for ethyl-acetate hydrolysis, $\Delta S^{\ddagger} = -100$ J K^{-1}. A mechanism consistent with these observations, the $A_{AC}2$ mechanism (Acid-catalysed, acyl-oxygen fission, bimolecular) is as follows: it differs from the $B_{AC}2$ mechanism in having two extra protons in the transition state.

$A_{AC}2$ Mechanism of Ester Hydrolysis

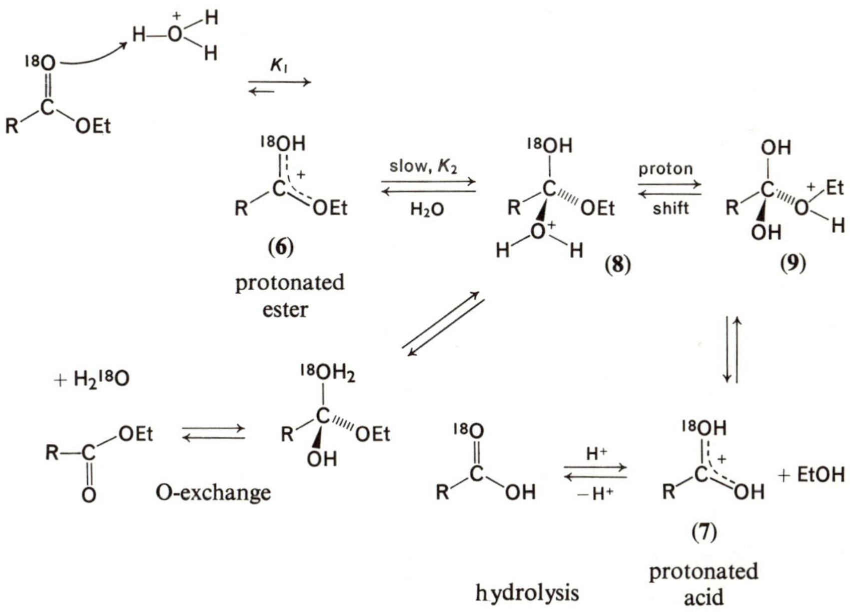

The initial coordination of a proton facilitates the attack by the weak nucleophile, H_2O. There are evidently a series of proton shifts required and as these are mostly very fast it is difficult to obtain more exact information on the real existence of some of the species involved. Protonated acid and ester with the formulae shown (6, 7) are known as stable species in highly acidic solutions (Section 2.19). The species 8 and 9 seem to be necessary intermediates since, in the hydrolysis, the expulsion of EtOH would be much more facile than EtO^-. It is likely that the sequence of changes is symmetrical from whichever end one starts. This reaction, unlike the $B_{AC}2$ mechanism, is reversible in each stage; one can therefore convert an acid to its ester by treating it with the alcohol in the presence of a strong acid, the normal method of esterification. By having an excess of either alcohol or water one is merely displacing a chain of equilibria to the left or to the right and the mechanism must be the same in either direction (principle of microscopic reversibility).

The effects of polar substituents are much less pronounced in the acid-catalysed reaction than in the base-catalysed process. Hammett ρ-factors for acid-catalysed hydrolysis of benzoic esters range from -0.2 to $+0.5$. It would be expected that an electron-donating group would facilitate the protonation of the ester (increase K_1) but retard the attack of water (reduce K_2), the two effects tending to cancel one another and vice versa for electron-withdrawing groups.

8.4   CARBOXYL HYDROLYSES WITHOUT TETRAHEDRAL INTERMEDIATES

It is appropriate at this point to contrast the hydrolytic mechanisms which are occasionally observed and which do not partake of tetrahedral intermediates, i.e. nucleophilic attack at carbonyl carbon.

Hydrolysis in Highly Acidic Media

The rate of hydrolysis of a simple ester in sulphuric acid increases rather irregularly with acid concentration or the acidity of the solution as measured by the acidity function H_0 (Section 1.22), Figure 1a. If we assume that the concentration of protonated ester increases with H_0* and correct for the increase in rate which is due to the higher concentration of this reactive intermediate, the rate of reaction is actually seen to fall up to about 85% concentrated H_2SO_4, after which it begins to increase (Figure 1b). This fall in rate constant of the slow step is due to the decreasing concentration of the

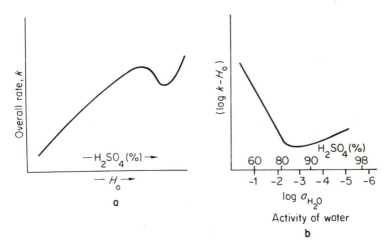

Figure 1. Ester-hydrolysis rates in strongly acid medium.

nucleophile, water. The magnitude of the slope in the first part of the curve is 2, indicating the participation of two water molecules in the transition state; i.e. one bearing the proton as H_3O^+ and one attacking the protonated ester. The abrupt change in rate constant at ca 85% sulphuric acid is due to a new mechanism of ester hydrolysis becoming predominant. This still occurs by acyl-oxygen fission, but now heterolysis of the protonated ester is the rate-determining step, not attack by the water molecule. An intermediate acyl cation (**10**) is formed, a species known to be formed in highly acidic media. This mechanism is designated $A_{AC}1$ (Acid-catalysed, acyl-oxygen fission, unimolecular).

---

* The acidity function, H_0, refers to protonation of nitrogenous bases and, in principle, a function appropriate to protonation of esters should be used. In practice the difference is slight.

$A_{AC}1$ Mechanisms of Ester Hydrolysis

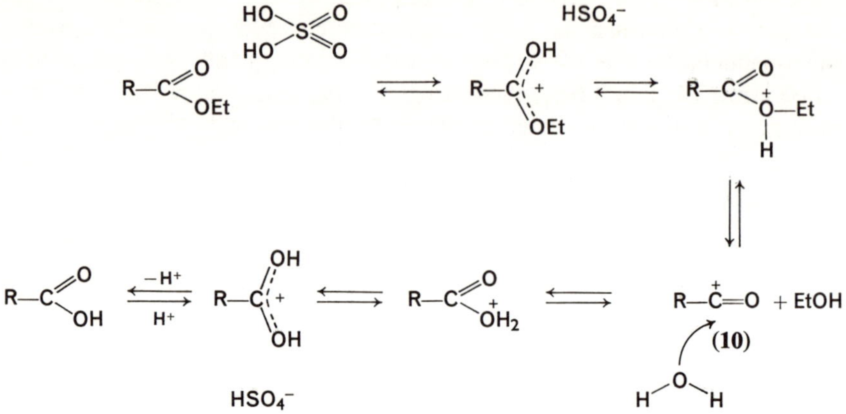

Hydrolysis by this mechanism can be useful when presented with very sterically hindered esters. For instance, the esters of mesitoic acid (**11**) are very resistant to hydrolysis under ordinary conditions of boiling aqueous acid or alkali ($A_{AC}2$ or $B_{AC}2$) on account of the *ortho*-methyl groups preventing the attack of the reagent. The esters are readily hydrolysed by dissolving in concentrated sulphuric acid and diluting with water. The highly stable mesitoyl cation is formed in the sulphuric acid solution. Since all steps are reversible, esteri-

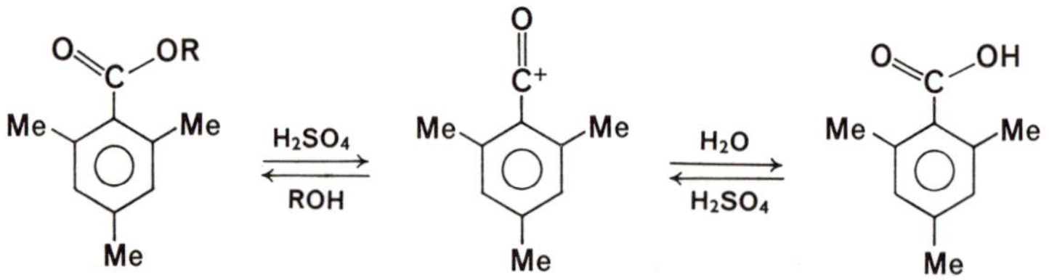

(**11**)   mesitoyl cation   (**12**)

fication is achieved by dissolving the acid (**12**) in sulphuric acid and diluting with alcohol.

Alkyl-Oxygen Fission in Ester Hydrolysis

Under certain circumstances, the alkyl-oxygen bond of the ester breaks; this is particularly likely when the alkyl group can form a highly stable carbonium ion and a highly polar medium is used. The hydrolysis in water of triphenylmethyl acetate is an example of the mechanism, designated $A_{AL}1$ (acid-catalysed, alkyl-oxygen fission, unimolecular). It may be regarded as an S_N1 displacement reaction on the triphenylmethyl group with acetate as the leaving group.

$A_{AL}1$ Mechanism of Ester Hydrolysis

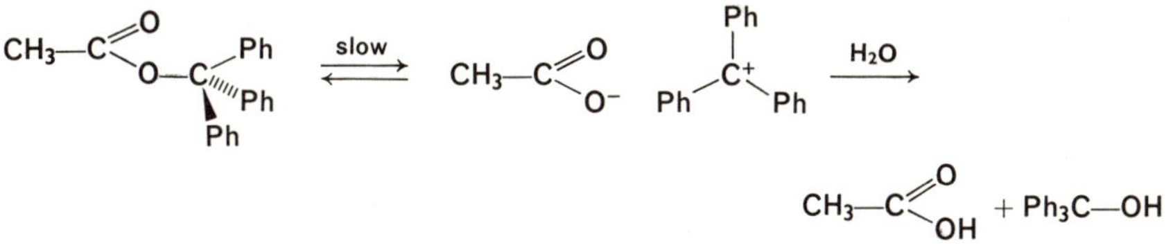

Elimination–Addition Processes

Net hydrolysis occurs if the leaving group and a β-proton are first eliminated, followed by addition (of water, etc.) to the intermediate ketene or isocyanate, thus:

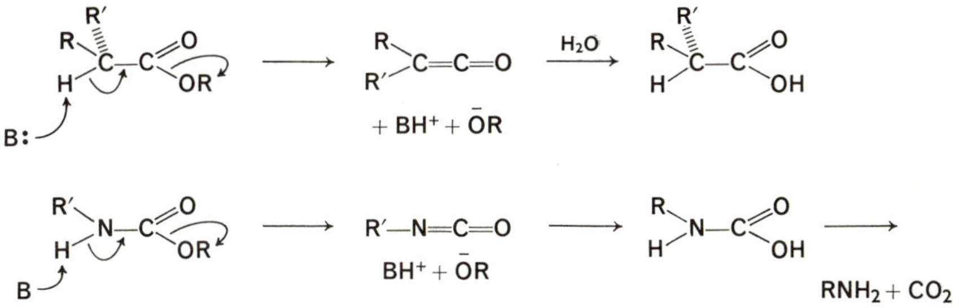

An example of the latter is the hydrolysis of carbamates of o-phenylene-diamine.

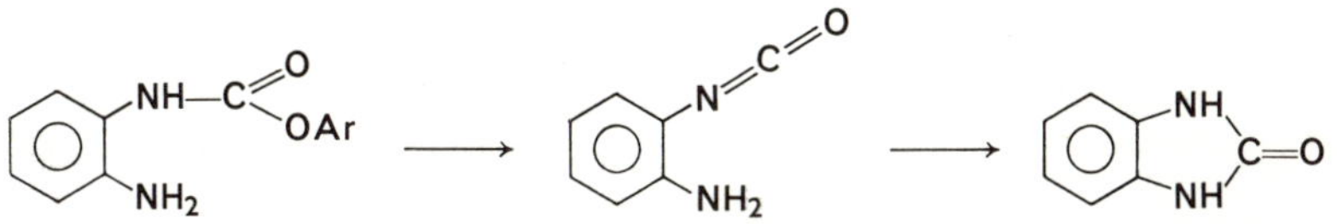

8.5   **NUCLEOPHILIC AND GENERAL-BASE CATALYSIS OF ESTER HYDROLYSIS**

The hydrolysis rates of esters of strongly acidic phenols (e.g. 2,4-dinitrophenyl acetate) are strongly catalysed by added nucleophiles such as acetate ion. If aniline is added to the hydrolysing solution, acetanilide may be isolated from the products. It is inferred that acetic anhydride is formed as an intermediate by the two-stage displacement of aryloxide ion by acetate. The rate of this

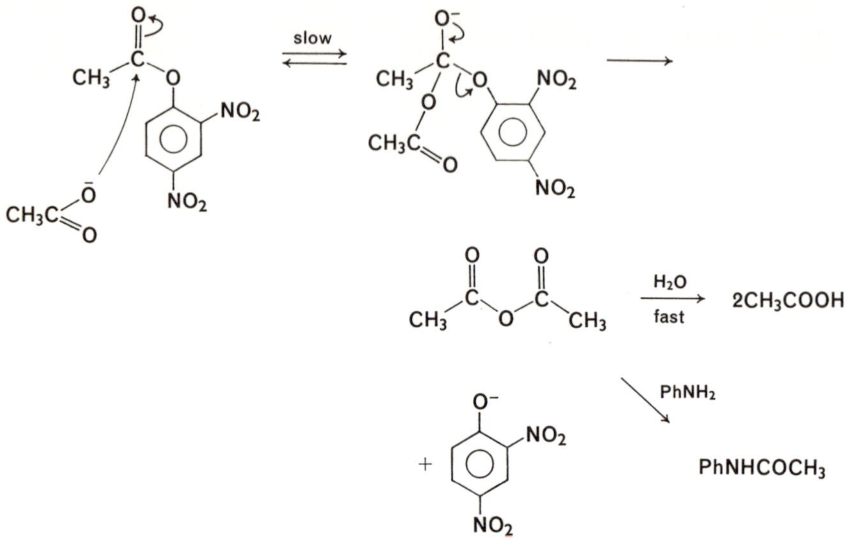

nucleophilically catalysed hydrolysis is ~10^5 times greater than hydrolysis by water, and the entropy of activation (-45 J K^{-1}) is consistent with a bimolecular reaction.

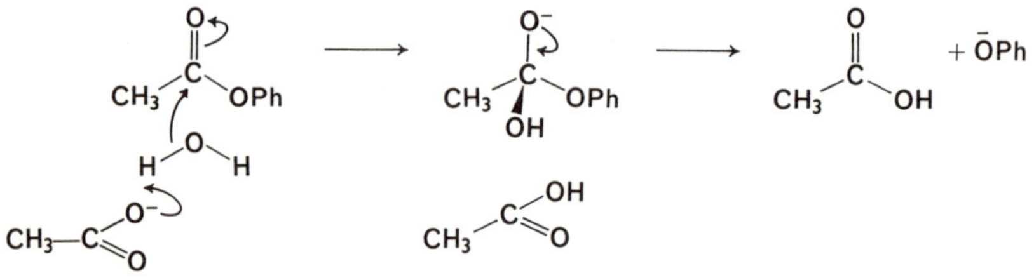

Esters of less acidic phenols (e.g. phenyl acetate) do not partake of this catalysis, the unactivated phenoxide ion being a much poorer leaving group, but instead hydrolyse in neutral solution by a general-base catalysed mechanism whereby a weak base or nucleophile such as acetate ion helps to remove a proton from a water molecule which then behaves like OH$^-$ in a $B_{AC}2$ reaction Catalysis rates are much less (about 30 times that of the uncatalysed reaction) and entropies of activation are more negative (-125 J K^{-1}) consistent with the participation of three molecules in the transition state.

The aspirin anion (*o*-acetylsalicylate) also hydrolyses by this type of mechanism between pH 4 and 8, but the general base is the neighbouring carboxyl group. The reaction is characterized by a large negative entropy (-130 J K^{-1}) and no effects of substituents at positions 4 and 5. An electron-

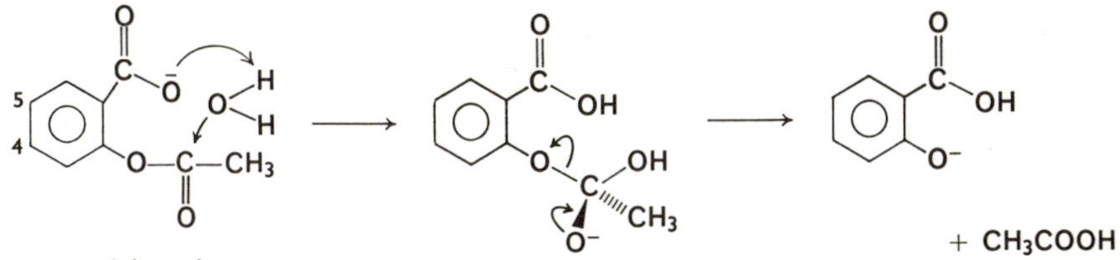

aspirin anion

withdrawing group at these positions will weaken the basic strength of the carboxyl and facilitate attack at the acetate carbonyl, the two effects being opposed and cancelling.

Imidazole Catalysis of Ester Hydrolysis

Imidazole and its derivatives have also been shown to be effective in catalysing the hydrolysis of certain esters such as *p*-nitrophenyl acetate. It is certain that

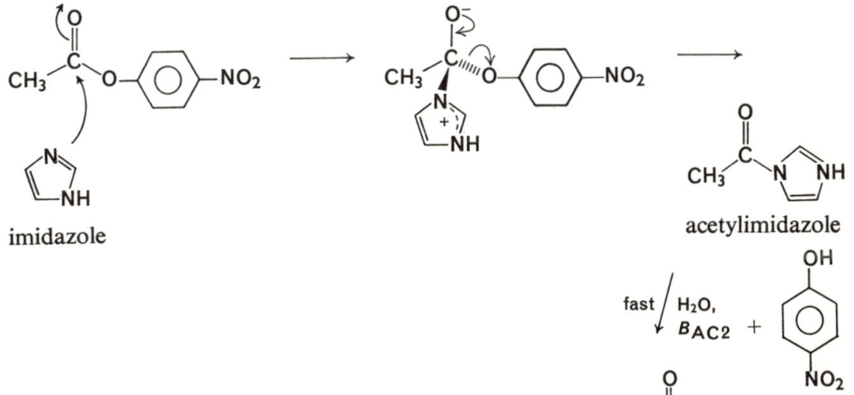

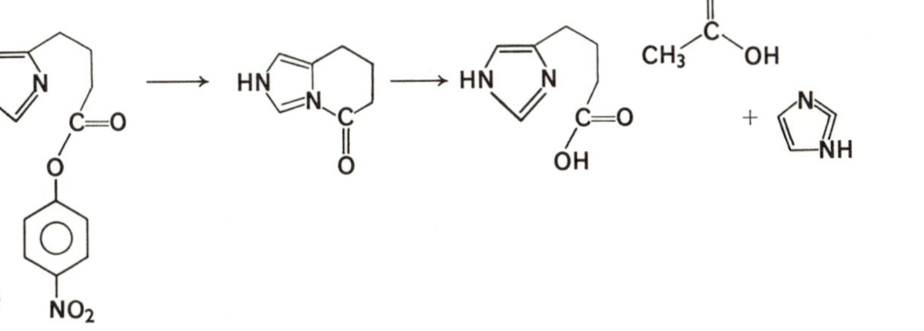

imidazole

acetylimidazole

(13)

the imidazole is involved in nucleophilic attack on the ester and subsequently itself suffers displacement. Acetylimidazole is an intermediate in the reaction and is known to be rapidly hydrolysed by water. In the intramolecularly catalysed hydrolysis of **13**, the intermediate acylimidazole may be detected.

Metal Catalysis of Ester Hydrolysis

Amino-acid esters hydrolyse by acid- or base-catalysed mechanisms and in addition are very efficiently catalysed by certain metal ions such as copper. The rates of hydrolysis are as follows:

$$PhCH_2CH—COOEt \quad \rightarrow \quad PhCH_2CH—COOH$$
$$\underset{NH_2}{|} \qquad\qquad\qquad \underset{NH_2}{|}$$

phenylalanine
ethylester

Catalyst	k (s^{-1})
H_3O^+	1.46×10^{-14}
OH^-	$5.8 \ \times 10^{-9}$
Cu^{2+}	2.67×10^{-3}

The copper ion is believed to coordinate to the amino group and act as an electrophile towards the carboxylate group which is then more susceptible to attack by water.

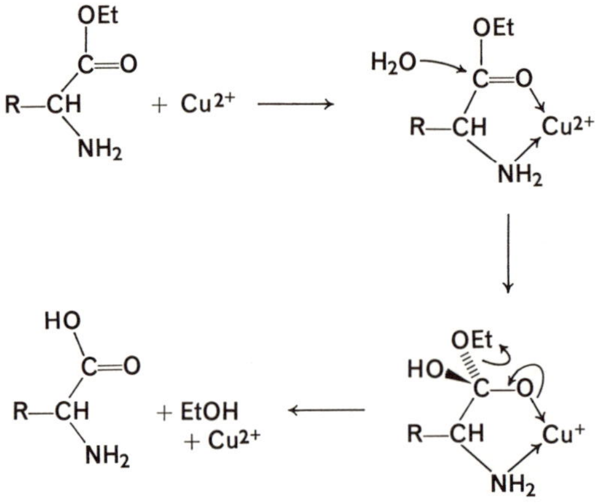

8.6   OTHER CARBONYL DISPLACEMENT REACTIONS

The mechanisms of ester hydrolysis discussed above are believed to apply to many other displacements occurring at the carbonyl group which may be summed up in the following table.

Table 2. Summary of carbonyl displacement reactions

$$R-C{\overset{O}{\underset{X}{\lessgtr}}} + :Y \longrightarrow R-C{\overset{O}{\underset{Y}{\lessgtr}}} + :X$$

Reaction	Nucleophile Y	Displaced group X	Mechanism
Amide hydrolysis	OH^-	NH_2^-	$B_{AC}2$
	H_2O	NH_3	$A_{AC}2$
Ester exchange	ROH	$R'OH$	$A_{AC}2, A_{AC}1$
Hydrolysis of acyl halides	H_2O	Cl^-, Br^-	$B_{AC}2$ $(A_{AC}2?)$
Ammonolysis of acyl halides	$NH_3, RNH_2,$ R_2NH etc.	Cl^-, Br^-	$B_{AC}2$
Alcoholysis of acyl halides	ROH	Cl^-, Br^-	$B_{AC}2$
Hydrolysis of anhydrides	H_2O, OH^-	$RCO\cdot O^-$	$B_{AC}2$
Alcoholysis of anhydrides	ROH, RO^-	$RCO\cdot O^-$	$B_{AC}2$
Ammonolysis of anhydrides	$NH_3, RNH_2,$ R_2NH	$RCO\cdot O^-$	$B_{AC}2$
Ammonolysis of esters	$NH_3, RNH_2,$ R_2NH	RO^-	$B_{AC}2$

An amide, for instance, is formed by the action of an amine upon an anhydride (or an ester or acyl halide),

while it is hydrolysed by heating in either acidic or basic solution. Reactions of acyl halides and anhydrides are much faster than those of esters and amides since they have better leaving groups (Cl^-, ^-OCOR, rather than OR^-, NH_2^-). For these, little evidence is available as to whether they undergo the normal $B_{AC}2$ type of reaction or whether the displacement is concerted, but the former may be presumed by analogy with ester reactions until evidence is available to suggest otherwise. In the presence of metal ions, e.g. Hg^{2+}, acyl halides may hydrolyse by a metal-catalysed ionization route, since large increases in rate are observed, most likely due to electrophilic assistance to C—hal bond breaking.

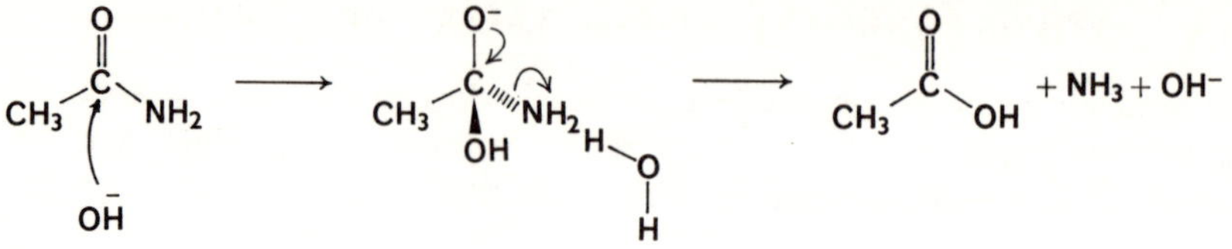

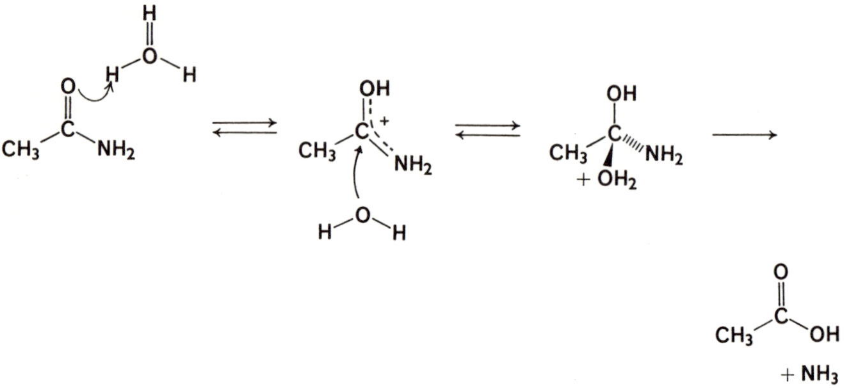

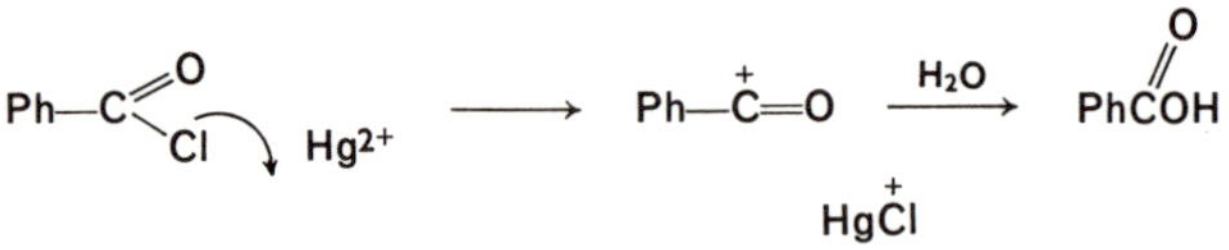

8.7   CONDENSATION REACTIONS

The term 'condensation' refers to a combination of two reagents with the simultaneous loss of a small molecule, usually water. Condensations at the carbonyl group are very familiar reactions and may be regarded as a nucleophilic addition to give a tetrahedral intermediate which then undergoes a β-elimination of water. A typical sequence is the formation of a ketone phenylhydrazone,

$$\begin{array}{c}\underset{R'}{\overset{R}{\diagdown}}C{=}O \end{array} \longrightarrow \underset{R'}{\overset{R\text{\tiny///}}{\diagdown}}\underset{NH}{\overset{OH}{\diagup}}C \qquad \overset{-H_2O}{\underset{(E2)}{\longrightarrow}} \quad \underset{R}{\overset{R}{\diagdown}}C{=}NNHPh$$

$$\underset{NHPh}{\big|}$$

Ph—NH—N̈H₂

phenylhydrazine

phenylhydrazone

A number of condensations of nitrogenous compounds, X—NH₂, are used in the preparation of crystalline derivatives of aldehydes and ketones:

$$R_2C{=}O + H_2N{-}X \rightarrow R_2C{=}N{-}X + H_2O$$

Compound	X	Product	
Primary amine	R, Ar	Anil (imine)	$R_2C{=}NH$
Hydrazine or substituted hydrazine	–NH₂, NHR	Hydrazone	$R_2C{=}N{-}NHR$
Semicarbazide	–NHCONH₂	Semicarbazone	$R_2C{=}N{-}NHCONH_2$
Hydroxylamine	–OH	Oxime	$R_2C{=}N{-}OH$

Oxygen exchange in a ketone or ester can be represented as a condensation reaction:

$$\underset{R'(O)}{\overset{R}{\diagdown}}C{=}^{18}O \underset{\beta\text{-elimination}}{\overset{\longrightarrow}{\longleftarrow}} \underset{R(O)}{\overset{{}^{18}OH}{R'{\diagdown}\underset{|}{C}{\diagup}OH}} \underset{}{\overset{\beta\text{-elimination}}{\overset{\longrightarrow}{\longleftarrow}}} \underset{R(O)}{\overset{R}{\diagdown}}C{=}O + H_2{}^{18}O$$

with H—O—H added.

Essentially similar processes are involved in the aldol family of condensations (Section 4.16) between a carbonyl compound and an enolate ion (14). In this case, the intermediate tetravalent alcohol is frequently isolable and requires further treatment to cause the elimination.

It will be noticed that condensation reactions are almost entirely limited to aldehydes and ketones. These are unable to undergo the displacements which are characteristic of esters and other carboxylic derivatives since they have no readily displaceable group attached to the carbonyl centre.

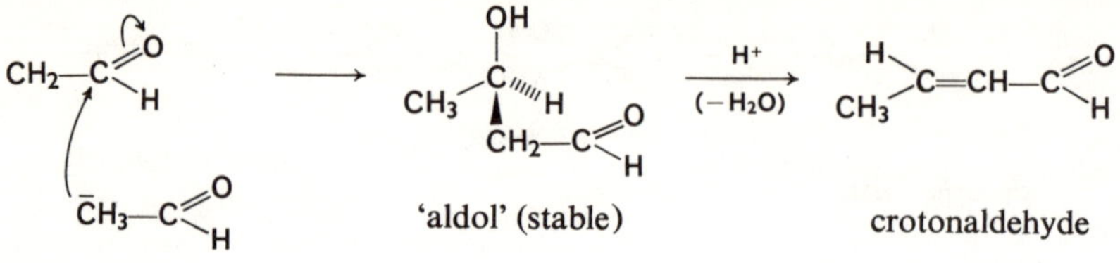

'aldol' (stable)

crotonaldehyde

(14)

8.8 ISOLATION OF TETRAHEDRAL INTERMEDIATES

The transient intermediates discussed above have two or three hetero atoms, oxygen or nitrogen, attached to the central carbon and unless these groups are alkylated, the compounds are usually highly unstable,

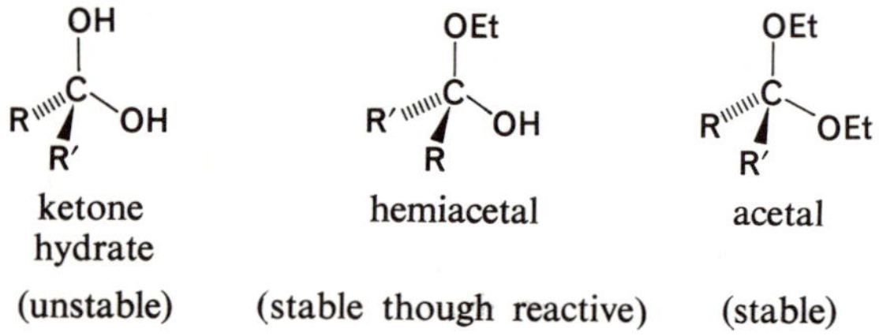

ketone hydrate	hemiacetal	acetal
(unstable)	(stable though reactive)	(stable)

A few instances are known in which this type of intermediate is stable. They are mostly derived from highly halogenated carbonyl compounds. The electron-withdrawing substituent tends to retard the breaking of the C—O bond which is necessary for decomposition. The isolation of these compounds strengthens the arguments for their transient formation elsewhere.

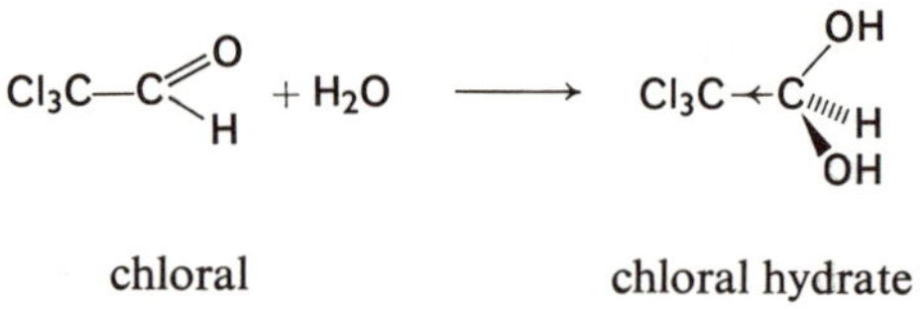

chloral

chloral hydrate

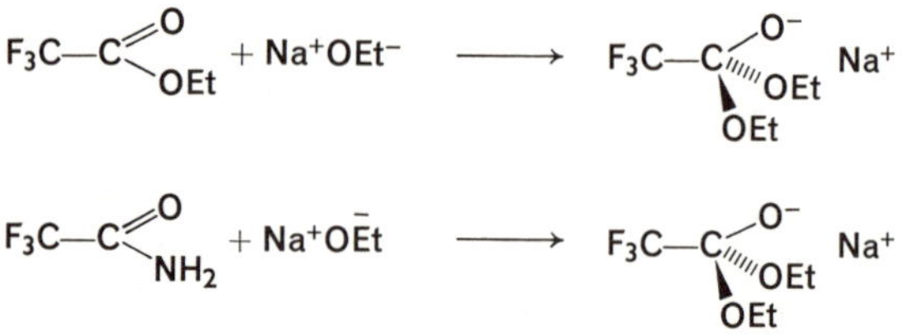

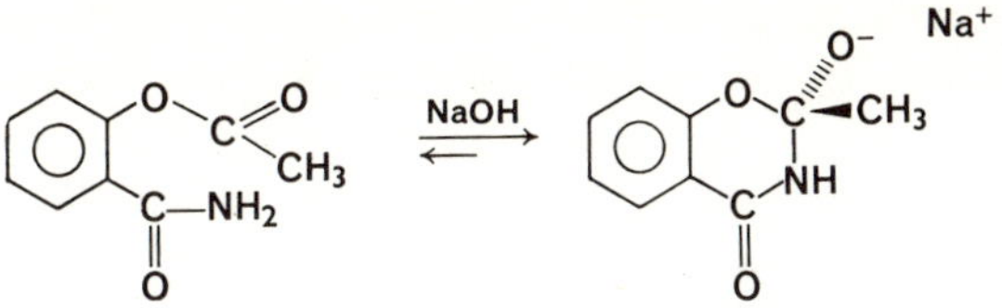

8.9   ADDITIONS TO THE CARBONYL GROUP

Numerous important reactions lead to stable tetrahedral derivatives of a carbonyl compound. The addition of Grignard reagents and other organo-metallic compounds—essentially carbanions and so related to the aldol reaction mentioned above and in Section 4.16, are familiar examples. Addition

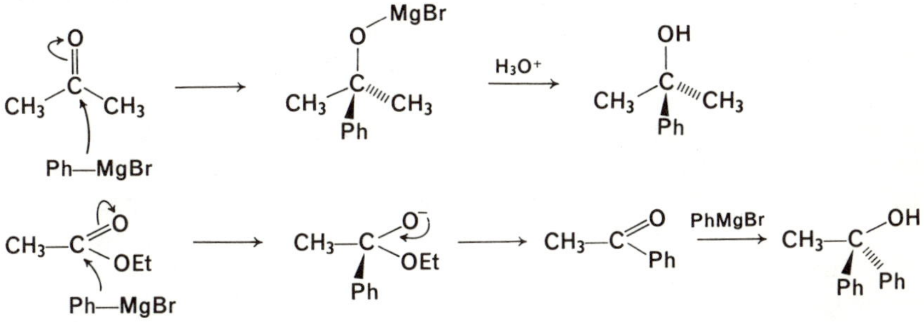

to ketones and aldehydes occurs since no displaceable group is present, nor any proton on the incoming nucleophile which can very readily take part in an elimination. An ester will undergo initial displacement of the alkoxide group and react with a second molecule of Grignard reagent. Reduction of the carbonyl function invariably leads to a saturated addition product—an alcohol. Of the homogeneous methods of reduction, complex metal hydrides provide clean and satisfactory conversion to alcohols. Lithium aluminium hydride, $Li^+AlH_4^-$, and sodium borohydride, $Na^+BH_4^-$, salt-like compounds, both act as donors of the hydride ion, $H^-:$, a powerful nucleophile. In each case all four hydrogens are available for reduction.

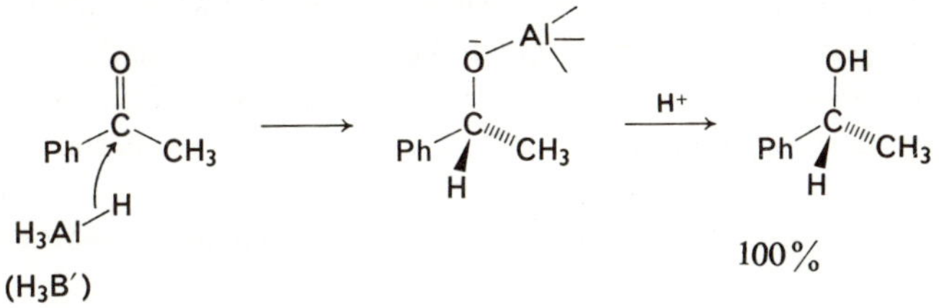

100%

An ester has a displaceable group (an alkoxide ion) and may be considered to form an intermediate ketone or aldehyde. This is immediately reduced by further hydride

The aluminohydrides are much more powerful reagents than the borohydrides. The former, used in anhydrous conditions in ether solvents, will reduce ketones, aldehydes, esters, amides, nitriles, etc., to their lowest oxidation state (alcohols, primary amines) while borohydrides are more selective and will reduce ketones and aldehydes to the corresponding alcohols in alcoholic or aqueous solution in the presence of ester groups.

Another effective source of the hydride ion is the aluminium 'salt' of a secondary alcohol such as isopropanol. This will undergo mutual oxidation and reduction with a ketone in a highly selective manner by exchange of H^- in a cyclic process. The reaction is an equilibrium and can be used for reduction of the ketone (Verley–Meerwein–Pondorff reduction) or oxidation of the secondary alcohol (Pondorff–Oppenauer oxidation), depending on whether isopropanol or acetone is in excess. A hydride exchange between two molecules

Al isopropoxide   transition state

of aromatic aldehyde is catalysed by strong base and known as the Cannizzaro reaction. The products are the alcohol and carboxylate, hence this is a mutual oxidation–reduction reaction. The reaction will occur with aliphatic aldehydes in the presence of aluminium ethoxide and is then known as the Tischenko reaction. If carried out in D_2O, the reaction yields benzyl alcohol containing no deuterium, showing that the hydrogen gained originated from a benzaldehyde molecule.

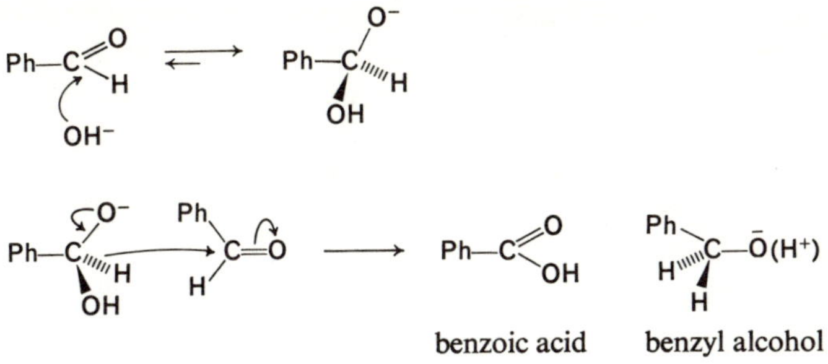

benzoic acid   benzyl alcohol

8.10   HYDROLYSES OF OTHER ESTERS

Sulphonic esters RSO_2OR', are potentially able to undergo $B_{AC}2$ and $A_{AC}2$
hydrolytic mechanisms and, indeed, relatively unreactive esters such as phenyl
benzenesulphonate and p-toluenesulphonate do so, as shown by uptake of ^{18}O
into the acid but not the phenol moiety on hydrolysis in $H_2^{18}O$. Sulphonate

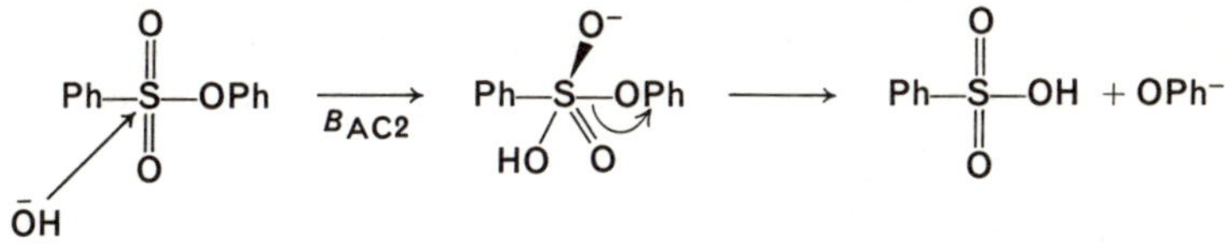

groups, however, are much better leaving groups than carboxylate (sulphonic
acids are stronger than carboxylic) and the intermediates may be less long lived,
or formed irreversibly. Also alkyl–oxygen bond fission is a much commoner
mode of hydrolysis—and solvolysis generally—as is illustrated by numerous
examples in Chapter 2. The hydrolysis of phosphate esters is biologically

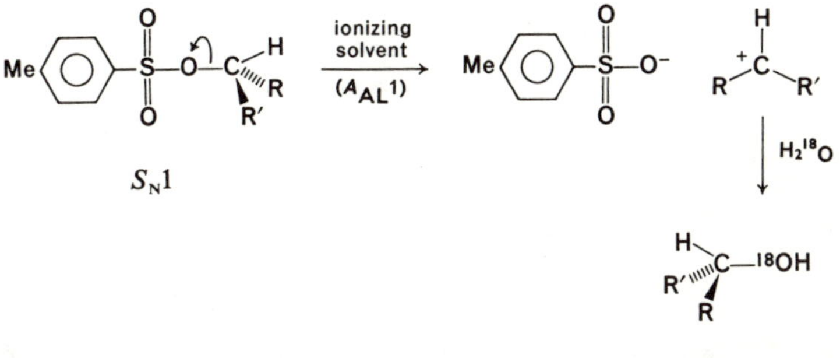

significant, although *in vivo* these reactions are mediated by enzymes. The
situation is complicated by the tribasic nature of phosphoric acid and the

tendency to form polyphosphates. In the following examples there is frequently some doubt as to the state of protonation of the reacting entities.

Simple alkyl dihydrogen phosphates such as the methyl ester (15) hydrolyse at very low rates in alkaline solution but the rate increases by some 10^8 between pH 8 and pH 5 and diminishes at more acidic values, to increase again in highly acidic solution (pH 1). The main reaction by which the hydrolysis proceeds is of the $B_{AC}1$ type—a decomposition of the monoanion as the slow step, and independent of OH^- or other nucleophiles. At higher pH, the less

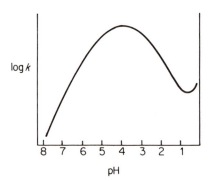

(15)   + MeOH   detectable

reactive dianion is formed. The incorporation of no ^{18}O from solvent into the alcohol indicates that phosphorus–oxygen bond breaking occurs, which is in agreement with the totally retained configuration of an optically active alcohol moiety. The rate–pH profile is parallel to the concentration of monoanion (Figure 2). At the highly acidic end of the range the reactive species is evidently a protonated phosphoric acid since there is an increase in rate in D_2O over

Figure 2. pH dependence of the hydrolysis of methyl dihydrogen phosphate.

H_2O and the rate is proportional to pH (rather than h_0). However, evidence from incorporation of ^{18}O is not clear-cut; the extent of incorporation into the alcohol produced is consistent with two pathways, 35 % requiring P—O fission and 65% requiring C—O fission. These are presumed to be $A_{AC}2$ and $A_{AL}2$ mechanisms, respectively.

$A_{AL}2$ Mechanism

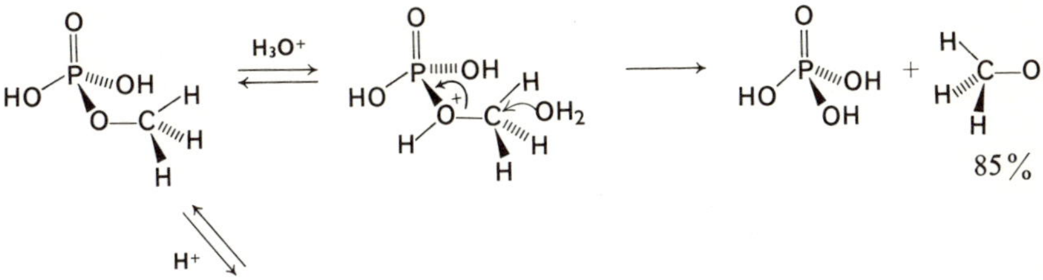

$A_{AL}2$ Mechanism

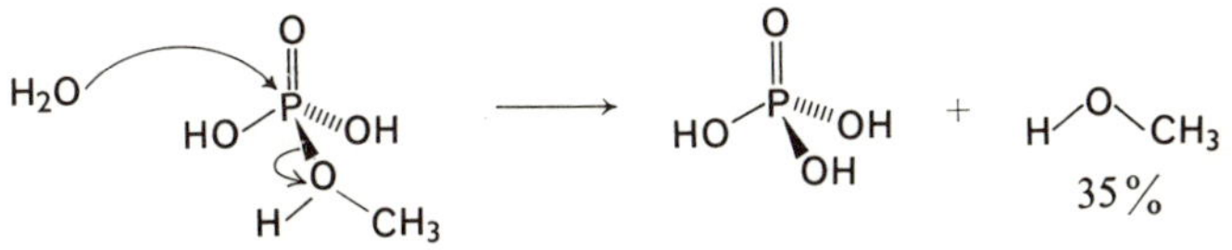

[$S_N2(P)$ process]

With the relatively few examples studied, secondary and tertiary phosphate hydrolysis can be accounted for by bimolecular (S_N2) displacements upon carbon or upon phosphorus by hydroxide ion. The former predominates in

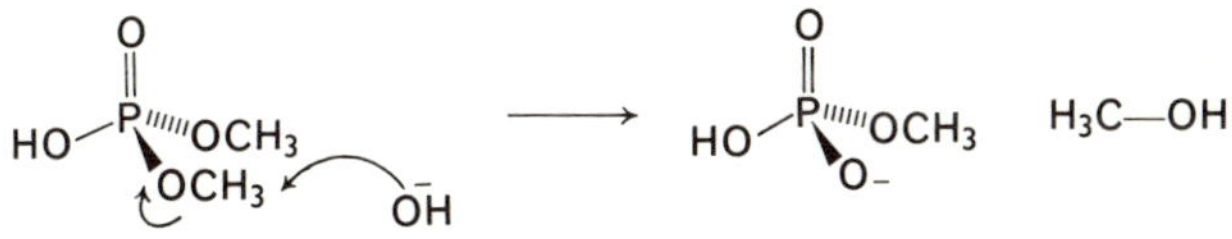

(16)

the hydrolysis of dimethyl phosphate (16), (90:10) while the latter is the exclusive route for trimethyl phosphate (17). Magnesium ion greatly catalyses

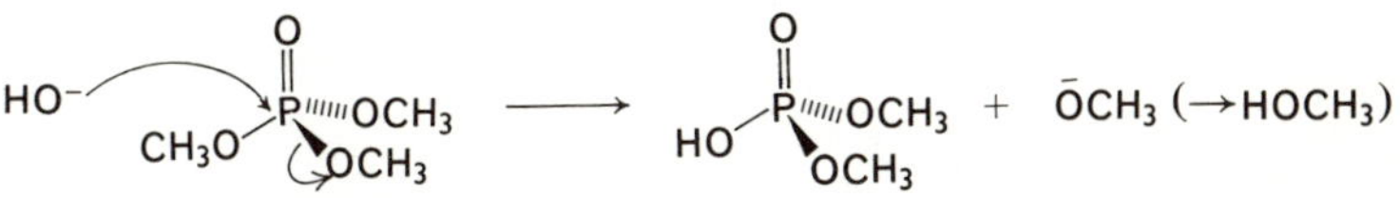

(17)

the hydrolysis of phosphate esters, the maximum efficiency for acetyl-phosphate hydrolysis being at pH 7.7, which corresponds to the dianion as the reactive species. Coordination of the metal ion probably renders the acetyl group more susceptible to the nucleophilic attack.

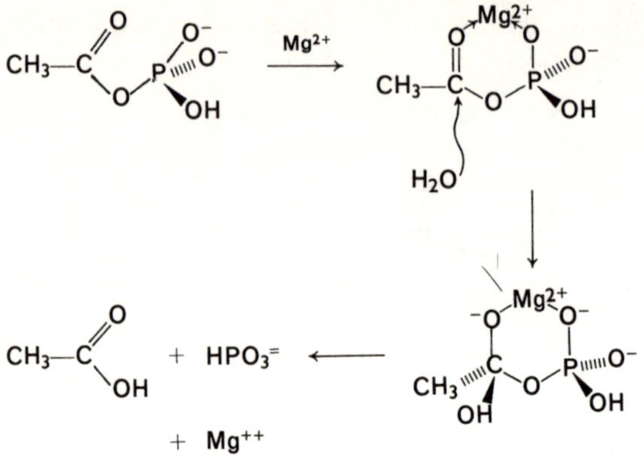

Chapter 9

Intermediates in Oxidation

The oxidation and reduction of organic compounds encompass a wide variety of mechanisms. The wide use of metals in high or low oxidation states as oxidizing and reducing agents, and the uncertainty as to the exact nature of these compounds in solution and of their valence states immediately after reaction, makes the subject one of the most complicated. The examples chosen for discussion here are those whose mechanisms are best understood and which constitute preparatively useful reactions and are thus important in organic chemistry.

9.1   OXIDATION OF ALCOHOLS

The oxidation of secondary alcohols to ketones is the most studied and usually most successful; primary alcohols, while readily able to be oxidized to aldehydes, frequently present a more complicated picture as these products are further converted to carboxylic acids. Tertiary alcohols are not readily oxidized but under forcing conditions deep-seated degradation may occur. Alcohol oxidation, in principle, involves loss of a proton from oxygen and either a hydride ion from the α-carbon or a hydrogen atom followed by an electron in two steps. It is frequently difficult to decide between these alternatives. Chromic acid, H_2CrO_4 in water, or as CrO_3 in acetic acid, is an

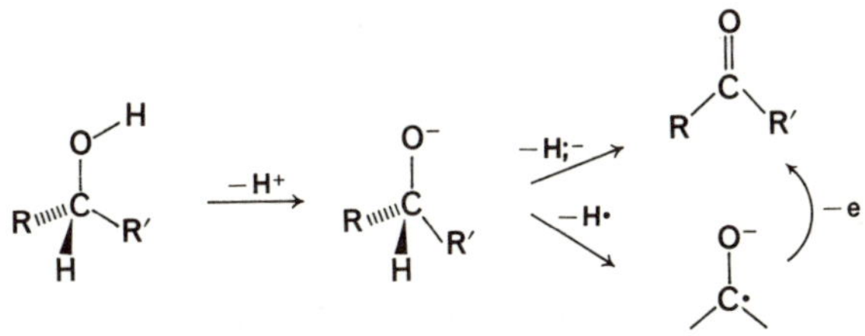

efficient reagent for effecting this conversion. The rate of oxidation in either solvent is accelerated by added strong acid and is proportional to the Hammett

445

acidity function, H_0 (h_0) (Section 1.10). In general, the rate expression, equation 9.1, is obeyed.

$$\text{rate} = [\text{alcohol}]\,[\text{CrO}_3]\,h_0 \tag{9.1}$$

Hence the transition state contains one molecule each of alcohol and chromic acid together with a proton. By careful analysis of the rate dependence on acidity and on the equilibrium,

$$2\,\text{HCrO}_4{}^{2-} \rightleftharpoons \text{Cr}_2\text{O}_7{}^{2-} + \text{H}_2\text{O}$$

Westheimer concluded that these were not parallel and that the attacking species was a monomeric, not a dimeric, chromate ion. A full primary deuterium isotope effect is found; the oxidations of isopropanol and iso-propanol-α-d have $k_H/k_D = 6.4$. Consequently, the α-C—H bond is being

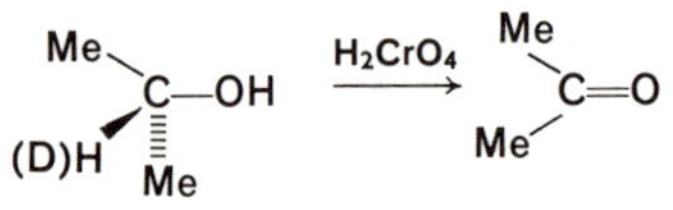

$$k_H/k_D = 6.4$$

broken in the slow step. The following mechanism, due to Westheimer, accommodates most of the facts. The first step is the formation of alkyl

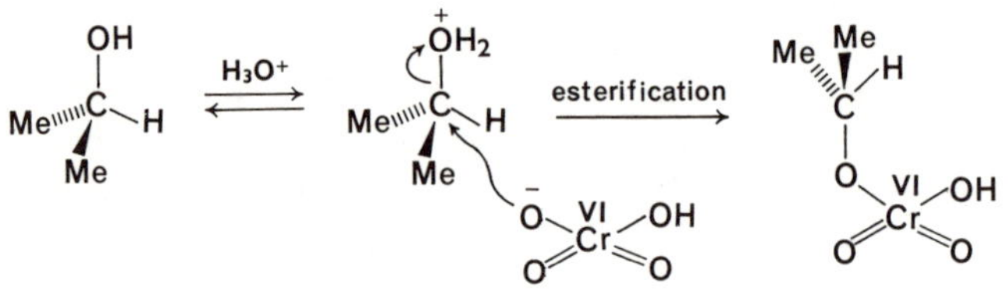

(1)

isopropyl chromate

chromate **(1)** by an ordinary acid-catalysed esterification (Section 8.3). Then, in a pre-equilibrium step, the chromate is protonated **(2)** rendering it a better leaving group (and explaining the dependence upon H_0); in the slow step this group is expelled together with the α-hydrogen in a β-elimination. The

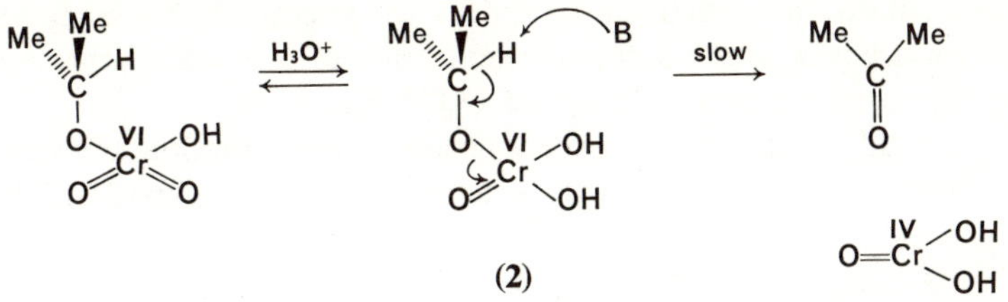

(2)

two-electron transfer to chromium in this step constitutes a reduction from the +6 to the +4 state. A disproportionation of this valence state occurs,

$$Cr^{IV} + Cr^{VI} \rightarrow 2\, Cr^{V}$$

and the more stable chromium (V) compound is capable of forming an ester for a second oxidation step, at which it drops to the stable +3 state, the final state of the chromium. The slow step of the oxidation therefore resembles an

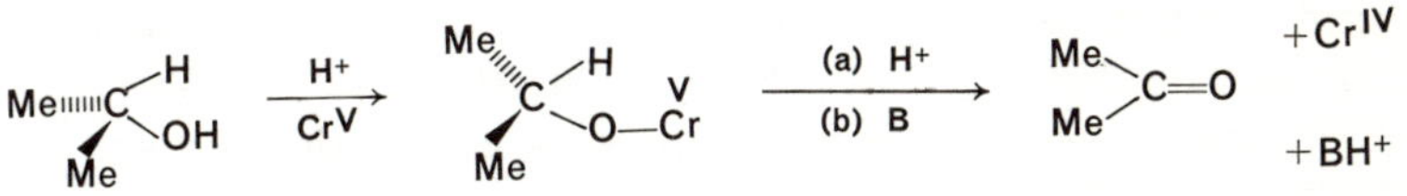

$E2$ elimination such as is commonly associated with the formation of olefin from an alkyl halide. Electronic influences on the rate are rather slight, the

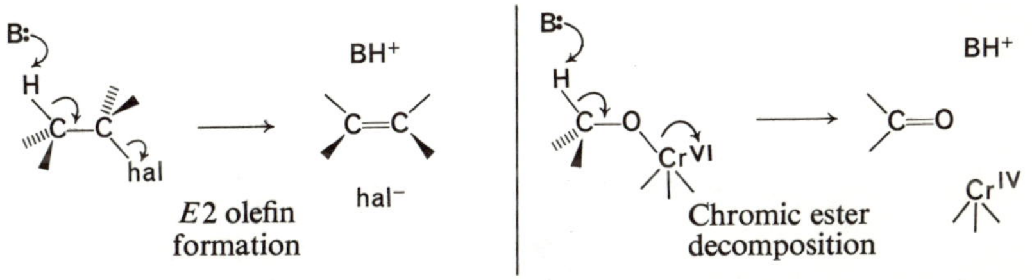

$E2$ olefin formation

Chromic ester decomposition

benzylic alcohols (3) react faster with electron-donating groups present ($\rho = -1.01$). This small value is not unreasonable since the electronic demands on C—H and O—Cr bond breaking are in opposite directions; nevertheless, the negative ρ indicates that the latter is the more advanced in the transition state,

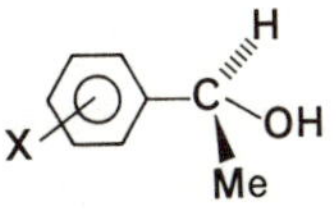

(3)

There is a substantial solvent-isotope effect, the rates in D_2O being faster than in H_2O on account of the higher concentration of protonated chromic ester consequent upon the greater acidity of D_3O^+ than H_3O^+.

An essentially similar mechanism is proposed for oxidation of a secondary alcohol by nitric acid. In this case the reaction is aided by electron-withdrawing

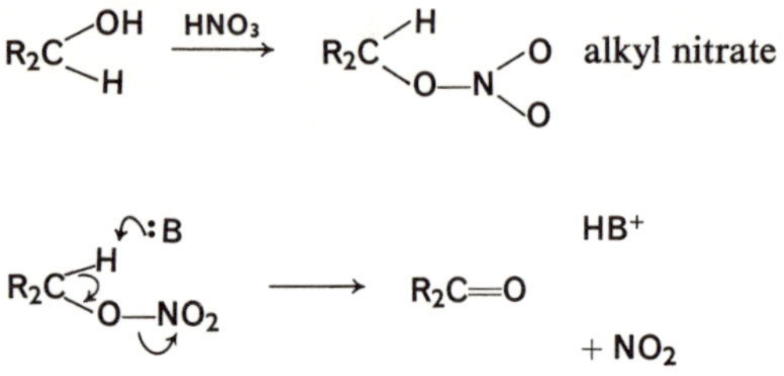

groups, hence proton removal is more advanced than O—N bond fission in the transition state.

Permanganate, especially in alkaline solution, is often employed in the oxidation of alcohols. The rate increases with pH and it is probable that attack of permanganate on alkoxide ion occurs with removal of either a hydride ion or hydrogen atom. The effect of substituents is not easy to assess since the

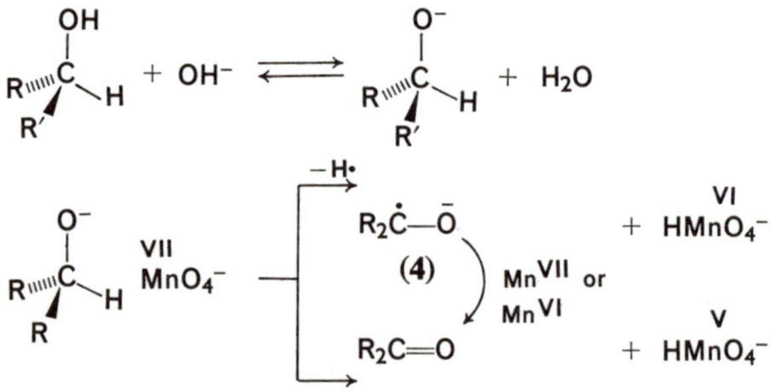

position of the pre-equilibrium and hence concentration of alkoxide ion is also affected. However, in the series

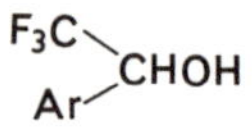

at high pH it is believed that complete ionization has occurred and rates are increased by electron withdrawal, $\rho = +1.0$. It would be expected that hydride loss would be characterized by a negative ρ—assistance from electron

donation—hence this tends to favour the radical mechanism. The intermediate would indeed be a ketyl (**4**) (Section 5.5), and such species are known
to have some stability and to be further stabilized by such groups as nitro
(which would lead to a positive value of ρ). Further, this intermediate may
explain why the alkoxide ion is oxidized so much faster than the neutral
alcohol. It is known that protonated ketyls are much more acidic than the
corresponding alcohols (by about 10^6), i.e. the ketyl is much more stable (with
respect to its conjugate acid) than the alkoxide (with respect to alcohol).
Oxidation of the alcohol leads to protonated ketyl while oxidation of alkoxide
ion leads directly to the much more stable ketyl, and thus should be favoured.

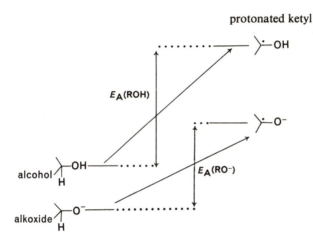

As with chromic acid, there is a full primary-isotope effect. In fact, k_H/k_D
can reach values as high as 19 for $(CF_3)_2C$—OH. This is much larger than the
theoretical maximum for loss of a C—H stretching vibration in the transition
state and may implicate changes in other vibrational modes, but quantum-
mechanical tunnelling (Section 1.11) through the potential barrier is the
likely explanation. We can be sure that the α-CH bond is greatly weakened in
the transition state.

'Active' manganese dioxide is a form of MnO_2 made by careful precipitation
and subsequent treatment under controlled conditions. It is likely that this
affects the surface properties since it acts as a useful heterogeneous oxidant
which will convert alcohols to ketones or aldehydes, often in good yield and
without causing remote double bonds to move into conjugation. Being a

$$\text{Cyclohexanol derivative} \xrightarrow[\text{3 days, } CH_3CN]{MnO_2} \text{ketone} \quad 70\%$$

$$C_3H_7CH_2OH \xrightarrow[\text{bed of } MnO_2]{\text{filter through}} C_3H_7CHO$$

$$70\%$$

heterogeneous reaction, the mechanism is difficult to investigate but a radical process similar to that utilized by permanganate is likely.

A similar radical mechanism is also proposed for alcohol oxidation by vanadium (V) (pervanadyl cation, VO_2^+). A red complex forms, on mixing with an alcohol, which is assumed to be pervanadyl ester (**5**). The system also initiates polymerization of acrylonitrile, which is diagnostic of the presence of free radicals. Cerium (IV) and cobalt (III) are powerful one-electron oxidants

$$R_2C\overset{OH}{\underset{H}{<}} \xrightarrow[H_2O]{\overset{V}{VO_2^+}} R_2C\overset{O-V\overset{OH}{\underset{OH}{\lessgtr}}}{\underset{H}{<}} \longrightarrow R_2\overset{\cdot}{C}-O-V(OH)_2 \xrightarrow{V^V} R_2C=O$$

$$(5) \qquad\qquad + H_2O \qquad\qquad + V^{IV}$$

which will readily convert alcohols to ketones. It is believed that coordination of the alcohol to the metal is followed by an internal hydrogen transfer.

$$\longrightarrow \qquad \xrightarrow[(-H\cdot)]{Co^{II}} O=C\overset{R}{\underset{R}{<}} \quad + Co^{II}$$

Molecular bromine, on the other hand, appears to operate by hydride-ion abstraction. There is a moderate primary isotope effect ($k_H/k_D \sim 3$) at the

$$\longrightarrow \qquad H-Br \xrightarrow{-H^+} R\overset{O}{\underset{R'}{\overset{\|}{C}}} $$

α-CH and also at the hydroxyl ($k_H/k_D = 1.5$), although the latter could be mainly a secondary effect due to changes in charge and hybridization taking place at oxygen. The rate is accelerated by electron donation.

9.2   OXIDATION OF HYDROCARBONS

Saturated Positions

Although the oxidation of saturated hydrocarbons at discrete positions is much less easily achieved than at a functional group such as an alcohol, certain compounds lend themselves to facile and specific oxidation by chromic acid. In general, tertiary positions are oxidized to alcohols and secondary, less readily, to ketones. The reactions are facilitated if the sites are in benzylic positions. The following examples occur in good yield.

(a) $(CH_3)_3C—CH_2CH_3 \xrightarrow{\;CrO_3\;} (CH_3)_3C—\overset{\overset{\displaystyle O}{\|}}{C}—CH_3$

(b)

adamantane   70%   10%

(c) $Ph_3CH \xrightarrow{\;CrO_3\;} Ph_3C—OH$

(d) $Ph_2CH_2 \xrightarrow{\;CrO_3\;} Ph_2C{=}O$

The conversion of diphenylmethane to benzophenone has been studied in detail; there is a large isotope effect ($k_H/k_D = 6.6$) and moderate acceleration by electron-donating groups ($\rho = -1.4$). The rate law is similar to that of alcohol oxidation,

$$\text{rate} = [Ph_2CH_2][CrO_3]h_0$$

It may be inferred that one molecule each of the reagents, and a proton but no water molecule, are involved in the transition state. The problem is to decide whether the oxidizable hydrogen is removed as a proton or a hydrogen atom. Rates of oxidation are found to parallel rates of hydrogen abstraction by radicals in a series of substrates rather than solvolytic rates of the corresponding chlorides, for example:

Substrate	$PhCH_3$	$PhCH_2Me$	Ph_2CH_2	Ph_3CH
k_{rel} (CrO_3 oxidation)	1.0	3.1	6.3	8.1
k_{rel} (H· abstraction by ·CCl_3)	1.0	3.0	8.0	16.7
k_{rel} (solvolysis of RCl)	1.0	10^2	2×10^3	10^7

But on the other hand, added azide ion in chromic-acid oxidations may give a good yield of alkyl azide. The azide ion is known to be an efficient carbonium-ion trap. The problem is further complicated since several optically active

$$Ph_3CH \xrightarrow[NaN_3]{H_2CrO_4} (Ph_3C^+?) \longrightarrow \underset{97\%}{Ph_3CN_3}$$

tertiary substrates (6) have been shown to be oxidized with retention of configuration. To accommodate this it is inferred that a carbonium ion or

(6)

radical intermediate is not free but is formed within a solvent cage and acquires oxygen from the same chromium species that removed the hydrogen. The following mechanism allows for the various possibilities.

Methylene groups adjacent to a carbonyl or allylic position may be conveniently oxidized to a carbonyl function by heating with selenium dioxide in water or acetic acid (as selenous acid, H_3SeO_3). It has been established that the reaction

$$PhCOCH_2Ph \xrightarrow{SeO_2} PhCOCOPh + Se + H_2O$$

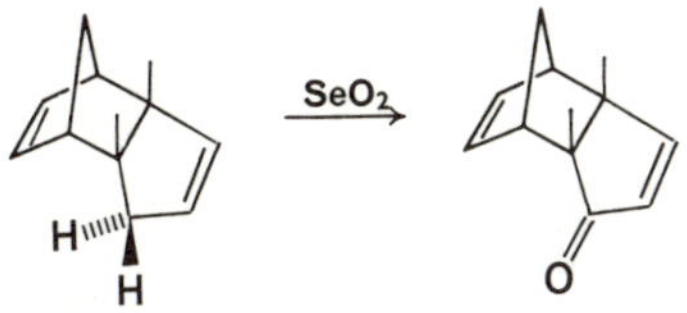

dicyclopentadiene

is catalysed both by acid and by base, but both routes proceed by way of an intermediate enol selenite (7) which subsequently decomposes. There is a full isotope effect ($k_H/k_D = 6$–7) due to the formation of enol, or its ester being the rate-determining step. The selenite is believed to undergo an internal re-

Acid-catalysed mechanism

$$H_2SeO_3 + H_3O^+ \rightleftarrows H_3SeO_3^+ + H_2O$$

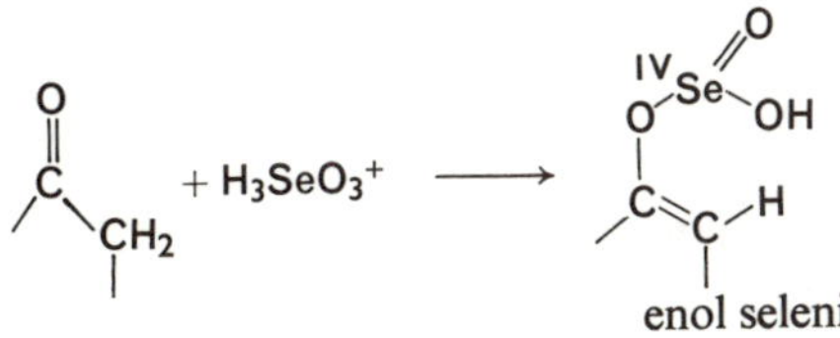

enol selenite (7)

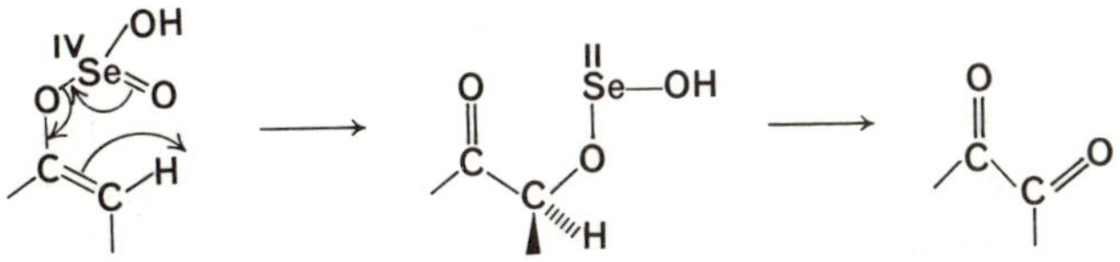

(8) $+ Se + H_2O$

arrangement with an effective two-electron reduction and a transient ester of Se^{II} (8) which decomposes with further reduction forming the diketone and elementary selenium.

The SeO_2 oxidation of olefins leads to allylic alcohols.

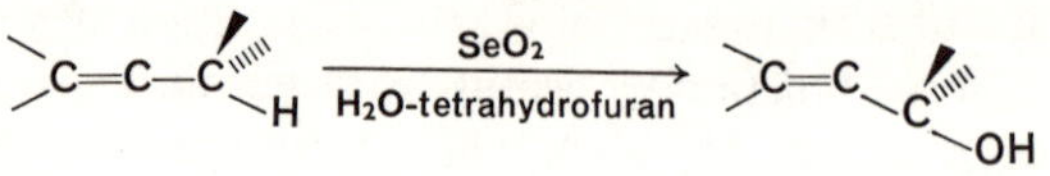

$$\begin{array}{c} \diagdown \\ C=C-C \\ \diagup \qquad\qquad H \end{array} \xrightarrow[\text{H}_2\text{O-tetrahydrofuran}]{\text{SeO}_2} \begin{array}{c} \diagdown \\ C=C \\ \diagup \qquad\qquad C \\ \qquad\qquad OH \end{array}$$

While the mechanism is somewhat controversial, the following scheme is plausible.

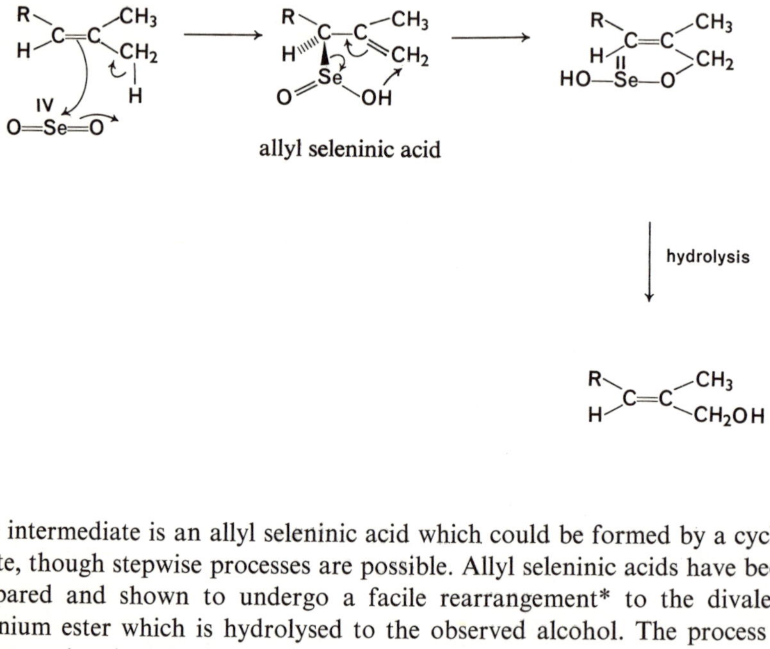

$$\begin{array}{c}
R \diagdown \quad\quad CH_3 \\
\quad C{=}C \\
H \diagup \quad\quad CH_2 \\
\\
IV \\
O{=}Se{=}O
\end{array}
\longrightarrow
\begin{array}{c}
R \diagdown \quad\quad CH_3 \\
\quad C{-}C \\
H \diagup \quad\quad CH_2 \\
\quad Se \\
O \quad OH
\end{array}
\longrightarrow
\begin{array}{c}
R \diagdown \quad\quad CH_3 \\
\quad C{=}C \\
H \diagup \quad\quad CH_2 \\
HO{-}Se{-}O
\end{array}$$

allyl seleninic acid

hydrolysis

$$\begin{array}{c}
R \diagdown \quad\quad CH_3 \\
\quad C{=}C \\
H \diagup \quad\quad CH_2OH
\end{array}$$

The intermediate is an allyl seleninic acid which could be formed by a cyclic route, though stepwise processes are possible. Allyl seleninic acids have been prepared and shown to undergo a facile rearrangement* to the divalent selenium ester which is hydrolysed to the observed alcohol. The process is known to involve attack at the less-hindered side of the double bond to give a highly selective oxidation.

Aromatic methyl groups may be oxidized to the aldehyde level only, by chromyl chloride, CrO_2Cl_2 (the Etard reaction), and methylene groups go to the ketone. An initial 2:1 complex of the reagents is formed (9) which is isolable as an amorphous solid. E.s.r. studies show that it contains one chromium atom in the +6 oxidation state and one in the +4. On adding water this is believed to break down to give a carbonium ion and the alcohol which is then further oxidized. The carbonium ion may be intercepted by azide ion.

---

* This is a 2,3-sigmatropic rearrangement, permitted by orbital-symmetry conservation.

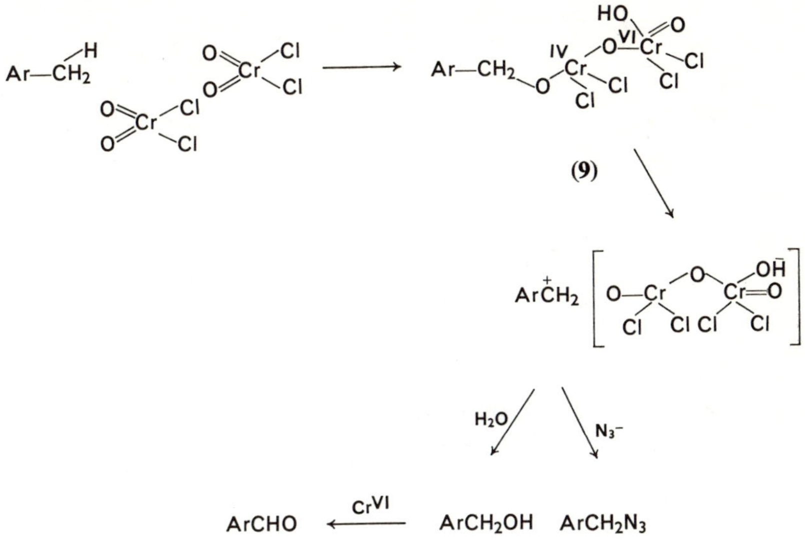

(9)

$$ArCHO \xleftarrow{Cr^{VI}} ArCH_2OH \quad ArCH_2N_3$$

Unsaturated Positions

Oxidation of olefinic double bonds by the common oxidizing agents, Cr^{VI}, Mn^{VII}, etc., occurs with the formation of ketones, carboxylic acids, rearrangement and cleavage products. The reactions are rarely clean and are of limited practical use. For example,

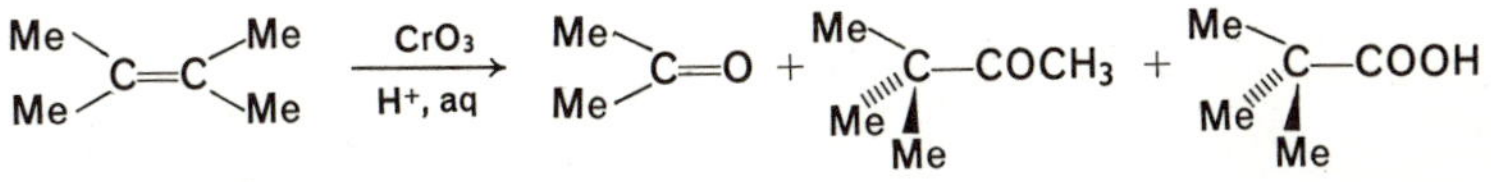

It is likely that carbonium-ion intermediates are formed since Wagner–Meerwein rearrangements (Section 2.29) are often observed (**10** → **11** and **12**). It is probable that an electrophilic attack on the double bond occurs. An intermediate epoxide (**13**) is quite probable; chromic-acid oxidation of tetraphenylethylene (**14**) leads to the corresponding epoxide (**15**) in good yield and further oxidation converts it to benzophenone. Furthermore, the same

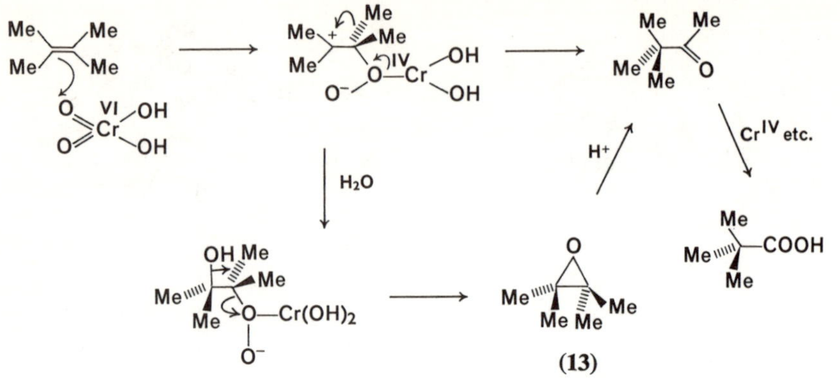

(13)

ratios of products are obtained from chromic-acid oxidation of cyclohexene in acetic acid as from the acetolysis of cyclohexene oxide. Aromatic rings are

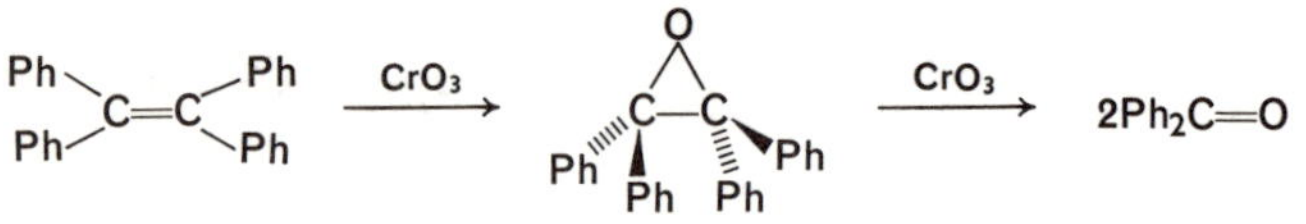

(14)   (15)

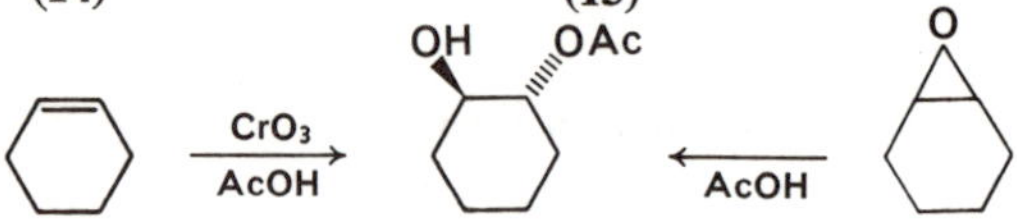

difficult to oxidize; fused-ring systems are more accessible to oxidation by chromic acid and will yield varying amounts of quinones. Little is known of the mechanism, but it is likely that the reaction commences with an electrophilic attack to form an intermediate benzenonium ion:

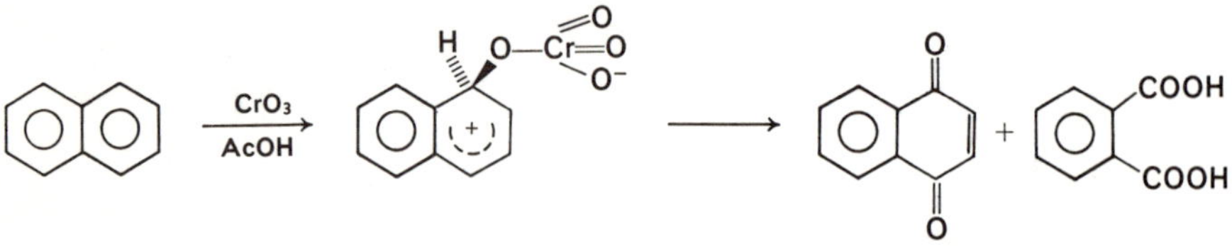

naphthalene   1,4-naphthoquinone

9.3   HYDROXYLATION AND GLYCOL-CLEAVAGE

A highly useful reaction sequence for preparative, analytical and structure-proving studies consists of the oxidation of an olefin to the 1,2-glycol (16), which is in turn cleaved to two carbonyl fragments. Permanganate, or better osmium tetroxide, will bring about hydroxylation: lead (IV) acetate or periodic ion causes the cleavage of the glycol.

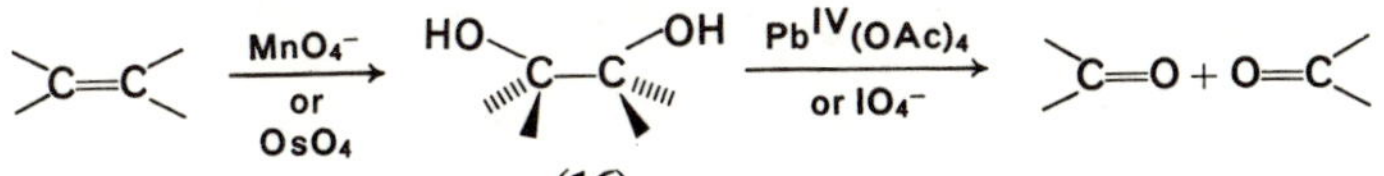

(16)

Hydroxylation by aqueous-ethanolic permanganate has been long used as a test for unsaturation (Baeyer test); the permanganate colour disappears in the presence of an olefin (in the absence of other reducing agents). Under carefully controlled conditions good yields of glycols can be obtained, though under other than optimum conditions, further oxidation may severely limit the yield. If a cyclic olefin is hydroxylated, the glycol is found to have the *cis* stereochemistry and gives a clue to the mechanism which undoubtedly occurs via a cyclic manganese ester **(17)**. Further, both glycol oxygens are derived from permanganate, not from water.

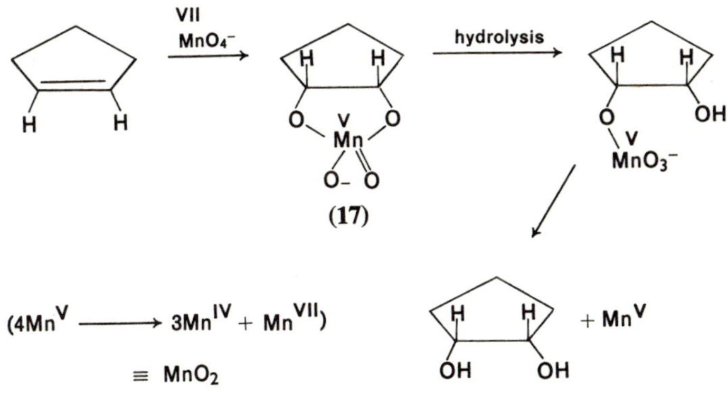

(17)

An intermediate of this type has been observed by its ultraviolet absorption during the hydroxylation of cinnamic acid:

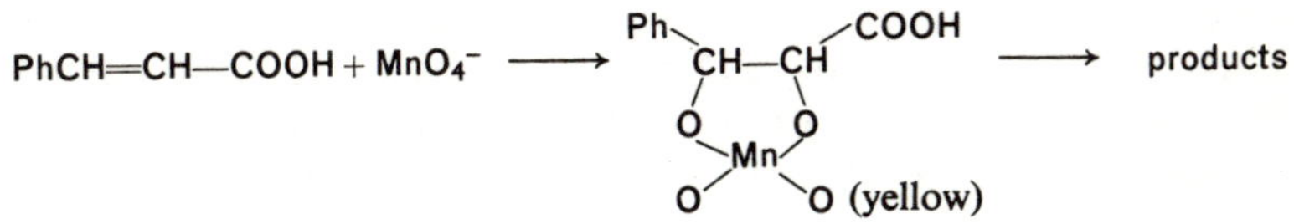

A very similar and generally cleaner reaction occurs with osmium tetroxide; the intermediate cyclic osmic ester (18) may be isolated. The expensive reagent

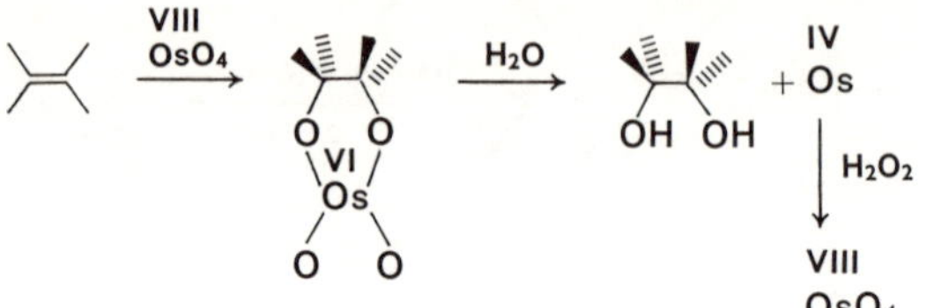

(18)

is economized by using only a catalytic quantity in conjunction with an oxidizing agent such as hydrogen peroxide which will continually regenerate it.

In an interesting reversal of this reaction, tungsten (IV) compounds will deoxygenate a glycol to form the olefin. A cyclic tungstic ester is presumed to be intermediately formed. *Trans*-diol gives *trans*-olefin with complete stereospecificity.

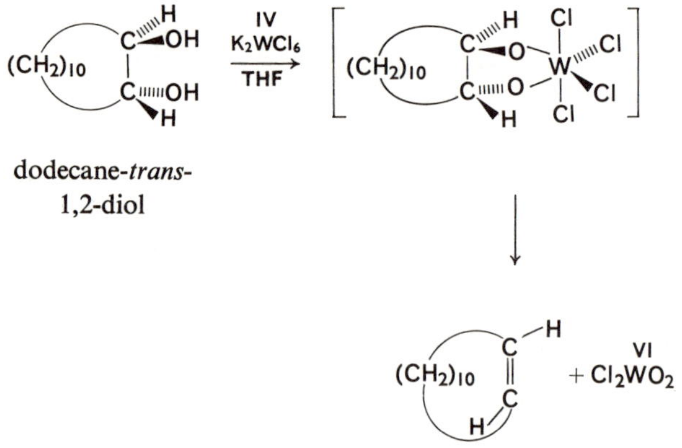

dodecane-*trans*-
1,2-diol

trans-cyclododecene

Cleavage of glycols is best accomplished by aqueous periodate which is a selective reagent and will only attack diols in which the hydroxyl groups can achieve a *cis* conformation. Initially, a cyclic periodic ester (19) forms, which is only possible if the hydroxyl groups are spatially adjacent. This then breaks down, possibly by a concerted reaction, to two carbonyl fragments. The original oxygens remain attached to carbon throughout. The α-diketone (20)

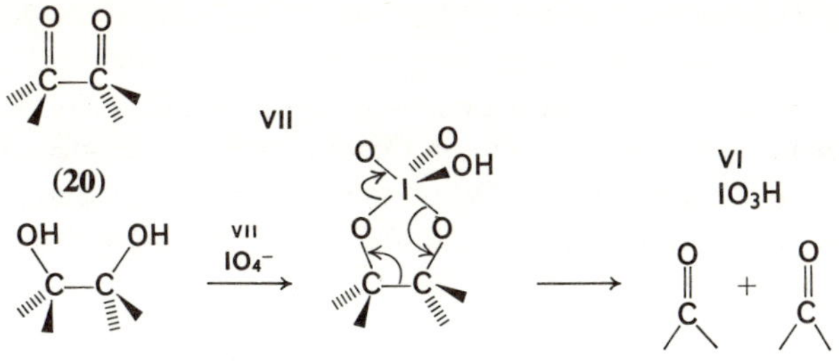

(20)

VII

VI
IO₃H

(19)

cannot be an intermediate since it may be shown to be cleaved to two carboxylic-acid fragments under the above conditions. Periodate has found great
use in determining carbohydrate structures: the number of *cis* glycol units in
the molecule may be obtained by titration. For example, methyl α-D-riboside
(22) consumes one mole of periodate and forms one mole of 21, showing that it
has only one pair of *cis* hydroxyl groups. Hydroxylation and glycol cleavage

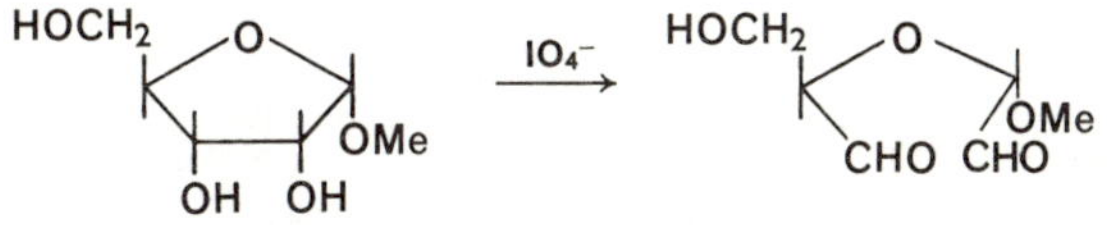

(22)   (21)

may elegantly be combined by using a reagent consisting of periodate with a
catalytic quantity of permanganate. The latter is continually replenished by
reoxidation by the periodate, but being in low concentration is less capable of
extensively degrading the substrate.

Lead tetraacetate is also used for glycol cleaving; a cyclic plumbic ester (23)
is believed to be an intermediate. This reagent is less selective than periodate,

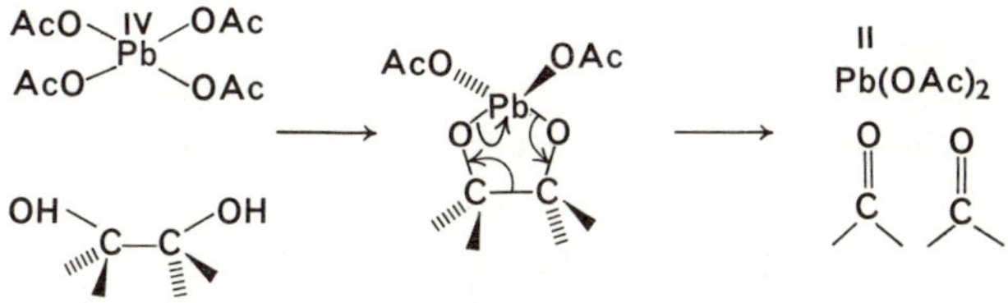

(23)

+ 2AcOH

and *trans* diols are also oxidized though usually at lower rates than *cis*. This may occur by a similar mechanism to that above in cases such as *trans*-cyclohexane-1,2-diol in which the cyclic ester is still geometrically possible. However, a cyclic intermediate is not possible in the cleavage of *trans*-decalin-9,10-diol (**24**); possibly two moles of lead (IV) acetate are required leading initially to a lead (III) product. This oxidation state of lead has previously been implicated

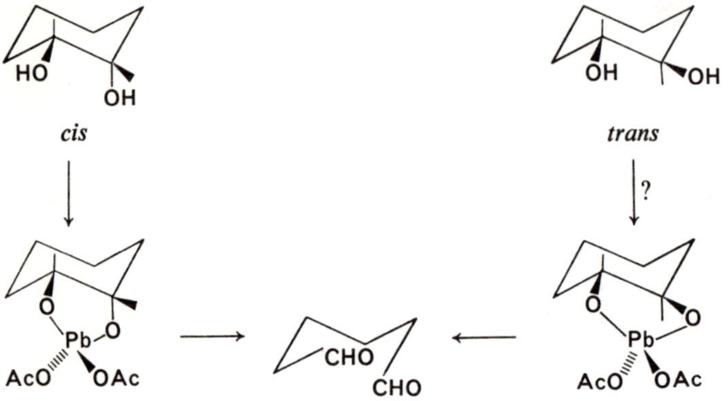

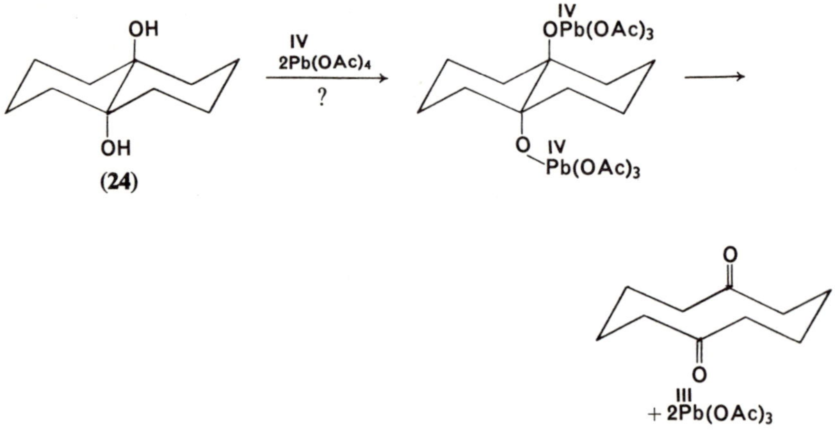

in oxidations and can itself initiate further oxidation. Thallium (III) acetate, $Tl^{III}(OAc)_3$, will also act in a similar way to lead (IV) acetate.

9.4   **OZONOLYSIS OF OLEFINS**

The oxidation and subsequent cleavage of an olefin by ozone, followed by reduction, affords a classical method of structure determination. Ozone, mixed

with oxygen, is allowed to add to the olefin at low temperature and in an inert
solvent when an isolable addition product, the ozonide (25), is formed. This is
usually somewhat unstable and is cleaved to two olefinic or alcoholic fragments
by zinc and acid, or lithium aluminium hydride, respectively. By-products of

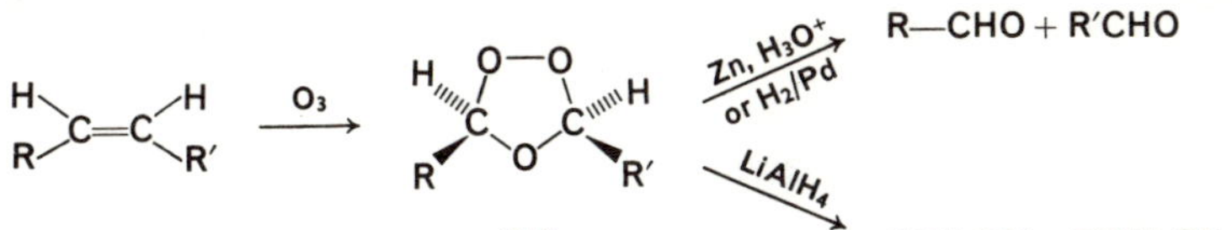

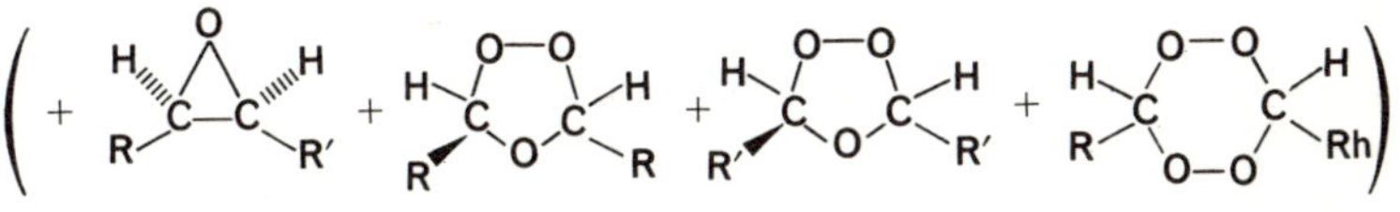

the reaction include the epoxide (26), 'crossed' ozonides (27 and 28), and
bis-peroxides (29). The reaction is quite complex but has been extensively
investigated. The scheme proposed by Story probably best accounts for the
course of the reaction.

The primary ozone adduct may well be the peroxy epoxide (30), which
would plausibly lead to epoxide by the loss of an oxygen molecule. Certain
tetrasubstituted ethylenes such as tetramethylethylene yield epoxide as the
major product. This primary ozonide then rearranges to a 'molozonide',
depicted as 31 (a structure originally proposed by Staudinger for the stable
ozonide). This in turn decomposes to the isolable ozonide (32) by several
competing routes. *Route A*, proposed by Criegee, proceeds by dissociation of
the molozonide into a carbonyl compound and a peroxy zwitterion (33), which
then recombine in the opposite sense.

Story's Scheme

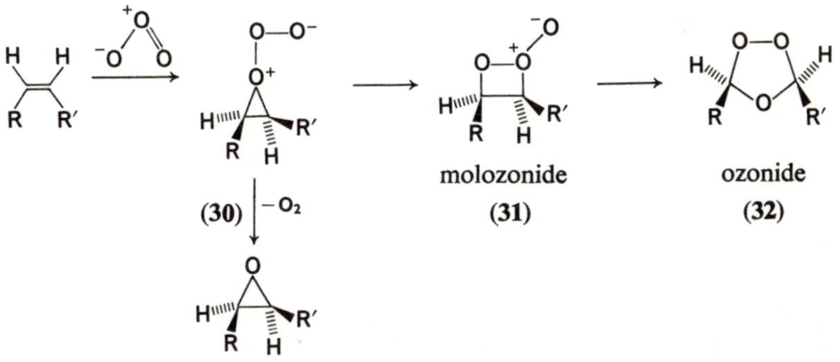

Routes for the decomposition of the molozonide:

Route A

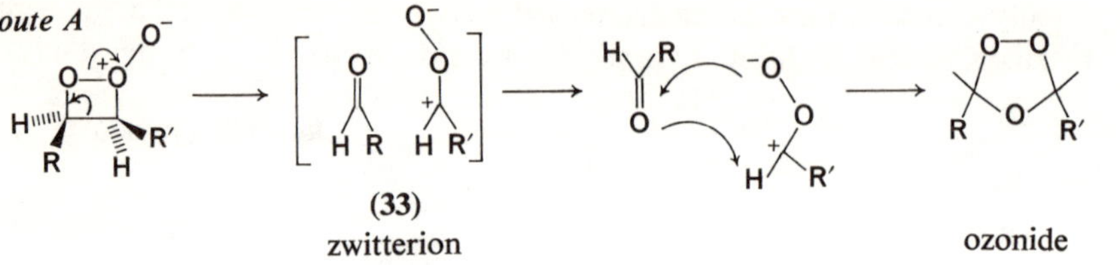

(33)

zwitterion   ozonide

Route B

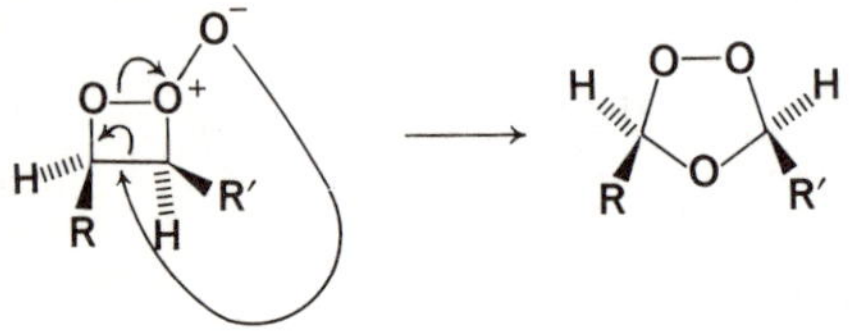

Route C

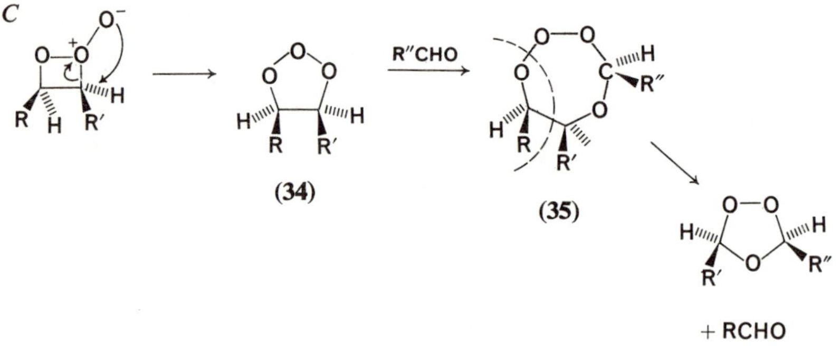

(34)

(35)

+ RCHO

Story has produced evidence for *Route B*, a rearrangement analogous to the Baeyer–Villiger reaction (Section 9.5) in which the essential step may be depicted thus:

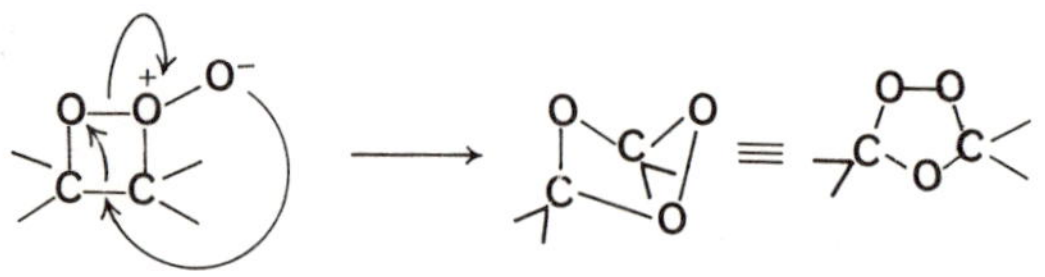

An isomeric molozonide, the trioxolan (**34**), may also be involved (*Route C*) and inserts an aldehyde molecule (which either originates from *Route A* or may be a 'foreign' aldehyde which is introduced). The resulting seven-membered ring (**35**) then expels a molecule of an aldehyde, which obviously explains the formation of the crossed product, although this can be well accommodated by the Criegee mechanism (*Route A*). Thus the observed formation in good yield of isobutene ozonide (**36**) from the ozonolysis of 2,3-dimethylbut-2-ene (**37**) in the presence of formaldehyde may be explained as follows:

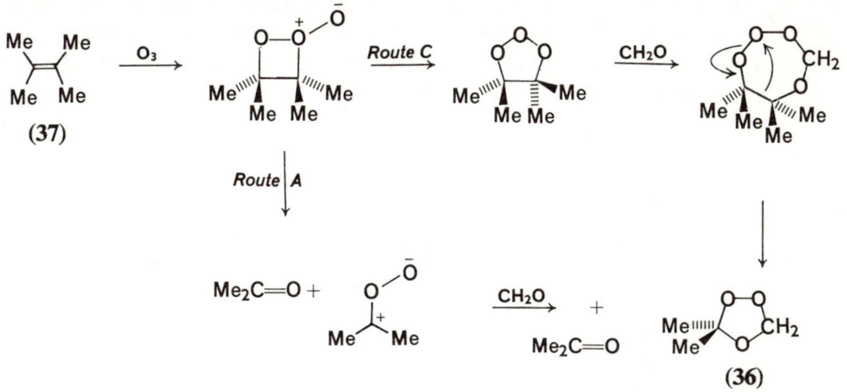

Me Me
$\overset{\text{Me Me}}{\underset{\text{Me Me}}{>=<}}$ $\xrightarrow{O_3}$ ... *Route C* ... $\xrightarrow{CH_2O}$ ...
(**37**)

Route A

$Me_2C\!=\!O +$ ... $\xrightarrow{CH_2O}$ + ...
$Me_2C\!=\!O$
(**36**)

The bis-peroxides observed also arise plausibly by dimerization of the zwitterion:

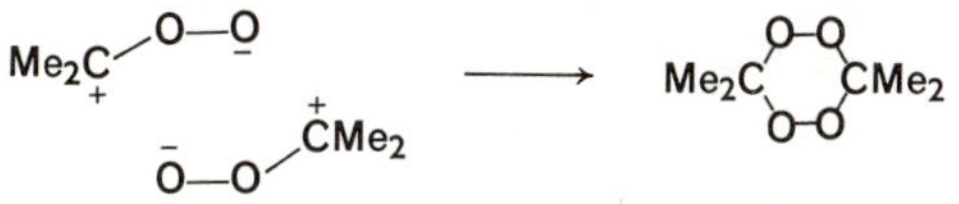

Two further lines of evidence are worth mentioning as they require at least two pathways leading from **31** to the ozonide. First, the formation of crossed ozonide by ozonization in the presence of a 'foreign' aldehyde which is labelled with ^{18}O should place the label at the ether oxygen by *Route C* and at peroxide oxygen by *Route A*. It is found that the ozonolysis of diisopropylethylene in the presence of acetaldehyde—^{18}O yields crossed ozonide which has 68 % of the label at the ether oxygen and 32 % at a peroxide oxygen, which implies that the

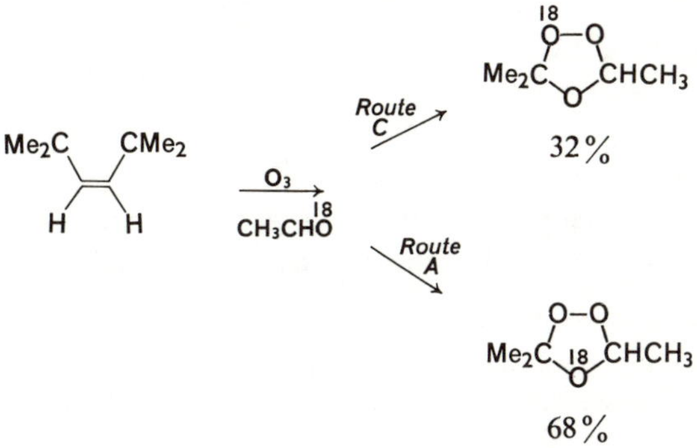

Route C → 32 %

$\overset{Me_2C\quad CMe_2}{\underset{H\quad\quad H}{>=<}}$ $\xrightarrow[CH_3\overset{18}{C}HO]{O_3}$

Route A → 68 %

compound is formed by both *Routes C* and *A* in the ratio 68:32.* Secondly, the amounts of the *cis* and *trans* isomers of either the normal or a crossed ozonide should be the same whether one started from *cis* or *trans* isomers of an olefin *if the only pathway is Route A*. Whatever the starting material, the separation of zwitterion and carbonyl compound causes all distinction between *cis* and *trans* to be lost; yet the yield of stereoisomeric ozonides *is* found to depend upon the stereochemistry of the original olefin. For example,

cis	$\longrightarrow$	1 : 1
trans	$\longrightarrow$	1 : 2

It follows that there must be at least one other route to the ozonide; *Routes B* and *C* are both capable of explaining the discrepancy.

In certain cases at least, the olefin appears to complex initially with ozone by charge transfer or π-overlap. Thus, the highly hindered and strongly donor olefin, 1-mesityl-1-phenylethylene (38), gives with ozone at −150° a deep-red complex, $\lambda_{max} = 450$ nm, which on warming leads to the normal products of ozonolysis (in this case, epoxide) (39).

The reactions of acetylenes with ozone are less understood and also less clean; both α-diketones and carboxylic acids can result, though yields are often low, e.g.

---

* It should also be mentioned, however, that the only crossed ozonide found in the ozonolysis of ethylene in the presence of formaldehyde–^{18}O is that with label in the ether position.

$$R-C\equiv C-R' \xrightarrow{O_3} R-COCOR' + RCOOH + R'COOH$$

Aromatic rings are cleaved by ozone; numerous products are formed and the reaction has no practical application. For example,

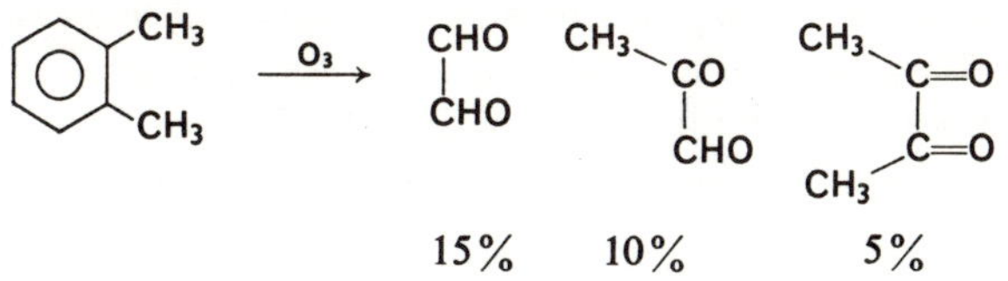

15%   10%   5%

9.5   THE BAEYER–VILLIGER REACTION

Ketones are oxidized to esters by peroxyacids, e.g. perbenzoic, peracetic and persulphuric (Caro's) acids. The yields vary but may be very high. The net result is the insertion of oxygen next to the carbonyl group.

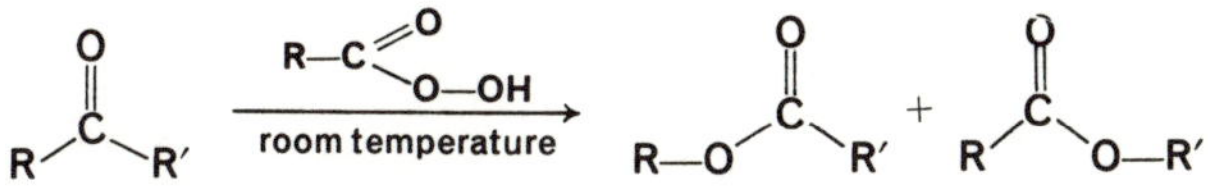

Cyclic ketones form lactones and α,β-unsaturated ketones are converted to enol esters. Aldehydes react less readily; benzaldehydes will only take part in the Baeyer–Villiger reaction if activated by donor groups. The mechanism of

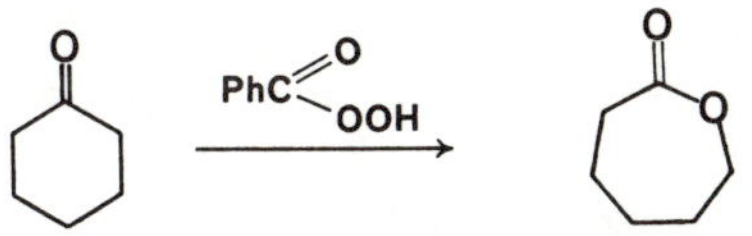

5-hydroxyhexanoic acid lactone

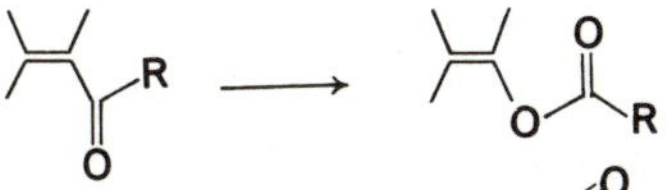

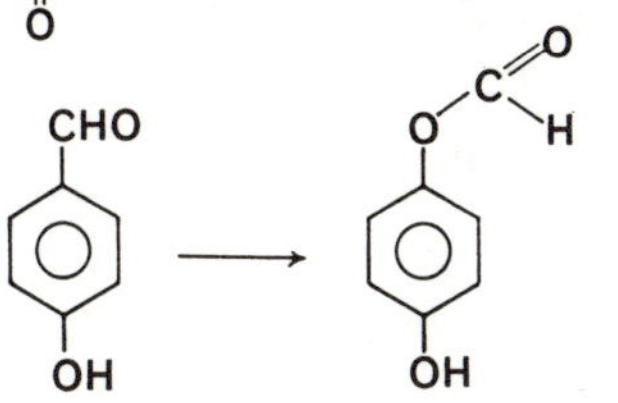

hydroquinone formate

this reaction is intriguing and not entirely understood. The oxidation of acetophenones, $ArCOCH_3$, shows a rather slight acceleration by electron-donating groups in the ring ($\rho = -0.5$). The kinetics are usually first order in both ketone and peroxy acid though substrates of low reactivity may show no dependence on peroxy-acid concentration. There is usually a large negative entropy of activation indicating a compact transition state with loss of internal degrees of freedom. Several mechanisms have been suggested. Doering proposed an attack of the peroxy-acid anion on the ketone to give a tetrahedral intermediate (39) with several possible decomposition pathways which must always include the migration of one of the alkyl groups from carbon onto oxygen. Either the tetrahedral intermediate decomposes by expulsion of a

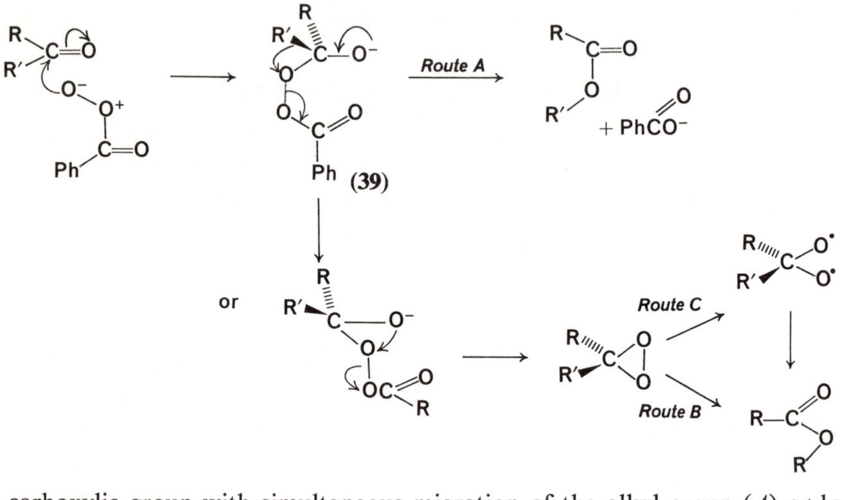

carboxylic group with simultaneous migration of the alkyl group (A) or by cyclization to a three-membered cyclic peroxide (39). This could reach the product structure by a concerted rearrangement (B) or via a diradical followed by alkyl migration (C).

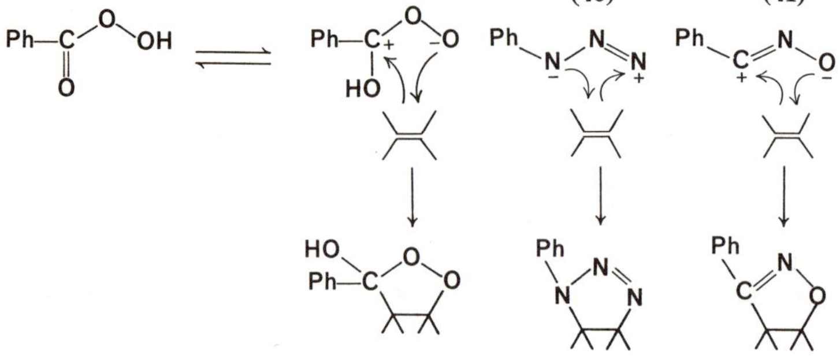

A quite different mechanism has been proposed by Kwart; the peroxy acid is supposed to act as a 1,3-dipolar reagent, analogous to well-authenticated examples of this type such as azide ion (40) or nitrile oxides (41). The peroxy acid adds across the double bond in the usual way to give a five-membered ring containing three oxygens (a 1,2,4-trioxolan (**42**)), reminiscent of an ozonide, or a transition state resembling it, from which carboxylic acid splits out with concomitant alkyl migration. A curious feature of this reaction is the preference

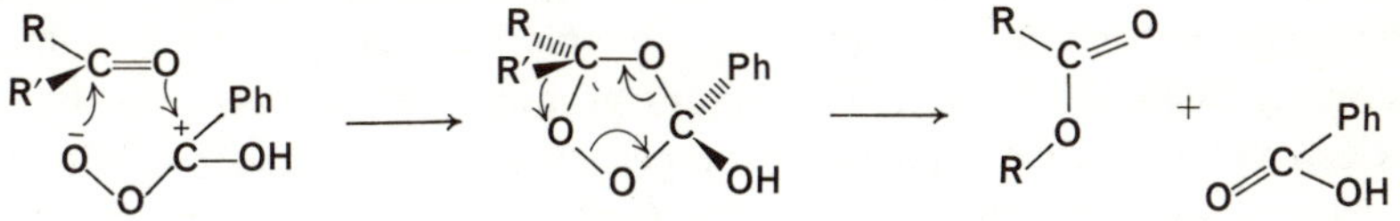

(**42**)

intermediate
or

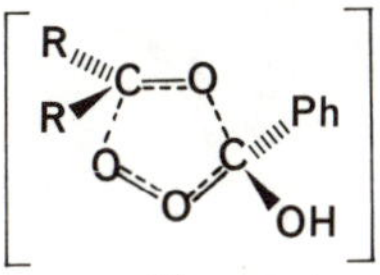

transition state

for migration exhibited by different groups. While secondary and tertiary alkyl groups migrate more readily than primary, aryl groups may migrate less readily than alkyl; thus,

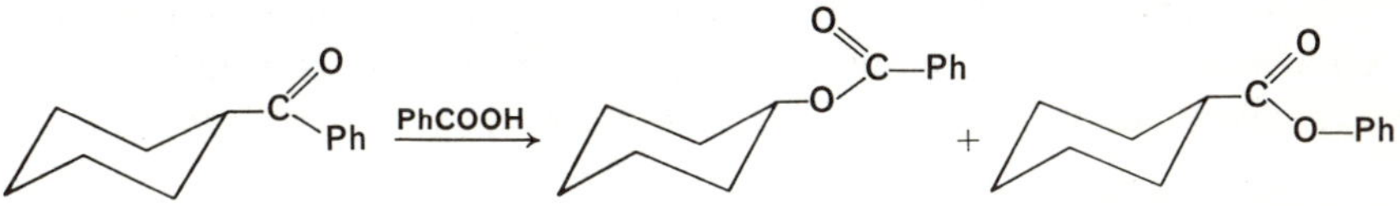

80%   20%

Normally in migrations to electron-deficient carbon, aryl groups are greatly preferred over alkyl. Little is known about migrations to electron-deficient oxygen; possibly differences exist between this and carbon.

9.6   **EPOXIDATION OF ALKENES**

A very important reaction for functionalizing an alkene double bond consists

of treatment with a peroxy acid ($R\!-\!\overset{\overset{O}{\|}}{C}\!-\!OOH$), peracetic, perbenzoic or best

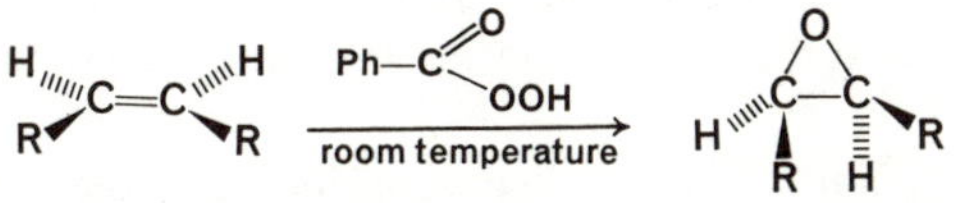

(**43**)

of all, trifluoroperacetic acid. The epoxide (43) is formed in a single operation, and usually in excellent yield, with retained stereochemistry. The reactions show first-order dependence upon olefin and peroxy-acid concentrations, are not acid catalysed, and show no salt effect. By studying epoxidation by perbenzoic acid of substituted stilbenes, and epoxidation of stilbenes (44) by substituted perbenzoic acids (45). Lynch and Pausacker showed that the reaction is facilitated by electron-donating groups in the olefin ($\rho = -1.6$) and by electron-withdrawing groups in the peroxy acid ($\rho = 1.5$). The following mechanism is compatible with these facts, which indicate that the olefin must be acting as a nucleophile and the peroxy acid as an electrophile:

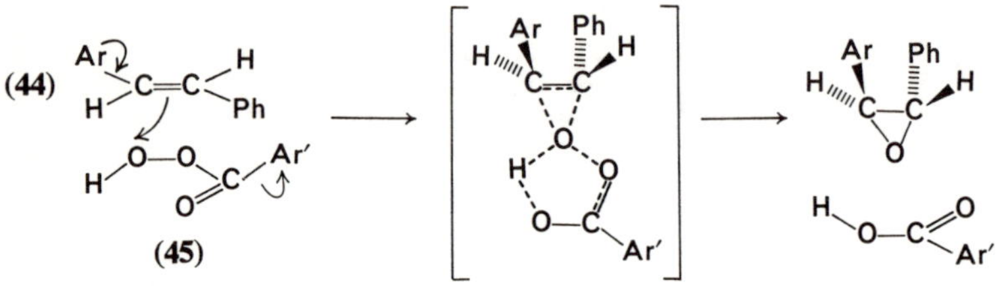

The epoxidation can be considered to take place by attack of an *incipient* hydroxyl cation, $\overset{+}{O}H$. However, the intermediacy of a free hydroxyl cation is *not* in accordance with the strictly *cis* addition mode, since it would lead to a carbonium ion (46) in which rotation should occur before final ring closure.

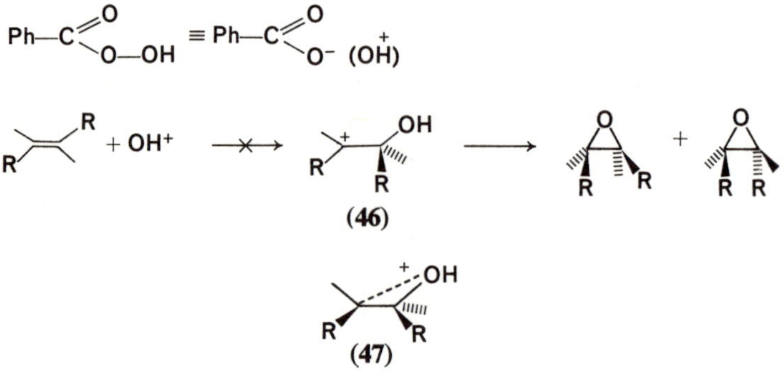

If 46 is indeed an intermediate, rotation is effectively prevented from occurring. This must be due to some interaction between the oxygen and the cationic centre, 47.

A further mechanism which has been considered is the concerted dipolar addition of peroxy acid (cf. the Baeyer–Villiger reaction, Section 9.5). The corresponding steps for addition to an olefin are,

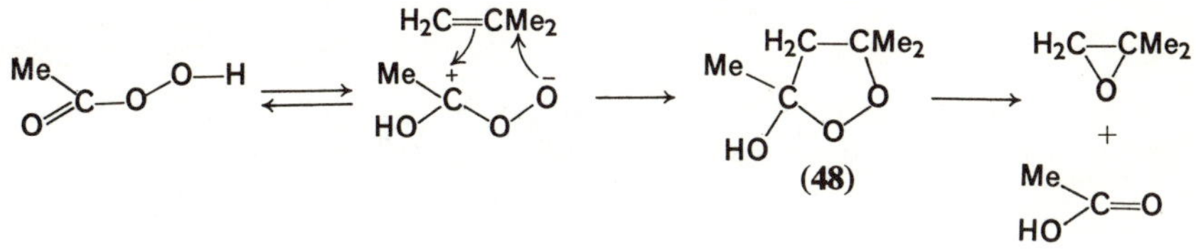

The mechanism as such evidently does not occur since the intermediate dioxolan (48) has been separately prepared and shown to be reasonably stable under the experimental conditions and then to decompose to give only a minor amount of epoxide. A modified version of this scheme may however

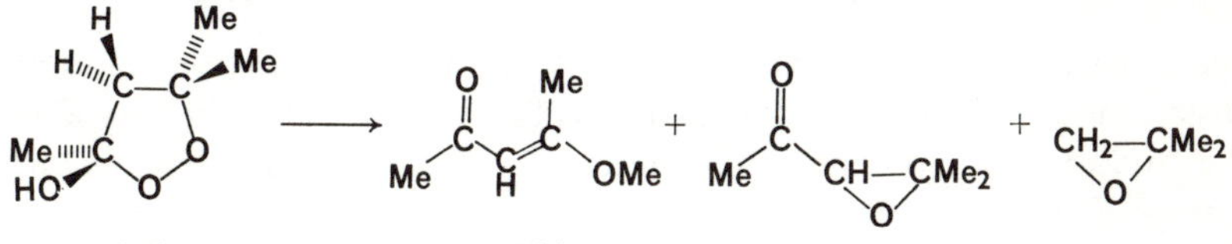

(48)   55%   36%   9%

still be considered; the approach of the reagent is as for a dipolar addition but rearrangement direct to epoxide occurs from the transition state. It appears

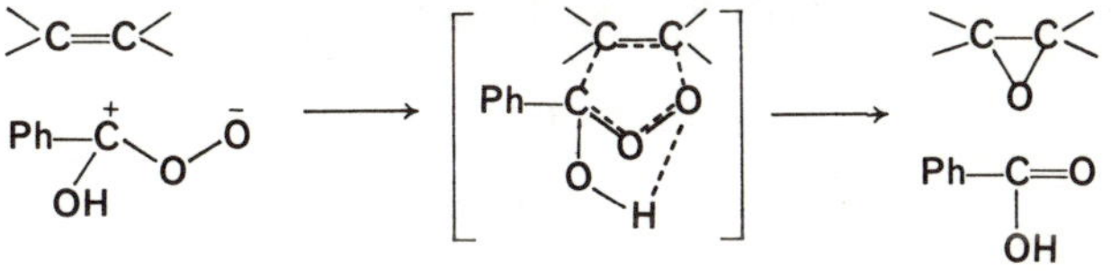

that contrary to previous suggestions, epoxidation of olefins proceeds along a pathway in which oxygen is transferred to olefin without the formation of discrete intermediates.

Little is known also of the mechanism of epoxidation by pertungstic acid. The reagent is formed by the equilibrium between tungstate and H_2O_2:

$$WO_4^{2-} + 2H_2O_2 \rightleftharpoons HWO_6^- + OH^- + H_2O$$
$$\text{and } WO_4^{2-} + 4H_2O_2 \rightleftharpoons WO_8^{2-} + 4H_2O$$

The pertungstate ions HWO_6^- and WO_8^{2-} are good epoxidizing agents and will react readily with double bonds which are flanked by electron-withdrawing

groups, compounds such as maleic acid (49), which are not epoxidized readily
by percarboxylic acids. The reaction generates base which must be continuously

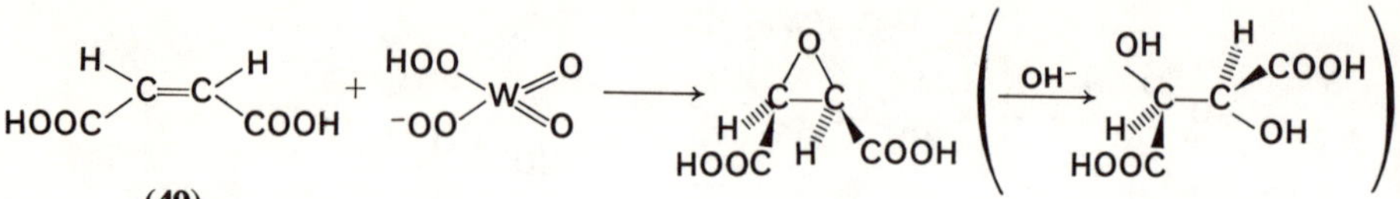

(49)

$+WO_4^{2-}$

(50)

neutralized or will react with the epoxide to give glycol (50). The tungstate need
only be present in catalytic amounts in the presence of H_2O_2 which continuously
reoxidizes it.

9.7   CATALYTIC OXIDATION OF AROMATIC SIDE-CHAINS

Chain reactions involving metal catalysts in different oxidation states make
feasible industrially important oxidations of alkyl benzenes to carboxylic acids.
The oxidation of *p*-xylene to terephthalic acid (used in the manufacture of
'*Terylene*' or '*Dacron*') is such an example. This may be accomplished by
oxygen in the presence of soluble cobalt catalysts such as **51**. The reaction is
carried out in acetic acid as solvent at a temperature of 150° and pressure
around 20 to 50 atmospheres.

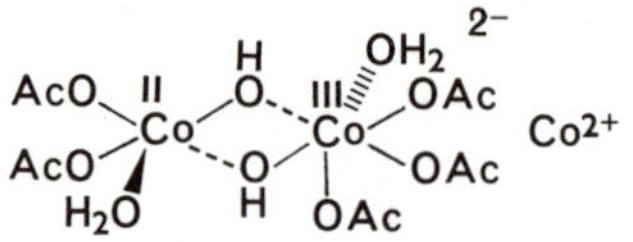

(51)

The stages which appear to occur are as follows:

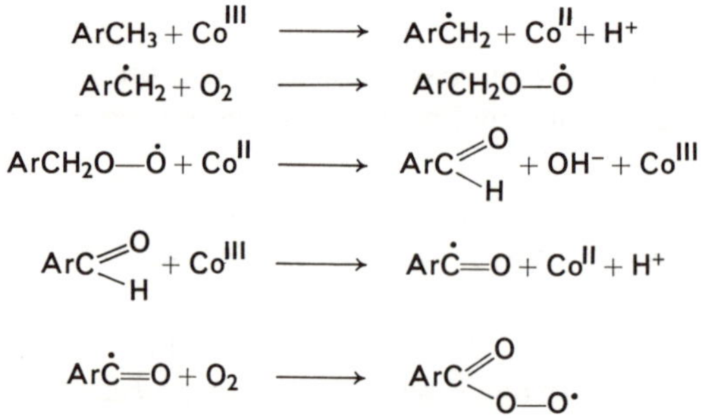

$$ArCH_3 + Co^{III} \longrightarrow Ar\dot{C}H_2 + Co^{II} + H^+$$

$$Ar\dot{C}H_2 + O_2 \longrightarrow ArCH_2O-\dot{O}$$

$$ArCH_2O-\dot{O} + Co^{II} \longrightarrow ArC\!\!\begin{smallmatrix}O\\\\H\end{smallmatrix} + OH^- + Co^{III}$$

$$ArC\!\!\begin{smallmatrix}O\\\\H\end{smallmatrix} + Co^{III} \longrightarrow Ar\dot{C}\!\!=\!\!O + Co^{II} + H^+$$

$$Ar\dot{C}\!\!=\!\!O + O_2 \longrightarrow ArC\!\!\begin{smallmatrix}O\\\\O-O^\bullet\end{smallmatrix}$$

$$ArC\underset{O-O^{\cdot}}{\overset{O}{\diagdown}} + \underset{H}{\overset{O}{\diagup}}C-Ar \longrightarrow ArC\underset{O-OH}{\overset{O}{\diagdown}} + Ar\overset{\cdot}{C}O$$

$$ArC\underset{O-OH}{\overset{O}{\diagdown}} + Co^{II} \longrightarrow ArC\underset{O^{\cdot}}{\overset{O}{\diagdown}} + Co^{III} + OH^{-}$$

$$ArC\underset{O^{\cdot}}{\overset{O}{\diagdown}} + \underset{H}{\overset{O}{\diagup}}CAr \longrightarrow ArC\underset{OH}{\overset{O}{\diagdown}} + Ar\overset{\cdot}{C}=O$$

The reactions in which the metal is involved presumably occur within its coordination sphere, although the details of intermediate organometallic compounds are not known with any certainty.

9.8 OXIDATIONS WITH SINGLET OXYGEN

The ground state of molecular oxygen is a triplet (two unpaired-electron spins); two excited states lie 93 and 155 kJ mol^{-1}, respectively, above the ground state and both are singlet species (no unpaired spins). The excited

Second excited state ($^1\Sigma_g^+$) $\uparrow$ $\downarrow$ 155 kJ mol^{-1}

First excited state   ($^1\Delta_g$) $\uparrow\!\uparrow$ — 93 kJ mol^{-1}

Ground state   ($^3\Sigma_g^-$) $\uparrow$ $\uparrow$

states are quite readily accessible, especially the first; oxygen in these states has properties very different from the ground state and is a useful and selective oxidant.

The first excited state has only paired electrons and would be expected to resemble ethylene in chemistry, but be more electrophilic, undergoing ionic or two-electron processes; the higher excited state is more akin to the ground state electronically and would be more prone to radical or one-electron processes.

Singlet oxygen may be prepared either photochemically or chemically. The photochemical method involves passing ordinary oxygen through a solution containing a suitable photosensitizing dye and also the substrate. On irradiating the solution, excited oxygen forms by one or other of the mechanisms below and reacts *in situ* with the substrate.

sens = photosensitizing dye,

    e.g. Rose bengal

S = substrate

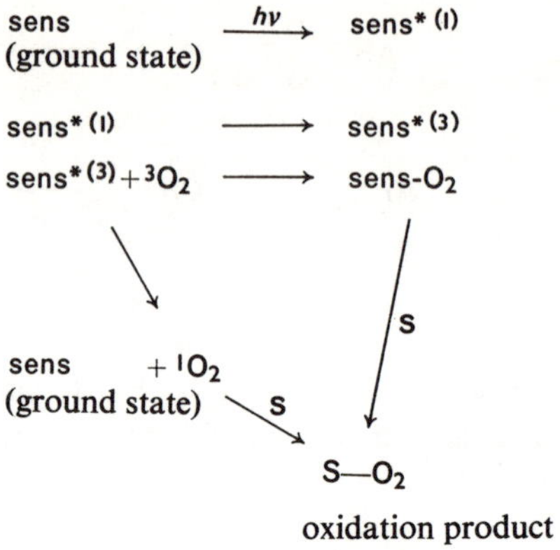

$$\text{sens (ground state)} \xrightarrow{h\nu} \text{sens* (1)}$$

$$\text{sens* (1)} \longrightarrow \text{sens* (3)}$$

$$\text{sens* (3)} + {}^3O_2 \longrightarrow \text{sens-}O_2$$

sens (ground state) + ${}^1O_2$

S—O_2

oxidation product

The superscript in these symbols for the states of oxygen describe them as singlets (1)—i.e. no unpaired electrons or triplets, (3)—two unpaired electrons; * indicates an excited state.

The chemical production of singlet oxygen uses the reaction between peroxide and hypochlorite. Although the mechanism of production is not well understood, it is likely that a four-centre reaction such as,

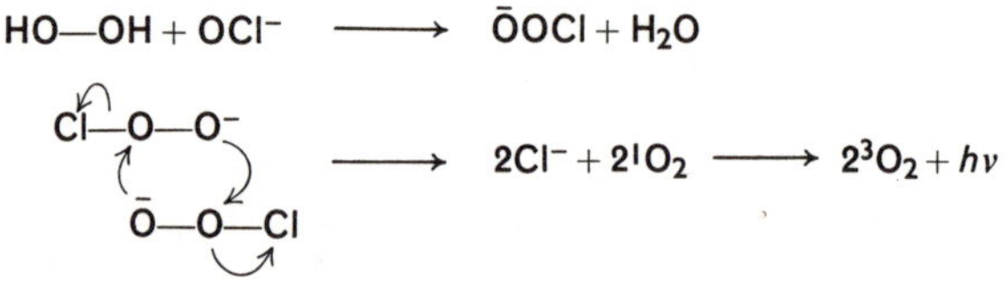

$$HO—OH + OCl^- \longrightarrow \bar{O}OCl + H_2O$$

$$2Cl^- + 2{}^1O_2 \longrightarrow 2{}^3O_2 + h\nu$$

is involved.

The system is chemiluminescent (Section 10.8) emitting red light (634, 703 nm) which corresponds to the transition, $O_2\ ({}^1\Delta_g) \rightarrow O_2\ ({}^3\Sigma_g^-)$.

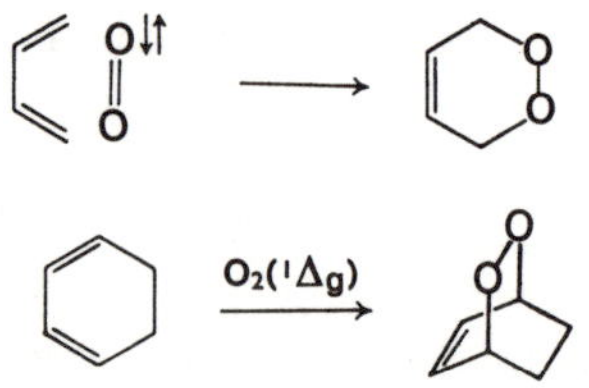

Diels–Alder type cycloaddition to dienes are symmetry allowed (Section 1.31). Olefins are attacked to give allylic alcohols with a shift of the double bond. This reaction is likely to be via a diradical (52), since the concerted cycloaddition of O_2 ($^1\Delta_g$) to an olefin is symmetry forbidden.

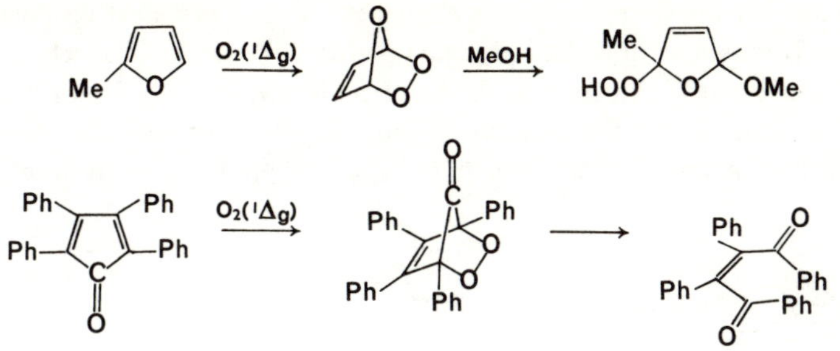

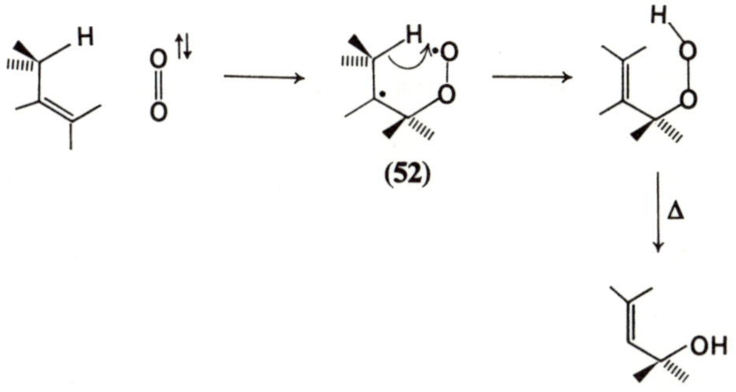

(52)

Some typical examples are,

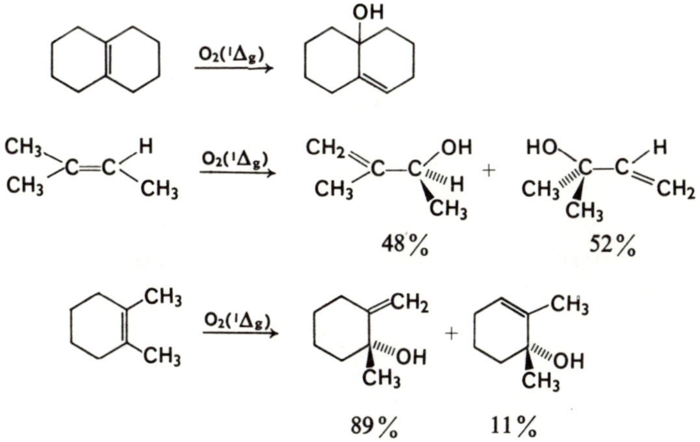

As a rule, all possible isomers are formed, so for preparative purposes the reaction is most useful when only one product is possible. The relative rates of reaction of olefins indicate that the reaction is favoured by donor character of the olefin and hence the oxygen is behaving electrophilically. Thus, in the oxidation of a series of styrenes (53), $\rho = -0.9$, there is no marked effect of

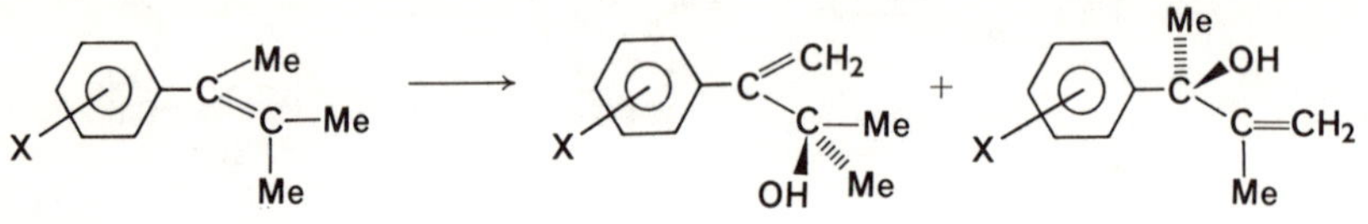

(53)

change of solvent upon rate, which tends to substantiate the suggested non-ionic reaction mechanisms.

Acetoxylation via Oxymercuration

Olefins react with mercuric acetate with the production of allylic acetates, a net oxidation of the molecule. There is a migration of the double bond as is

$$-CH-CH=CH- + Hg^{II}(OAc)_2 \longrightarrow -CH=CH-\underset{\underset{OAc}{|}}{CH} + Hg^{I}(OAc)$$

25–70%

obvious from the product from α-cyclogeraniol, 2,4,4-trimethylcyclohexene (54). Olefins are known to add mercury (II) acetate, and in the presence of

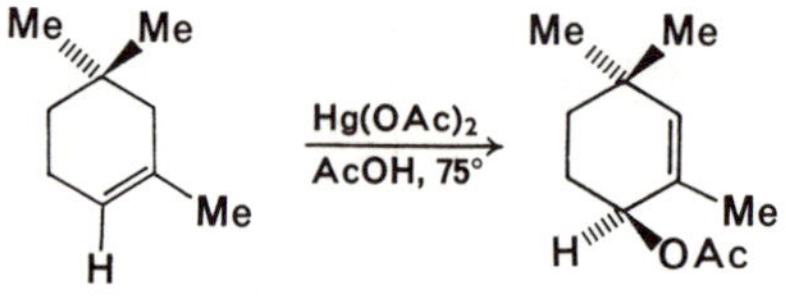

(54)

protic solvents oxymercurial compounds, e.g. 55, can be isolated:

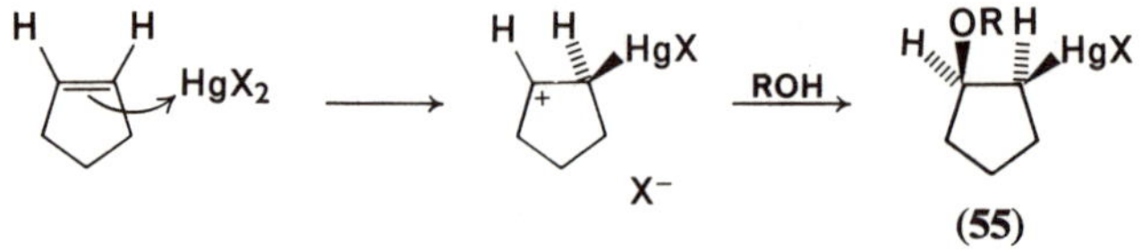

(55)

In the presence of acetic acid it appears that coordination of mercury is followed by loss of the β-proton and the mercury-containing group is subsequently displaced by acetate:

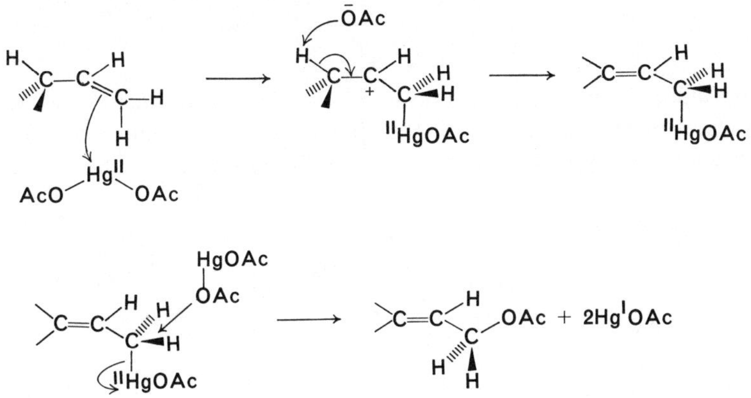

Oxidations by Thallium Compounds

Thallium occurs in Group III of the Periodic table, hence its compounds should be acceptors (Lewis acids) similar to those of boron and aluminium. However, in addition a monovalent state is stable, hence trivalent thallium can act as an oxidizing agent rather similar to Hg^{II} and Pb^{IV}, its redox potential lying between those of mercury and lead.

Olefins react with thallium nitrate in aqueous solution to give ketones and aldehydes in which a skeletal rearrangement has occurred:

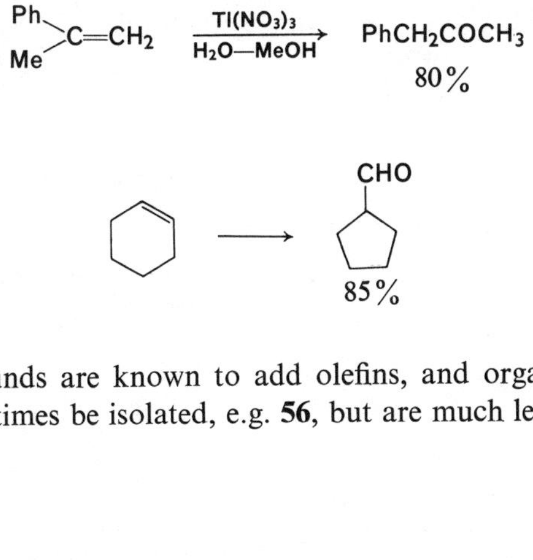

Thallium compounds are known to add olefins, and organothallium compounds can sometimes be isolated, e.g. **56**, but are much less stable than the

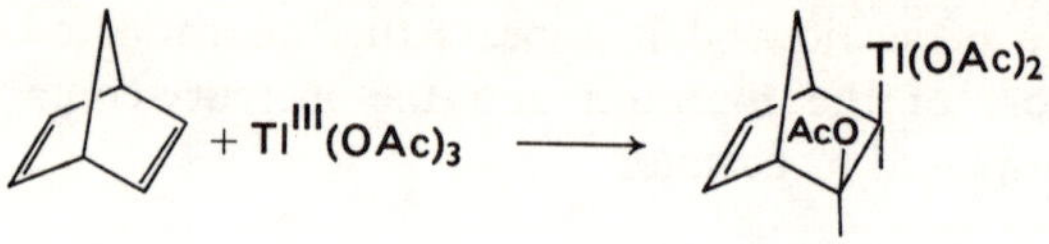

(56)

corresponding mercury compounds. The oxidation of olefins no doubt commences by such an addition; the expulsion of the thallium group (as Tl^I) occurs concurrently with a Wagner–Meerwein rearrangement (Section 2.29).

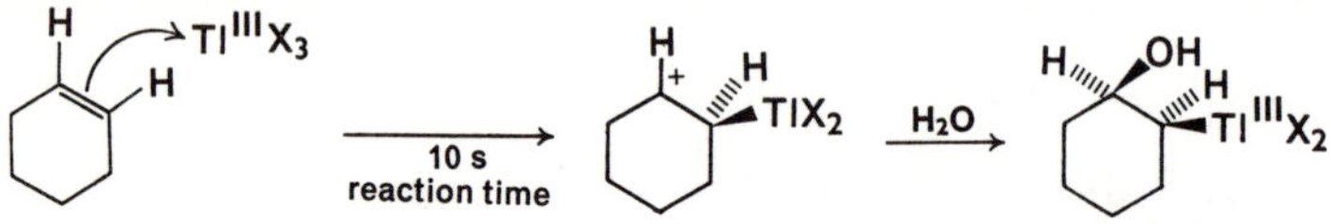

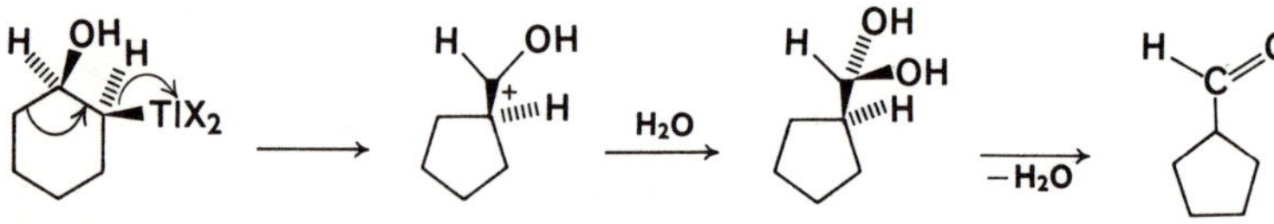

In general $RCH=CH_2$ $\xrightarrow{Tl(NO_3)_3}$ $R-C\overset{\displaystyle O}{\underset{\displaystyle CH_3}{\big\langle}}$ 90+%

Aliphatic acyclic olefins frequently give a great deal of hydroxylation (or alkoxylation) product (57):

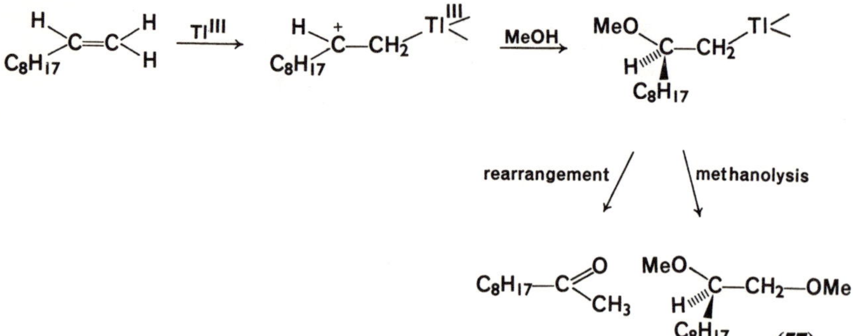

$C_8H_{17}-C\overset{\displaystyle O}{\underset{\displaystyle CH_3}{\big\langle}}$

28%

$\underset{\displaystyle C_8H_{17}}{\overset{\displaystyle MeO}{\underset{\displaystyle H^{\backslash\backslash\backslash}}{\big\rangle}}}C-CH_2-OMe$

52%

(57)

The thallium reagent also attacks acetylenes; monoalkyl acetylenes yield carboxylic acids with loss of the terminal carbon:

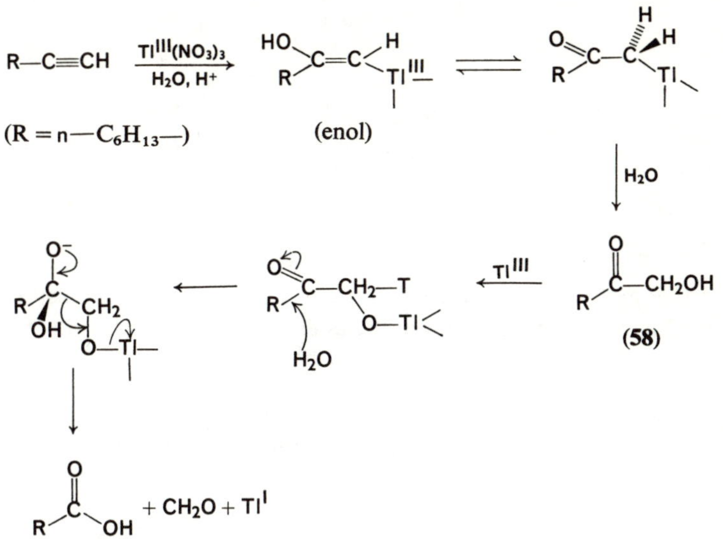

$(R = n\text{—}C_6H_{13}\text{—})$ (enol)

(58)

It is believed that the reaction takes place via an intermediate hydroxyketone (58) which complexes with more thallium and renders the terminal carbon susceptible to nucleophilic displacement.

Similar mechanisms account for the formation of 1,2-diones (59) from diarylacetylenes:

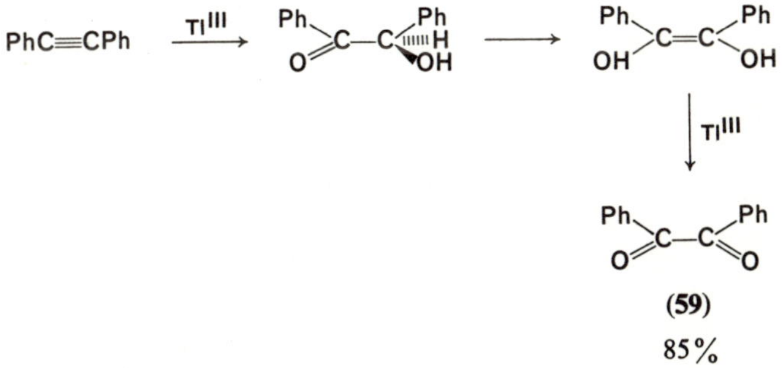

(59)

85%

and for the formation of esters (60) from alkylacetylenes. In this case a migration takes place and a keten acetal (61) may be an intermediate. This would add methanol and hydrolyse to form the product.

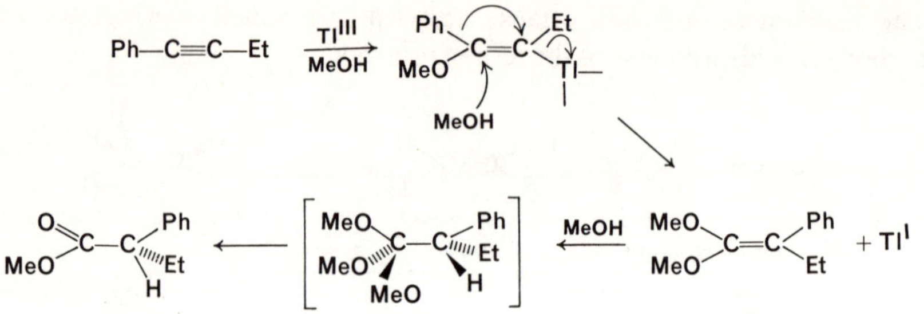

$$Ph-C\equiv C-Et \xrightarrow[MeOH]{Tl^{III}}$$

(60) ← [ ] ← (61)

95%

A further molecular rearrangement occurs in the oxidation of acetophenone (presumably via the enols) to arylacetic esters. The high yields, mild conditions

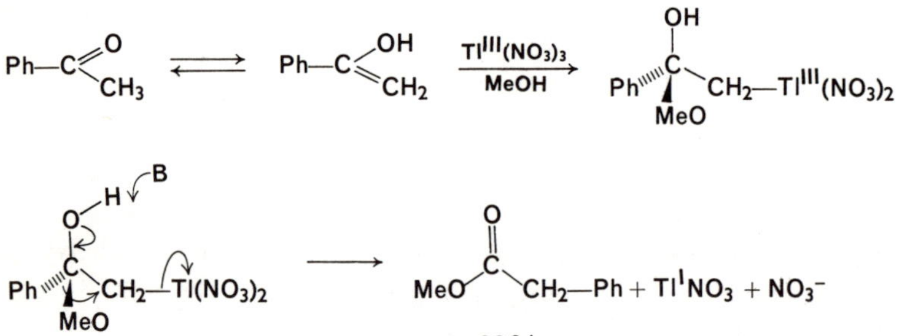

~90%

and unusual reactions make thallium oxidations very useful preparatively.

Chapter 10

Miscellaneous Intermediates

10.1 AROMATIC SUBSTITUTION VIA QUINONES

A number of polyphenols undergo substitution in the ring by nitrous acid leading to a *nitro* compound; e.g. catechol (**1**) gives the 4-nitro derivative (**2**).

(40%)

(**1**) (**2**)

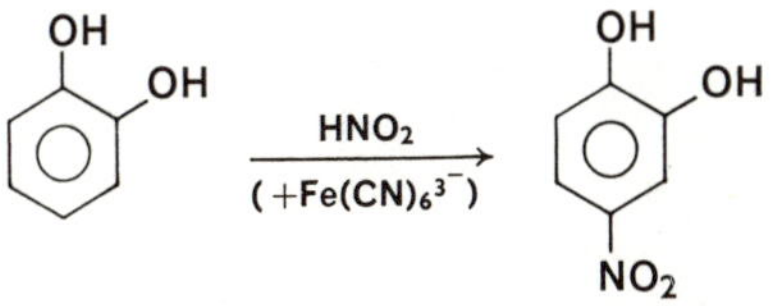

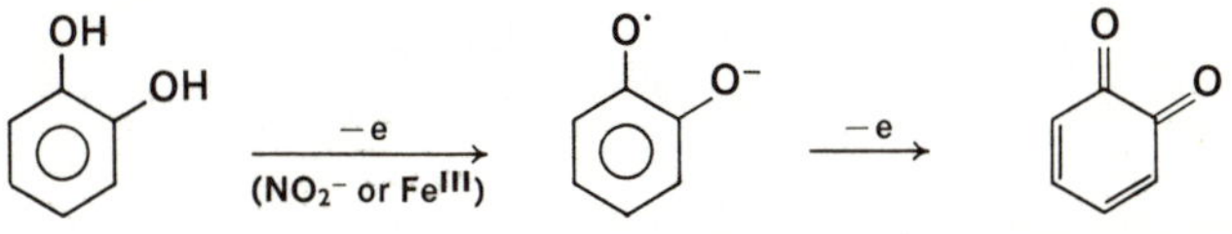

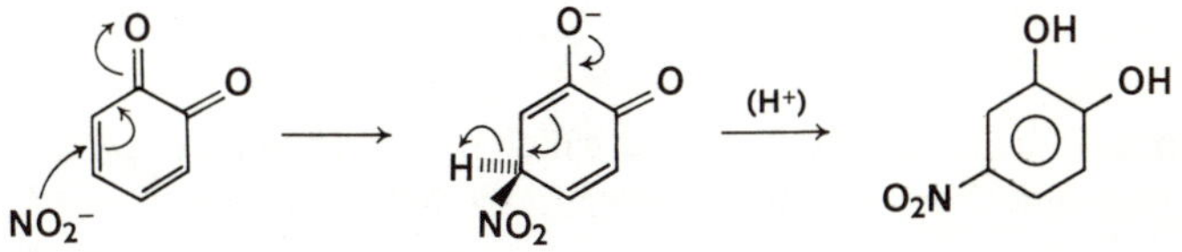

Phenols would not be expected to undergo nucleophilic substitution (Section 4.12) and anyway, the reaction does not resemble electrophilic substitution (Section 2.14) by which nitrous acid might be expected to attack such reactive substrates with the production of a nitroso compound. Furthermore, the yield is greatly improved by the addition of an oxidizing agent such as potassium ferricyanide. It is believed that this is an example of a substitution via an intermediate quinone (**3**) formed through the oxidizing action of the nitrous

semiquinone (+ NO_2, Fe^{II})

o-benzoquinone (**3**)

acid (or ferricyanide); the intermediate is susceptible to nucleophilic attack by the nitrite ion. Benzenesulphinate ion (4) also efficiently substitutes in poly-phenols in the presence of the oxidant and will give the same product when added to preformed quinone (5):

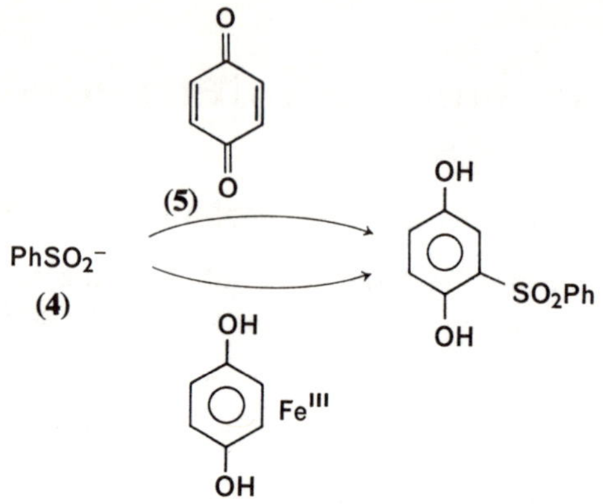

In some examples, the product retains the quinone structure,

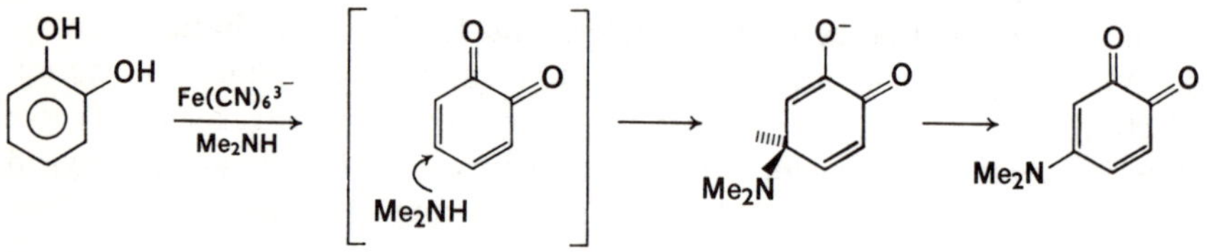

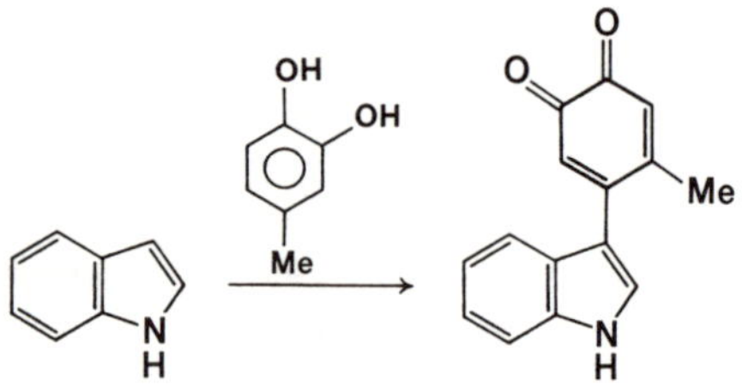

indole

In principle, any aromatic compound which possesses *ortho* or *para* dihydroxy or hydroxyamino functions may take part in these reactions.

10.2   **SOME ACID-CATALYSED REACTIONS**

Reactions of Diazo Compounds with Acids

Diazo compounds, which include the parent compound diazomethane, CH_2N_2, and its derivatives (**6**), have been mentioned previously as precursors of carbenes which form on thermal or photochemical decomposition (Chapter 6). In the presence of acids an entirely different set of products may form. Since

$$\underset{R'}{\overset{R}{\diagdown}}C=\overset{+}{N}=\overset{-}{N} \longrightarrow \underset{R'}{\overset{R}{\diagdown}}C\colon + N_2$$

(**6**)

diazoalkanes such as diazomethane are rather unstable, most investigations of this reaction have used either diphenyldiazomethane, Ph_2CN_2, or ethyl diazoacetate, EtO_2CCHN_2, both of which may be obtained pure. These two related compounds exhibit surprisingly different behaviour.

Diphenyldiazomethane (**7**), a dark-red crystalline solid, reacts in alcoholic solution in the presence of a strong acid catalyst to give the benzhydryl ether (**8**) and nitrogen with a small amount of tetraphenylethylene (**9**) as by-product. If the diazo compound is allowed to react with a carboxylic acid in alcoholic solution, a mixture of benzhydryl ether (**8**) and ester (**10**) is obtained, while in an aprotic solvent the ester is the sole product. These products may be seen as arising from a carbonium ion derived by proton transfer to the diazo compound. If the reaction is carried out in a deuterated-alcohol solvent, either with

$$\underset{Ph}{\overset{Ph}{\diagdown}}C=\overset{+}{N}=\overset{-}{N} \underset{\longleftarrow}{\overset{H^+}{\rightleftharpoons}} \left[\underset{Ph}{\overset{Ph}{\diagdown}}\underset{\underset{N\equiv N}{\overset{+}{|}}}{C}\diagdown^{H} \right] \longrightarrow \underset{Ph}{\overset{Ph}{\diagdown}}\overset{+}{C}{-}H + N_2$$

(**7**)

$$\underset{Ph}{\overset{Ph}{\diagdown}}\overset{+}{C}{-}H$$

ROH → $\underset{Ph}{\overset{Ph}{\diagdown}}\overset{\text{\tiny|}}{\underset{H}{C}}{-}OR$ (**8**)

Ph_2CN_2 → $Ph_2CH{-}\underset{+}{C}Ph_2 + N_2$ $\overset{-H^+}{\longrightarrow}$ $Ph_2C{=}CPh_2$ (**9**)

$R'COO^-$ → $\underset{Ph}{\overset{Ph}{\diagdown}}\underset{H}{\overset{\text{\tiny|}}{C}}{-}O{-}C\overset{\diagup O}{\diagdown}_R$ (**10**) a strong or a weak acid, the rate is reduced by a factor of ca 3.5 compared to the ordinary solvent. This is a significant observation and suggests that the proton transfer is rate determining, i.e. that a bond to hydrogen in the acid is being broken in the transition state. Furthermore, the intermediate carbonium ion is not available for trapping by added nucleophiles, which indicates that it is not free but exists transiently as an ion pair with the carboxylic ion which then collapses to ester products or is captured by solvent with ether formation. The following scheme is consistent with these observations.

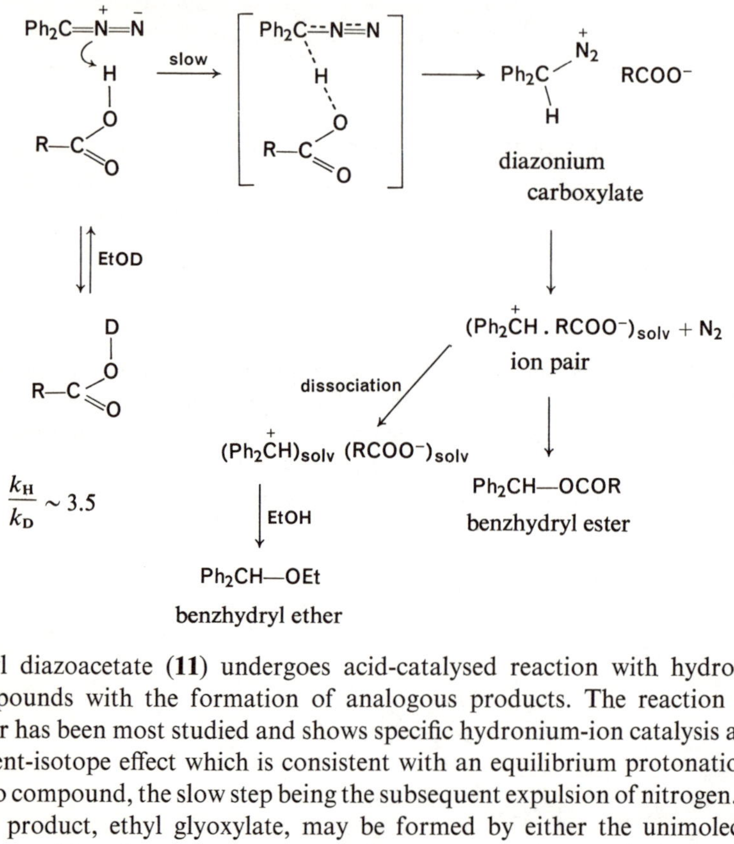

Ethyl diazoacetate (11) undergoes acid-catalysed reaction with hydroxylic compounds with the formation of analogous products. The reaction with water has been most studied and shows specific hydronium-ion catalysis and a solvent-isotope effect which is consistent with an equilibrium protonation of diazo compound, the slow step being the subsequent expulsion of nitrogen. The final product, ethyl glyoxylate, may be formed by either the unimolecular decomposition, k_c, of the protonated diazo compound or by displacement of nitrogen by water, k_s. The first route looks unlikely since the carbonium ion formed would not only be primary but further destabilized by the adjacent carboxyl group. However, $-N_2^+$ is about the best leaving group known so this should not be excluded. The S_N2 route is perhaps more plausible on the grounds that this process appears to operate with other nucleophiles such as Cl^-

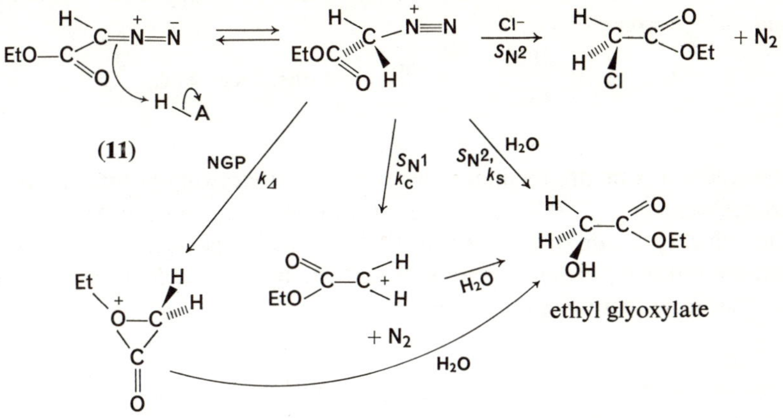

(11)

present. The reaction is faster in hydrochloric acid than in perchloric of the same acidity (as measured by H_0, Section 1.22). Furthermore, the nucleophilic displacement of N_2 could be intramolecular (k_A) by the carboxylate group (Section 3.1).

It has been pointed out that this difference in behaviour between diphenyl-diazomethane and the diazoacetate lies in a difference in the magnitudes of the activation energies for the loss of nitrogen from the protonated compound. The diazo ester, in the extreme case, forms a highly destabilized cation, its formation becoming rate determining; the diphenyldiazomethane forms a highly stable benzhydryl cation hence C—N bond cleavage is very fast and the proton transfer becomes rate determining. The latter shows the more negative entropy of activation, -65 J K^{-1}, compared with -7 for the diazoacetate, which is consistent with this picture, since proton transfer would create ions which would increase the solvation, while dissociation of the diazonium ion would not.

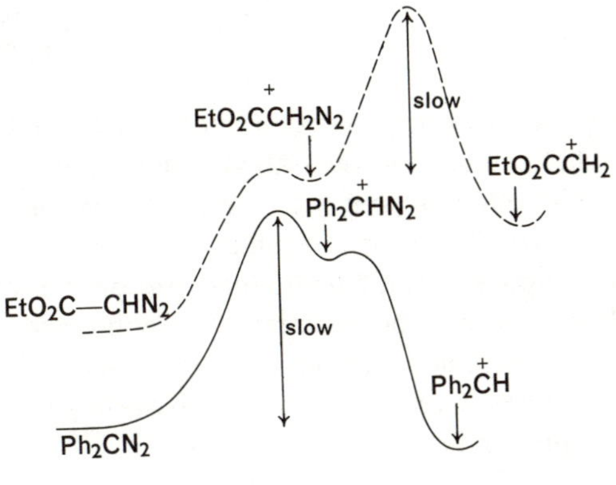

The decomposition of arylalkyltriazenes (**12**) by carboxylic acids is a useful route to the esters:

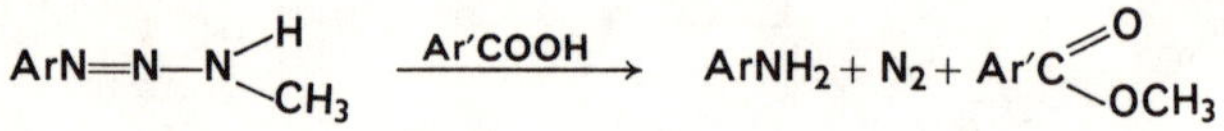

$$ArN{=}N{-}N\underset{CH_3}{\overset{H}{\diagup}} \xrightarrow{Ar'COOH} ArNH_2 + N_2 + Ar'C\underset{OCH_3}{\overset{O}{\diagup}}$$

(**12**)

The reaction is mildly facilitated by electron-withdrawing groups in Ar' (i.e. stronger acids), $\rho = -0.92$, and by electron-donating groups in Ar ($\rho = 1.20$). The n.m.r. spectrum of the reagents (before reaction has occurred) shows the triazene and acid protons to be rapidly exchanging. The mechanism is probably via a pre-protonation equilibrium:

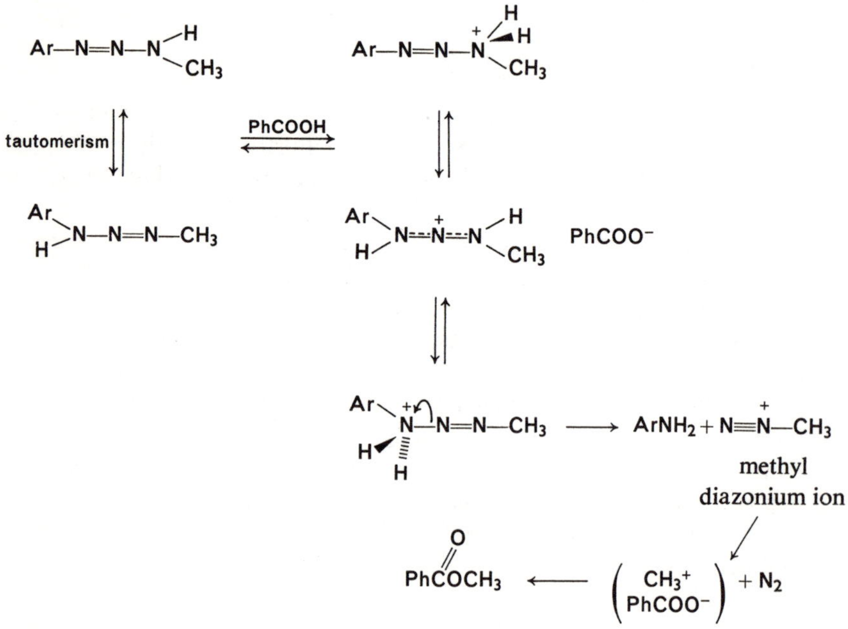

There are probably a number of protonated intermediates and the alkyl group is expelled as alkyldiazonium ion then alkyl cation in an ion-pair with the carboxylate ion. This collapses rapidly to the ester. The solvent effects upon this and the reactions of diphenyldiazomethane are quite characteristic: solvents which are good electron-pair donors (irrespective of their polarity properties) markedly retard the reaction. The observed rate order is: $CH_3NO_2 > CH_2Cl_2 >$ benzene $>$ acetone $>$ ether $> (CH_3)_2SO$. This is also the order of acid strengths in these solvents. Consequently the rate of reaction is controlled by the pre-equilibrium protonation.

The Benzidine Rearrangement

In strongly acid solution, diphenylhydrazines (hydrazobenzenes, **13a**) rearrange to the isomeric diaminobiphenyls (benzidines, **13b**), together with greater or lesser amounts of the diphenylines (**13c**), *ortho*-benzidines (**13d**) and semidines (**13e**, **13f**). The latter are particularly important products from 4-substituted hydrazobenzenes.

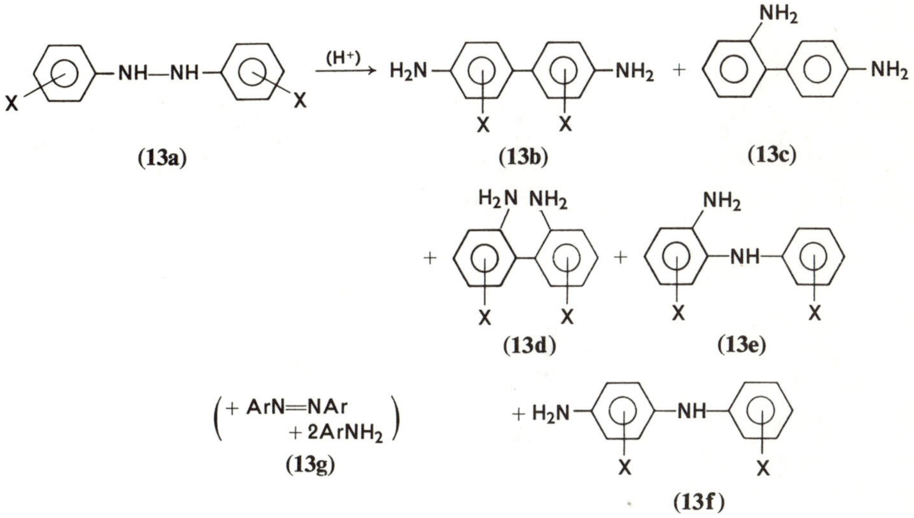

			Per cent yield			
X	13b	13c	13d	13e	13f	13g
H	75	25	—	—	—	—
2-MeO	100	—	—	—	—	—
4-MeO	—	—	—	55	25	20
4-Cl	—	—	—	20	30	30
2-I	100	—	—	—	—	—

The kinetics are frequently third order,

$$\text{rate} = k_3[\text{hydrazobenzene}][\text{acid}]^2$$

and consequently a doubly protonated intermediate is presumed to be the species which undergoes the rearrangement. Isotopic labelling shows conclusively that the rearrangement is intramolecular and a small solvent-isotope effect ($k_{H_3O^+}/k_{D_3O^+} \sim 2$)—and zero *para*-D kinetic-isotope effect—is consistent with an equilibrium protonation. The following sequence accounts for these observations:

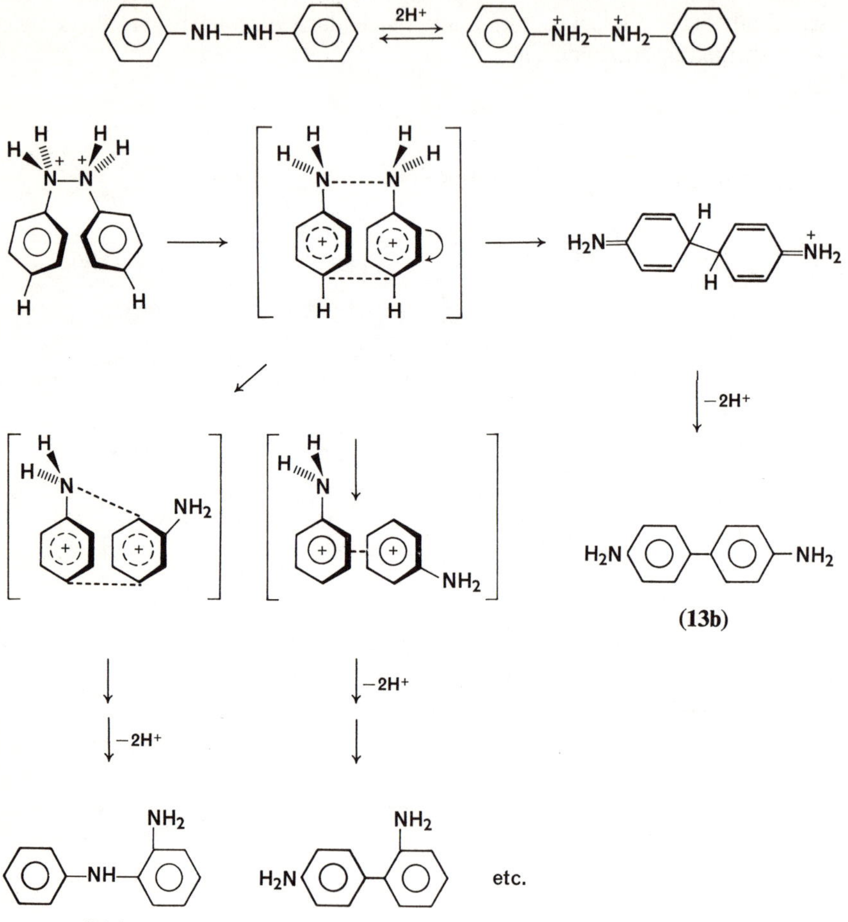

(13b)

(13e)   **(13c)**   etc.

The dication apparently undergoes fission at the N—N bond giving rise to an intermediate in which the two rings, face to face, are held together but are able to rotate relative to one another to give, after loss of the protons, the various observed products. The reaction is strictly intramolecular and is greatly accelerated by electron-donating substituents, which obviously make protonation easier. The rates for the methoxy-substituted hydrazobenzenes are proportional to the first power of the hydrogen-ion concentration, showing that only one proton is required in these cases. Other products include the azo compound and the aniline, formed by disproportionation. Tetraphenyl-hydrazine undergoes rearrangement via radical ions:

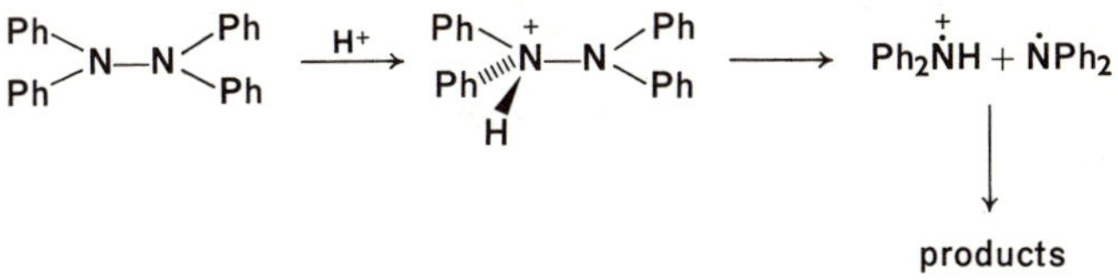

The radical cation can be detected during reaction by its characteristic visible absorption ($\lambda_{max} = 465$ nm) and also by its e.s.r. spectrum. This demonstrates a second type of mechanism which may be used only when circumstances are especially favourable but may turn out to be much more common or indeed quite general when further investigations have been made.

The Beckmann Rearrangement

When an oxime is treated with a strong acid, a rearrangement to an amide occurs. This involves migration of the alkyl or aryl group from carbon to nitrogen and it is found that only the group which is *trans* to the leaving hydroxyl function migrates. The function of the acid is to assist the departure

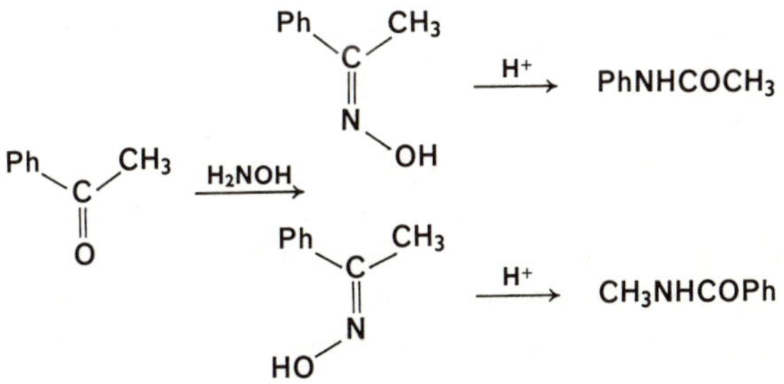

of the hydroxyl group. This can equally well be achieved by using a hydrazone and treating with nitrous acid (Schmidt reaction). The reaction is facilitated by electron release from substituents in the aromatic group ($\rho = -2.0$) and it appears that initial protonation is followed by heterolysis of the leaving group (H_2O) and simultaneous rearrangement to give, initially, the enolic form of an amide.

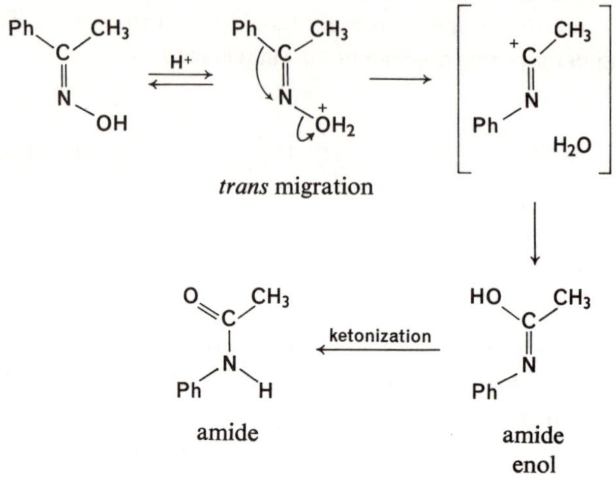

trans migration

ketonization

amide

amide
enol

10.3   REACTIONS CATALYSED BY ENZYMES

Enzymes are protein molecules, often of very high molecular weight and sometimes containing in their structure relatively small functional molecules (coenzymes). Their function essentially is to reduce the activation energy required for a given chemical reaction so that it will take place at a suitable rate under conditions such as are found in the living cell—dilute aqueous media at 20–35°. The reaction which is mediated by the enzyme takes place at a specific point on its surface (the 'active site') where are to be found the amino-acid residues and perhaps the coenzyme which assists the reaction. It is generally supposed that the substrate is able to bind to the active site specifically by a combination of electrostatic forces and hydrogen bonds. The active site usually involves only a minute fraction of the whole enzyme; the remaining protein chain is necessary for the maintenance of the exact stereochemical alignment of the components of the active site mainly through hydrogen-bonding forces. This picture of enzyme action is substantiated by the peculiar

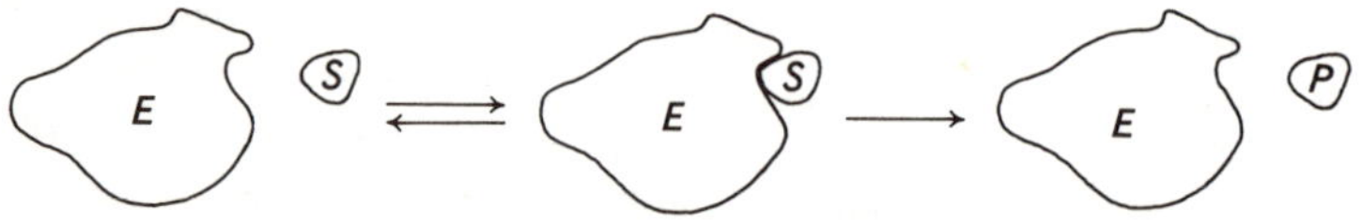

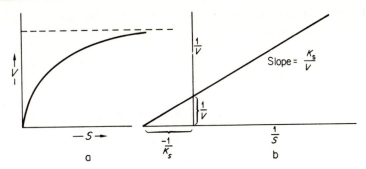

Figure 1. (a) Initial velocity of an enzymic reaction as a function of substrate concentration; (b) Lineweaver–Burk plot for an enzymic reaction.

kinetic form exhibited by most enzymic reactions, which may be deduced from the scheme,

$$E + S \underset{}{\overset{K_s}{\rightleftharpoons}} ES$$

$$ES \xrightarrow{k} \text{products}, p$$

Writing E, S and ES, for the concentrations of total enzyme, free substrate and enzyme–substrate complex, respectively,

$$K_s = \frac{(E - ES)\,S}{ES} \quad \text{whence} \quad ES = \frac{E.S}{K_s + S}$$

the rate of reaction,

$$v = kES = \frac{kE.S}{K_s + S} = \frac{kE}{1 + K_s/S}$$

A plot of v against S (Figure 1a) shows that the reaction velocity tends towards a maximum as S increases, after which the reaction becomes zero order, which indicates that the enzyme is saturated with respect to substrate. Thus, if S is large in comparison with K_s, $v = V = kE$; therefore

$$v = \frac{V}{1 + K_s/S}$$

where V is the maximum velocity. This expression may be written in the form,

$$(V - v)(K_s + S) = VK_s$$

and is known as the Michaelis equation; V and K_s characterize the kinetic properties of the reaction. There are several ways of testing the validity of this expression. If $1/v$ is plotted against $1/S$, a straight line results of slope K_s/V (Lineweaver–Burk plot)—Figure 1b. That such a plot is indeed linear for very many enzymic reactions adds support to the proposed mechanism.

The Mechanism of Action of Chymotrypsin

In order to understand the molecular basis of enzyme action it is first necessary to know the structure of the enzyme or at least the active site. This is a formidable task which must be attacked by obtaining an X-ray crystallographic structure of the (crystalline) enzyme. Chemical studies can, in principle, determine the number and sequence of the amino-acid sub-units present but cannot give information on their stereochemical arrangement. Today, the complete structures of several enzymes have been determined and this is illustrated by considering the proteolytic enzyme, chymotrypsin.

Chymotrypsin is a digestive enzyme whose function is to hydrolyse the amide linkages of protein. While it shows a preference for the peptide links of aromatic amino acids, chymotrypsin is relatively unspecific and can be made to hydrolyse a wide variety of substrates, both natural and artificial, including amino-acid esters. The enzyme is made up of three separate peptide chains of 13, 131 and 97, amino-acid units respectively, linked together by the disulphide bonds of cystine residues. It specifically hydrolyses the L-enantiomers of these substrates (**14a**, i.e. those related to natural amino acids) the D compounds being bound but not hydrolysed (i.e. they act as inhibitors).

Classical kinetic studies have given some information as to the nature of the active site, since it must fit a substrate with the general structure **15**. There are

(15)

N-acetyl-L-alanine ethyl ester

(14)

L series

a

D series

b

likely to be three loci for interactions between enzyme and substrate: the amide group, the carboxyl function and a rather large 'pocket' for the hydro-

Substrate (reaction site underlined)	K_s (mol)	10^3 rate V (mol min^{-1} ml^{-1} mg^{-1})

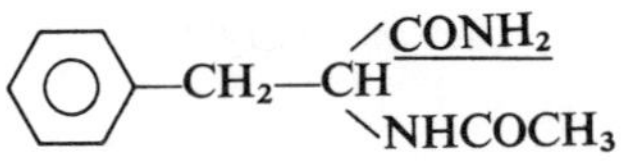

N-acetylphenylalaninamide

| | 31 | 0.8 |

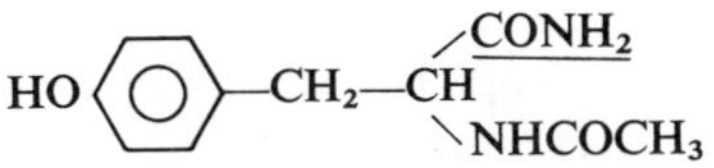

N-acetyltyrosinamide

| | 32 | 2.4 |

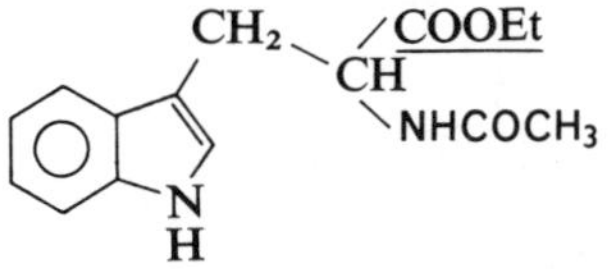

N-acetyltryptophan ethyl ester

| | 4.8 | 360 |

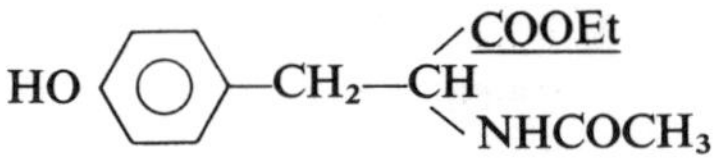

N-acetyltyrosine ethyl ester

| | 74 | 4500 |

phobic aromatic side-chain. The rates of hydrolysis (V) of the following artificial substrates illustrate the importance of structure. The basic mechanism of hydrolysis involves transfer of an acyl group to the enzyme, releasing amine or alcohol depending on whether an amide or ester is being hydrolysed. The acyl group is then displaced in turn to release the other moiety of the substrate and free the enzyme for further reaction. It was early considered likely that the acylation occurred at a serine hydroxyl. The full three-dimensional structure of chymotrypsin is now known and the active site identified. The conformation of the surrounding protein is such as to allow a good fit to the typical substrates of the enzyme. The mode of action depends upon a 'charge-relay' system whereby a carboxylic ion (in the aspartic acid residue No. 102) is able to supply electrons through the imidazole ring of histidine No. 57 thus effectively making

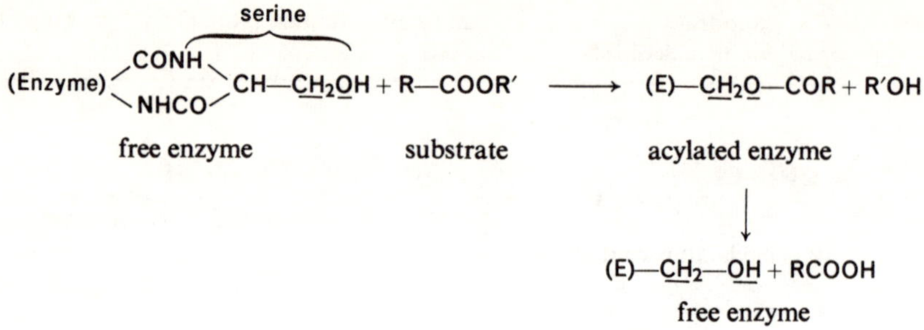

serine

$$(\text{Enzyme})\underset{\text{NHCO}}{\overset{\text{CONH}}{\diagdown\diagup}}\text{CH}-\text{CH}_2\text{OH} + \text{R}-\text{COOR}' \longrightarrow (\text{E})-\text{CH}_2\text{O}-\text{COR} + \text{R}'\text{OH}$$

free enzyme   substrate   acylated enzyme

$$\downarrow$$

$$(\text{E})-\text{CH}_2-\text{OH} + \text{RCOOH}$$

free enzyme

the serine oxygen (serine No. 195) much more nucleophilic and able to attack the substrate carbonyl which is lying in the correct position.

Aminoacids in a protein chain

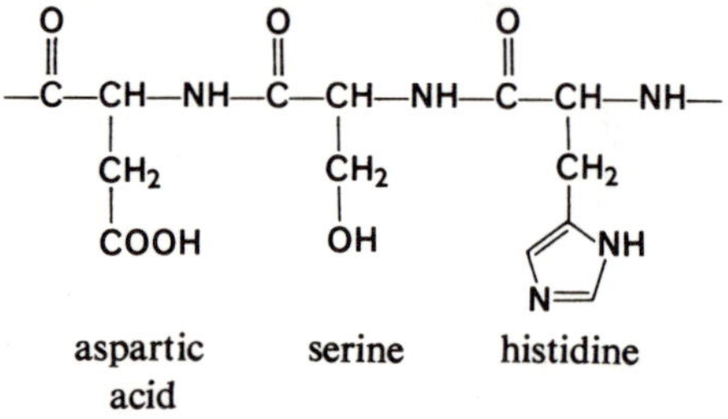

aspartic acid   serine   histidine

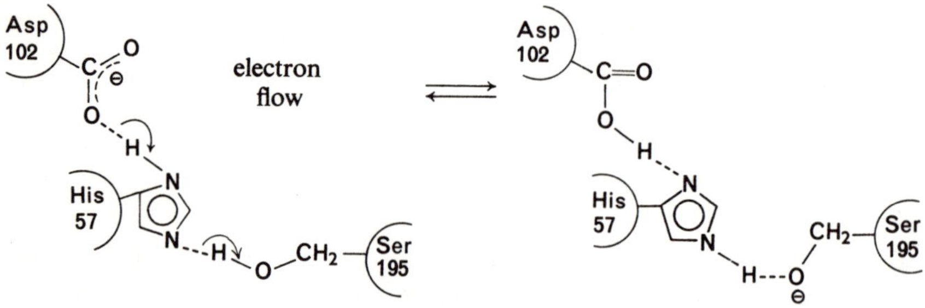

Charge-relay system of chymotrypsin
(dotted lines are hydrogen bonds)

The hydrolytic mechanism may be depicted as in Figure 2. The effect of this charge-relay system is to provide electrons at the serine oxygen on demand during the part of the reaction which normally requires the greatest activation energy, and which thereby becomes considerably reduced. It is likely that other proteolytic enzymes such as trypsin, papain, ficin and rennin use similar mechanisms, though the environments of their active sites are different and accordingly their substrate specificities.

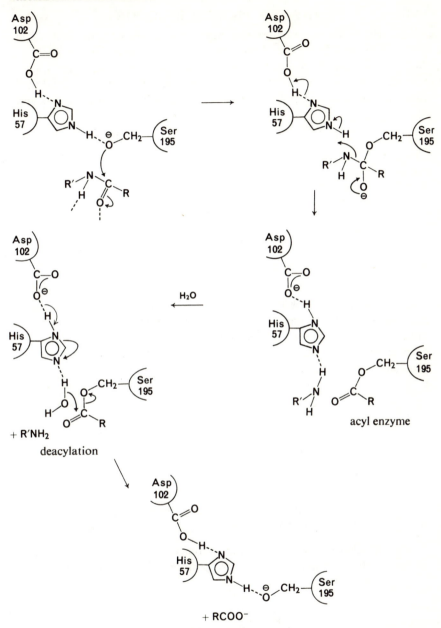

Figure 2. Mechanism of action of chymotrypsin (adapted from J. J. Birktoft et. al., *Phil. Trans. Roy. Soc.*, **B257**, 67, (1970)).

10.4   **THE VON RICHTER REACTION**

When an aromatic nitro compound is heated with alcoholic potassium cyanide, a carboxylic acid is formed. This reaction was discovered by von Richter in 1871 and it was soon realized, through the work of Griess and others, that it was highly unusual in that the carboxyl function entered the ring *ortho* to the original nitro group. The reaction has been reexamined in recent years and a

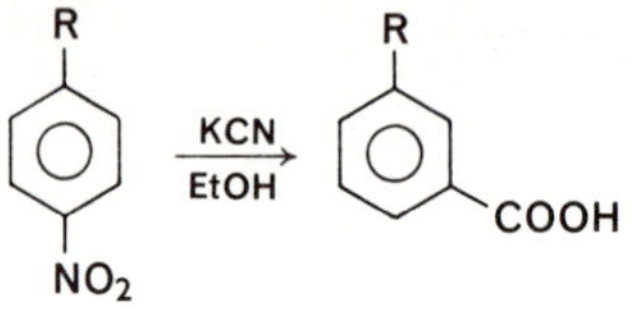

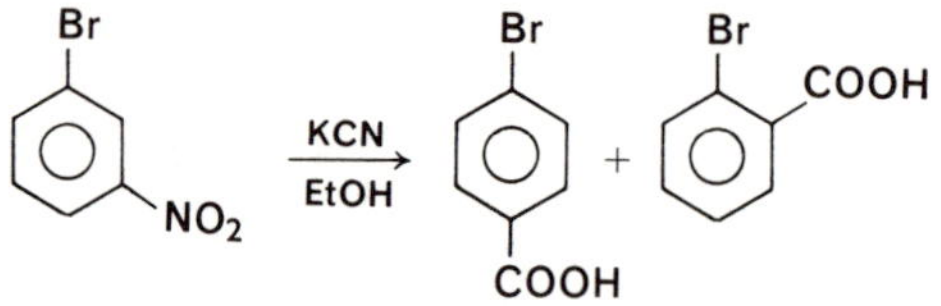

plausible mechanism has been deduced. It is found that unreacted starting material is recovered unchanged, i.e. there is no prior isomerization of the nitro compound, and also the reaction does not proceed via an intermediate cyanide or amide since these, when added, do not lead to the observed products.

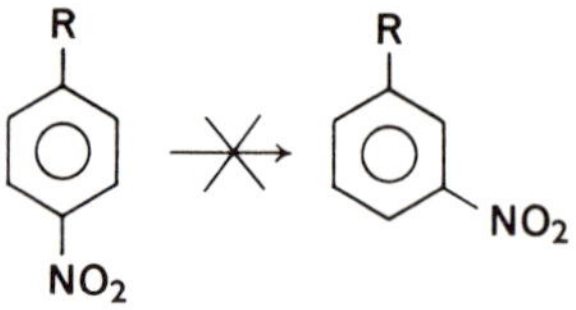

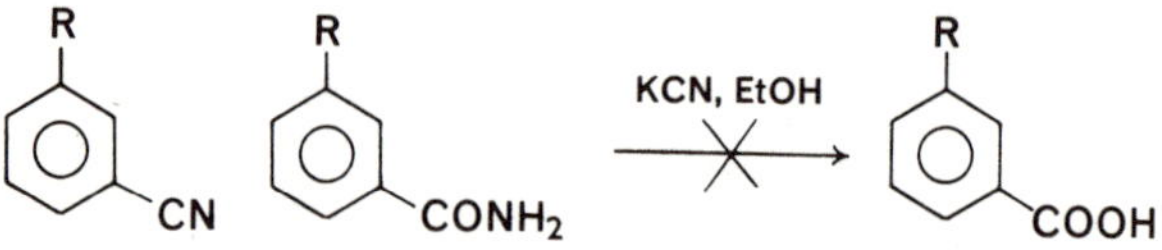

When the reaction is carried out in isotopically labelled protic solvent, the proton which results at the original nitro position is seen to come from solvent.

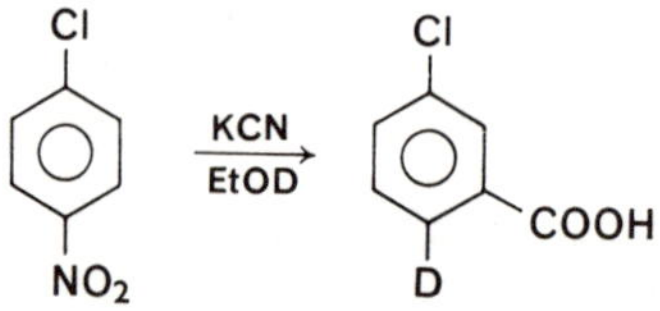

A product of the reaction is nitrogen gas, formed in approximately the same yield as the carboxylic acid and furthermore, isotopic labelling shows that one nitrogen atom of the N_2 comes from the nitro group and the other from cyanide and that neither have existed intermediately as ammonia.

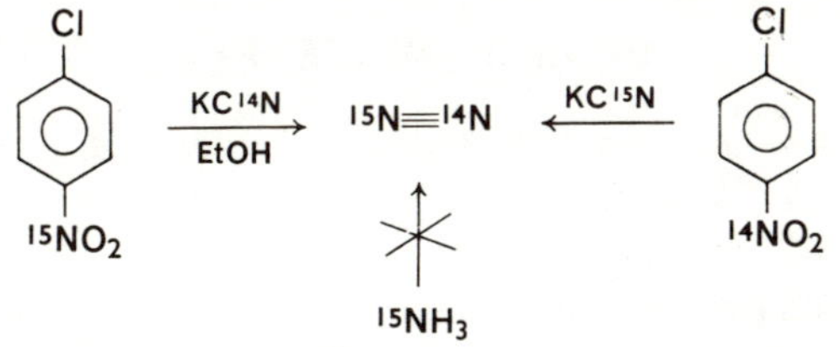

A further labelling experiment with oxygen shows that one carboxyl oxygen comes from the nitro group and one from the solvent.

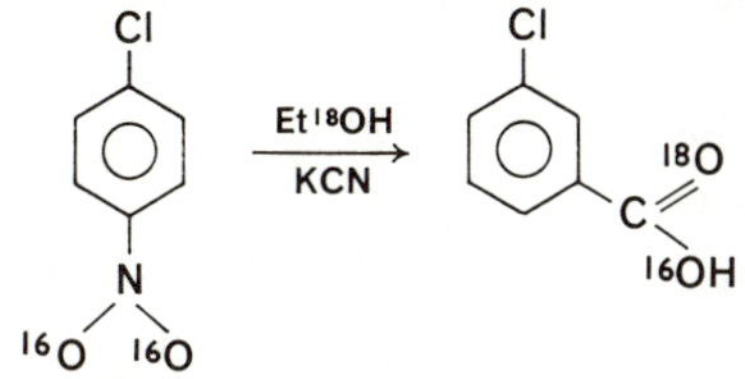

It is likely that the cyanide ion, a strong nucleophile, attacks the aromatic ring *ortho* to the nitro group (cf. aromatic nucleophilic substitution, Section 4.12); it is clear that the nitro group must then interact with the complex which is formed (18). The following mechanism (due to Rosenblum) is consistent with the facts.

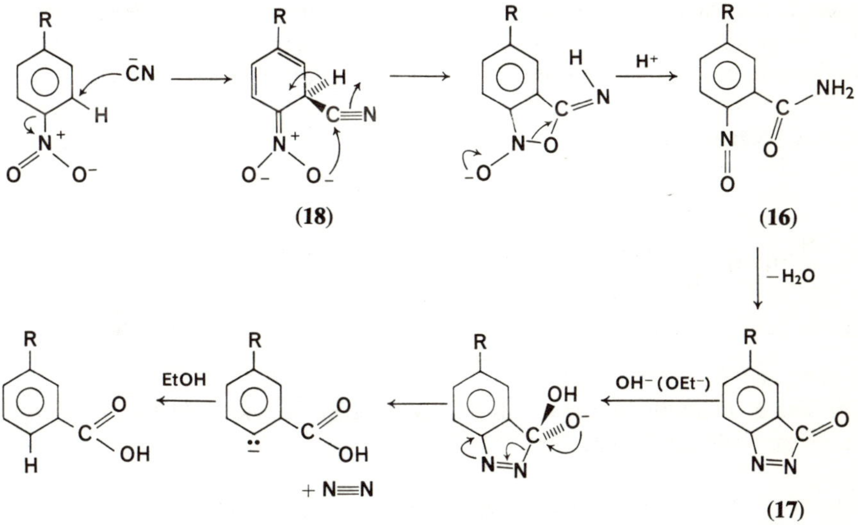

In support of this mechanism, both the nitrosoamide (**16**) and the indazolone (**17**) have been isolated and shown to be converted to the appropriate carboxylic acid under the conditions of the von Richter reaction.

10.5   DIIMINE (DIAZENE)

The series **19** makes an interesting study in stability. Azo compounds (**19a**), especially aromatic derivatives, are well known and extremely stable (PhN=NPh to 300°); monosubstituted diazenes (**19b**) are quite reasonably stable though much less so than azo compounds and have only recently been discovered, while the parent compound (**19c**), usually known as diimine, is highly unstable and normally exists only as a transient intermediate with, however, interesting and useful chemical properties.

$$RN=NR \qquad RN=NH$$
$$\qquad\qquad\qquad\qquad\qquad HN=NH$$
$$ArN=NR \qquad ArN=NH$$

$$\text{(a)} \qquad\qquad \text{(b)} \qquad\qquad \text{(c)}$$

$$\text{(19)}$$

Generation of Diimine

(a) It has been known since about 1930 that hydrazine can act as a reducing agent towards olefinic double bonds *when in the presence of an oxidizing agent.* It is now known that the function of the oxidizing agent is to convert hydrazine to diimine which acts as a specific hydrogen-transfer agent to olefinic centres. The most useful oxidizing agents for this purpose are ferricyanide or chloramine T, but oxygen, peroxide with catalytic amounts of Cu^{II} and mercuric oxide, are also suitable.

(b) The acid-catalysed decomposition of azodicarboxylic acid and its salts leads smoothly to diimine.

$$HOOC-N=N-COOH \xrightarrow{\;H^+\;} HN=NH + 2CO_2$$

(c) The decomposition of a number of other nitrogenous compounds furnishes convenient routes to diimine, e.g. **20** and **21**.

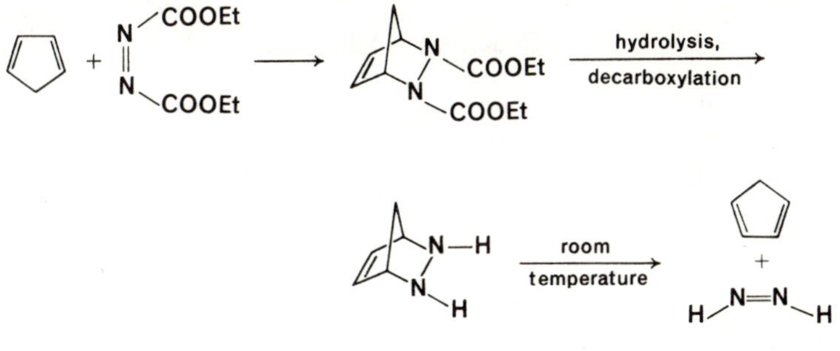

(20)

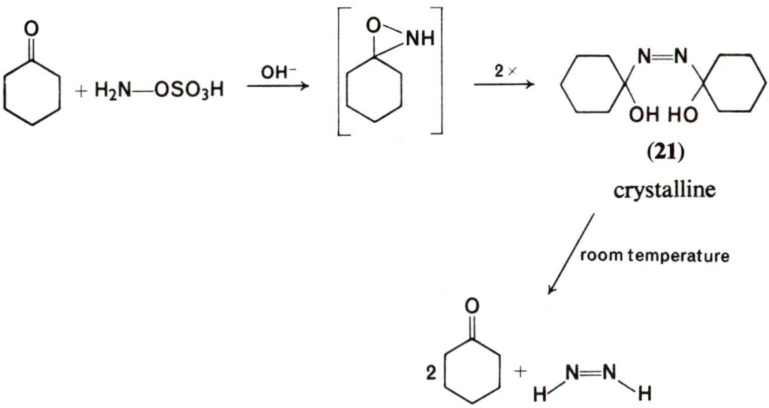

(21)

crystalline

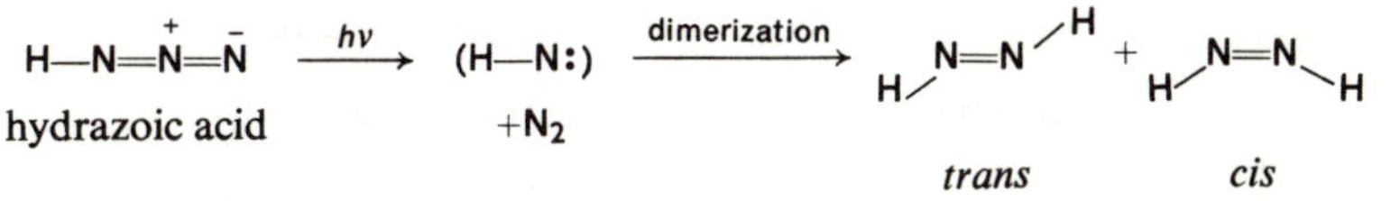

2

Diimine has been directly detected at very low temperature. When hydrazoic acid was photolysed in a solid nitrogen matrix, new infrared absorption bands were detected and assigned as follows:

λ_{max} 1286 1279, 3074 cm^{-1}

Indeed, recently diimine has been prepared pure and shown to be fairly stable at $-80°$. As expected, diimine can exist as *cis* or *trans* geometrical isomers.

When trapped in the nitrogen matrix, no reaction of this very labile species is possible.

Reactions of Diimine

Diimine may undergo disproportionation with itself yielding hydrazine and nitrogen by a hydrogen-transfer mechanism essentially similar to its reaction with olefins. In the presence of base, ionization of the rather acidic hydrogen

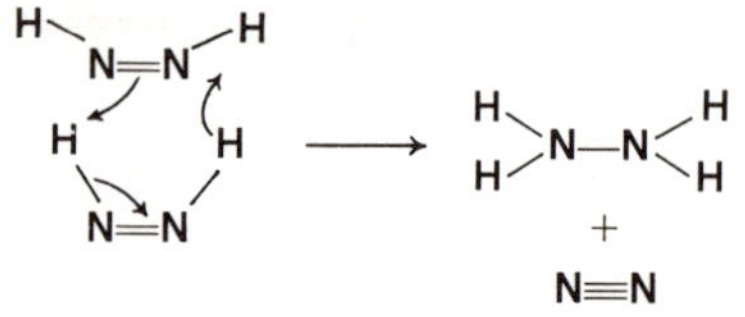

occurs followed by loss of a hydride ion to water and the production of nitrogen.

$$HN—NH + B \longrightarrow HN=\bar{N} \xrightarrow{H_2O} N_2 + H_2 + OH^-$$

The most characteristic property of diimine is its ability to transfer hydrogen to olefinic double bonds by a concerted mechanism which leads to specifically *cis* hydrogenation. The symmetry of the transition state is reflected in the observation that reduction is confined to homonuclear double bonds such as $C=C$, $C\equiv C$ and $N=N$, but not heteronuclear bonds such as $N=O$, $C=O$ and $C=N$, which are unaffected. These hydrogen transfers are symmetry-allowed concerted reactions (Section 1.31). There is another useful property of

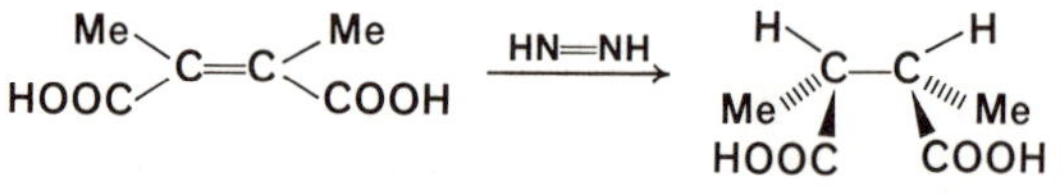

dimethylmaleic acid

meso-1,2-dimethyl-succinic acid

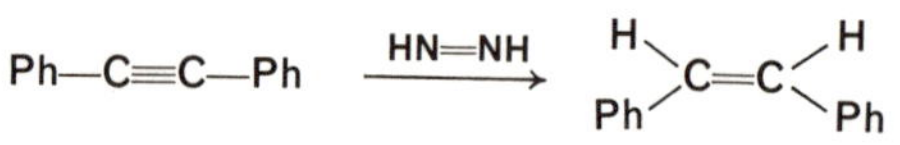

diphenylacetylene

cis-stilbene

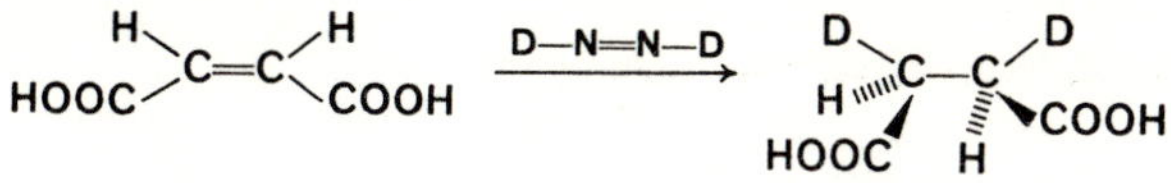

maleic acid

meso-succinic
acid-1,2-d_2

the reduction in that *trans* olefins are usually attacked faster (3–10 times) than *cis*, which may allow specific reductions to be carried out, e.g.

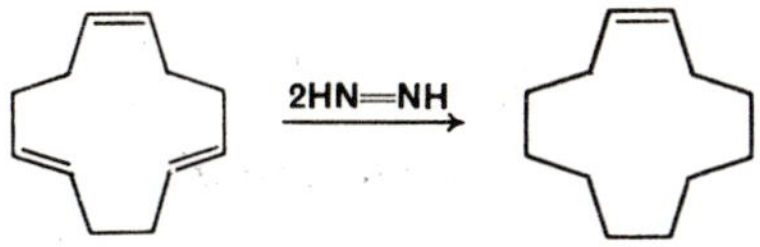

cis,trans,trans-dodecatriene

cis-dodecene
95%

Monosubstituted Diazines

The Wolff–Kishner reduction of aldehydes and ketones to alkanes has been interpreted in terms of the rearrangement of a hydrazone to a diazene (**22**) which, under the basic conditions of the experiment, decomposes to the hydrocarbon via a carbanion. A number of monosubstituted diazenes have now been

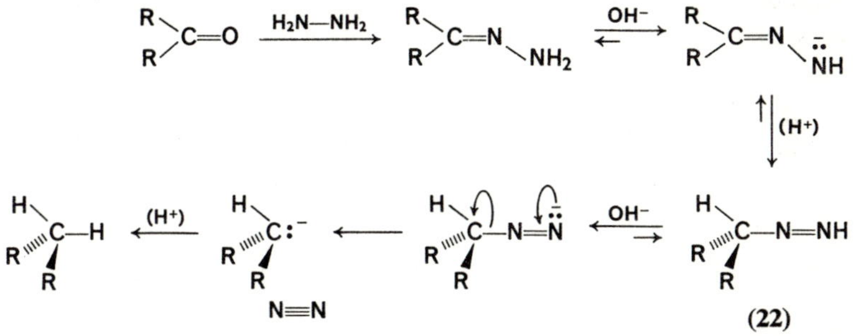

(22)

prepared and their chemical properties shown to support this mechanism. A convenient route to many such compounds is the decarboxylation of appropriate azocarboxylic acids.

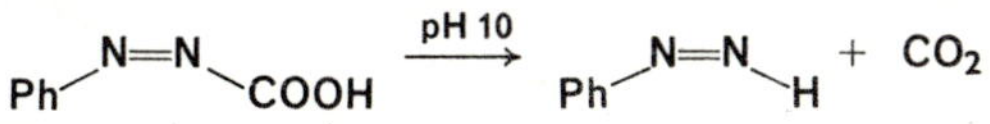

phenylazocarboxylic acid $\lambda_{max} = 417$ nm

Elimination reactions have also been used,

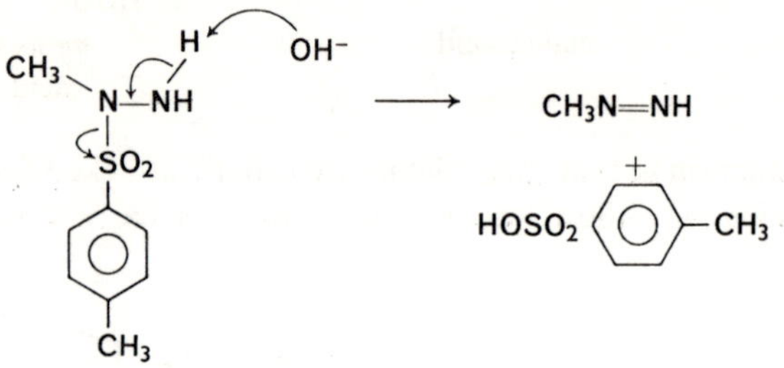

$$
\begin{array}{c}
CH_3N{=}NH \\[4pt]
\overset{+}{} \\
HOSO_2\text{—}\langle O \rangle\text{—}CH_3
\end{array}
$$

N-methyl-*p*-toluenesulphonyl hydrazide

and also oxidations of hydrazines,

$$PhNH\text{—}NH_2 \xrightarrow{\ Fe(CN)_6{}^{3-}\ } PhN{=}NH$$

Diazenes decompose in basic media, producing carbanions,

$$PhN{=}NH + OH^- \longrightarrow PhN{=}N:^- \longrightarrow Ph:^- + N{\equiv}N$$

and also undergo a bimolecular mode of decomposition which may proceed via radicals: the products are the hydrazine and hydrocarbon corresponding to the diazene.

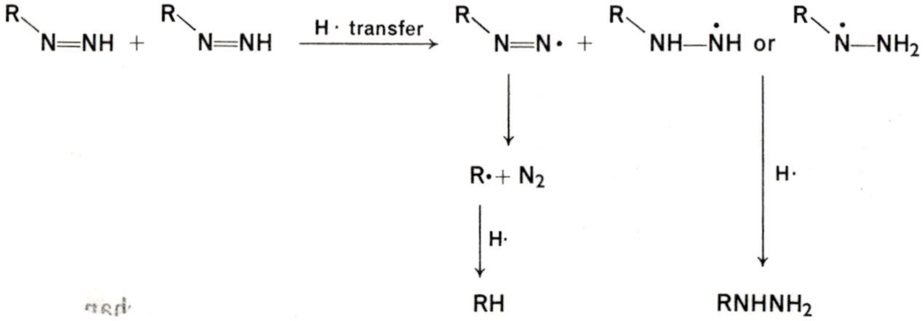

$$
\begin{array}{ccc}
\underset{N{=}NH}{R} + \underset{N{=}NH}{R} \xrightarrow{\ H\cdot\ transfer\ } & \underset{N{=}N\cdot}{R} + & \underset{NH{-}\dot{N}H}{R}\ or\ \underset{\dot{N}{-}NH_2}{R}
\end{array}
$$

$$R\cdot + N_2$$

$$\Big\downarrow H\cdot$$

$$RH \qquad\qquad RNHNH_2$$

10.6   TRANSITION-METAL COMPLEX INTERMEDIATES

Many reactions are now known to be mediated by transition metals in homogeneous media and a number of these form the bases of important large-scale processes.

The transition metals are characterized by possessing a set of d-orbitals which are of comparable energy to the p-orbitals of the next higher group of the periodic table. They are therefore capable of coordinating to a wide variety of molecules or ligands which act in most instances as electron donors and form a coordinate bond, e.g. **23**. Carbon ligands are also possible and in this

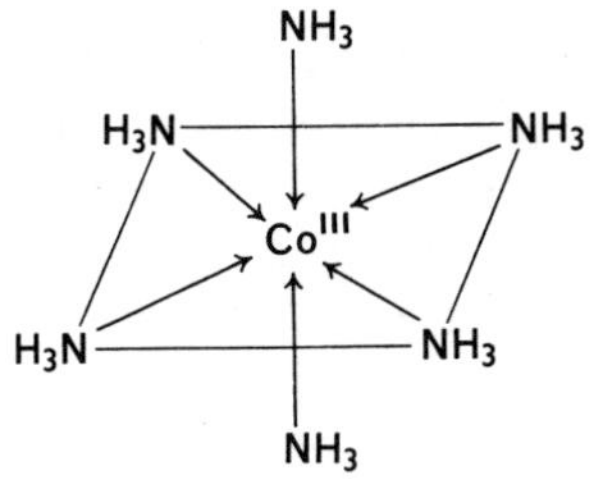

(23)

octahedral coordination at
Co^{III}

instance coordination to the metal may be by π-electrons or by a formal σ-bond. The two may also be in equilibrium (**24a, b**) and antibonding π-orbitals may also be involved (**25**). A transition metal may therefore be seen as

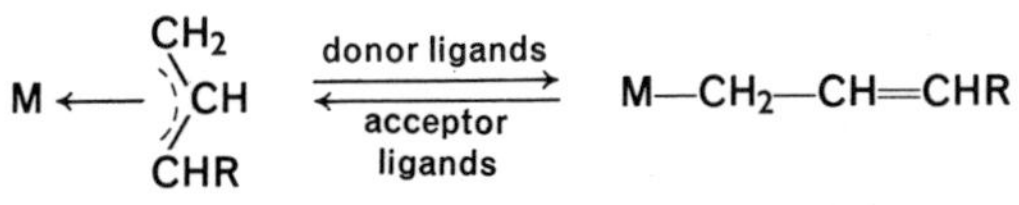

π-allyl σ-propenyl

(a)   **(24)**   **(b)**

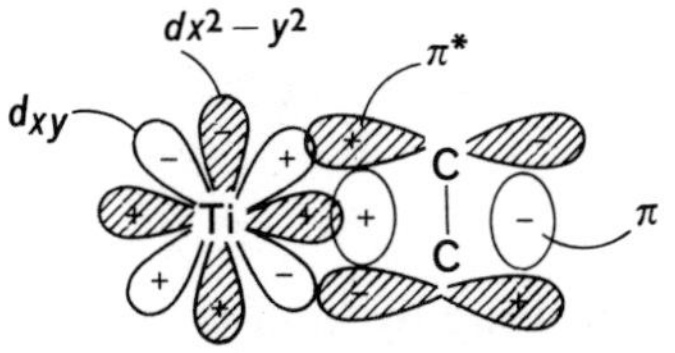

π-bonding of ethylene

to titanium

(25)

a site for coordinating organic groups which may subsequently react with themselves or external reagents. Whether a given transition-metal compound will act as a catalyst in an organic reaction is almost impossible to predict, and small differences in structure can affect the system profoundly through a variety of variables which may include the strength of metal–carbon bond of both reagent and product, the ability to coordinate a reagent, if necessary, and favourable thermodynamics of the reaction. These in turn may be affected by such features as the donor ability of both reacting and non-reacting ligands, inductive and resonance effects, electron density at the metal and ligand effects upon orbital energies. It is often difficult to separate these factors but the basic mechanisms of many metal-catalysed reactions are now understood at least in outline. The subject is very large but the following examples will illustrate some important concepts.

The Oxo Process

Terminal olefins were found by Roelen (1938) to react with carbon monoxide and hydrogen in the presence of a soluble cobalt–carbonyl complex to give aldehydes. Various soluble complexed forms of cobalt can be used as catalyst

$$RCH{=}CH_2 + CO + H_2 \xrightarrow{\text{Co}_2(CO)_3} RCH_2CH_2CHO$$

precursors but it is believed that the active catalyst is cobalt tricarbonyl hydride (26) formed in the following reaction:

$$Co + 8CO \longrightarrow Co(CO)_8$$
$$\text{(or complex)}$$

$$Co(CO)_8 + H_2 \longrightarrow 2HCo(CO)_4 \longrightarrow 2HCo(CO)_3 + CO$$
$$\textbf{(26)}$$

This is capable of coordinating by both π- or σ-bonds a molecule of olefin which then undergoes an insertion reaction with a carbonyl group, a very common and important process in these metal-catalysed reactions. The acylated cobalt is then cleaved by hydrogen to form the aldehyde and regenerate the cobalt hydride.

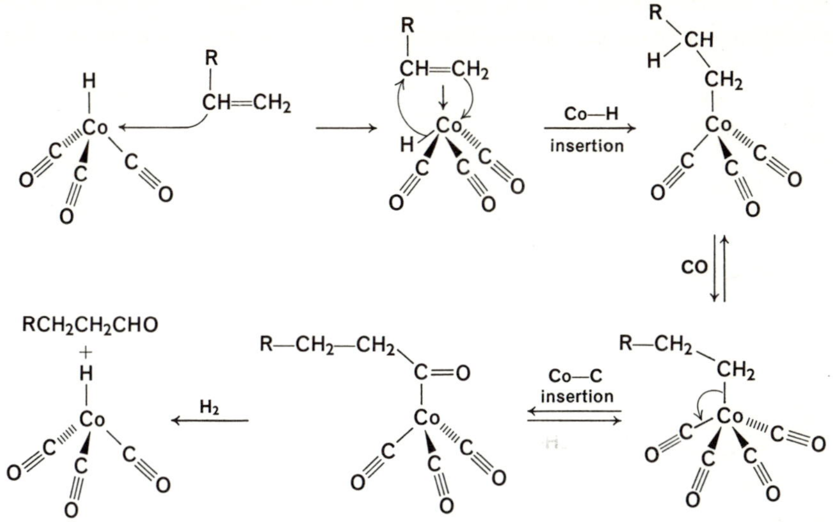

This process is used extensively for the production of butanols, for instance, from propylene:

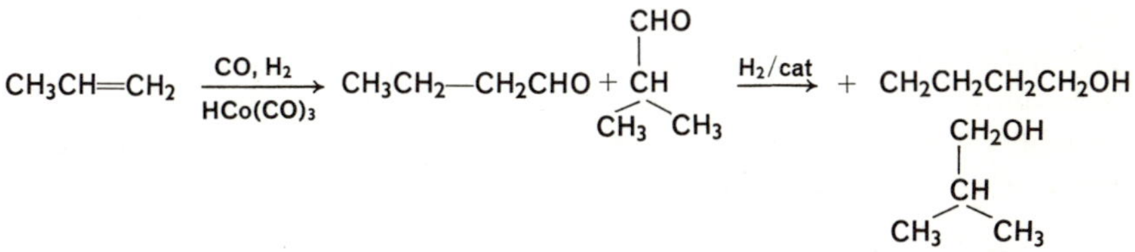

$$CH_3CH{=}CH_2 \xrightarrow[HCo(CO)_3]{CO,\ H_2} CH_3CH_2{-}CH_2CHO + \underset{\underset{CH_3\quad CH_3}{\diagdown\diagup}}{\overset{\overset{CHO}{|}}{CH}} \xrightarrow{H_2/cat} + CH_2CH_2CH_2CH_2OH$$

$$\underset{\underset{CH_3\quad CH_3}{\diagdown\diagup}}{\overset{\overset{CH_2OH}{|}}{CH}}$$

Ziegler–Natta Polymerization Catalysts

The polymerization of ethylene and propylene to useful plastics originally was carried out at extremely high pressures. Around 1955 Ziegler and Natta discovered a range of catalysts which would accomplish this reaction at ordinary pressure and even at quite low temperatures. Typical Ziegler–Natta catalysts contain two components, a titanium halide and an aluminium alkyl, although various other additives may be used in practice. The catalysts are species containing titanium and aluminium atoms bridged by chlorine or oxygen (27); the titanium atom is five-coordinate which means that there is a

$$nR{-}CH{=}CH_2 \xrightarrow{TiCl_4{-}AlEt_3} RCH_2{-}CH_2{-}\overset{\overset{R}{\diagup}}{(CH}{-}CH_2){-}_{n-1}$$

vacant site for further coordination of an olefin molecule. The polymerization step then consists of an insertion of the olefin into the metal–carbon bond. One

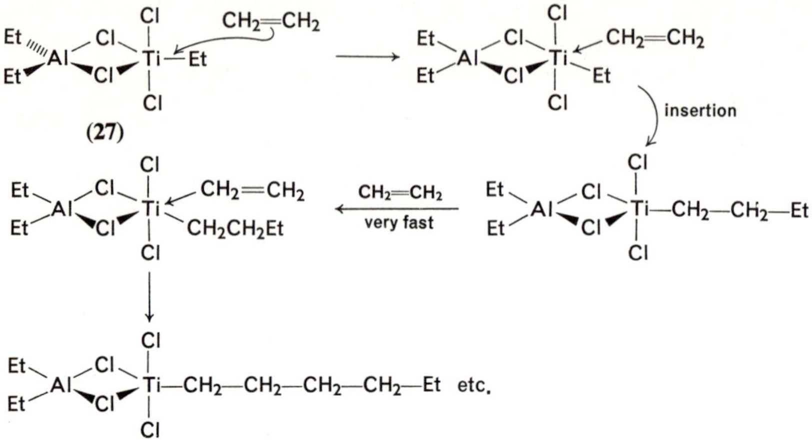

(27)

of the important consequences of this mode of polymerization is the stereo-regularity which is imparted to the polymer of a substituted ethylene. Each monomer-addition step creates a new asymmetric centre. In a given polymer chain, all centres have the same configuration but the material as a whole will not be optically active since there will be equal numbers of enantiomeric polymer molecules. A polymer of this type is 'isotactic' and is formed as a consequence of a steric preference for the approach of the olefin and the inability of the olefinic double bond to rotate during the addition (cf. radical polymerization, Chapter 5):

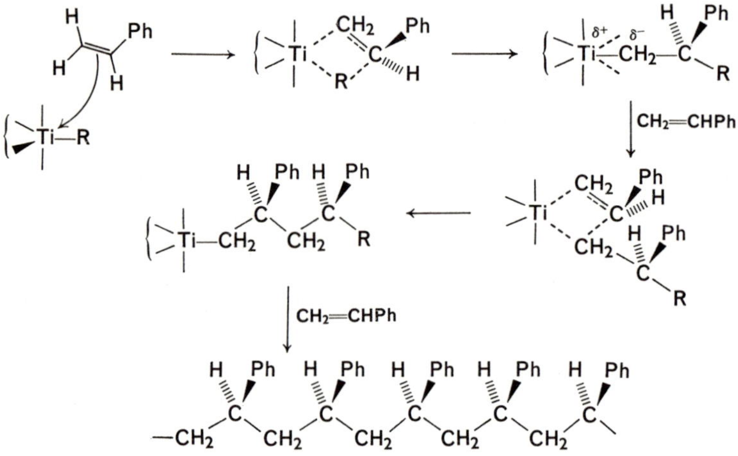

isotactic polymer

Since the metal–carbon bond is polarized, the growing end of the polymer chain has a residual negative charge and the process therefore has some relationship to anionic polymerization (Section 4.24). Ziegler–Natta catalysts, since they contain strong Lewis acids, may initiate ordinary cationic (atactic) polymerization concurrently with the above process. The Lewis acidity of the catalyst may be reduced and hence the stereoregularity of the polymer improved by suppressing the cationic process, if electron-donating ligands are placed on the metal. For example, the proportion of isotactic polymer increases from about 30% to over 90% (i.e. the proportion of cationic to coordinative polymerization) when the catalyst **27** is replaced by **28** or **29**. Only

$$
\begin{array}{c}
\text{Cl} \quad \text{Cl} \quad \overset{\displaystyle \text{Cl}}{\underset{\displaystyle \text{Cl}}{\text{Al}}} \quad \text{Ti}-\text{Et}
\end{array}
$$

TiCl$_4$.AlEtCl$_2$

(28)

$$
\begin{array}{c}
\text{Cl} \quad \text{Cl} \quad \overset{\displaystyle \text{OEt}}{\underset{\displaystyle \text{Cl}}{\text{Ti}}} \quad \text{Et}
\end{array}
$$

TiCl$_3$(OEt).AlEt$_2$Cl

(29)

ethylene and some terminal olefins are polymerized, but many Ziegler-type catalysts have been prepared which permit the reaction to occur for a limited number of steps (oligomerization) with other olefins. No theory is yet available to predict with accuracy the effect of these catalysts on a given olefin. Thus the compound MeTiCl$_3$MeAlCl$_2$ causes rapid oligomerization of ethylene and numerous 1-alkenes to dimers, trimers and small amounts of higher molecular weight products. The products are often complicated since chain branching is possible (Figure 3). Many other systems will cause oligomerization of α-olefins. The following data are of interest as they suggest that the products are determined by a combination of electronic and steric effects which operate upon the catalyst when the ligands are changed in a systematic way:

$$
2\text{CH}_3\text{CH}{=}\text{CH}_2 \xrightarrow[\text{PY}_3]{\text{NiCl}_2-\text{AlCl}_3}
$$

	hexenes +	$\left\{\begin{array}{c}\text{2-methyl-}\\ \text{pentenes}\end{array}\right\}$ +	$\left\{\begin{array}{c}\text{dimethyl-}\\ \text{butenes}\end{array}\right\}$
Y	% molar	% molar	% molar
Ph	22	74	4
Me	10	80	10
cyclo-C$_6$H$_{11}$	3	38	59
t-Bu	0	70	30

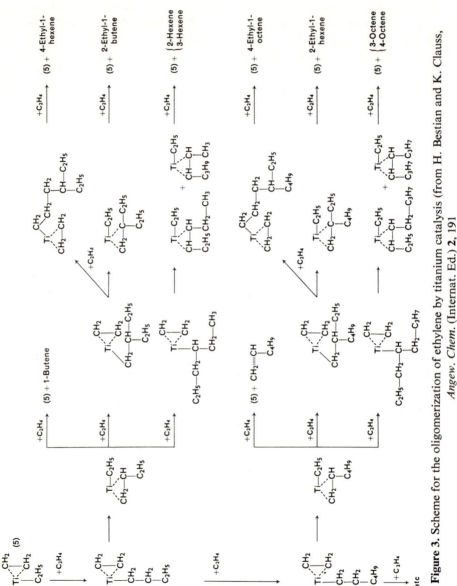

Figure 3. Scheme for the oligomerization of ethylene by titanium catalysis (from H. Bestian and K. Clauss, *Angew. Chem.* (Internat. Ed.) **2**, 191

Cyclo-Oligomerization of Butadiene

Butadiene behaves atypically towards Ziegler–Natta catalysts; for example, the system tetrabutyl titanate–triethyl aluminium (30) will dimerize ethylene but polymerize butadiene by a 1,2-addition:

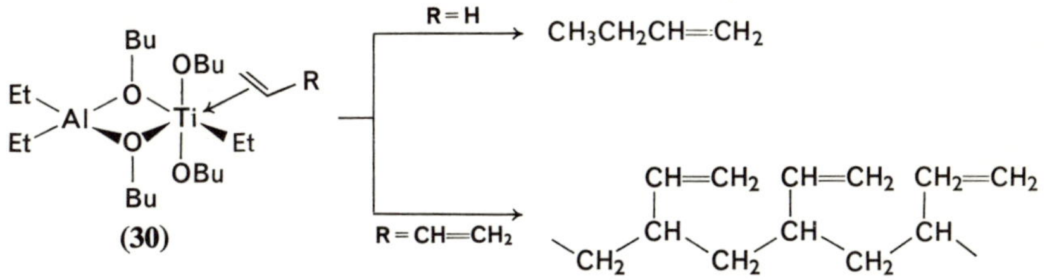

TiCl$_4$–Et$_2$AlCl, a typical polymerization catalyst for ethylene, was found by Wilke to initiate a surprising reaction with butadiene leading to *trans,trans,cis*-1,5,9-cyclodecatriene, the cyclic trimer (31), in excellent yield. Another

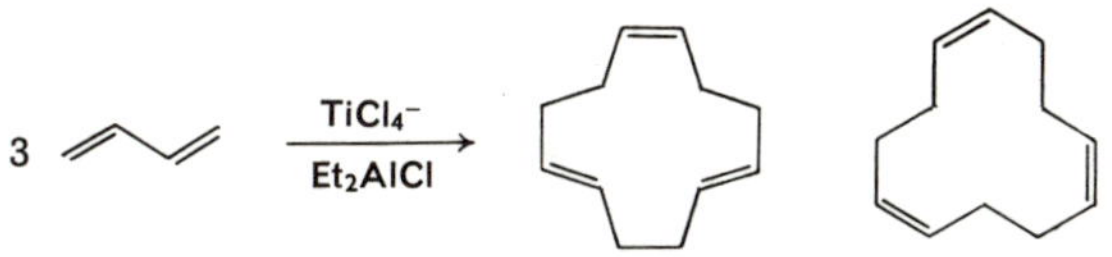

excellent catalyst for this transformation is formed by adding TiCl$_4$ to a mixture of diethylaluminium hydride and aluminium chloride; a catalyst consisting of chromyl chloride and triethyl aluminium, however, leads to a mixture of the *trans,trans,cis* and *trans,trans,trans* (32) isomers in the ratio 40:60.

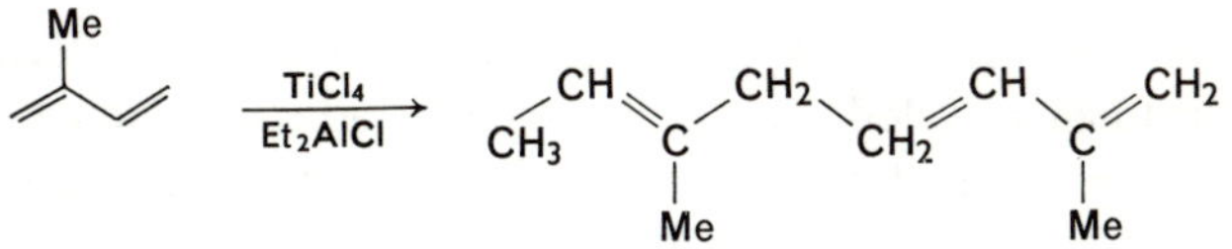

The reaction is much less facile with isoprene, although some trimethyl analogues of 31 have been isolated; the open-chain dimer 33 is mainly formed here instead. The mechanism of cyclotrimerization has been investigated in some detail. Two possible pathways might be considered; the reaction may commence as a Ziegler–Natta polymerization (1,4) in which the growing end of the polymer is held by a metal–allyl π-bond and cyclization by radical dimerization occurs after addition of three units (**34a**). On the other hand, three molecules of butadiene may be able to coordinate simultaneously and a concerted cycloaddition then occur (**34b**). An indication that the latter

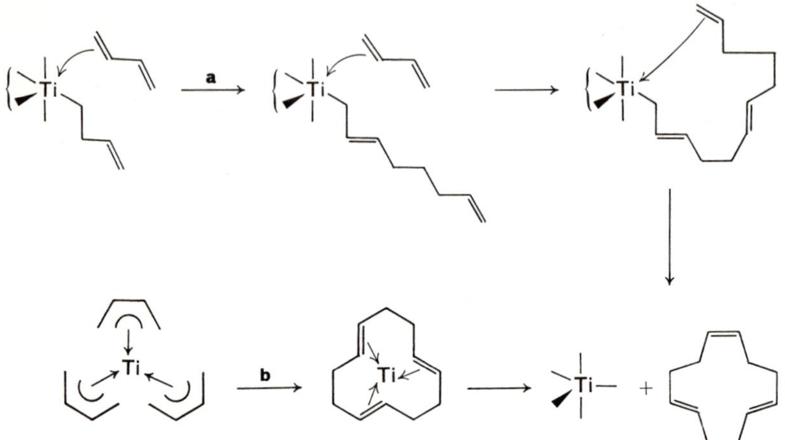

(34)

mechanism is correct comes from the observation that certain chromium catalysts will convert 2-butyne to the dibenzene chromium cation (**35**); evidently six alkyne molecules must interact with one chromium atom.

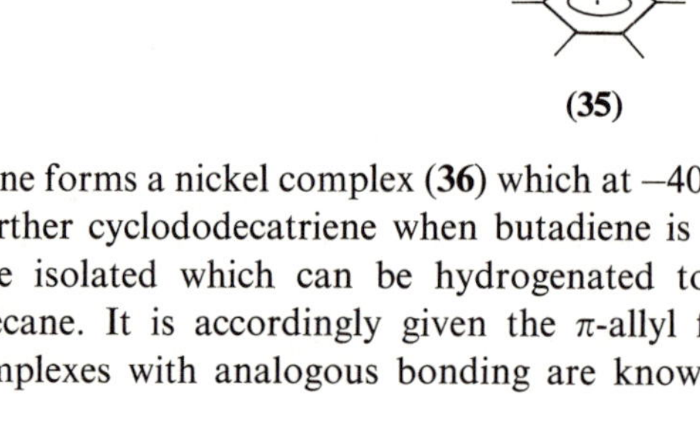

(35)

Cyclododecatriene forms a nickel complex (**36**) which at −40° will catalyse the formation of further cyclododecatriene when butadiene is added. An intermediate may be isolated which can be hydrogenated to the open-chain paraffin, n-dodecane. It is accordingly given the π-allyl formulation (**37**). Nickel–allyl complexes with analogous bonding are known as stable compounds, e.g. **38**. Yet other nickel compounds, including various nickel⁰ carbonyl nitrile phosphines and phosphites (**39**) catalyse the *cyclodimerization* of butadiene, presumably by an essentially similar type of mechanism to the trimerization. Linear trienes are formed by cobalt and iron complexes or by

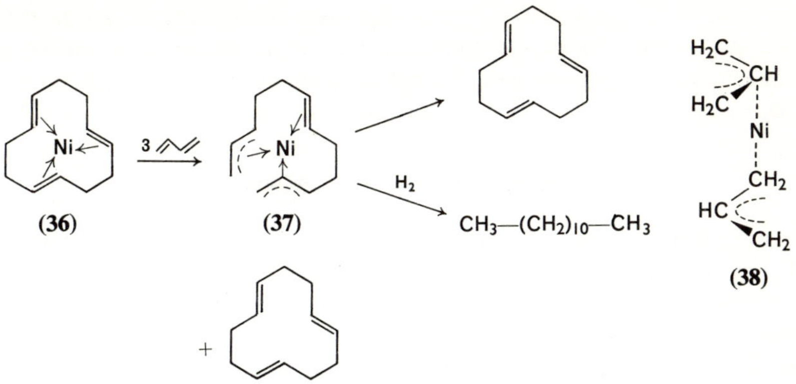

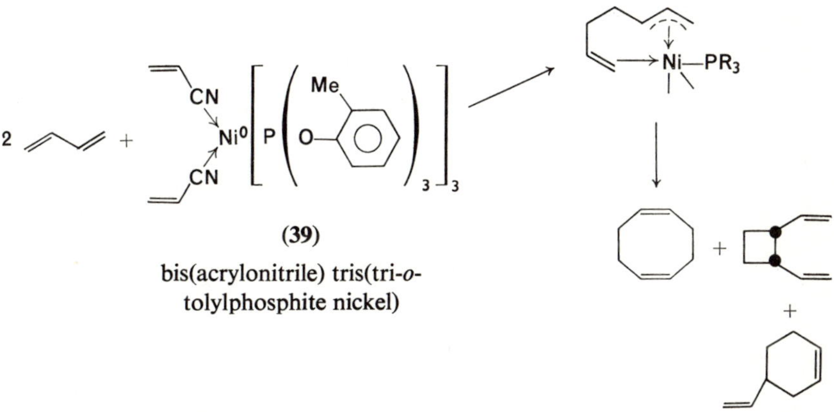

(**39**)

bis(acrylonitrile) tris(tri-*o*-tolylphosphite nickel)

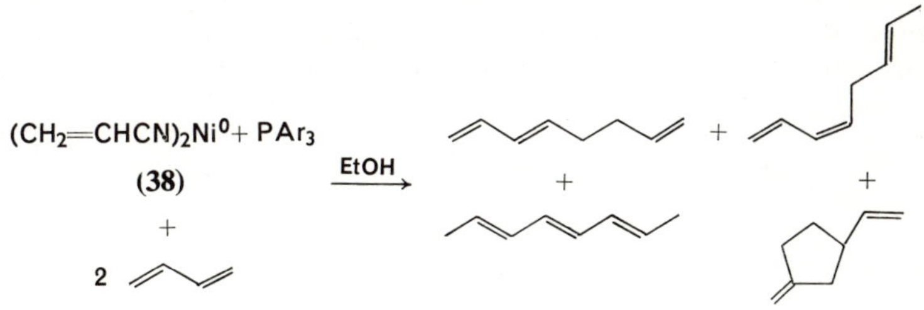

the nickel compounds used for cyclodimerization, but with better donor ligands, or in the presence of alcohols. The wide variety of the products obtainable from the inexpensive starting material, butadiene, makes these processes of great industrial potential. The subtle influences of structural change on the catalysts increase this potential and make 'catalyst tailoring' to achieve a particular reaction a realistic goal.

Homogeneous Hydrogenation

Rhodium forms a series of soluble hydride complexes which are capable of reducing an olefinic double bond in the presence of molecular hydrogen. Wilkinson has examined the series, $(R_3Y)_3RhX$ **(40)** as hydrogenation catalysts, where Y is a Group V element, P, As, Sb, and X a halogen or CO. The reaction is assisted by electron-donating groups R (e.g. the series of p-substituted phenyl groups); phosphines are more reactive than arsines while stibines are practically non-active and the halogen affects the rate in the order, $Cl < Br < I$. The following mechanism has been proposed.

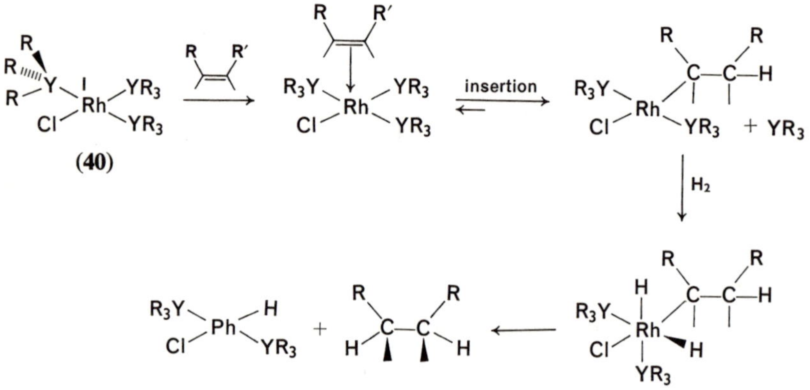

There are consecutive insertions of coordinated olefin into metal–hydrogen, followed by hydrogen into metal–carbon bonds. As with the insertions which lead to olefin polymerization, this reaction is stereospecific and gives the cis-dihydro compound. Terminal olefins are reduced some 10^4 times faster than non-terminal, so that specific reductions of dienes can be achieved, e.g.

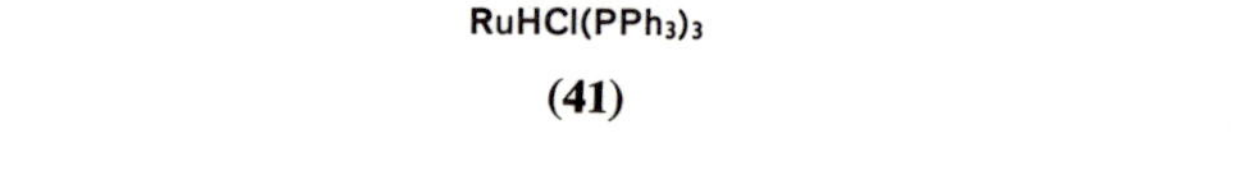

(41)

Some analogous ruthenium complexes are active in homogeneous hydrogenation, and also certain iridium and cobalt compounds.

Olefin Metathesis

Certain transition-metal halides, notably tungsten (VI) chloride, catalyse the interchange of two olefins at the double bond thus, pent-2-ene is equilibrated with but-2-ene and hex-3-ene. Complex hydrides make the reaction more

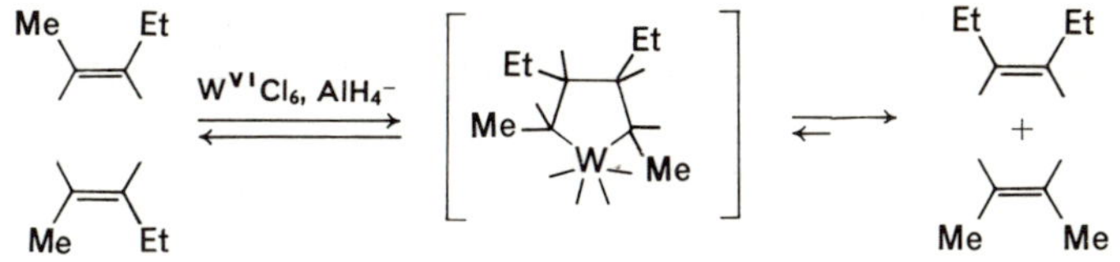

facile. Cyclic tungsten-containing intermediates are likely and indeed have even been isolated.

Palladium-catalysed Olefin Oxidation; the Wacker Process

A useful industrial oxidation of ethylene to acetaldehyde employs a trace of palladium chloride in the presence of oxygen and copper. The following sequence of reactions has been shown to occur:

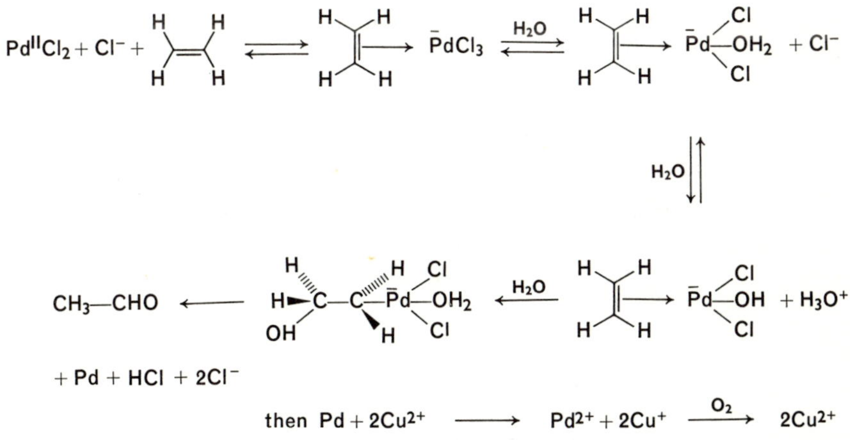

10.7   CATALYTIC HYDROGENATION AND HYDROGEN EXCHANGE ON METAL SURFACES

The addition of hydrogen to olefinic and acetylenic π-bonds is a standard chemical technique which has been of importance for at least a century.

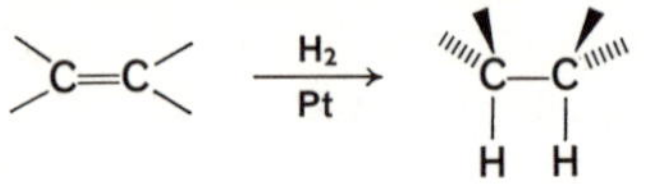

Less well known, but of great importance in the preparation of deuterium-labelled compounds, is the exchange reaction of a saturated compound.

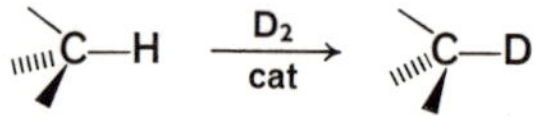

These reactions clearly take place on the surface of the metal and recent work has done much to elucidate the processes which are occurring. For practical purposes the metals which are of most use are palladium, platinum and nickel, but many other transition metals are active to some extent, including iron, molybdenum and rhodium. The nature of adsorbed hydrogen is not well understood but it is likely that hydrogen atoms become σ-bonded to the surface metal atoms.

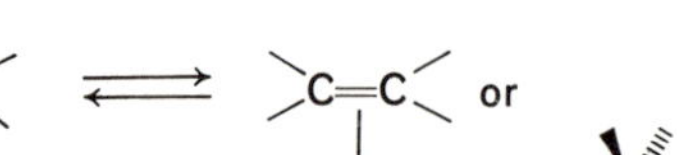

Similarly it is fairly certain that olefins and acetylenes may complex to the metal by σ- or π-bonds. The process of catalytic reduction may be pictured as follows:

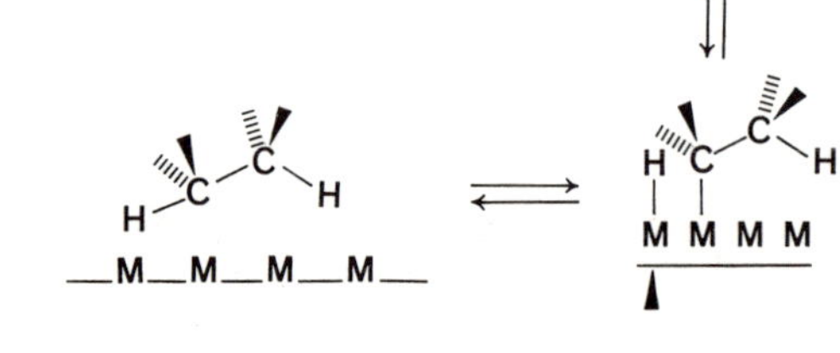

The hydrogens are assumed to be transferred singly; the intermediate may be described as a σ-bonded radical.

Evidence for this type of mechanism may be obtained from a study of the more complicated hydrogen-exchange reaction, but one consequence is the mainly *cis* nature of the addition observable when a cyclic olefin or acetylene is used. Several processes have been identified in exchange at saturated carbon.

Perhaps the most important is the reverse of the hydrogenation reaction in which initially the alkane C—H bond is broken and the two fragments σ-bonded to the metal surface. This process, an α,β exchange, requires two

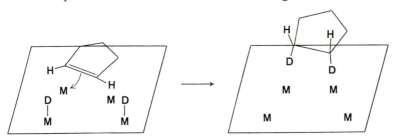

adjacent *cis* CH bonds. If the ratio of rates $k_3:k_{-2}$ is known, statistical predictions can be made as to the isotopic composition of the substrate at the beginning of the reaction (assuming that once desorbed, no further exchange occurs) and these predictions agree well with the experimental values. For example, deuterium exchange of ethane on palladium (for which $k_3/k_2 = 18$) shows the following isotopic distribution early in the reaction:

$$CH_3CH_3 \xrightarrow[Pd]{D_2} CH_3CH_2D \longrightarrow CH_3CHD_2 \longrightarrow \longrightarrow CD_3CD_3$$

	d_1	d_2	d_3	d_4	d_5	d_6
Calculated	5.3	7.3	8.7	11.2	16.9	50.6
Observed	5	6	8	11	19	51

If this mechanism alone operates, a cyclic olefin such as cyclopentane should only be able to exchange hydrogens on one side and there should therefore be a maximum of five deuterium atoms exchanged in the early stages of the reaction. This is not found to be the case. At $0°$ the ratio of d_5 to d_{10} compounds

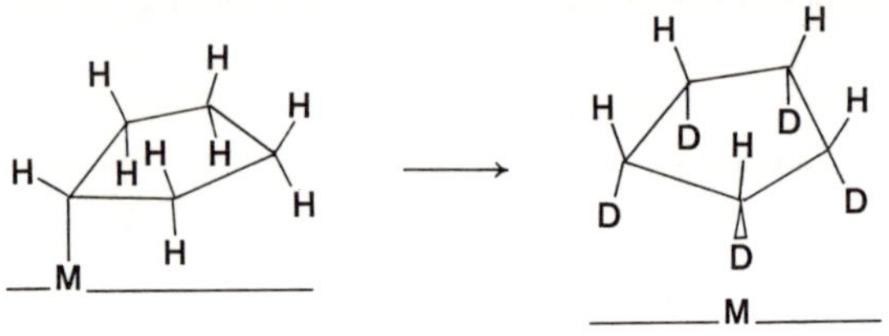

is $26:17$, while at $25°$ this ratio becomes $15:32$. Thus there must be a mechanism for the compound to 'turn over' on the catalyst surface without being desorbed and this has a higher activation energy than the exchange step. Methane has been shown to exchange by a process involving the formation of a metal methylene (presumably both σ- and $p\pi–d\pi$ bonding, as in metal carbonyls is involved here)—an α,α-exchange mechanism. It is possible that a similar

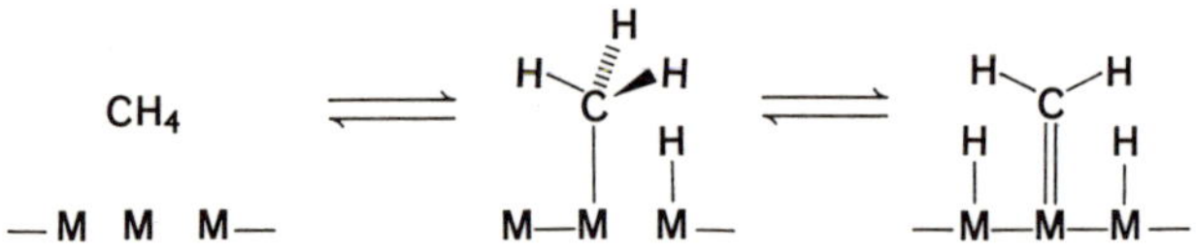

mechanism permits the cyclic olefin to invert on the catalyst surface and so form the perdeutero compound:

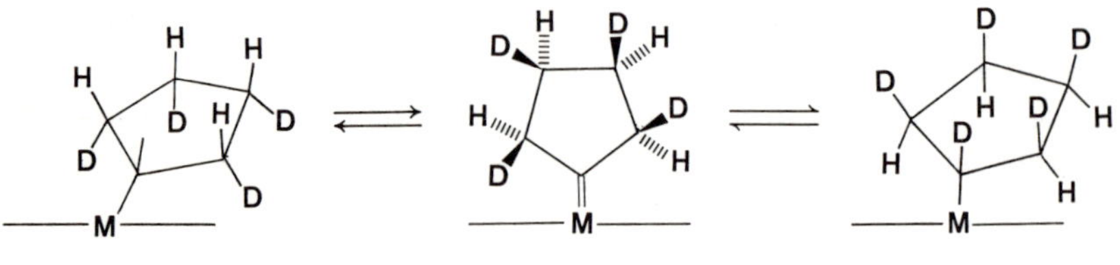

It is also possible that π-bonding from the metal to a more delocalized π-system could occur, such as an allyl residue and in its formation or reversal hydrogen transfer to the top side of the molecule may occur (π-exchange). Such a process requires a three-carbon system, only the outer ones being capable of exchange.

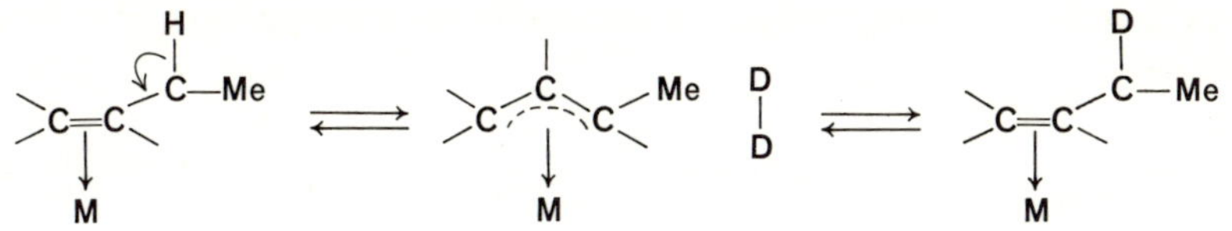

Excellent support for this theory comes from exchange studies on the hydrocarbons **42**, **43** and **44**; the exchangeable positions are ringed.

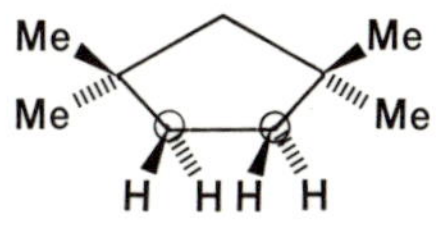

(42)

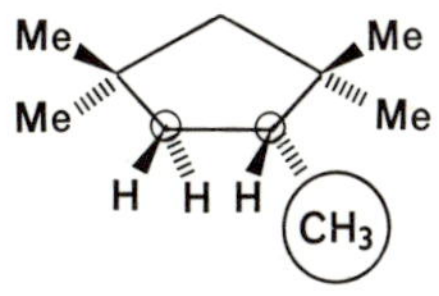

d_1	43
d_2	57
d_3	1

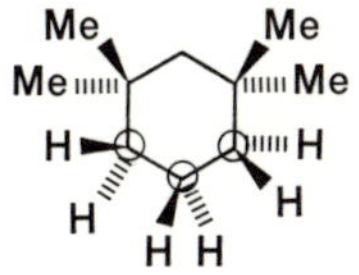

(43)

d_1	55
d_1–d_5	5
d_6	40

(44)

d_1	40
d_2	3
d_3	13
d_4	8
d_5	37
d_6	<1

The *gem*-dimethyl groups limit the carbon chain in which exchange may occur (a position flanked by two such groups is sterically hindered towards exchange). Compound **42** has only a two-carbon chain and hence the initial exchange can occur only by an α,β-process leading to a *cis-d_2* compound as observed. The extra methyl in **43** provides a third centre for π-exchange in addition to α,β. All six protons are now rapidly exchanged including the central one which must be assumed to approach on the side towards the catalyst. However in **44** the process can permit exchange of three *cis* protons and π-exchange the 1,3-*trans* protons, but there is no mechanism for exchange of the *trans* central proton, H′. It is found that only five are rapidly exchanged. Other processes which can

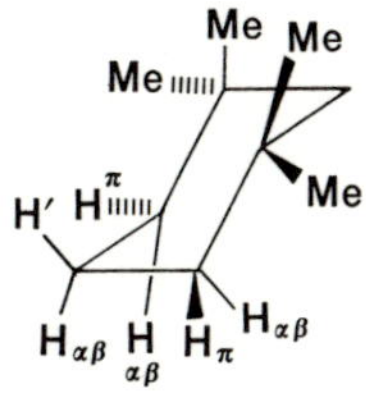

$H_{\alpha\beta}$—α,β-exchanged
H_{π}—π-exchanged
H′—not exchanged

be explained on similar mechanisms are double-bond migration and *cis–trans* isomerization:

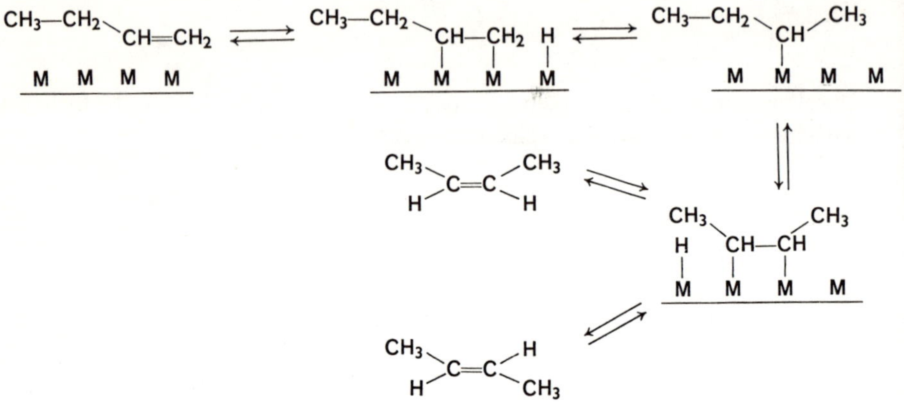

The catalytic disproportionation of cyclohexene and cyclohexadiene to cyclohexane and benzene,

$$3C_6H_{10} \underset{}{\overset{Pd}{\rightleftarrows}} 2C_6H_{12} + C_6H_6$$

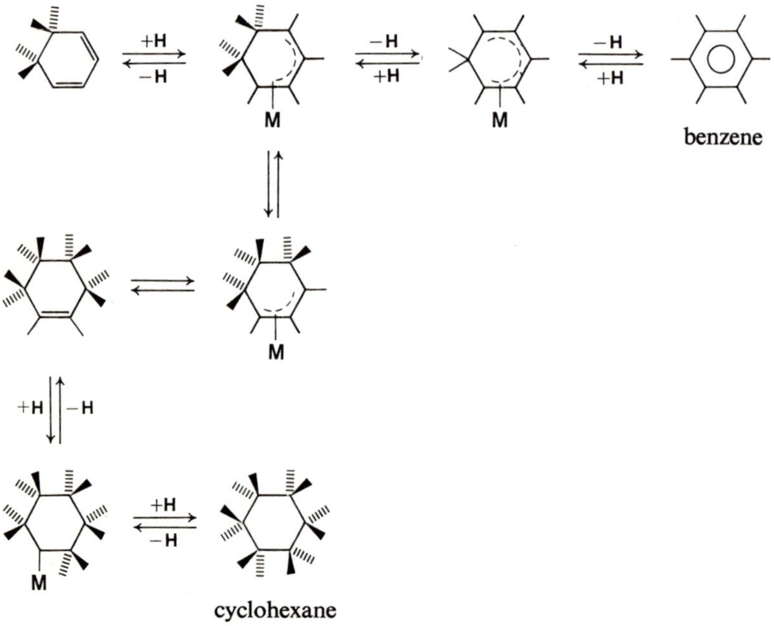

and dehydrogenation reactions, typically those which lead to an aromatic system such as 9,10-dihydroanthracene (45).

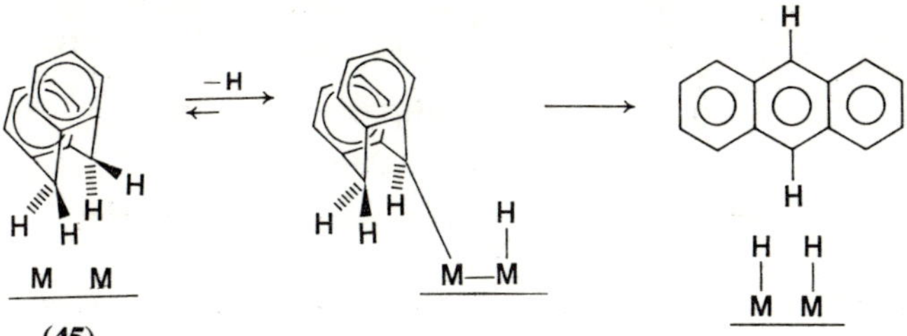

(45)

10.8 CHEMILUMINESCENT INTERMEDIATES

There are many fascinating examples of chemical reactions which are chemiluminescent or emit light without appreciable heat. In nature there are the examples of the fire-fly and glow-worm and of micro-organisms involved in the decay of vegetable matter. It has recently become apparent that the basic reason for the emission of chemiluminescence is the formation of an intermediate in an electronically excited state which fluoresces as soon as it is formed emitting its characteristic excitation energy in the form of quanta of visible radiation. Since many of these reactions take place with little activation energy, the excited state of the emitting species cannot lie very high in energy, relative to the reagents.

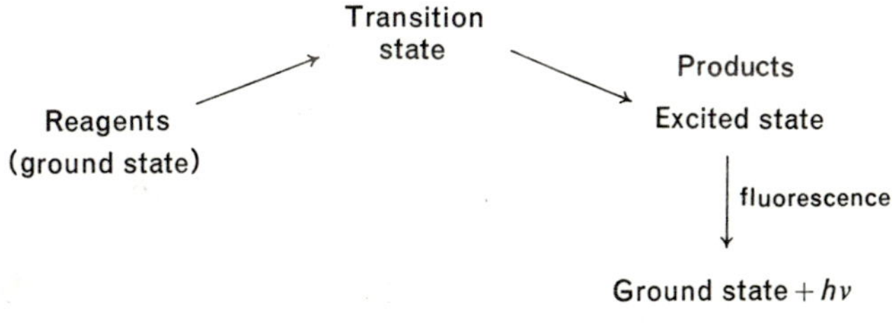

Cyclic Hydrazides

Some of the earliest investigated artificial chemiluminescent systems were cyclic hydrazides typified by luminol (4-aminophthalic hydrazide, 46); some other members of this class are the compounds 47 and 48. All undergo reaction with the emission of light in the presence of alkaline H_2O_2 or oxygen. The oxidation of luminol leads to 4-aminophthalate (49), initially in the excited

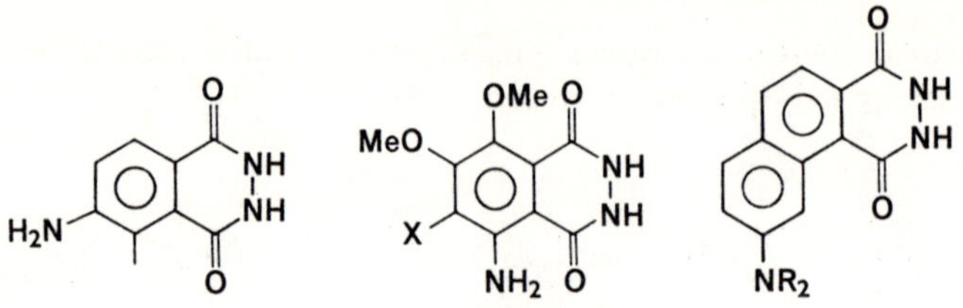

(46)   (47)   (48)

X = H, OMe

state, together with nitrogen. Kinetic studies show that the reaction is first
order in luminol and in oxygen and that the monosodium salt is stable, but
that light emission occurs on the further addition of strong base. It is therefore
deduced that the dianion of luminol (50) is a critical intermediate. Labelling
studies with ^{18}O show that the oxygen is completely incorporated into the
phthalate product. The following scheme is consistent with these findings.

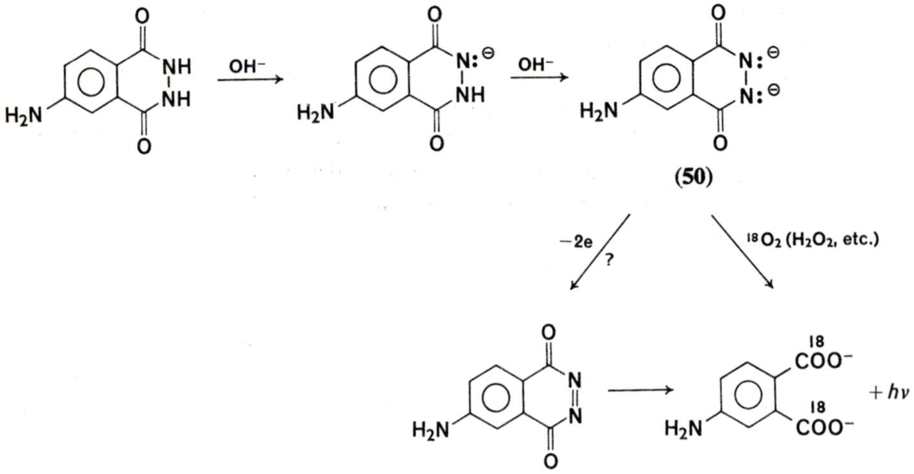

(50)

(51)   (49)

It has been confirmed that the light evolved is the fluorescence radiation from
(singlet-) excited 5-aminophthalate. The problem is to decide how oxygen (or
the peroxide ion) adds to the system and permits the loss of nitrogen. It is
likely that a prior oxidation of the dianion causes the formation of the diazo-
quinone (51) as a reactive intermediate. An analogous compound (52) has been
isolated from the oxidation of the naphthalene derivative (53). The azaquinone

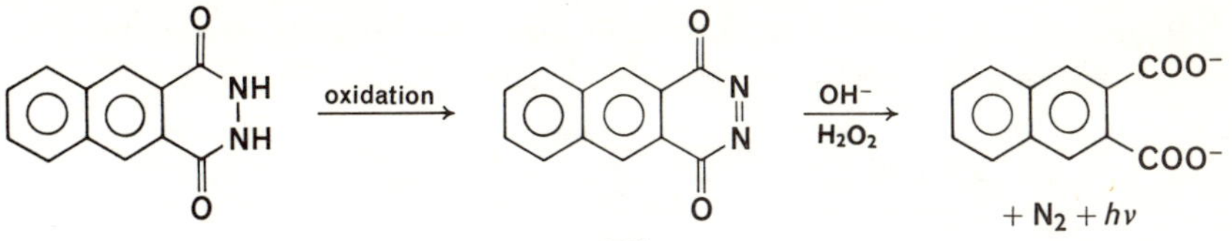

(53)   (52)

                                    (isolable)

(51) presumably adds oxygen or peroxide to give a bicyclic intermediate such as 54 which then breaks down in a symmetry-disallowed process (though, on account of the weakness of the bonds, energetically favourable) to give the observed products. Since the latter is 'forbidden' in the ground state, the product appears in an excited state initially (55) and then fluoresces.

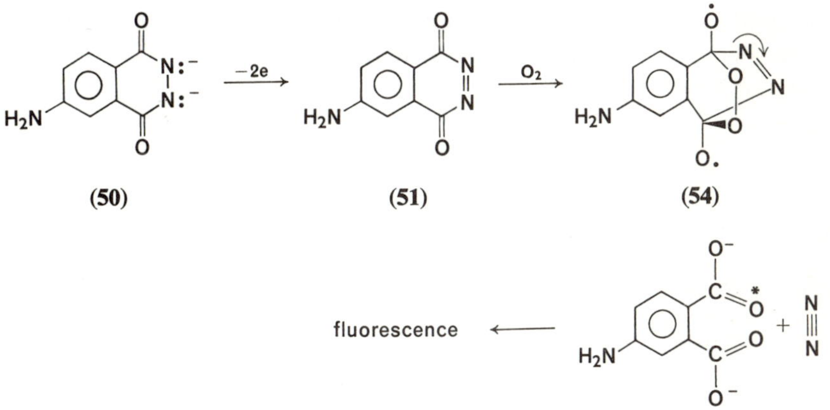

(50)   (51)   (54)

(55)

(triplet) state

Fire-fly Chemiluminescence

It seems likely that all organic chemiluminescent reactions depend upon the formation of an intermediate which can decompose by a pathway which is feasible energetically but which is 'forbidden' on grounds of orbital-symmetry factors (Section 1.31). This means that the products (or at least one species) will be produced in the first instance in an excited state, whereupon fluorescence will immediately occur with the emission of a quantum of radiation, and conversion to the ground state.

Reactions which would seem to fulfil these requirements include the concerted decomposition of four-membered cyclic peroxides **56** and **57**:

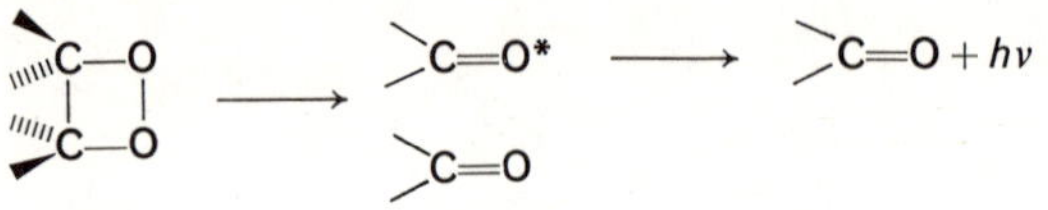

(56)

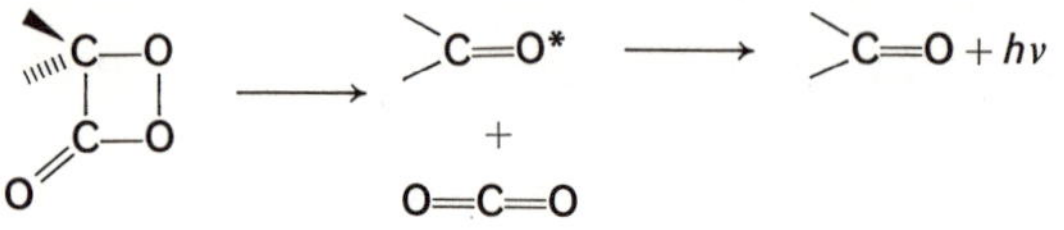

(57)

It is probable that the latter basic system occurs in nature. There are a number of 'luciferins' produced by various insect species, e.g. fire-fly luciferin (from *Photinus pyralis*).

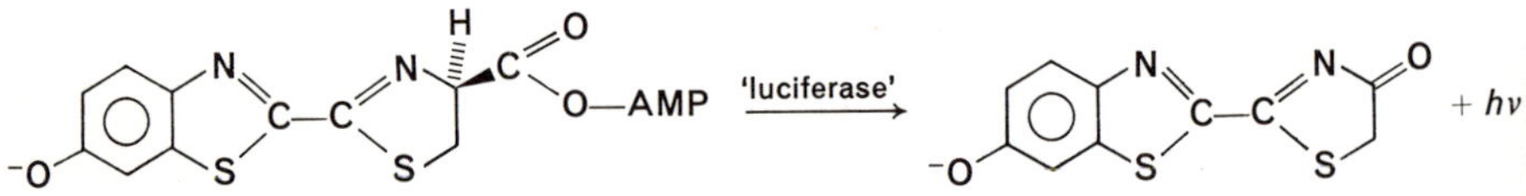

(AMP = adenosine monophosphate)

Luciferin from *Cypridina hilgendorfii*

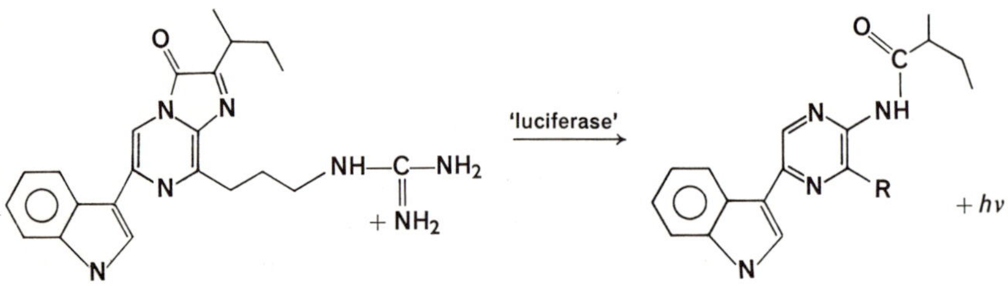

Luciferin from *Latia neritoides*

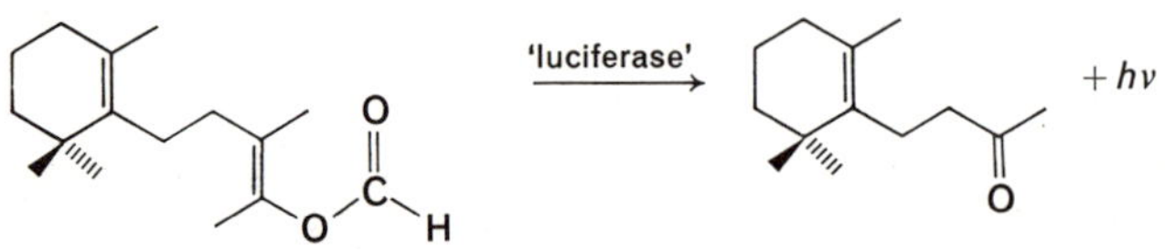

The chemiluminescence reactions are, in all cases, oxidations mediated by enzymes ('luciferases'). A basic scheme which may be applied to these systems leads to the formation of an oxodioxetan (58); this decomposes to the excited

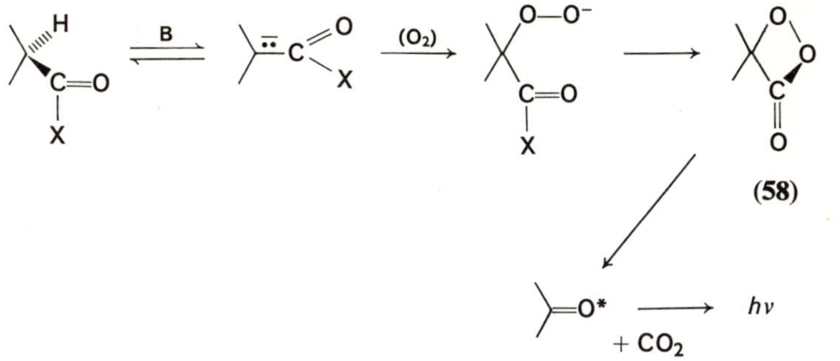

(58)

hv
+ CO$_2$

product. Thus, the fire-fly luciferin oxidation may be written,

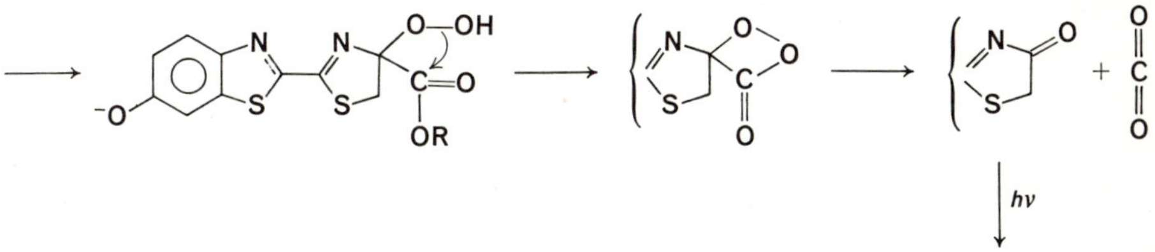

hv

The light emitted at 562 nm (yellow) represents energy of about 230 kJ mol^{-1} which is feasible from the estimated bond strengths. Virtually quantitative efficiency of light emission occurs in the natural system but this is much reduced *in vitro*.

Other Cyclic Peroxides

The decomposition of six-membered bis-peroxides (59) has been shown to be a forbidden process and therefore if it occurs must lead to one excited carbonyl product. Chemiluminescence has been observed from a number of compounds

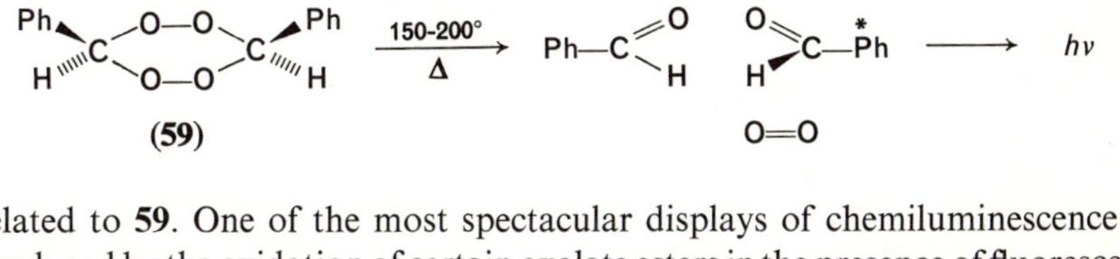

(59)

O=O

related to 59. One of the most spectacular displays of chemiluminescence is produced by the oxidation of certain oxalate esters in the presence of fluorescent dyes. When bis-(2,4-dinitrophenyl) oxalate (60) in dimethoxyethane is treated with concentrated hydrogen peroxide the dioxetan (61)—a dimer of CO_2—is produced transiently. This decomposes in the presence of dye to give the excited dye which then fluoresces. The colour of the chemiluminescence is

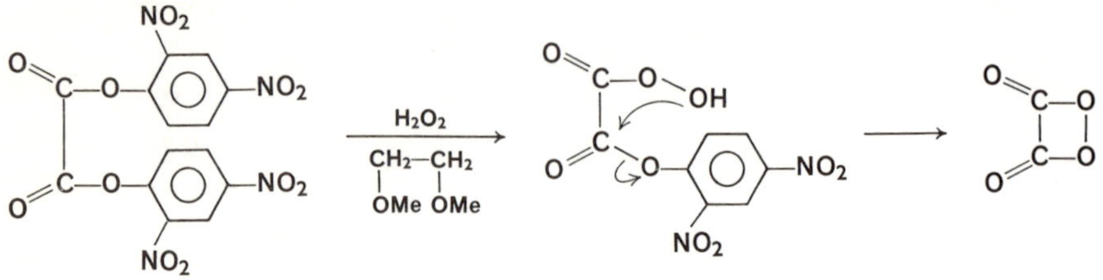

(60)   (61)

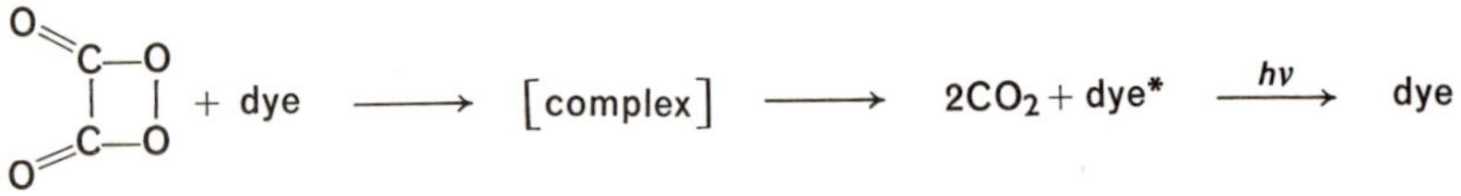

determined by the dye—eosin produces yellow light, 9,10-diphenylanthracene blue and Rose Bengal, orange.

10.9   **CHARGE-TRANSFER COMPLEXES**

It is well known that the colour of molecular iodine in the gas phase and in solvents such as carbon tetrachloride is magenta ($\lambda_{max} = 520$ nm), but that in oxygenated solvents such as ethanol the solution is yellow-brown due to absorption at 460 nm. The explanation lies in the species present in these solutions; in carbon tetrachloride light is absorbed by iodine molecules while in alcoholic solution the absorbing species is a complex of iodine and an alcohol molecule. Similar changes in absorption spectra are apparent in

solutions of tetracyanoethylene, TCNE (62) in aromatic solvents. The colourless components often yield coloured solutions (in toluene, yellow; in xylene, orange; in anisole, red; in aniline, blue).

(62)

colourless   red

These are examples of interactions between molecules which are of wide generality. In each case, one of the interacting species has readily available π- or unshared electrons (i.e. the alcohol or the aromatic compounds) while the other (I_2 or TCNE) has low-lying vacant molecular orbitals which will accept electrons. The result is an energetically favourable interaction between the electrons of the *donor* molecule and the vacant orbital of the *acceptor*. The resulting complex can be expressed in valence-bond terms as a combination of the structures, **63** and **64**. Mulliken has developed the theory of these complexes in terms of a relatively small contribution of **62** to the structure of the

$$D: + A \rightleftharpoons [D:A] \longleftrightarrow [\overset{+}{D}\cdot\cdot\overset{-}{A}]$$

(63)   (64)

ground state, but the combination of the interacting orbitals produces a new pair of *molecular* orbitals the upper of which is vacant but accessible by excitation due to light absorption. This new excitation process, a property of the complex, is often of lower energy than any available in either component alone and is responsible for the colours developed on mixing. The excited complex is postulated to partake of a much greater proportion of the dipolar canonical form **64** than the ground state and hence is formed essentially by the transfer of *one electron* from donor to acceptor. From this, the new absorption band is properly termed a 'charge-transfer' band and for similar reasons, though less soundly, the complexes themselves are often known as 'charge-transfer complexes', though 'donor–acceptor complex' is perhaps a better term.

Charge-transfer complexes are likely to be formed in solutions containing any combinations of the donors and acceptors set out in Table 1.

Table 1. Some typical donor and acceptor molecules which will enter into complex formation

Donors. Aromatic molecules, especially if containing donor groups (R_2N-, $RO-$, $RS-$, alkyl), amines, alcohols, ethers.

Acceptors.

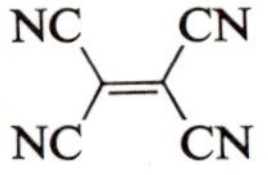

tetracyanoethylene

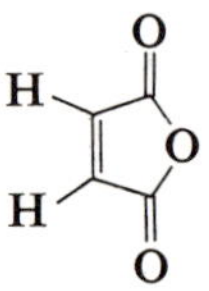

maleic
anhydride

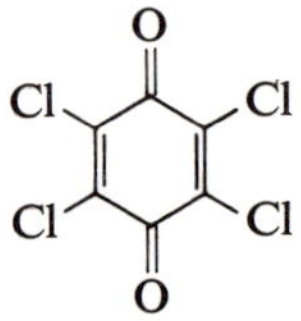

chloranil
(and other quinones)

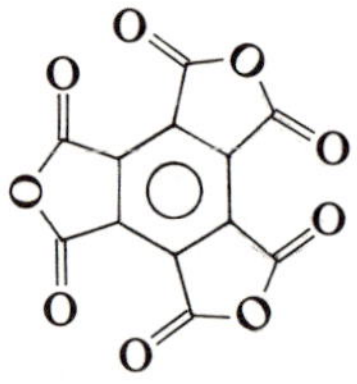

pyromellitic
trianhydride

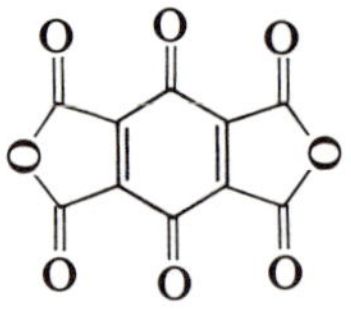

benzoquinone
tetracarboxylic acid dianhydride

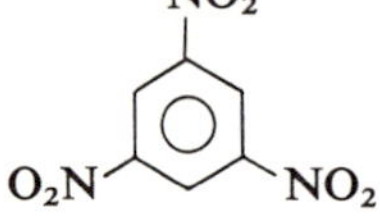

trinitrobenzene

also Lewis acids in general, Cl_2, Br_2, SO_2, carbonium ions, $AlCl_3$, etc.

Stoichiometry of Charge-Transfer Complexes

If it is assumed that donor and acceptor molecules are in equilibrium in solution with a 1:1 complex,

$$D + A \underset{\longleftarrow}{\overset{K}{\longrightarrow}} DA$$

and if D and A are the initial concentrations of the two reagents and DA the equilibrium concentration of complex,

$$k = \frac{[DA]}{([A] - [DA])([D] - [DA])}$$

If the donor is allowed to be in large excess (e.g. the solvent), then $([D] - [DA]) \ll [D]$; the expression may be rewritten,

$$K' = \frac{[DA]}{([A] - [DA]) N_D}$$

where N_D is the mole fraction of donor, but

$$[DA] = \mathscr{A}/l\varepsilon$$

where $\mathscr{A}$ is the absorbance due to the complex, ε, its molar extinction coefficient and l the path length. Substituting and rearranging,

$$\frac{Al}{\mathscr{A}} = \frac{1}{N_D K' \varepsilon} + \frac{1}{\varepsilon}$$

which is known as the Benesi–Hildebrand equation. By plotting $[A]l$ against $1/N_D$, a straight line should be observed with slope $= 1/K'\varepsilon$ and intercept $= 1/\varepsilon$.

If complexes of stoichiometry other than 1:1 were formed, the above expression would not hold. The observation that a large number of donor–acceptor systems obey the Benesi–Hildebrand law well is evidence that 1:1 complexes are the rule.

Lifetime of Charge-Transfer Complexes

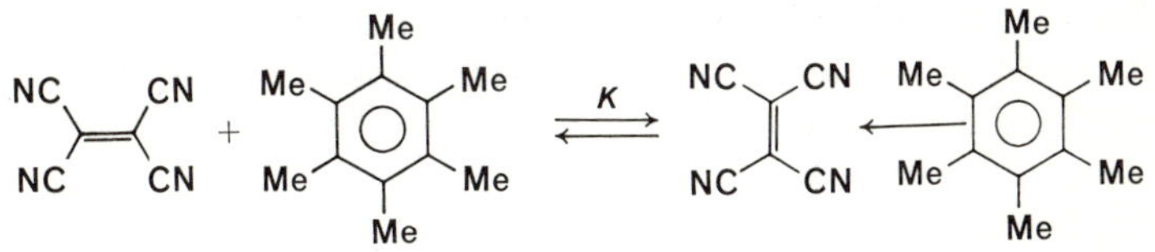

Values of equilibrium constants obtained from Benesi–Hildebrand plots are often quite high; e.g. for TCNE and hexamethylbenzene, K has been variously estimated as 263 and 1530. Solutions containing mixtures of donors and acceptors of the kind discussed here give little physical evidence for association other than spectra. For instance, a solution containing chloranil and tetra-

(65)

methyl-*p*-phenylenediamine (65) is a deep blue-black yet vapour pressure measurements of the solution which could detect less than 1 % association of the two fail to do so. It would appear that the lifetime of the complex which is evidently present, must be extremely short, of the order of a few vibrational periods of the molecules, and could be described as 'sticky collisions' between the donor and acceptor molecules.

The Role of Charge-Transfer Complex as Reactive Intermediates

Many reactions take place between donor and acceptor types of reagents and it is natural to speculate whether complexes of a charge-transfer nature are actually involved on the reaction pathway. The availability of interacting orbitals between the reagents would appear to provide a potential energy well, which would serve to bring them into close proximity and from which reaction might proceed. The following examples will illustrate this point.

Nucleophilic Aromatic Substitution

Nucleophiles are capable of displacing chloride from 2,4-dinitrochlorobenzene (Chapter 4) by an addition–elimination mechanism. The reaction exhibits second-order kinetics but the measured rate constant depends upon the initial concentrations of the reagents. Thus, at a fixed initial concentration of dinitrochlorobenzene of 0.05 M, k_2 has the value of 0.262 mol^{-1} h^{-1} when the initial aniline concentration is 0.2 M, and 0.192 when it is raised to 1.0 M. It is suggested that the added aniline is more effectively complexing the dinitro-chlorobenzene but the reduced rate constant must mean that the π-complex (66) is not on the reaction coordinate leading to products. Tetracyanoethylene reacts with dimethylaniline in a rather striking way; the solution on first mixing is deep blue and this, on standing, eventually becomes bright red. The

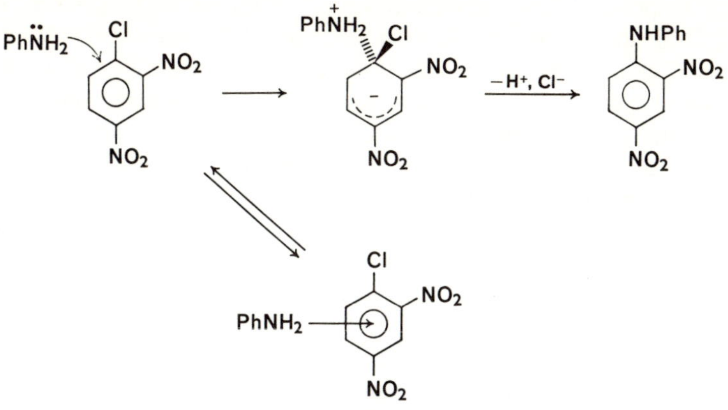

(66)

blue colour is evidently due to charge transfer and the red is the colour of the final product, 4-(tricyanovinyl)-dimethylaniline **(67)**, formed by electrophilic substitution by TCNE in the aniline ring. While it is tempting to speculate that the blue charge-transfer complex is an intermediate in the reaction, there is no

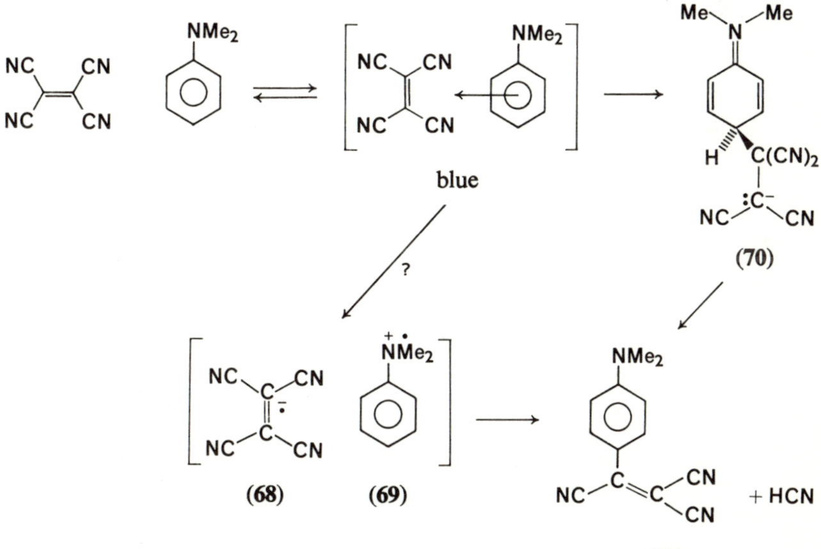

(70)

(68)   **(69)**   **(67)**

definite evidence that it lies on the reaction coordinate. However, solutions containing the reagents are found to give electron-spin resonance signals showing the presence of unpaired electrons, and the species responsible is the stable TCNE radical anion, **68** (Section 5.4), which has a characteristic nine-line spectrum. The presence of the much more reactive dimethylaniline radical cation (**69**) is much more difficult to establish but presumably must be the other product. It seems plausible that these two species would rapidly co-ordinate to form the σ-complex (**70**) and the product. The rate of reaction is found to be very sensitive to solvent, though this would be expected in any event since a dipolar intermediate must be formed.

Photochemical Reactions

Photochemical additions of a donor to an acceptor are of frequent occurrence. If the mixture of reagents gives rise to an absorption band at longer wavelength than either component alone and if the photochemical reaction is initiated by light which can only excite the charge-transfer transition, then we may conclude that the reaction passes through the charge-transfer complex, both in the ground and excited states. This is the case with the cycloaddition of benzene to maleimides (**71**), which are initiated by irradiation in the charge-transfer band ($\lambda = 300$ nm). The intermediate diene (**72**) adds a further molecule of maleimide or may be diverted to form **73** by the addition of TCNE, a powerful dienophile.

Similarly, charge-transfer complexes are inferred in the photo-cycloaddition of cyclohexene to maleonitrile and maleic anhydride, which is stereoselective and contrasts with the photoaddition of diethyl maleate (**74**) which forms no

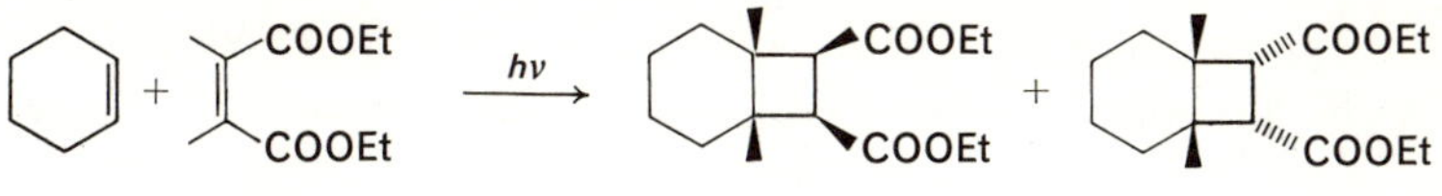

(74)

diethyl maleate

55%   25%

20%

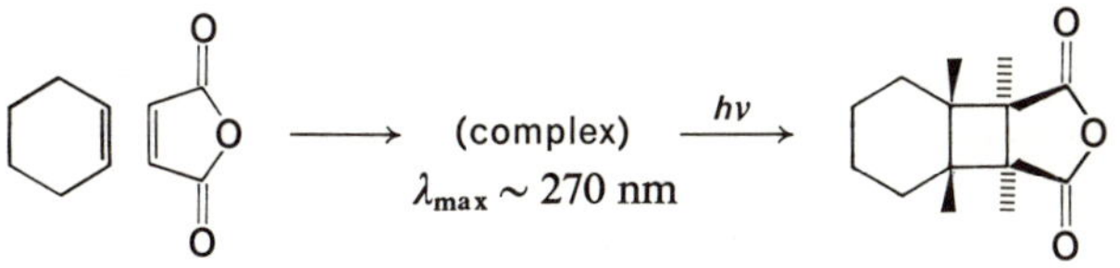

maleic anhydride

detectable complex and is stereo-unselective. Excited monomeric species may themselves complex with reagents to form a so-called 'exiplex'; charge-transfer intermediates may be very general in photochemical reactions but it is difficult to obtain unambiguous evidence for their existence.

Other Reactions

Vinyl ethers, e.g. **75**, add to TCNE and also form coloured species when mixed in solution. It may be supposed that the products, tetracyanocyclo-butanes, result from the complexes. Aromatic electrophilic substitution may

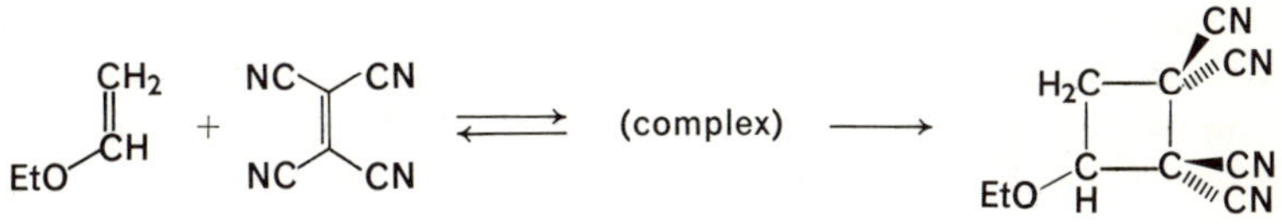

(75)   orange, $\lambda_{max} = 410$ nm

ethyl vinyl ether

also commence by interaction of the aromatic molecule and the electrophile via a π-complex. Diels–Alder and other cycloaddition reactions are also likely candidates for the intermediacy of complexes. In many instances,

π-complex   σ-complex

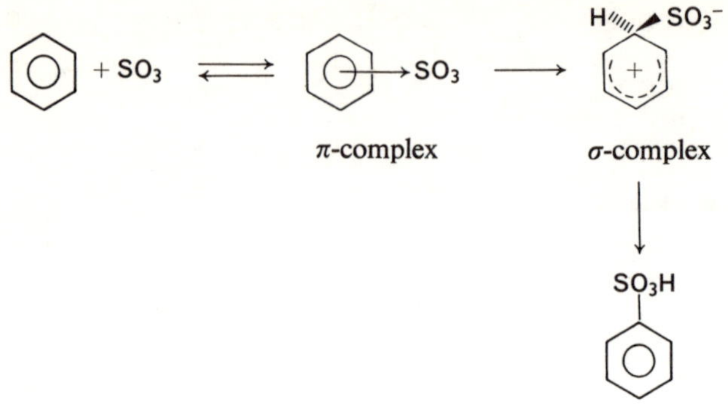

these are detectable spectroscopically but, as before, no unambiguous evidence exists that these species lie on the reaction path.

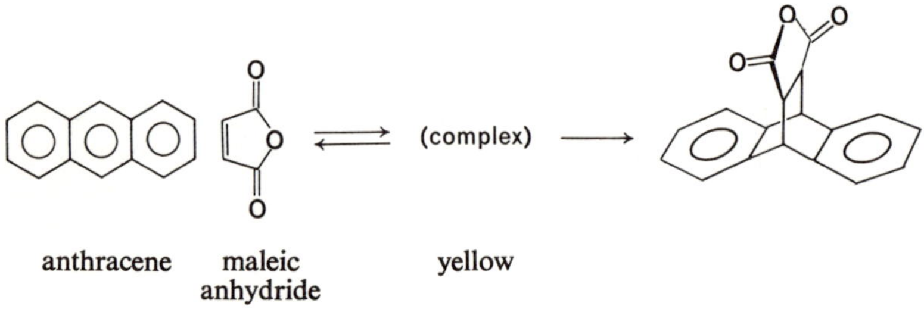

anthracene   maleic   yellow
               anhydride

Solid Charge-Transfer Complexes

Many donor–acceptor pairs will form crystalline complexes with a definite (usually 1:1) stoichiometry. The fact that these compounds will pack into a crystal implies some kind of favourable attractive force between them. It does not necessarily imply that the complex has any existence in solution.

Examples are the bromine–dioxan complex (**76**), the naphthalene–picric

(**76**)

acid complex (77) and the chloranil–hexamethylbenzene (78).

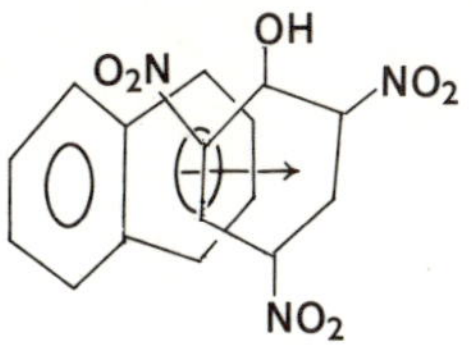

(77) 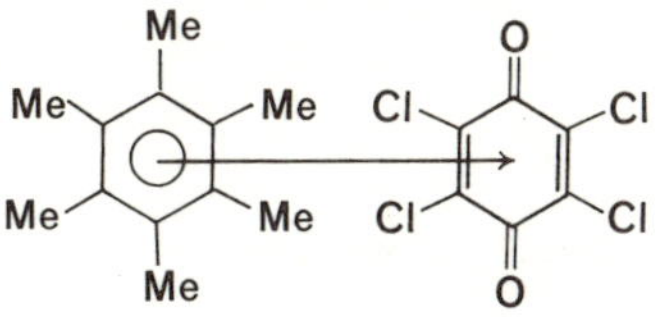   (78)

10.10   **THE RAMBERG–BÄCKLAND REACTION**

Formally, the sulphonate analogue of the Favorskii rearrangement, the intramolecular elimination of HCl from an α-halosulphone (79), leads eventually to an olefin. An intermediate three-membered cyclic sulphone (an episulphone, 80), is postulated but loses SO_2 rather than undergo base-catalysed ring opening as does the carbonyl analogue. The episulphone may be

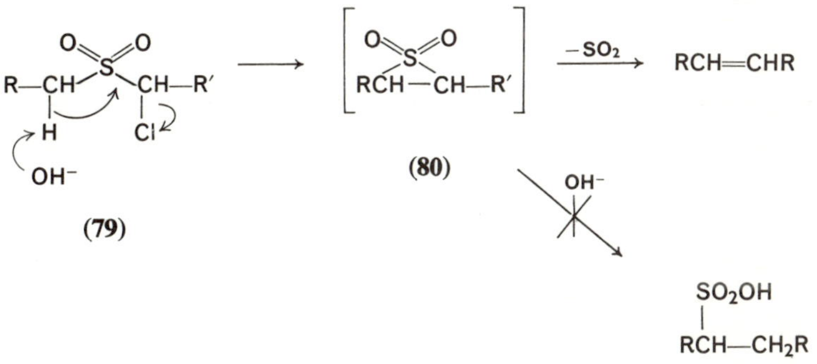

prepared by a separate route (81) and shown to decompose in the expected manner. This step is an orbital-symmetry permitted reaction and is highly stereospecific, i.e. *cis*-episulphone gives only *cis* olefin and *trans* likewise. However as an olefin-forming reaction the Ramberg–Bäckland reaction is not stereospecific, since in the first place two stereoisomeric episulphones may form, e.g. 82 and 83, and these in turn may be interconvertible in the presence of base via a stabilized carbanion (84); the olefin obtained reflects the relative stabilities of the two episulphones, usually *trans* > *cis* (e.g. for R = Me, the products are *trans*- and *cis*-2-butenes in the ratio 78 % to 22 % respectively).

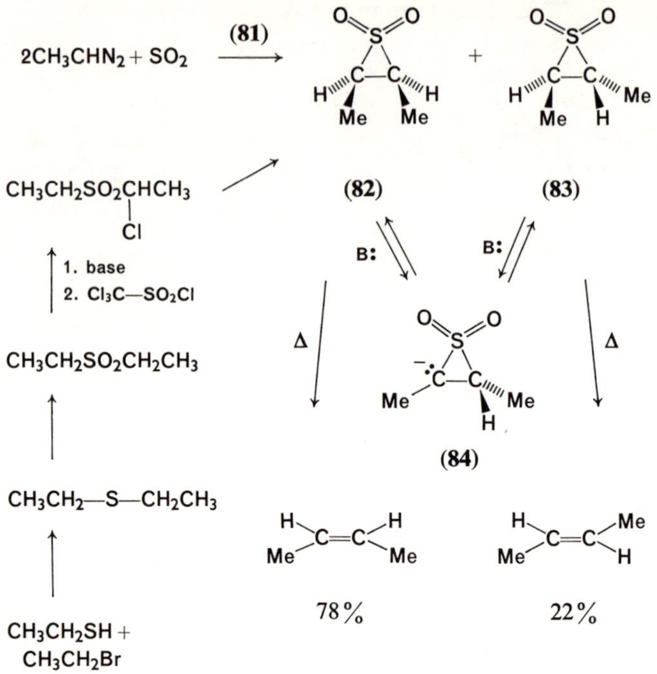

(81)
2CH3CHN2 + SO2
CH3CH2SO2CHCH3
Cl
1. base
2. Cl3C—SO2Cl
CH3CH2SO2CH2CH3
CH3CH2—S—CH2CH3
CH3CH2SH +
CH3CH2Br
(82)
(83)
B:
B:
Δ
Δ
(84)
78%
22%

The reaction has been used in the synthesis of phenanthrene derivatives (**85**). α,α′-Dichlorosulphones (**86**) undergo the Ramberg–Bäckland reaction with

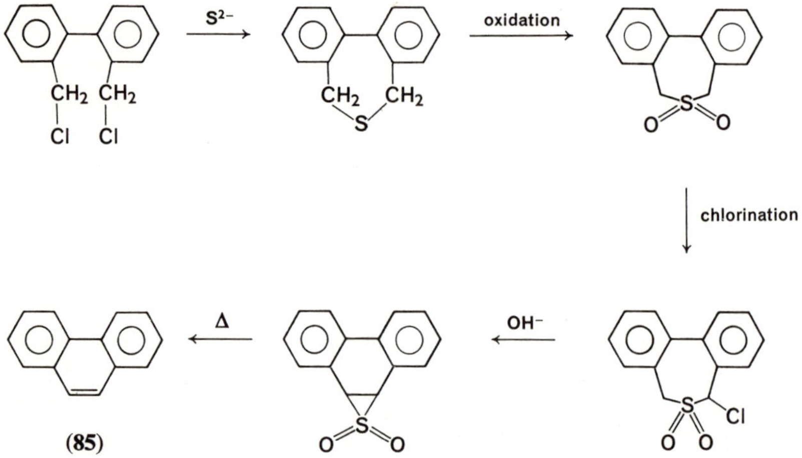

S2−
oxidation
chlorination
OH−
Δ

(**85**)

the formation of an unsaturated sulphonic acid. This may be rationalized as being due to the solvolysis, rather than decomposition, of the intermediate episulphone.

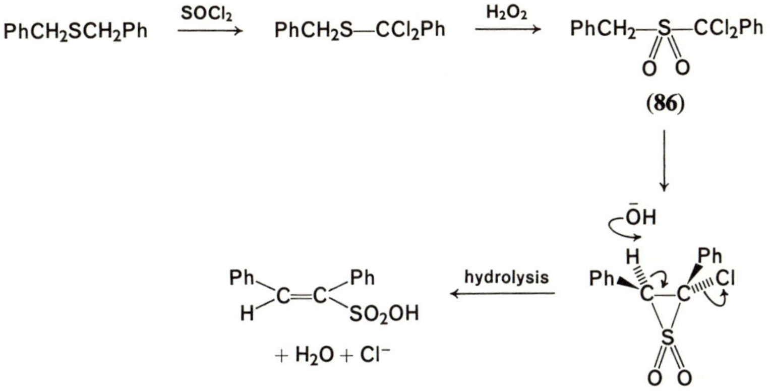

(86)

10.11 THE COREY–WINTER OLEFIN SYNTHESIS

A carbene intermediate (87) is suggested in the reaction of cyclic carbonates and thiocarbonates (88) with phosphines (89). This then loses CO_2 to form an olefin (90) with the same stereochemistry as the cis-1,2-diol which forms the starting material. Preparation of the thiocarbonate is accomplished by the use of thiocarbonyldiimidazole (91). Since the 1,2-diols are most conveniently

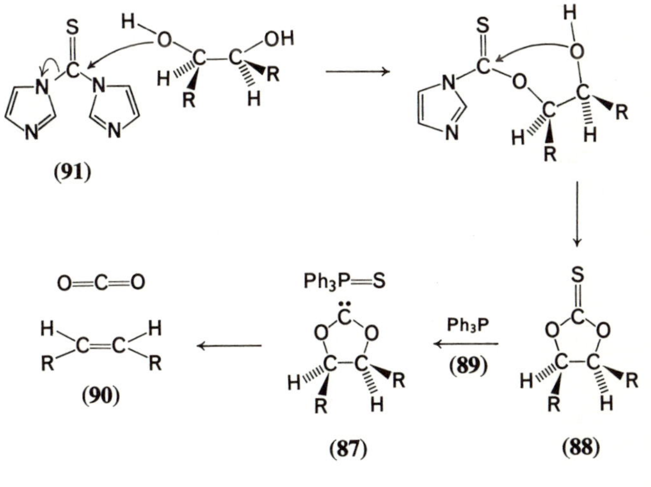

(91)

(90)   (87)   (89)   (88)

prepared from the olefin, the reaction is mainly of value in the conversion of a
cis olefin to a *trans* or vice versa. *trans*-Cyclooctene for example, which has a
very strained double bond, may be prepared in this way.

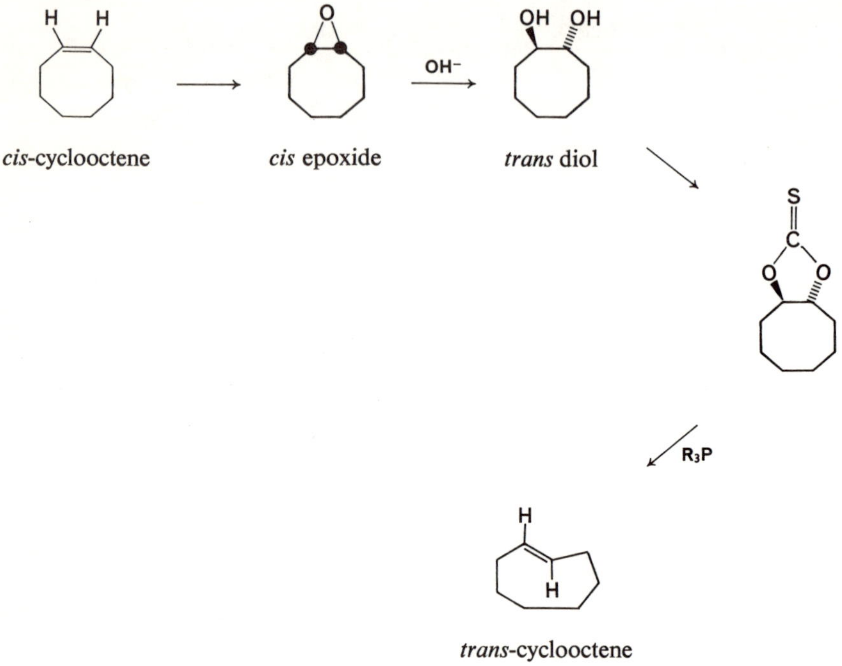

cis-cyclooctene   *cis* epoxide   *trans* diol

OH⁻ over the second arrow

trans-cyclooctene

R₃P

10.12   'TETRAVALENT-SULPHUR' INTERMEDIATES

The expansion of the sulphur valence shell in sulphonium compounds may be
forced by a powerful nucleophilic reagent such as an alkyl lithium. The
resulting species, postulated to be a tetravalent-sulphur compound (**92**), then
fragments. The vinylation of the triphenylsulphonium ion (**93**) occurs by this
mechanism rather than by direct nucleophilic displacement at the aromatic
ring:

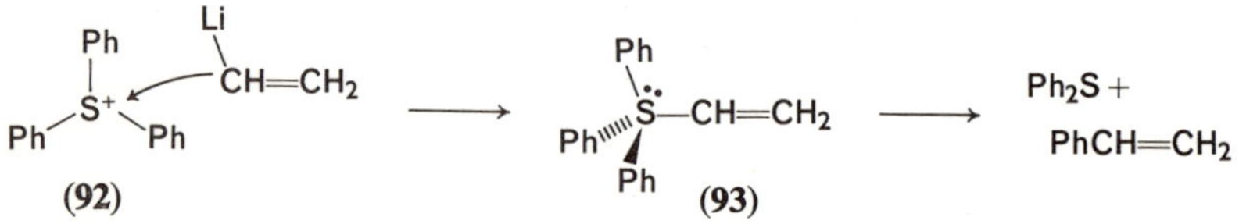

(**92**)   (**93**)

$Ph_2S +$
$PhCH{=}CH_2$

The analogous reaction with butyl lithium gives a variety of products but no butylbenzene, which again is evidence against the nucleophilic substitution reaction.

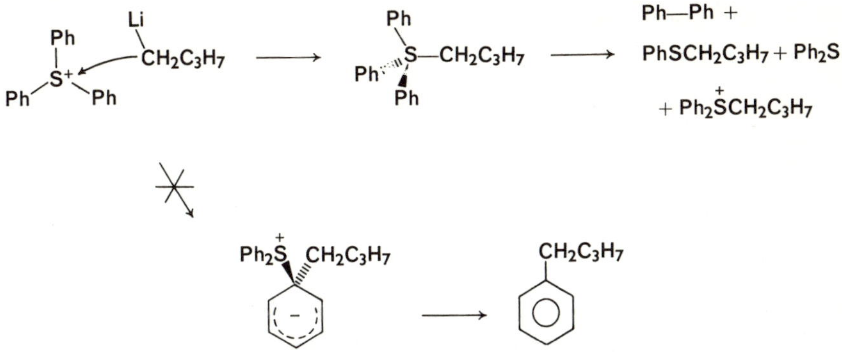

This reaction may be of use in the preparation of cyclopropanes from thietans (**94**):

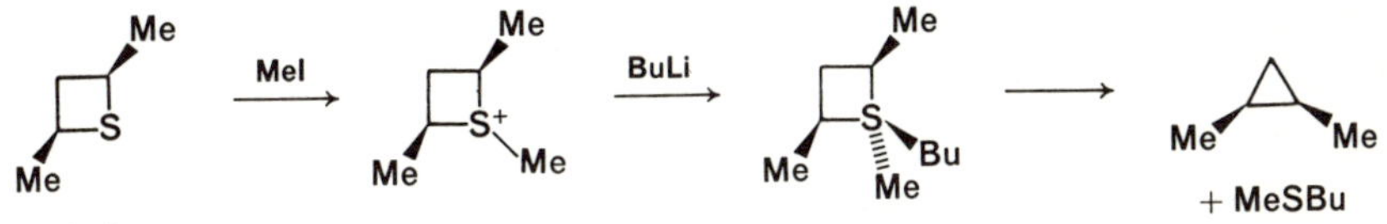

10.13   ORGANOBORANE INTERMEDIATES

When diborane, B_2H_6, is allowed to react with an olefin, addition occurs to give a trialkyl boron (**95**). There is a succession of additions of B—H across the double bonds during which the dimeric borane fragments into monoboron species going through the mono- di- and finally trialkylboranes. The reaction may be stopped at any stage by adjustment of the quantities, and mixed boranes may be made. A concerted mechanism of addition should be symmetry forbidden hence it probably occurs via a betaine (**96**) in two steps; however the addition is stereospecific, so the nature of **94** is such as to prohibit rotation about the C—C bond. The alkylboranes are quite stable entities which may be isolated and characterized in the usual way. Their use for organic chemists lies in the variety of subsequent transformations which may be made to occur and

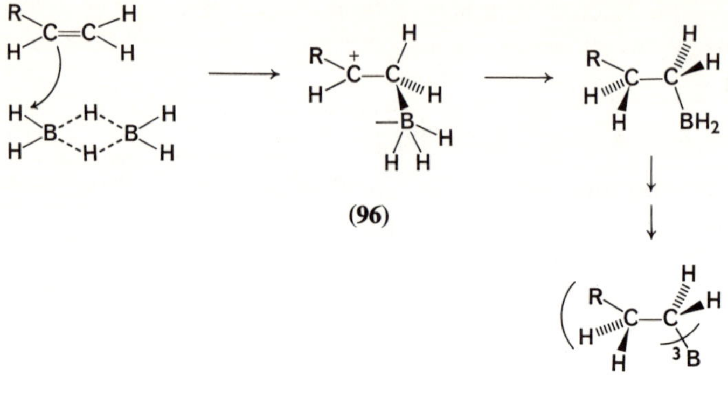

(96)

(95)

therefore they behave as versatile preparative intermediates. Displacements of boron by electrophilic reagents occur with retention of configuration,

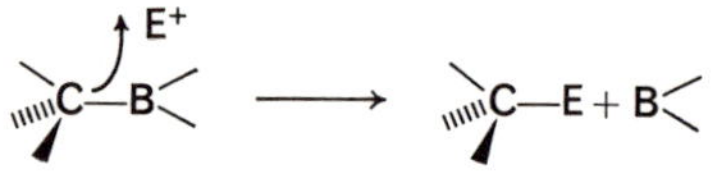

The discovery and development of these reagents is due almost entirely to H. C. Brown.

Generation of Borane

The most convenient procedure is to treat a solution of sodium borohydride (or lithium aluminium hydride) in an ethereal solvent with the boron trifluoride–ether complex. B_2H_6 may be passed into the reaction vessel or the reagent added to the generating system and the addition of borane accomplished *in situ*.

$$3RCH{=}CH_2 \quad \xrightarrow[\quad B_2H_6 \quad]{\overset{NaBH_4 + BF_3}{\downarrow}} \quad (RCH_2{-}CH_2{-})_3B$$

Orientation of Addition

1-Alkenes add borane to give about 94% primary borane and 6% secondary. Greater specificity is achieved by using a sterically hindered dialkyl borane (99% primary attack with diisopentylborane). The addition is reversible at

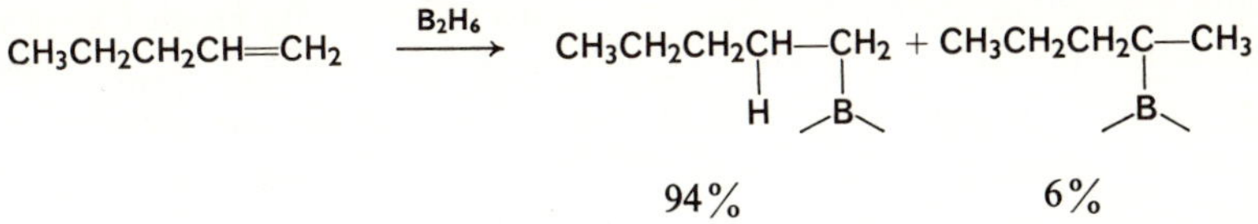
CH3CH2CH2CH=CH2
B2H6
CH3CH2CH2CH—CH2 + CH3CH2CH2C—CH3
H
B
B
94%
6%

higher temperatures and thermodynamic control allows the formation of the most stable borane which is usually the least sterically hindered; the stability order is primary > secondary > tertiary, in general. The same dissociative

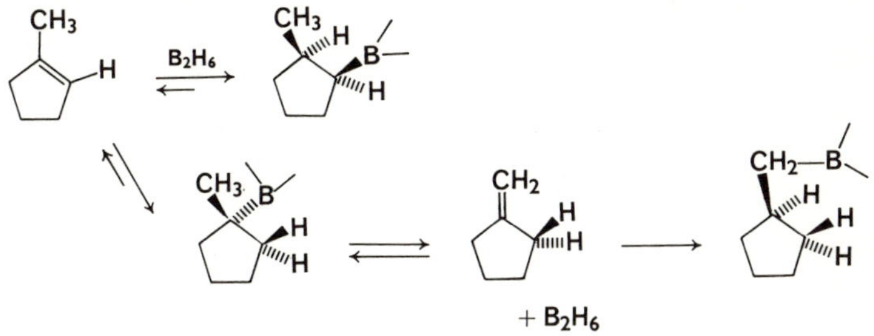
CH3
H
B2H6
B
+ B2H6
CH2
CH2—B

mechanism accounts for borane exchange. If an alkylborane is distilled with a relatively high-boiling olefin, a more volatile olefin may be released from the alkyl groups. There are several uses to this reaction, one being the isomerization of olefins to the thermodynamically *less stable* isomers, e.g. **97** → **98**.

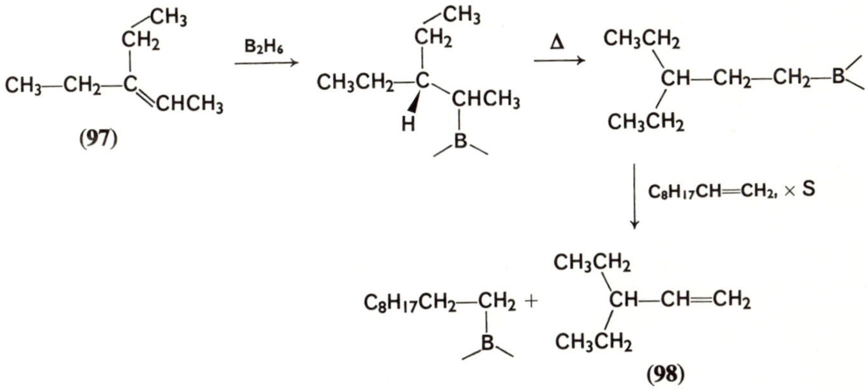
CH3—CH2—C=CHCH3
CH2
CH3
(97)
B2H6
CH3CH2—C—CHCH3
Δ
CH3CH2
CH—CH2—CH2—B
CH3CH2
C8H17CH=CH2, × S
C8H17CH2—CH2 +
CH—CH=CH2
(98)

Oxidation of the Boranes

Probably the principal preparative use of boranes is their oxidation by hydrogen peroxide in alkali to alcohols ('hydroboration'). This provides a simple, efficient and stereospecific (*cis*) means of hydrating an olefinic double bond, and, moreover, this occurs in the 'anti-Markownikov' sense (Section 2.13) to give the isomer which is not normally available via carbonium-ion routes.

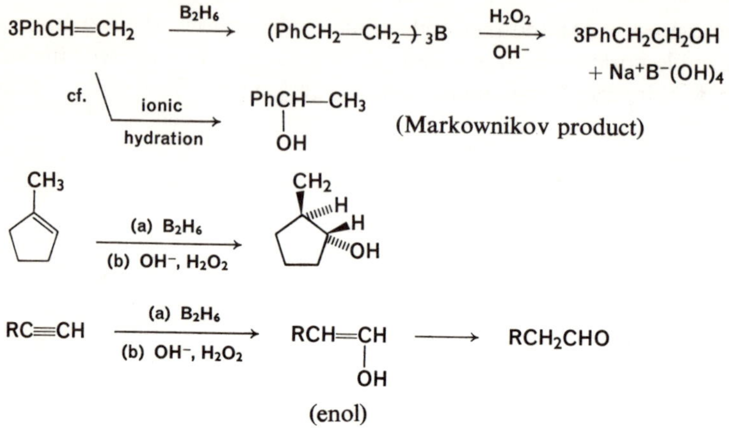

$$3PhCH{=}CH_2 \xrightarrow{B_2H_6} (PhCH_2{-}CH_2{\rightarrow}_3B \xrightarrow[OH^-]{H_2O_2} 3PhCH_2CH_2OH$$

$$+ Na^+B^-(OH)_4$$

cf. $\xrightarrow[\text{hydration}]{\text{ionic}}$ $PhCH{-}CH_3$ (Markownikov product)
$\qquad\qquad\qquad\qquad\quad$ |
$\qquad\qquad\qquad\qquad\quad$ OH

$$RC{\equiv}CH \xrightarrow[\text{(b) OH}^-\text{, H}_2\text{O}_2]{\text{(a) B}_2\text{H}_6} RCH{=}CH \longrightarrow RCH_2CHO$$
$$\qquad\qquad\qquad\qquad\qquad\quad |$$
$$\qquad\qquad\qquad\qquad\quad OH$$
$$\qquad\qquad\qquad\qquad\text{(enol)}$$

Chromic-acid oxidation of the boranes leads directly to ketones:

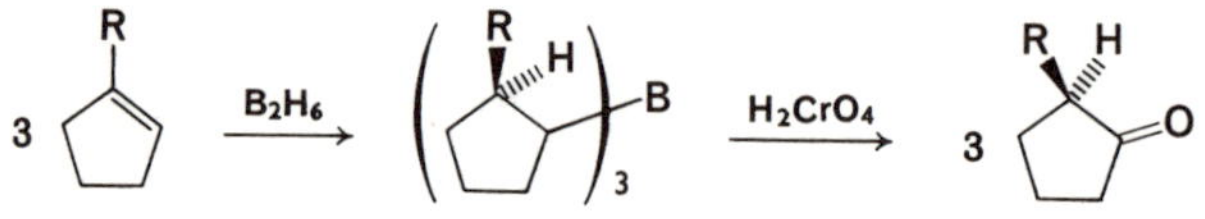

while with molecular oxygen, hydroperoxides or alcohols can be obtained in good yield, if the amount of oxygen is controlled. A radical chain reaction is involved here.

$$R_3B + O{=}O \longrightarrow R\cdot + R_2B{-}O{-}\overset{\cdot}{O}$$

$$R\cdot + O_2 \longrightarrow R{-}O{-}O\cdot \xrightarrow{(H\cdot)} R{-}O{-}OH$$

$$\downarrow R_3B$$

$$R{-}O{-}O{-}BR_2 + R\cdot$$

$$\downarrow RO_2\cdot$$

$$2ROOH + ROH \xleftarrow{H_2O_2} (RO_2)_2BR + R\cdot$$

Using an optically active borane, e.g. **99**, optical asymmetry may be induced in the hydroboration of suitable olefins.

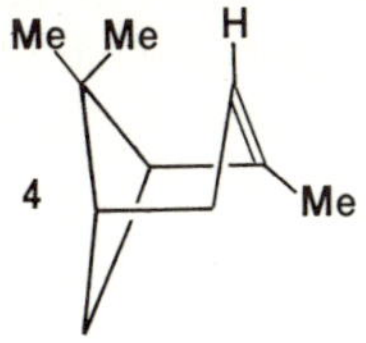

α-pinene

(optically active olefin)

(99)

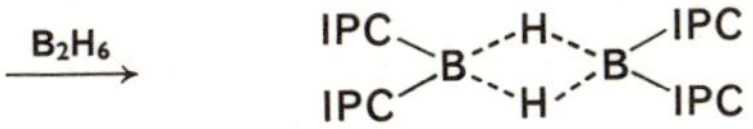

diisopinocampheylborane (dimer)

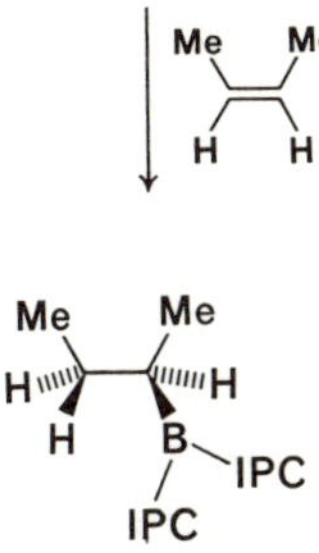

IPC = isopinocampheyl

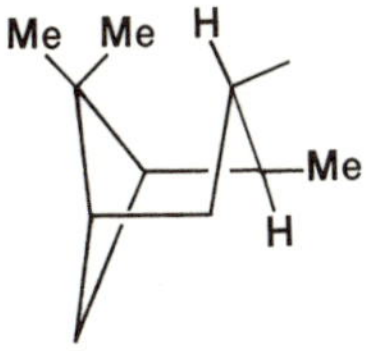

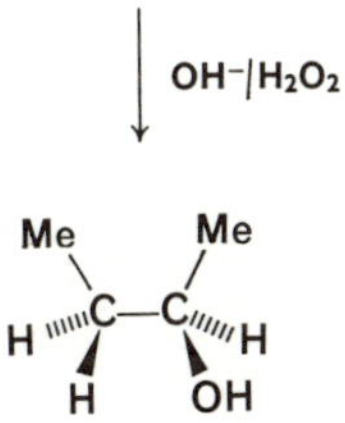

(+) 2-butanol, 80% optical purity

Halogenation of Boranes

Bromine and iodine in alkaline solution cleave the alkylboranes with the formation of alkyl halides. Inversion apparently occurs in this step.

norbornene

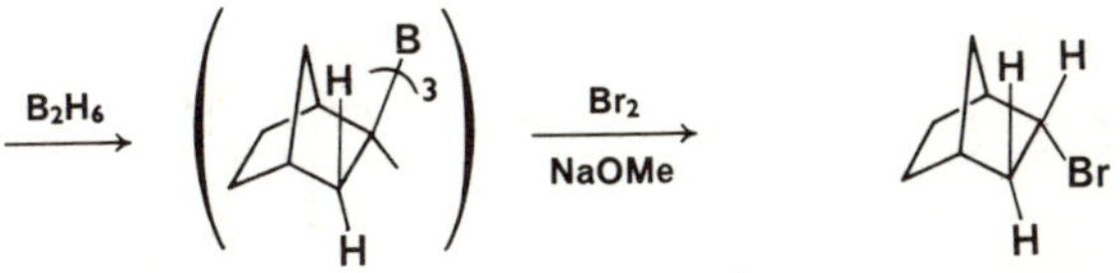

endo-2-bromonorbornane

Amination of Boranes

Cleavage by 'positive-nitrogen' compounds such as chloramines or O-hydroxylaminosulphonic acid (100) leads to the primary amine. As in hydration, the reaction is cleanly a net *cis* addition to the olefin although the mechanism is believed to be a radical chain type.

$$PhCH{=}CH_2 \xrightarrow{B_2H_6} (PhCH_2CH_2{\longrightarrow})_3B \xrightarrow[\substack{\overset{\delta^+ \;\; \delta^-}{H_2N{-}OSO_3H} \\ (100)}]{} PhCH_2CH_2NH_2$$

$$\xrightarrow[\text{(b) } H_2N{-}OSO_3H]{\text{(a) } B_2H_6}$$

$$\xrightarrow[\text{(b) } Cl{-}NMe_2]{\text{(a) } B_2H_6}$$

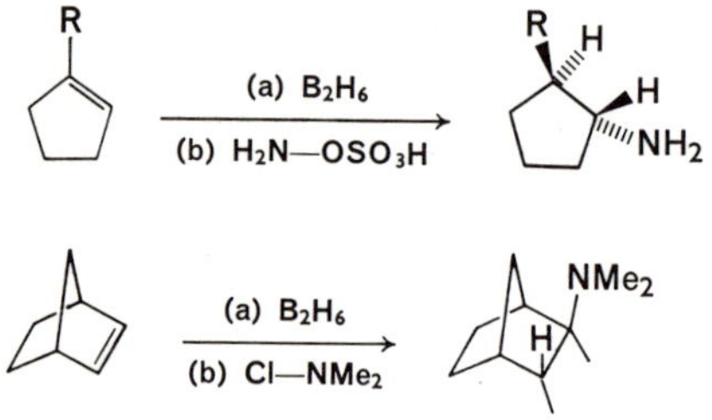

Acid Cleavage of Boranes

Cleavage of the carbon–boron bond occurs in the presence of carboxylic acids though not by water or mineral acids (H_3O^+). The products are paraffins.

$$(RCH_2{=}CH_2)_3B \xrightarrow{CH_3COOH} RCH_2{-}CH_3$$

Alkyl Coupling in Boranes

Alkaline silver nitrate converts a borane to the dimeric paraffin, probably via an intermediate silver alkyl (101) which is known to react thus—

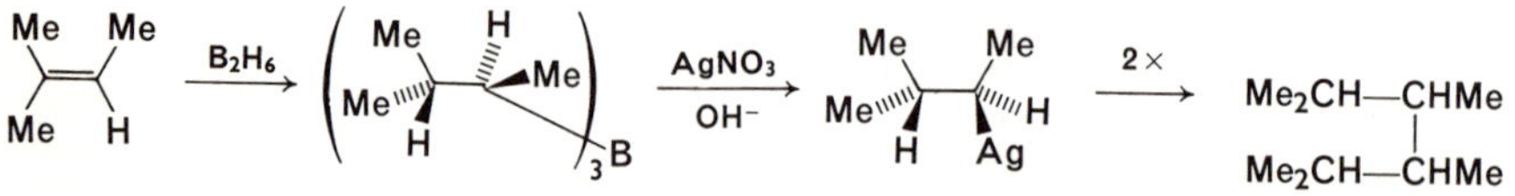

(101)

Mixed boranes or mixtures of boranes will give the crossed product in essentially the statistical proportions.

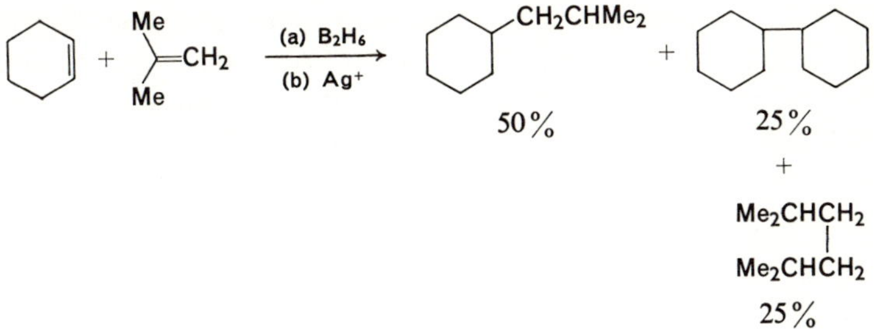

$$\text{Me}_2\text{CHCH}_2$$
$$\text{Me}_2\text{CHCH}_2$$
25%

Cyclization via Boranes

An allylic halide may be hydroborated and the γ-haloborane then undergoes cyclization to a cyclopropane. The preferred reagent is the secondary borane, 9-borabicyclo[3,3,1]nonane (102). Analogous reactions may yield four-, five-

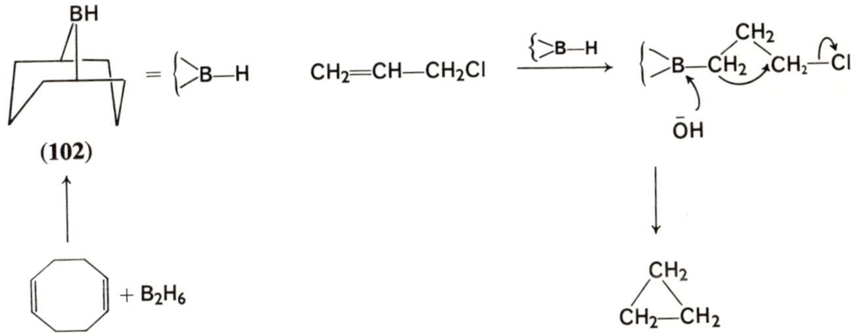

or six-membered rings. Intramolecular alkylations may be accomplished with these reagents, particularly of α-haloketones and esters.

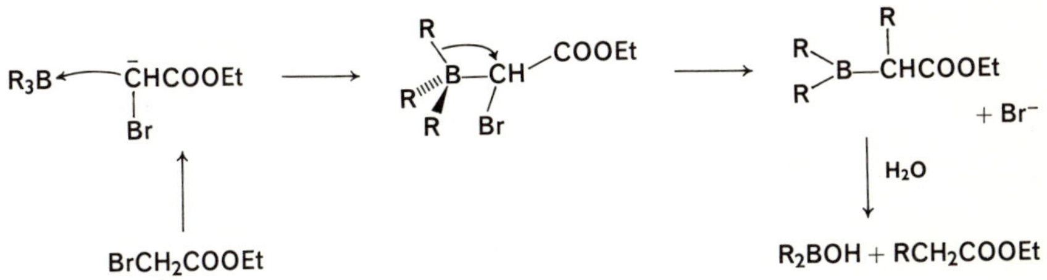

Carbonylation of Boranes

The addition of carbon monoxide to a borane in the presence of a hydride donor of only moderate power, e.g. **103**, leads to the formation of aldehydes (**104**) or homologous alcohols (**105**) in excellent yield. Functional groups such as carboxylic esters are not affected by the hydride reagent. As with other electrophilic substitutions, the stereochemistry involves strict retention of configuration. In the presence of water, the carbon monoxide inserts into two

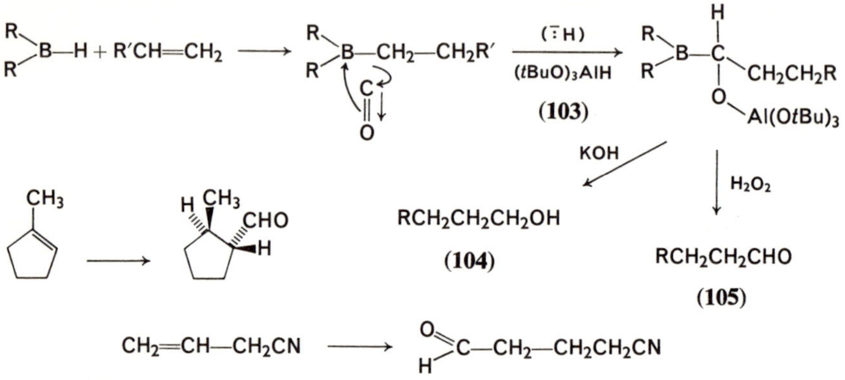

C—B bonds. The product can then be oxidized to a ketone. In order to avoid the variety of products which would be formed in making unsymmetrical ketones, *t*-hexylborane is used as a starting material; this is sterically hindered and does not take part in the insertion reaction.

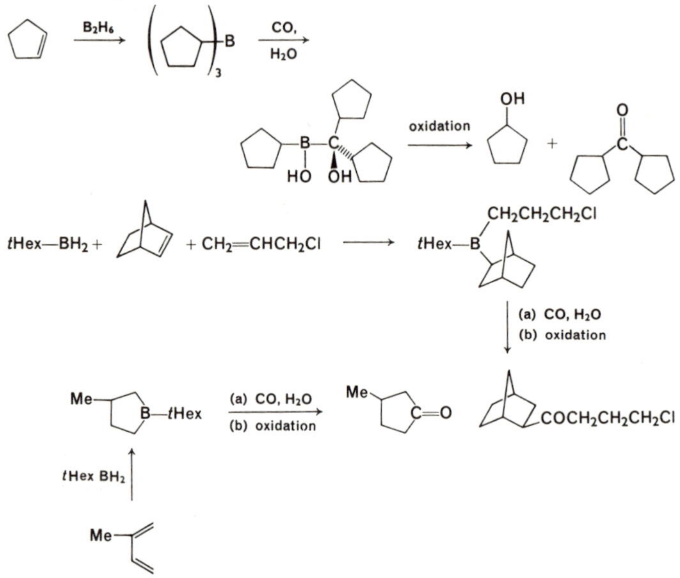

Carbonylation in the presence of oxidizing agents leads to tertiary alcohols including sterically hindered molecules difficult to obtain by alternative routes.

SUGGESTIONS FOR FURTHER READING

Enzyme Reactions

J. J. Birktoft, D. M. Blow, R. Henderson, and T. A. Steitz, 'The structure of α-chymotrypsin', *Phil. Trans. Roy. Soc.*, **B257**, 67 (1970).

M. Dixon and E. C. Webb, *The Enzymes*, Longmans, London, 1958.

A. Williams, 'Mechanism of action and specificity of proteolytic enzymes', *Quart. Rev.*, **23**, 1 (1969).

Metal-catalysed Reactions

J. Boor, *Ind. Eng. Chem., Prod. Res. Develop.*, **9**, 437 (1970).

G. Henrici-Olivé and S. Olivé, *Angew. Chem.* (Internat. Ed.), **10**, 105, 776 (1971).

H. Muller, *Angew. Chem.* (Internat. Ed.), **4**, 327 (1965).

H. Weber and J. Falbe, *Ind. Eng. Chem.*, **62**, 4, 33 (1970).

G. Wilke, *Angew. Chem.* (Internat. Ed.), **2**, 105 (1963); **5**, 151 (1966).

Chemiluminescence

D. M. Hercules, 'Chemiluminescence from electron transfer reactions', *Accounts Chem. Res.*, **2**, 301 (1969).

D. R. Kearns, *J. Amer. Chem. Soc.*, **2**, 6554 (1969).

F. McCapra, 'The chemiluminescence of organic compounds', *Quart. Rev.*, **20**, 485 (1966); 'The Chemiluminescence of Organic Compounds'; IUPAC International Symposium, St. Moritz, 1970, p. 611 (Butterworth).

E. H. White and D. F. Roswell, 'The chemiluminescence of organic hydrazides', *Accounts Chem. Res.*, **3**, 54 (1970).

Borane Reactions

H. C. Brown, *Hydroboration*, Benjamin, San Francisco, 1962.

H. C. Brown, 'The versatile organoboranes', *Chem. Brit.*, **7**, 458 (1971).

D. S. Matteson, *Organomet. Chem. Rev.*, **5**, 35 (1969).

G. Zweifel and H. C. Brown, 'Hydroboration of olefins, dienes and acetylenes via hydroboration', *Organic Reactions*, **13**, 1 (1963).

Subject Index